Progress in Colloid & Polymer Science, Vol. 109 (1998)

PROGRESS IN COLLOID & POLYMER SCIENCE
Editors: F. Kremer (Leipzig) and G. Lagaly (Kiel)

Volume 109 (1998)

Horizons 2000 – aspects of colloid and interface science at the turn of the millenium

Editor:
G. Lagaly (Kiel)

SPRINGER-VERLAG BERLIN HEIDELBERG GMBH

Springer

ISBN 978-3-662-15706-0
ISSN 0340-255 X

Die Deutsche Bibliothek –
CIP-Einheitsaufnahme

Progress in colloid & polymer science. –
Darmstadt : Steinkopff ; New York :
Springer
Früher Schriftenreihe
Vol. 109. Horizons 2000 – aspects of
colloid and interface science at the turn
of the millenium. – 1998
Horizons 2000 – aspects of colloid and interface science at the turn of the millenium / guest ed.: G. Lagaly. –

(Progress in colloid & polymer science ;
Vol. 109)
ISBN 978-3-662-15706-0
ISBN 978-3-7985-1654-0 (eBook)
DOI 10.1007/978-3-7985-1654-0

Chemistry Editor:
Dr. Maria Magdalene Nabbe;
Production: Holger Frey, Ajit Vaidya.

Typesetting and Copy-Editing:
Macmillan Ltd., Bangalore, India

Prog Colloid Polym Sci (1998) V

PREFACE

The issue is dedicated to Professor Dr. Dr. h. c. Milan Schwuger on the occasion of his 60th birthday. A central topic of his scientific studies concerns the interaction of surface active agents with solid materials. Though most of the papers collected in this issue are related to this central theme, they comprise an impressing overview of the actual research in colloid science and reveal expected trends of this strongly developing field at the turn of the millennium.

G. Lagaly

The major part of the papers published in this volume has been presented during the 6th Wolfgang Ostwald Colloqium "Horizons 2000" from 4–6 June in Jülich. The meeting was sponsored by

BASF AG, Ludwigshafen
Bayer AG, Leverkusen
Bruker Analytische Meßtechnik GmbH, Karlsruhe
Chemische Betriebe Pluto GmbH, Herne
Chemische Fabrik Kreussler, Wiesbaden
Dr. Dietrich Steinkopff Verlag GmbH & Co. KG, Darmstadt
DSM Research, Geleen
Forschungszentrum Jülich GmbH
Henkel KGaA, Düsseldorf
Kolloid-Gesellschaft e.V., Köln
Merck KGaA, Darmstadt
Perkin Elmer, Düsseldorf
Quarzwerke GmbH, Frechen
Stuart GmbH, Wuppertal
Süd-Chemie AG, München
ThermoBioanalysis, Hemel Hempstead.

Prog Colloid Polym Sci (1998) VII

CONTENTS

Progr Colloid Polym Sci (1998) 109:1–2

LAUDATIO

W. von Rybinski

Honoring Milan Johann Schwuger on his 60th birthday

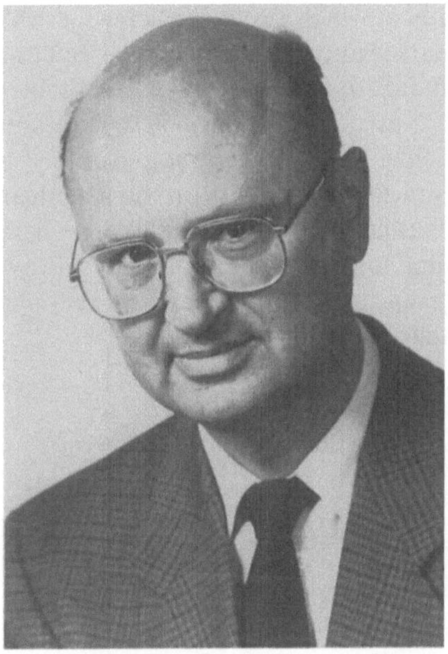

Prof. Dr. Dr. h.c. Milan Johann Schwuger

Dr. Wolfgang von Rybinski
Henkel KGaA
Henkelstraße 67
D-40191 Düsseldorf
Germany

On the occasion of his 60th birthday Professor Schwuger is honored as a scientist with an extraordinary career, which not only includes an unusually broad field of colloid and interface science, but also both basic science and application.

Starting as a student of Physical Chemistry in 1957 his diploma thesis "Comparison of BET-surface areas for different adsorbents" led into the field of interface science. His PhD thesis continued this topic with a successful defense at the University of Aachen in 1966. In the same year he joined the chemical company Henkel in Düsseldorf, Germany. In collaboration with Professor Hermann Lange he extended his scientific work into another field of interface science with a specific focus on applied research. During this time his first remarkable papers on the structure of association colloids were published. This early scientific success resulted in him becoming group leader within Corporate Research at Henkel in 1968. His main interest was the study of the interfacial properties of association and dispersion colloids in aqueous phases based on specific application problems. This period of research led to the replacement of phosphates in detergents with zeolite A.

Based on fundamental studies of the mode of action of complexing agents Professor Schwuger proposed the replacement of phosphates not by one, but by a mixture of at least two different components. Instead of choosing a water-soluble complexing agent, the water-insoluble zeolite A in combination with a carrier fulfilled the performance profile of phosphates in detergents and became the new technical standard for builders in detergents.

From 1972 to 1977 Professor Schwuger was the leader of a project team, which would eventually establish the scientific and technical basis and the ecological evaluation of this invention. All aspects ranging from technical problems to the environmental compatibility were covered in this project. Additionally, physicochemical studies of ion exchange, especially of heavy-metal ions, were conducted.

After the successful finalization of the project and the introduction of the first test detergent containing zeolite A, Professor Schwuger was appointed the head of Physical Chemical Research at Henkel. In recognition for his research in the replacement of phosphates in detergents he was honored by the Theodor–Steinkopff award of the German Colloid Society. Nowadays about 1.5 million tons of zeolite A and 5 million tons of detergents are produced according to the patents of Professor Schwuger. This invention is displayed in the Deutsche

Museum in Bonn, Germany, where the one hundred most important German inventions since 1945 are shown.

As head of Physical Chemical Research at Henkel from 1977 to 1989 Professor Schwuger succeeded in solving many problems of the worldwide activities of the company regarding colloid and interface science. In addition to his job at Henkel he became a lecturer at the University of Düsseldorf in 1982. In 1989 Professor Schwuger was appointed as the director of the Institute for Applied Physical Chemistry in Jülich, Germany, retaining his position as professor in Düsseldorf. Since then the physical chemistry of interfaces in the environment became his major field of research. He was able to build up a unique branch of research which deals with the physico-chemical interaction of surfactants in the aquatic and terrestrial environment, especially in multicomponent systems containing contaminants. The results regarding the kinetics and equilibrium during the adsorption of surfactant–contaminant mixtures onto clay minerals and biodegradable microemulsions for soil decontamination received awards from the American Oil Chemist's Society and the company Nordac. In 1989 Professor Schwuger also became director of the Environmental Specimen Bank of Germany, which belongs to the Institute of Applied Physical Chemistry in Jülich. In 1991 Professor Schwuger was elected as chairman of the German Colloid Society. He is a Dr Honoris Causa from Clarkson University in Potsdam, New York.

During a period of more than 30 years Professor Schwuger's scientific research impact has been tremendous. He has published about 130 original scientific papers in international journals and 15 contributions to books. In addition to this, there are more than 70 patent applications which show a unique combination of basic science and application in the research of Professor Schwuger.

I have had the priviledge to work together with Professor Schwuger at Henkel for seven years, when I joined Henkel in 1980. This time he convinced me that the study of the fundamentals in colloid and interface chemistry is of tremendous importance in many fields of applications. There was, and still is, a huge demand for further the understanding of technical processes regarding colloid and interface chemistry. Professor Schwuger is a brilliant scholar who in addition to this always had an open ear for his collegues and co-workers.

Looking at such a successful life's work, what remains for the future? Additional scientific achievements are a goal which Professor Schwuger will pursue and achieve of course, but one should not forget that there is also a private person in Professor Schwuger whose great passion is travelling, especially on the American continent. So, I would like to wish him and his wife, who has always given him strong support, enough spare time for travels.

Progr Colloid Polym Sci (1998) 109:3–12

H. Möhwald
U. Dahmen
K. de Meijere
G. Brezesinski

Lipid monolayers to understand and control properties of fluid interfaces

Received: 23 October 1997
Accepted: 27 October 1997

Prof. Dr. H. Möhwald (✉) · U. Dahmen
K. de Meijere · Dr. G. Brezesinski
Max-Planck-Institut für
Kolloid- und Grenzflächenforschung
Rudower Chaussee 5
D-12489 Berlin
Germany

Abstract Langmuir monolayers are probably the best defined interfacial systems since they enable variation of many parameters independently and since many techniques have been established in recent years to characterize them at the molecular level. In the first part of the paper the progress in the physical understanding of the polymorphism is given. Ordered mesophases are encountered, distinguished by aliphatic chain tilt, azimuth and the freedom of chain rotation about the long axis. The orientation of polar molecules at the interface in addition is responsible for long range electrostatic interactions causing peculiar domain shapes and superlattices. Having largely understood the polymorphism one can use structural studies to understand interactions of lipids with adjacent molecules. Samples presented are phospholipid molecules electrostatically interacting with polyelectrolytes, enzymatically attacked by phospholipases and partly solubilized by hydrocarbons. It is shown that these interactions specifically affect the monolayer structure, changing chain tilt and partly increasing the order. In addition it is shown that enzymatic reactions at interfaces are varied via structure and composition of the monolayer.

Key words Lipid monolayers – model membranes – interface structure

Introduction

Horizons 2000 – this must be the area in colloid and interfacial science between what we just faintly observe and what we may soon observe. It may also be the area that we greatly desire to see, and there may be different reasons for this desire: new applications and new demands, e.g. in environmental science, biotechnology or information technology; new principles, e.g. in general physics, biophysics or chemistry. These different motivations are not conflicting and one may ask: Where has been the most progress, what can one expect from it and how will one make use of it? Obviously answers on this will be highly subjective, and we will try to argue from our backgrounds in physics, chemistry and biology. In condensed matter physics much is known about macroscopic systems ($> \mu$m) and atoms and molecules, the "hot" areas being interfaces, clusters, soft matter and glasses. In chemistry the field of supramolecular chemistry is exploding where repetitive synthetic routes and self-organization principles are used to prepare functional macromolecules. In biology besides understanding the genetical code and the interplay between functional entities there are large efforts to resolve structure and dynamics of "small" units like proteins. The dimensions of these systems are between nm and μm, i.e. the colloidal domain, and it is therefore not surprising that colloid science is part of these larger areas and is thus per se also interdisciplinary. Interfacial science on the other hand is also essential to understand colloids, because the behaviour of colloidal systems largely depends on their interfaces. Thus colloids and interfaces are merged, e.g. in most relevant journals or in names of institutes like ours. In addition it has a virtue on its own because via suitable control of interfaces interactions can be controlled and understood and also processes like wetting, adsorption and detergency can be manipulated.

This work considers a very well-defined fluid interface, a Langmuir monolayer of amphiphiles at the air/water and oil/water interface. It enables an independent variation of parameters like molecular density, pressure or ionic conditions. The systems have been known for a century [1] and are of no direct practical use. Yet, they have recently achieved much attention because many techniques had been developed in the last 15 years to resolve the film structure at the molecular level [2]. Hence, these systems can now be used as models to understand interfacial interactions. Consequently, this paper is organized as follows: following the description of experiments we will describe general features of Langmuir monolayers which when first observed were unexpected or controversial. These features result from the fact that polar molecules are partially oriented at interfaces thus leading to long range electrostatic interactions and that aliphatic tails can assume many different configurations causing a rich polymorphism. Then we will give examples of the use of Langmuir layers to understand interactions of relevance in colloid and biosciences.

Experimental

Materials

The double-chain phospholipids 1,2-dipalmitoylphosphatidylcholine (DPPC), 1,2-dipalmitoylphosphatidylethanolamine (DPPE) and 1,2-dipalmitoylphosphatidic acid (DPPA) are commercially available and were spread from a 10^{-3} M solution in chloroform onto an ultrapure water subphase. The purification of the water in a Millipore desktop filtering system leads to a specific resistance of 18.2 MΩ cm.

As an example the chemical structures of a polymer coupled lipid (DPPA) [3] are shown in Fig. 1. The polyelectrolyte poly (diallyldimethylammonium chloride) (PDADMAC) was added to the water subphase in a concentration of 10^{-3} M (referring to the molecular weight of one monomer unit) before spreading the monolayer solution. After spreading the lipid monolayer, the polymer was given about 30 min to adsorp. Longer adsorption times up to 8 h did not change the isotherm.

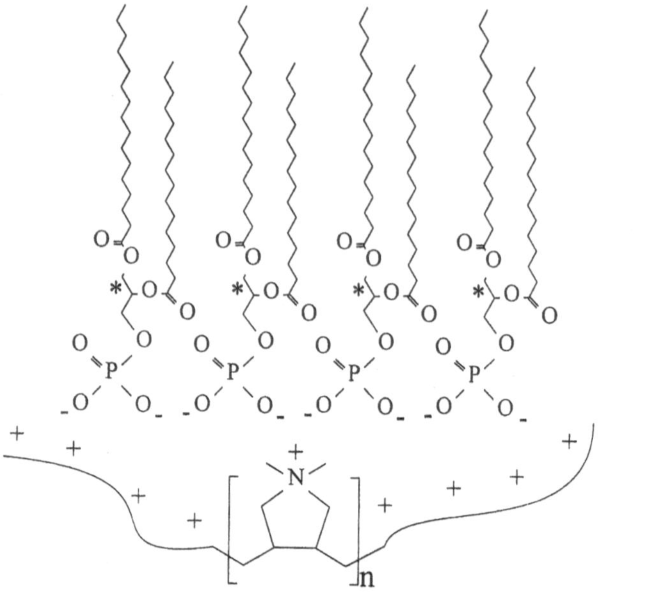

Fig. 1 Chemical structures of 1,2-dipalmitoylphosphatidic acid (*top*) and poly(diallyldimethylammonium chloride) (*bottom*). The asterisk marks the asymmetric carbon atom (chiral center)

Film balance and fluorescence microscopy

The pressure/area isotherms were measured on a thermostated film balance equiped with a Wilhelmy system to continuously record the surface pressure during the compression of the monolayer.

Fluorescence microscopic images were obtained with a Zeiss Axiotron microsocpe using a long distance objective ($d = 8$ mm). One mol% of the fluorescence dye L-α-phosphatidylcholine-β-[N-(7-nitrobenz-2-oxa-1,3-diazol-4-yl)aminohexanoyl]-γ-palmitoyl (Sigma, Taufkirchen, Germany) was added to the spreading solution. Its emission is excited by a 59-W Hg high-pressure lamp via a dichroic mirror and detected by means of a SIT camera (Hamamatsu C2400).

Grazing incidence X-ray diffraction (GIXD)

Synchrotron X-ray experiments at grazing incidence (GID) were carried out on the liquid surface diffractometer at the undulator beamline BW1 at HASYLAB, DESY (Hamburg, Germany) [4–7]. The Synchrotron beam was made monochromatic by Bragg reflection by a beryllium (0 0 2) crystal and was adjusted to strike the monolayer on the water surface at an angle of incidence $\alpha_i = 0.85\alpha_c$, where α_c is the critical angle of total external reflection. The intensity of the diffracted radiation is detected by a position sensitive detector (PSD) (OED-100-M, Braun,

Garching, Germany). The resolution of the horizontal scattering angle $2\Theta_{hor}$ is given by a Soller collimator located in front of the PSD. The scattering vector **Q** has an in-plane component $Q_{xy} = (4\pi/\lambda)\ \sin(2\Theta_{hor}/2)$ and an out-of-plane component $Q_z = (2\pi/\lambda) \sin \alpha_f$, where λ is the X-ray wave length and α_f the vertical scattering angle. The accumulated position-resolved scans were corrected for polarization, footprint-area and powder-averaging (Lorentz factor). The intensities of the diffraction peaks were least-squares fitted to model peaks as products of a Lorentzian parallel and a Gaussian normal to the water surface. The lattice spacings are obtained from the in-plane diffraction. The lattice parameters can be calculated from the lattice spacings and the unit cell area A_{xy} is calculated from the lattice parameters.

If two diffraction maxima occur, the monolayer forms a centred rectangular lattice where the tilt azimuth can be in one of the two symmetry directions, towards the nearest neighbour (NN) or towards the next nearest neighbour (NNN). Three non-degenerate diffraction maxima at $Q_z > 0$ Å^{-1} are measured if the unit cell has an oblique symmetry with three different lattice spacings.

Polarization-modulated infrared reflection absorption spectroscopy (IRRAS)

The goal of the polarization modulated FTIR technique used is to obtain a difference spectrum $(R_p - R_s)/(R_p + R_s)$ with R_p and R_s, the p- and s-reflectivity. Only anisotropic absorptions contribute to a difference spectrum, isotropic absorptions (from water vapor, etc) are not detected. Our setup consists of a Bruker IFS66 (Bruker, Karlsruhe, Germany) spectrometer, and an external reflection unit. At the exit of the Michelson interferometer, the IR-beam is directed into the external reflection unit, where it passes a polarizer in p- and a photoelastic modulator (ZnSe). Then the IR-beam is focused onto the water surface with an angle of incidence of 73°. The reflected intensity is focused onto a liquid-nitrogen-cooled MCS detector. The electronical setup used (two bandpass filters and a lock-in amplifier) is designed to obtain two signals $(R_p - R_s)$ and $(R_p + R_s)$. The signals are multiplexed and connected with the standard electronics of the spectrometer. Here the resulting interferogram is amplified, digitized (16 bit AD-board) and Fourier transformed (OPUS-software).

The IR spectra were collected using 200 scans at 4 cm^{-1} resolution with a scanning time of about 5 min.

General features

Figure 2 shows typical pressure/area isotherms as they are observed for phospholipids, fatty acids, fatty alcohols and many other compounds. For sufficiently high temperature or small attraction one observes a so-called liquid expanded (LE) state with no order and diffusivity like in a fluid oil. On compression one enters an ordered state and the many different types of order will be discussed below. Reducing the temperature or increasing the lateral interaction one may suppress the LE phase and by compression the system can be sublimed from a gaseous to an ordered state.

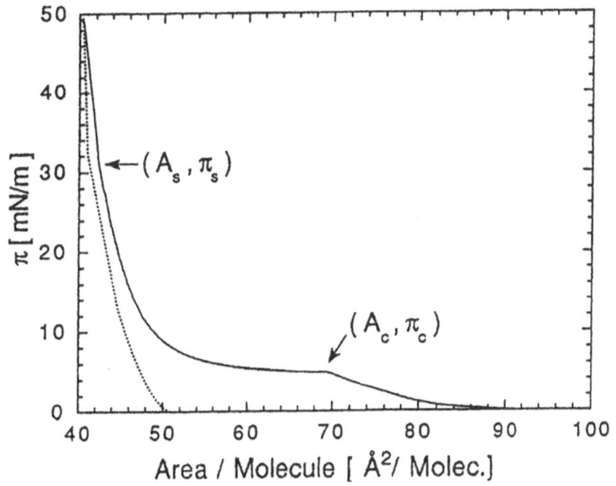

Fig. 2 Pressure–area isotherms of an 1,2-dimyristoylphosphatidylethanolamine (DMPE) monolayer at $T = 20\,°C$ (*solid line*) and an 1,2-dipalmitoylphosphatidylethanolamine (DPPE) mono-layer at $T = 25\,°C$ (*dotted line*)

There has been a long discussion as to whether the break in the slope of the isotherm corresponds to a first order phase transition and the clearest proof of this would be to observe two-phase coexistence. This has been achieved by staining the LE phase by a fluorescing dye and recording fluorescence micrographs of the monolayer upon compression. As an example Fig. 3 shows dark (ordered phase) domains in a continuous LE phase (bright) environment. These results on phase coexistence were later confirmed with techniques not relying on probes like Brewster angle microscopy or imaging ellipsometry [8–10]. In addition these showed that the ordered state is distinguished by a uniform tilt of aliphatic tails. Figure 2 also shows that there may be peculiar domain shapes and domain superlattices. These shapes are stable over observation times of hours, and this leads to a specific feature of the interfacial system. For a three-dimensional system one would expect that following a nucleation period large domains would have to grow on expense of small ones in order to minimize the boundary energy corresponding to the phases. For the interfacial system, however, the molecules carrying dipole moments are partially aligned at the interface and this leads to a long range

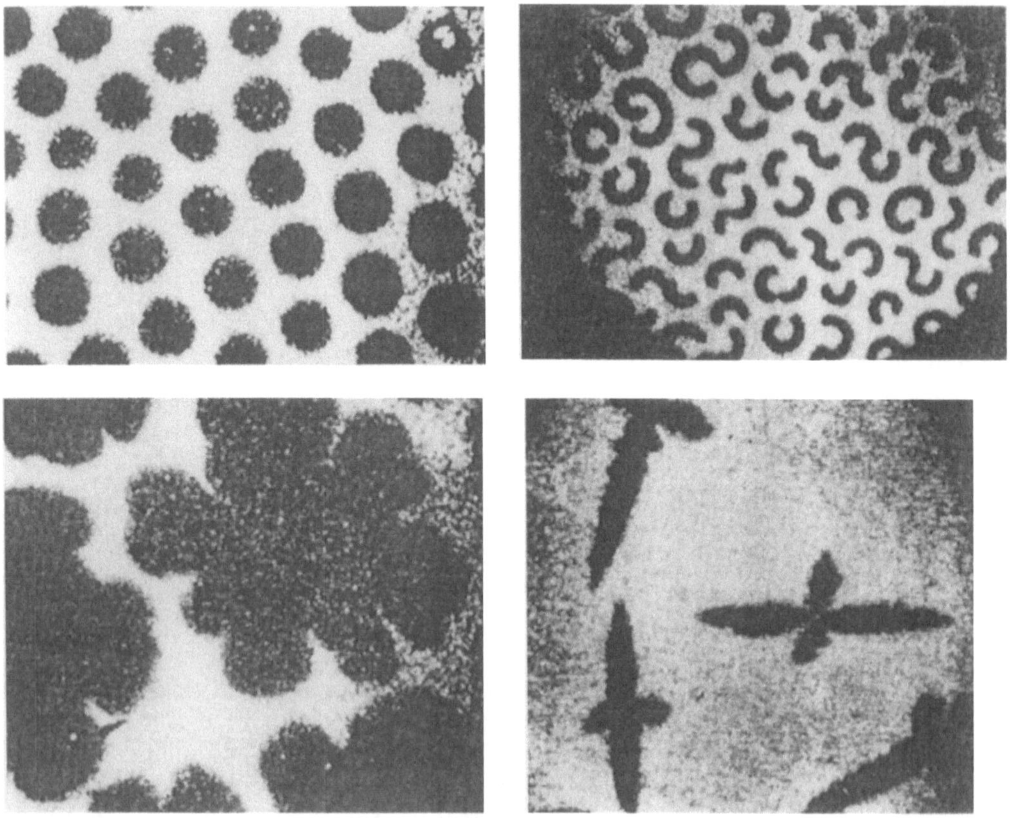

Fig. 3 Fluorescence micrographs of lipid monolayers in the LE/LC phase coexistence region. *Upper left*: 1,2-dimyristoylphosphatidic acid (1) at pH 5 and in the presence of monovalent ions. *Upper right*: compound (1) containing 1 mol% cholesterol at pH 11. *Lower left*: compound (1) at pH 5, no other ions added. *Lower right*: a diacetylenic lipid in the unpolymerized state

repulsion. For a two-dimensional system this energy increases with domain size faster than the boundary energy [11]. Hence there is a critical size R_c above which electrostatic repulsion dominates. This force limits the domain size and causes domain repulsion leading to a superlattice. If the boundary energy is anisotropic, its interplay with electrostatics can cause lamellar domains and if there is additionally a chiral interaction, spiral shapes can be formed. In order to quantify the ideas elegant theories have been formulated [11] and the electrostatic energy can be measured via the surface potential difference of coexisting phases [12]. Controversies exist concerning the boundary energies which may be strongly system dependent. They may vary from zero near a critical point to $\sim 10^{-4}$ mN/m [13]. Although there are undoubtedly systems where equilibrium between the two types of energies exists there are other systems where equilibrium may not be established at all. In fact it has been shown that establishment of chemical equilibrium may require days due to the very small chemical potential gradient [14].

For the last ten years it has been possible to characterize the ordered state, previously called liquid condensed (LC) by X-ray scattering studies [15, 16]. The fact that one observes Bragg diffraction reveals that there is positional order in the plane. However, often the peaks are broad indicating order only over some 10 nm, typical for a liquid crystal. Measuring the diffraction intensity along the plane normal one can also derive the tilt angle t of the aliphatic chains and the tilt direction. Generally, t decreases upon compression and from a systematic measurement and analysis one deduces a phase diagram as given in Fig. 4 for behenic acid [17, 18]. One realizes that there are at least eight different ordered phases. Two phases existing at low temperatures are crystalline and exhibit very narrow diffraction peaks, the others are mesophases with low positional order. They are distinguished by a tilt to a nearest neighbour (NN), a next nearest neighbour (NNN) or no tilt and by a free rotation or a hindered rotation about the long molecular axes. The latter modes are also found for n-alkanes [19]. This polymorphism is not unique as it is

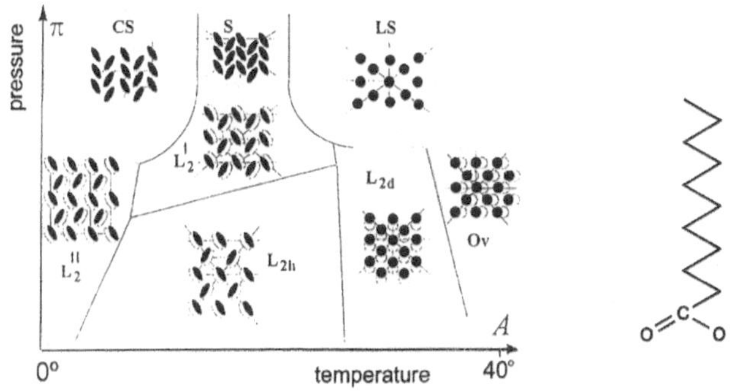

Fig. 4 Temperature-pressure phase diagram for a fatty acid monolayer displaying hexatic phases (LS, L_{2d} and Ov), hindered rotator phases (S, L_2' and L_{2h}) and crystalline phases (CS and L_2''). The temperature axis is roughly correct for a behenic acid and would have to be shifted for chain length variations. The chemical structure of a typical fatty acid molecule is shown on the right side

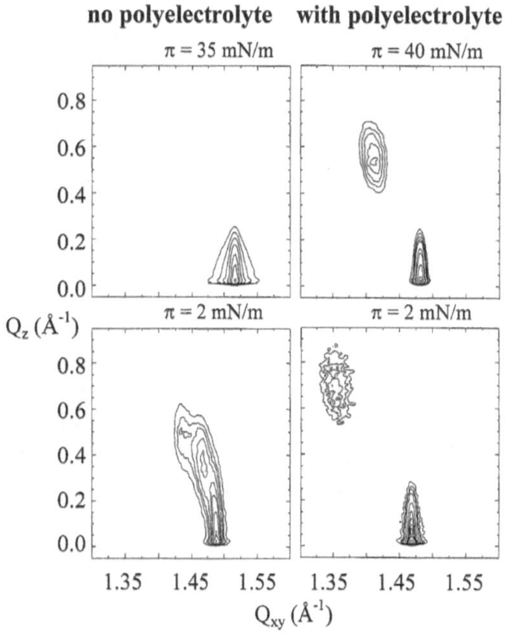

Fig. 5 Contour plots of the corrected X-ray intensities as a function of the in-plane component Q_{xy} and the out-of-plane component Q_z of the scattering vector **Q** of DPPA monolayers on water (*left*) and on a PDADMAC subphase (*right*) at different surface pressures (indicated)

also found, although at other pressures and temperatures, for amphiphiles with other chain lengths and head groups [20]. Mesophases can be suppressed for amphiphiles with strong and specific lateral interactions. Phospholipids as double-chain surfactants with an extended head group also exhibit mesophases, but crystalline monolayers have not yet been observed [21]. Probably the incommensurability of head and tail inhibits crystalline ordering. On the other hand it was possible to find head group chirality expressed in a chiral structure of the tail lattices [22, 23]. This among others indicates orientational order of the heads. This order is in addition reflected in the domain shapes. For example, the spirals are only observed for the pure enantiomer of the lipid, whereas the racemate exhibits linearly elongated domains [24].

Langmuir monolayers to model interactions

Polyelectrolyte/surfactant interactions

For various applications e.g. in separation and filtration technology stable lamellar structures existing of sheets of hydrophobic and hydrophilic regions are desired. One approach to this may be self-organized amphiphilic polymers or blockcopolymers, another approach, the biomimetic one, would be to design a lipid bilayer coupling to a polyelectrolyte [25]. In the latter case an important question concerns the influence of the polymer binding on the membrane structure. In the specific example the membrane exists of a negatively charged phospholipid, to which a positively charged polyelectrolyte is electrostatically bound. The question then arises: since binding reduces the electrostatic repulsion between the phospholipid heads does the membrane condense or does the disorder induced by the flexible polymer dominate and lead to a lateral expansion of the membrane?

This question can be answered studying a phospholipid monolayer with and without polyelectrolyte in the subphase [3]. Figure 5 shows contour plots of X-ray diffraction intensity of a monolayer of enantiomeric dipalmitoylphosphatidic acid (DPPA) in absence (left) and presence (right) of poly(diallyldimethylammonium chloride) (PDADMAC) at low and high surface pressures. For the monolayer without polyelectrolyte coupling one observes the typical behaviour of a phospholipid with small head: at low lateral pressure there are three diffraction spots indicative of an oblique lattice and the fact that all peaks are above the horizon ($Q_z > 0\,\text{Å}^{-1}$) proves that there is a chain tilt. On compression the peak maxima shift towards $Q_z = 0\,\text{Å}^{-1}$, and at high pressure there is no chain tilt. All reflections degenerate indicating a hexagonal lattice. The situation is qualitatively different for the polymer coupled lipid monolayer. At all pressures the lattice remains rectangular and the chains tilted. The positions of the two reflections change only slightly upon compression and differ from those in the absence of polymer. Therefore, the unit cell is larger compared to the uncoupled system. This means that the expansion effected by the polymer dominates the reduction of the Coulombic repulsion which would lead to a shrinking of the unit cell. It is also worth noting that the peaks with absence and presence of coupling are distinctly different and that coupling does not broaden the (merely) two reflections. Hence there are no (at least not ordered) areas of the film not coupling to the polymer underneath. In addition the uniformity of the reflections indicates that also the interactions averaged over the positional correlation lengths of some 100 Å are uniform.

Analyzing the line shapes in detail one realizes that the reflection perpendicular to the tilt azimuth has the smallest FWHM (full width at half maximum) corresponding to a large coherence length. This suggests that the polymer is oriented perpendicular to the tilt direction like a rigid rod. Although one can deduce the correct repeat distance (6.4 Å between charges on opposite sites) from molecular dynamics calculations [26] one may question whether this polymer can be considered as a rigid rod. Criticizing the above model one should keep in mind that the relevant lengths are merely 100 Å, that persistence lengths of this order have also frequently been found for flexible polymers [27] and that these lengths may be increased at the interface. Experiments systematically varying both the polymer and lipid charge density as well as polymer flexibility should help to conclude on this matter.

Enzymatic interfacial reactions

Many lipolytic enzyme reactions occur at interfaces and in one type of reaction, with much biophysical significance, phospholipids are cleaved by phospholipases. In the case considered here, the reactions with phospholipase A_2, the products are a fatty acid, presumably remaining in the membrane, and a lysolipid which is partially water soluble [28]. The reaction depends on the phase structure and the chemical structure of the phospholipid. Studies with well-defined monolayers should help to conclude on this in a precise way. The reaction is also enantio-selective, i.e. only the L-enantiomer is cleaved. However, using fluorescence microscopy one can show that the enzyme also binds to the monolayer of the D-enantiomer [29]. Hence the structural influence of the two steps, binding and reaction, can be separated.

The binding can again be studied via measurement of X-ray diffraction of a D-dipalmitoylphosphatidylcholine (D-DPPC) monolayer with phospholipase A_2 homogeneously distributed in the subphase (Fig. 6). In this case one realizes that all reflections can be shifted towards $Q_z = 0\,Å^{-1}$, i.e. the tilt decreases near zero. This is surprising because for the pure DPPC monolayer the tilt angle cannot be reduced below 25°. This is generally ascribed to the large hydrated choline head group and its orientation nearly parallel to the surface [30]. Hence we conclude that

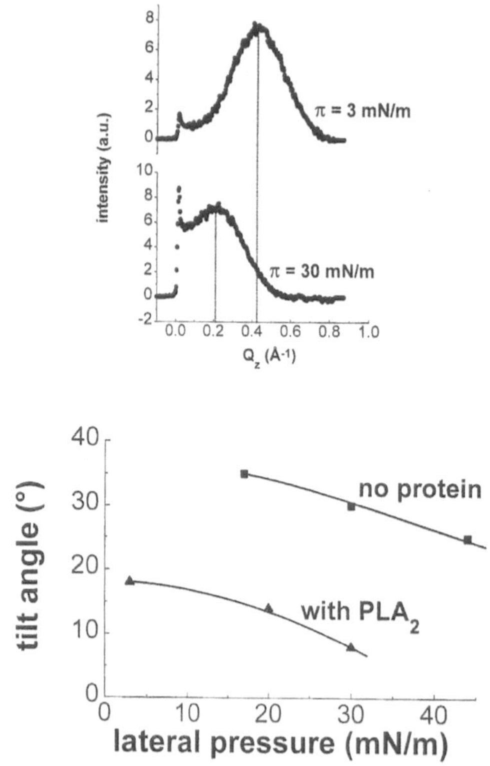

Fig. 6 *Top*: Diffraction intensity (Bragg rod) as a function of out-of-plane scattering vector component Q_z integrated over the Q_{xy} interval of the twofold degenerate peak ($\pi = 3$ mN/m: $1.435\,Å^{-1} < Q_{xy} < 1.465\,Å^{-1}$, $\pi = 30$ mN/m: $1.475\,Å^{-1} < Q_{xy} < 1.495\,Å^{-1}$) of a centered rectangular lattice of D-DPPC after adsorption of PLA_2. *Bottom*: Tilt angle versus lateral pressure for a D-DPPC monolayer on a buffer solution and after adsorption of PLA_2

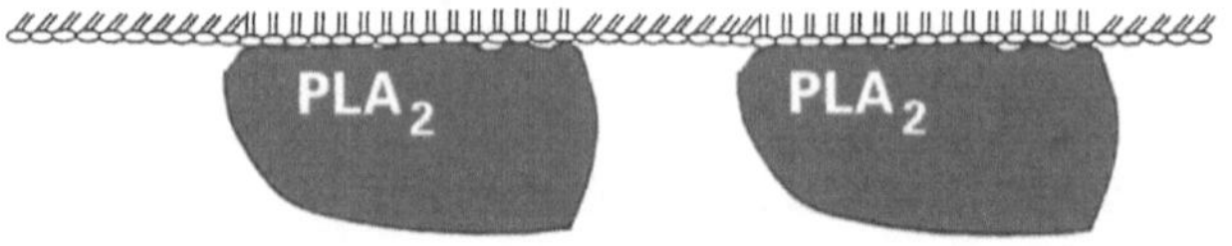

Fig. 7 Schematic diagram of the structural change induced upon PLA_2 adsorption

protein binding enforces a dehydration and reorientation of the head group. It is remarkable, that the protein binding does not only affect a single head which might be pulled into a binding pocket, but induces a cooperative change involving at least a hundred head groups (see Fig. 7). In fact from the line widths one can deduce the area over which positional order extends. It amounts to $180 \times 80\,Å^2$ which, incidentally, is close to be projected area of the enzyme [29].

We should also note that the extent of head group change strongly depends on the head group, e.g. it is much less pronounced if an ester linkage is replaced by an ether

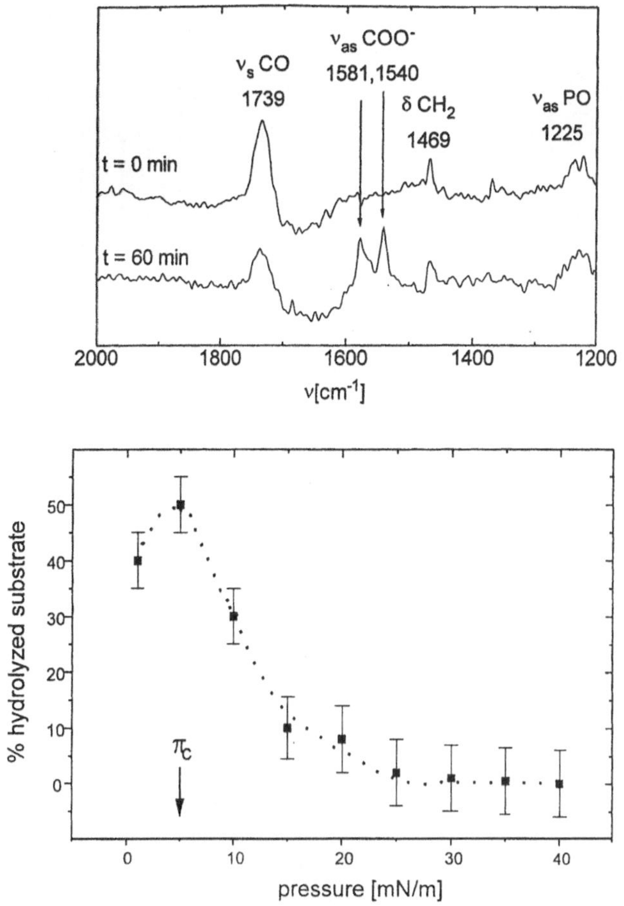

Fig. 8 *Top*: Spectra of L-DPPC before ($t = 0$ min) and after 60 min of hydrolysis at 5 mN/m by PLA_2 (0.3 ng enzyme/ml subphase; PLA_2 from *Crotalus Atrox* venom); spectra measured at 40 mN/m. The asymmetric stretching vibrations of the carboxylate (fatty acid) have been assigned to free carboxylate (1540 cm^{-1}) and enzyme associated carboxylate (1581 cm^{-1}), the shoulder at 1562 cm^{-1} is due to Ca^{2+} associated carboxylate. *Bottom*: Percentage (%) hydrolyzed substrate after 60 min vs. lateral pressure applied during hydrolysis

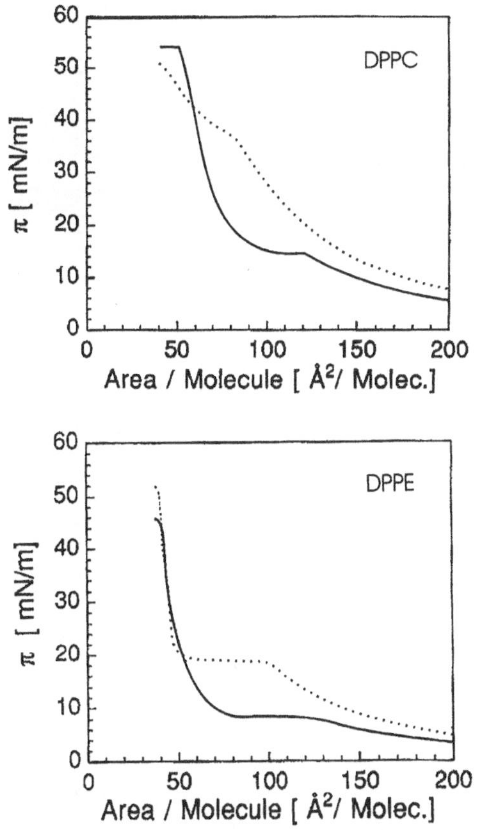

Fig. 9 Pressure–area isotherms of DPPC (*top*) and DPPE (*bottom*) monolayers at bicyclohexyl (BCH)–water ($T = 20\,°C$) (*dotted lines*) and hexadecane (C_{16})–water($T = 25\,°C$) (*solid lines*) interfaces

linkage near the head. In this case the larger change correlates with larger enzyme activity. Also there were experiments where reflections of protein bound and protein free monolayers were observed simultaneously. This means that two types of areas may coexist if the monolayer is not completely covered by the enzyme.

In order to follow the enzyme activity FTIR spectroscopy has proven to be extremely valuable since product and educt are distinguished among others by vibrations in the carbonyl and the phosphate region. Because the interface causes partial molecular orientation, this technique can be interface specific by using polarization modulation and comparing the p-(in-plane polarization) and s-(out-of-plane polarization) IR reflectivities [31]. An example of polarization modulated IR-reflection spectra of an L-DPPC monolayer before and after enzymatic reaction is given in Fig. 8. One clearly observes the appearance of peaks characteristic of fatty acid and the decrease of the ester vibrations. At this stage typical measurement times for spectra like in Fig. 8 are on the order of 5 min, and this determines the time resolution. From an integration of characteristic peaks one can determine reaction rates and it turns out that these are maximal for pressures corresponding to LE/LC phase coexistence [32]. This result agrees with fluorescence microscopic observations of maximum enzyme activity at the phase boundary [33].

Monolayers at the oil/water interface

Microemulsion stability is largely determined by the lowering of the oil/water interfacial energy by surfactant enrichment at the interface. Therefore understanding the surfactant layer at this interface is of utmost importance to control microemulsions. In this respect the planar monolayer is very promising because of its defined geometry and

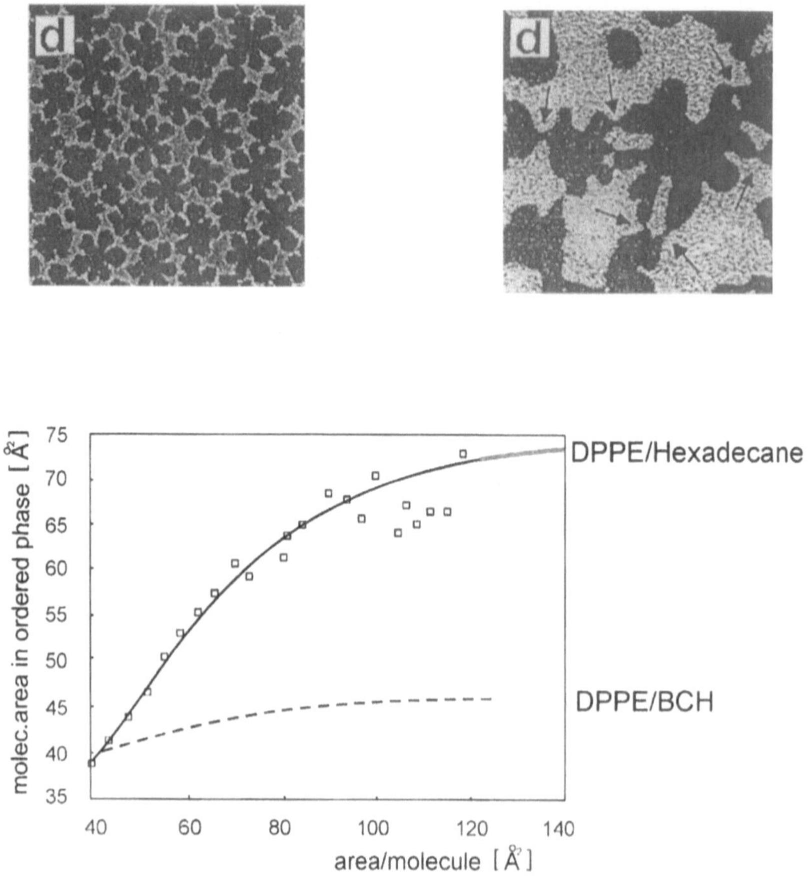

Fig. 10 *Top*: fluorescence micrographs from the phase coexistence region of a DPPE monolayer at the bicyclohexyl (BCH)/water interface (*left*) and at the hexadecane/water interface (*right*). *Bottom*: molecular area per lipid in the condensed phase as a function of the area per molecule for the two interfaces from above as derived from quantitative analysis of the fluorescence micrographs

suitability for optical, X-ray and Neutron reflection techniques. Preparing the monolayer with phospholipids that are insoluble in the water as well as the oil phase one can also change the surfactant density and state in a well-defined way.

The pressure/area isotherms (Fig. 9) reveal that also in this case the pronounced break in the slope typical for a first order phase transition is observed [34]. The area differences of LE and LC phases are much larger compared to the air/water interface and this as well as the flat isotherm in the LE phase can be ascribed to the entropic repulsion by oil penetrating the monolayer [34]. The change of transition pressure and limiting molecular area at high pressures depend on the lipid [35] as well as the type of oil [34]. Concerning the latter it is also informative to analyse quantitatively the fluorescence micrographs observed for the phase coexistence range (Fig. 10, top). Comparing the bulky oil bicyclohexyl (BCH) and the (linear) n-alkane hexadecane (C_{16}) in the first case, contrary to the latter, one finds a dendritic structure and domain repulsion. Determining the LC phase area fraction as a function of molecular area one can also derive the area per phospholipid A_p in the LC domains [36]. One derives values between 40 and 45 Å^2 in the case of BCH as expected for two nearly perpendicularly oriented tails of the phospholipid DPPE. In the case of C_{16}, however, A_p is well above 45 Å^2 and is decreased upon compression (Fig. 10, bottom). This indicates that commencing the transition the alkane partly remains in the ordered phase and is continuously squeezed out upon compression. To quantify this statement, since in the ordered state at low pressure the molecular area demand of the two tilted tails of the phospholipid is about 45 Å^2 and that of an ordered alkane similarly tilted is about 22 Å^2, one estimates 1–2 alkanes per phospholipid at the beginning of the transition.

The pressure/area isotherms are qualitatively much more expanded if a lipid like DPPC with a large head group is studied in contact with hexadecane [35]. In this region isotherms with oil contact and at the oil/water interface agree and there are good arguments that this also holds for the monolayer structure. The latter can again be studied by surface diffraction [37], and the analysis of the measurements in Fig. 11 shows that the tails can be aligned vertically. In this case, opposite to the situation of enzyme

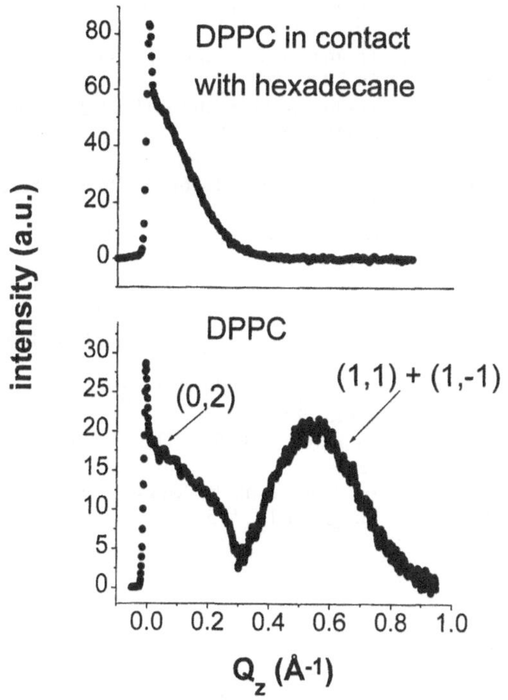

Fig. 11 Diffraction intensity (Bragg rod) as a function of the out-of-plane scattering vector component Q_z integrated over the Q_{xy} interval of the corresponding peaks. *Top*: threefold degenerate peak (at $Q_z = 0\,Å^{-1}$) of a hexagonal unit cell of untilted DPPC molecules in contact with hexadecane ($T = 20\,°C$, $\pi = 38$ mN/m, $1.465\,Å^{-1} \langle Q_{xy} \rangle\ 1.565\,Å^{-1}$). *Bottom*: DPPC monolayers on water exhibit a slightly distorted rectangular lattice of tilted molecules at all lateral pressures investigated ($T = 20\,°C$, $\pi = 41$ mN/m, $1.335\,Å^{-1} \langle Q_{xy} \langle 1.525\,Å^{-1}$)

binding, this cannot be ascribed primarily to a head group reorientation but to incorporation of the alkane into the ordered state. Judged from the isotherms the lipid/alkane stoichiometry is close to 1:1 but this may not be a regular arrangement which would explain the highly flexible domain shapes. Hence via the contact with oils via liquid or gas phases [38] the monolayer can be manipulated.

Conclusions and outlook

In this paper we have shown first that Langmuir monolayers possess unique physical features. These result from the anisotropic molecular orientation at interfaces causing long range electrostatic interaction and thus peculiar domain structures and superlattices. Since these features qualitatively do not depend on the type of lipid or interface one should expect them to also exist in many interfacial systems. Actually, domain superlattices have also been found for so called self-assembled monolayers like silanes, siloxanes or thiols on solid support [39]. A second specific feature is the rich polymorphism with elements from smectic liquid crystals and polymeric crystals. Thus the systems possess many internal and external, not necessarily coupled, order parameters leading to various mesophases. Consequently, one should expect that increasing the complexity of molecules even more phases will be encountered, and the situation will resemble that of the already better known thermotropic liquid crystals.

This paper due to lack of space did not touch other basic aspects of the systems related to two-dimensional molecular transport, defects and melting in 2D, isothermal nucleation and growth. Instead we intended to demonstrate that these are excellent model systems to study interactions at fluid interfaces and some modern methods were introduced that enable these studies at the molecular level. The examples were selected from one biophysical problem to control enzyme activity at a hydrophobic/hydrophilic interface and two problems with association colloids, to control polymer/lipid interactions and microemulsions. The list of examples could have been extended, e.g. to conclude on the organization in soluble monolayers [40], to show that these are best-defined systems for wetting studies [41], for molecular recognition [42, 43] or for latex coating. All these aspects are faintly seen at the horizon, and the examples are still weak hints into this direction. Hopefully they could raise interest to have more pathfinders into the different areas ahead.

References

1. Pockels A (1891) Nature 43:437
2. Möhwald H (1993) Rep Prog Phys 56:653
3. de Meijere K, Brezesinski G, Möhwald H (1997) Macromolecules 30:2337
4. Als-Nielsen J, Jaquemain D, Kjaer K, Lahav M, Leveiller F, Leiserowitz L (1994) Phys Rep 246:251
5. Kjaer K (1994) Physica B 198:100
6. Majewski J, Popovitz-Biro R, Bouwman WG, Kjaer K, Als-Nielsen J, Lahav M, Leiserowitz L (1995) Chem Eur J 1:304
7. Brezesinski G, Dietrich A, Struth B, Böhm C, Bouwman WG, Kjaer K, Möhwald H (1995) Chem Phys Lipids 76:145
8. Hönig D, Möbius D (1991) J Phys Chem 95:4590
9. Reiter R, Motschmann H, Orendi H, Nemetz A, Knoll W (1992) Langmuir 8:1784
10. Harke M, Teppner R, Schulz O, Motschmann H, Orendi H (1997) Rev Sci Instrum 68:3130
11. McConnell HM (1991) Annu Rev Phys Chem 42:171
12. Miller A, Helm CA, Möhwald H (1987) J Physique 48:693
13. Muller P, Gallet F (1991) Phys Rev Lett 67:1106
14. McConnell HM (1996) Proc Natl Acad Sci, USA 93, p 15001
15. Kjaer K, Als-Nielsen J, Helm CA, Laxhuber LA, Möhwald H (1987) Phys Rev Lett 58:2224

16. Dutta P, Peng JB, Lin M, Ketterson JB, Prakash M, Geogopoulos P, Ehrlich S (1987) Phys Rev Lett 58:2228
17. Kenn RM, Böhm C, Bibo AM, Peterson IR, Möhwald H, Kjaer K, Als-Nielsen J (1991) J Phys Chem 95:2092
18. Overbeck GA, Möbius D (1993) J Phys Chem 97:7999
19. Ewen B, Strobl GR, Richter D (1980) Faraday Discuss Chem Soc 69:19
20. Bibo AM, Peterson IR (1990) Adv Mater 2:309
21. Kenn RM, Kjaer K, Möhwald H (1996) Colloids Surfaces A 117:171
22. Böhm C, Möhwald H, Leiserowitz L, Als-Nielsen J, Kjaer K (1993) Biophys J 64:553
23. Bringezu F, Brezesinski G, Nuhn P, Möhwald HN (1996) Biophys J 70: 1789
24. Weis RM, McConnell HM (1984) Nature 310:47
25. Antonietti M, Kaul A, Thünemann A (1995) Langmuir 11:2633
26. Donath E, Walther D, unpublished
27. Förster S, Schmidt M (1995) Advances in Polymer Sci 120:51
28. Verger R, de Haas GH (1973) Chem Phys Lipids 10:127
29. Dahmen-Levison U, Brezesinski G, Möhwald H, Thin Solid Films, in press
30. Brumm T, Naumann C, Sackmann E, Rennie AR, Thomas RK, Kanellas D, Penfold J, Bayerl TM (1994) Eur Biophys J 23:289
31. Blaudez D, Buffeteau T, Cornut JC, Desbat B, Escafre N, Pezolet M, Turlet JM (1993) Appl Spectrosc 47:869
32. Dahmen-Levison U, Brezesinski G, Möhwald H, Progr Colloid Polym Sci, in press
33. Grainger DW, Reichert A, Ringsdorf H, Salesse C (1990) Biochim Biophys Acta 1023:365
34. Thoma M, Pfohl T, Möhwald H (1995) Langmuir 11:2881
35. Thoma M, Möhwald H (1995) Colloid Surf A 95:193
36. Thoma M, Möhwald H (1994) J Colloid Interface Sci 162:340
37. Brezesinski G, Thoma M, Struth B, Möhwald H (1996) J Phys Chem 100:3126
38. Harke M, Motschmann H (1998) Langmuir 14:313
39. Quint P, Möhwald H (1991) Makromol Chem Makromol Symp 46:329
40. Melzer V, Vollhardt D (1996) Phys Rev Lett 76:3770
41. Pfohl T, Möhwald H, Riegler H, Langmuir, submitted
42. Kunitake T (1996) Thin Solid Films 285:9
43. Weck M, Fink R, Ringsdorf H (1997) Langmuir 13:3515

Progr Colloid Polym Sci (1998) 109:13–20

H. Hoffmann
M. Bergmeier
M. Gradzielski
C. Thunig

Preparation of three morphologically different states of a lamellar phase

Received: 10 December 1997
Accepted: 12 December 1997

Prof. Dr. H. Hoffmann (✉) · M. Bergmeier
M. Gradzielski · C. Thunig
University of Bayreuth
Physical Chemistry I
D-95440 Bayreuth
Germany

Abstract An aqueous L_α-phase from uncharged zwitterionic surfactants alkyldimethylaminoxide and cosurfactants in which a few percent of the zwitterionic surfactants are protonated can exist in three well-defined structurally different states:

- a state with stacked bilayers (state I),
- a state with polydisperse multilamellar vesicles (state II),
- a state with monodisperse small unilamellar vesicles (state III).

State I can be prepared from a L_3-phase that is at rest and in which the bilayers are protonated with formic acid which is produced by hydrolysis methylformiate that is added to the of L_3-phase. State II can be obtained from state I by modestly shaking a sample in state I. State III is obtained if a sample in state II is sheared for about 1 h at high shear rates ($\sim 4000\ s^{-1}$). All three states have different macroscopic properties and can be kept without change for long times (days and weeks). State I has a domain-like birefringence and is a low viscous solution. State II is a viscoelastic fluid that shows stress birefringence and state III is a viscoelastic fluid with no birefringence.

Key words Vesicles – shear – morphological transformations – lamellar phases

Introduction

Lamellar phases are encountered in different regions of binary and ternary phase diagrams. They are observed in the dilute region of double-chain surfactants like didodecyldimethylammoniumbromide or AOT in water or in ternary systems with single chain surfactants and cosurfactants [1–3]. In the ternary systems the lamellar phases are formed only in solutions having a narrow cosurfactant/surfactant ratio. Lamellar phases can usually be recognized easily on their characteristic birefringence pattern (oily streaks). The birefringence results from the optical anisotropy of the bilayers that is due to the ordered surfactant molecules and their different polarizabilities parallel and perpendicular to their chain. Both uncharged and ionically charged L_α-phases can exist as highly swollen phases [4, 5]. In extreme situations, L_α-phases can contain as little as 1% of surfactant. Lamellar phases are usually sketched as stacks of flat bilayers. Uncharged bilayers of single-chain surfactants and cosurfactants are rather flexible and the bilayers are therefore not flat but wrinkled [6]. The lamellar phases in these situations are actually stabilized by repulsive undulation forces. In charged systems the undulations are suppressed and the large interlamellar spacings are supported by electrostatic repulsions [7].

Charged lamellar phases can also exist as vesicle phases. The vesicles in such phases usually have many shells and are therefore also called onion phases [8]. The vesicles usually have a large size distribution. An onion can be considered as a small volume of a L_α-phase. For charged systems [9] theoretical considerations on the basis of the Helfrich theory favor the formation of L_α-phases with vesicles in comparison to L_α-phases with flat bilayers. Several groups have recently prepared vesicle phases by adding a few percent of an ionic surfactant to lamellar phases that had been prepared from uncharged single-chain surfactants [7] and cosurfactants [10, 11]. Lamellar phases from double-chain ionic surfactants have also been shown to contain vesicles [12]. Very recently, it was shown that monodisperse unilamellar vesicles do form spontaneously in ternary systems of surfactant and cosurfactant when the ionic charge density on the bilayer is adjusted to a well defined characteristic value [13]. It also was shown recently that cubic phases from small unilamellar vesicles can form spontaneously when cosurfactant and ionic surfactants are mixed in the right concentration and ratio [14].

These results demonstrate that lamellar phases can be built up from flat-stacked bilayers or from small unilamellar vesicles or from polydisperse multilamellar vesicles. It is the purpose of this manuscript to demonstrate that it is even possible to prepare the three different states of the lamellar phase from one and the same system with the same composition and that the macroscopic properties of the three states are remarkably different. The three different states are demonstrated on the ternary system tetradecyldimethylamineoxide, hexanol, water in which a few percent of the tetradecyldimethylamineoxide are protonated with formic acid. In order to avoid mechanical mixing of the formic acid with the lamellar phase (made up of the three components) what would cause shear forces that could influence the structures, the formic acid is produced by hydrolysis of methylformiate that is added to the L_3-phase.

Experimental results and their discussion

Preparation of the samples

The three different L_α-phases are prepared from the single L_3-phase with the composition 100 mM TDMAO (tetradecyldimethyl amineoxide) and 240 mM hexanol [15]. The L_3-phase is a low viscous, optically isotropic, slightly turbid solution that shows flow birefringence under shear. The phase remains in the L_3-state when methylformiate is added to reach a concentration of 10 mM. Methylformiate is a hydrophilic compound and is not incorporated into the bilayer of the L_3-phase. The phase remains isotropic for a few minutes after the methylformiate is added to the original L_3-phase and then develops birefringence that can easily be observed in the solution at rest between crossed polarizers. The phase transformation is caused by hydrolysis of methylformiate to methanol and formic acid that protonates the alkyldimethylamineoxide. The protonation reaction charges up the surfactant bilayers and thereby stiffens the bilayer. The L_3-phase is thereby transformed into the L_α-phase. The reaction is complete in about one day [16].

Birefringence of the samples

Figure 1A shows a photograph of the thus produced sample between crossed polarizers (state I). The whole sample shows a typical domain-like birefringence pattern. Figure 1B shows the same sample after it has been moderately shaken a few times, and put at rest again. Now the birefringence pattern is very different: the domains have disappeared and the fluid has become viscoelastic. Small air bubbles which are dispersed in the fluid by the shaking of the sample do not rise (state II). They can, however, be removed by centrifugation of the samples what was done before Fig. 1C was photographed. Finally, the sample was filled into a rheometer with a Couette cell and sheared for about 1 h at a shear rate of $4000\ s^{-1}$. The resulting fluid (state III) is still viscoelastic with even increased elastic properties.

FF-TEM micrographs of the samples

The visual inspection of the samples with and without polarizers shows that the fluid must be in different states and that different microstructures must exist in the samples after they have been manipulated differently. In order to identify different microstructures, we prepared freeze fracture electron micrographs from the samples, the results are shown in Figs. 2A–C.

Figure 2A (state I) shows the typical pattern for stacked bilayers of a L_α-phase. The interlamellar distance between the bilayers is about 500 Å, a distance that is expected on the basis of the surfactant concentration and a bilayer thickness of about 22 Å. The micrograph thus clearly shows that the protonation of the bilayers of the L_3-phase in the absence of shearing forces leads to a normal L_α-phase. It is therefore not correct to assume that vesicles are formed instantaneously when bilayers are charged with ionic surfactants. Figure 2B (state II) shows multilamellar vesicles which have the same interlamellar distance as in Fig. 2A. Obviously, the onions have been prepared by shear forces in the shaking procedure and not by charging the membranes. As the shear rates in the shaking process are not well defined the size distribution of the onion is polydisperse. Finally, Fig. 2C shows small unilamellar vesicles of more or less the same size (state III).

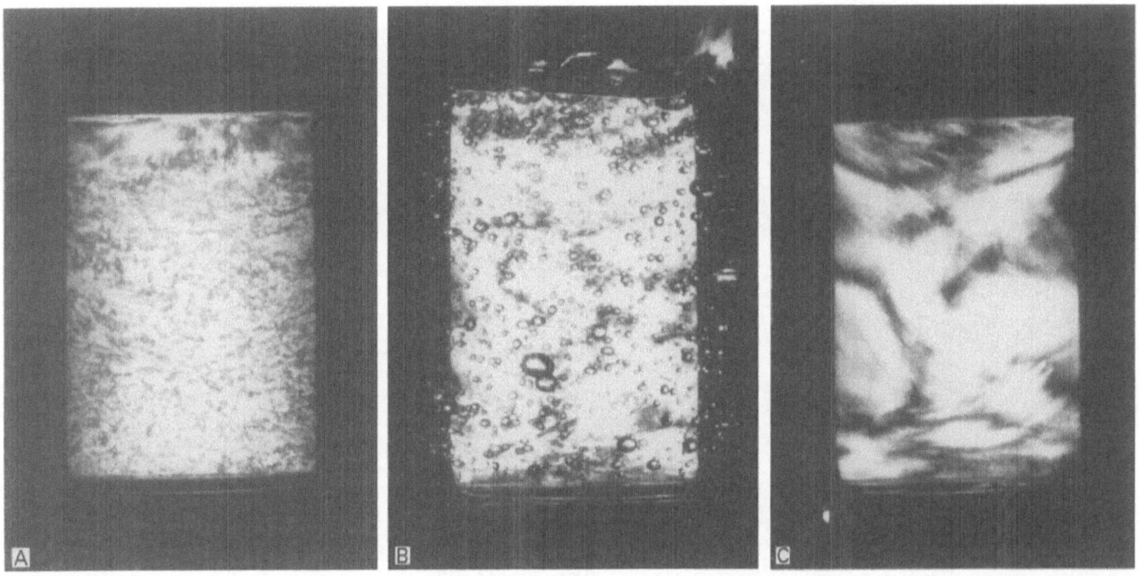

Fig. 1 (A) Sample after completion of the hydrolysis reaction (state I). Typical domain pattern of the stacked bilayer phase between crossed polarizers. (B) Streak like birefringence pattern of the sample that was shaken by hand after completion of the hydrolysis reaction (state II). The entrapped air bubbles demonstrate the yield stress. (C) The same sample as in Fig. 1B, but centrifuged

In comparison to the multilamellar vesicle system (Fig. 2B) the polydispersity is reduced drastically. The high shear now has peeled off all but the innermost shell of the vesicles. In the micrograph a few vesicles can be seen with a second open shell. This is a shell which is just being ripped off. Thus, the picture gives a nice idea of how the actual mechanism of the transformation to unilamellar vesicles takes place. The radius of some vesicles is of the order of the interlamellar distance between the bilayers as found in Fig. 2A.

Viscoelastic properties of the samples

The three states of the samples have different macroscopic properties. This is demonstrated on the rheological behavior of the samples. Figure 3A shows a rheogram for state I in which the complex viscosity (η^*), the loss (G'') and the storage (G') modulus are plotted against the frequency. The plot shows the unusual situation of a fluid with very weak viscoelatic properties where for low frequencies the storage modulus is higher than the loss modulus but where at higher frequencies the loss modulus is higher than the storage modulus. For many viscoelastic fluids the situation is the other way around like in Fig. 3B where the rheogram for state II is shown. This fluid behaves like a Bingham fluid with a yield stress value and a storage modulus that is frequency independent and always about an order of magnitude higher than the loss modulus. It is also noteworthy that the viscoelastic properties of state II are about two orders of magnitude larger than for state I. State III finally has even more pronounced viscoelastic properties as state II (Fig. 3C).

SANS-measurements

We also monitored the transformation of the L_3-phase to the L_α-phase by SANS measurements. For that purpose a sample in the L_3-phase (100 mM TDMAO/220 mM 1-hexanol) was prepared in D_2O. To this sample, 10 mM methyl formiate were added and the progress of the reaction monitored continuously by SANS measurements. After about 2 h the samples were taken out of the neutron beam and tempered for 24 h at 40 °C to ensure completeness of the reaction. This sample was measured again, and then the sample was shaken vigorously in order to transform it into the vesicle state and then measured again.

Scattering curves for the L_3-phase with and without methylformiate are shown in Fig. 4A. The scattering curve in the presence of methylformiate was recorded about 5 min after the methylformiate had been added to the L_3-phase. Both scattering curves for the L_3-phase are very similar and show the typical scattering features of a L_3-phase, namely a weak correlation peak from which the mean distance d of the bilayers can be determined and a slope of -2 in the double log plot of I against q in the

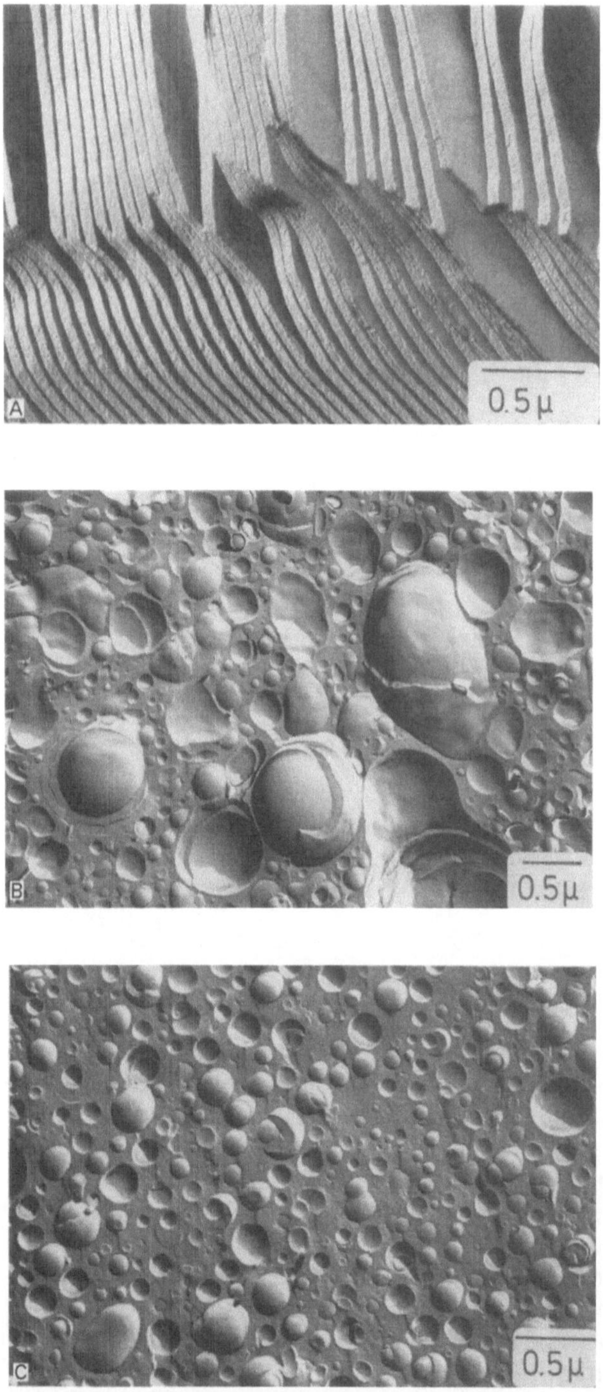

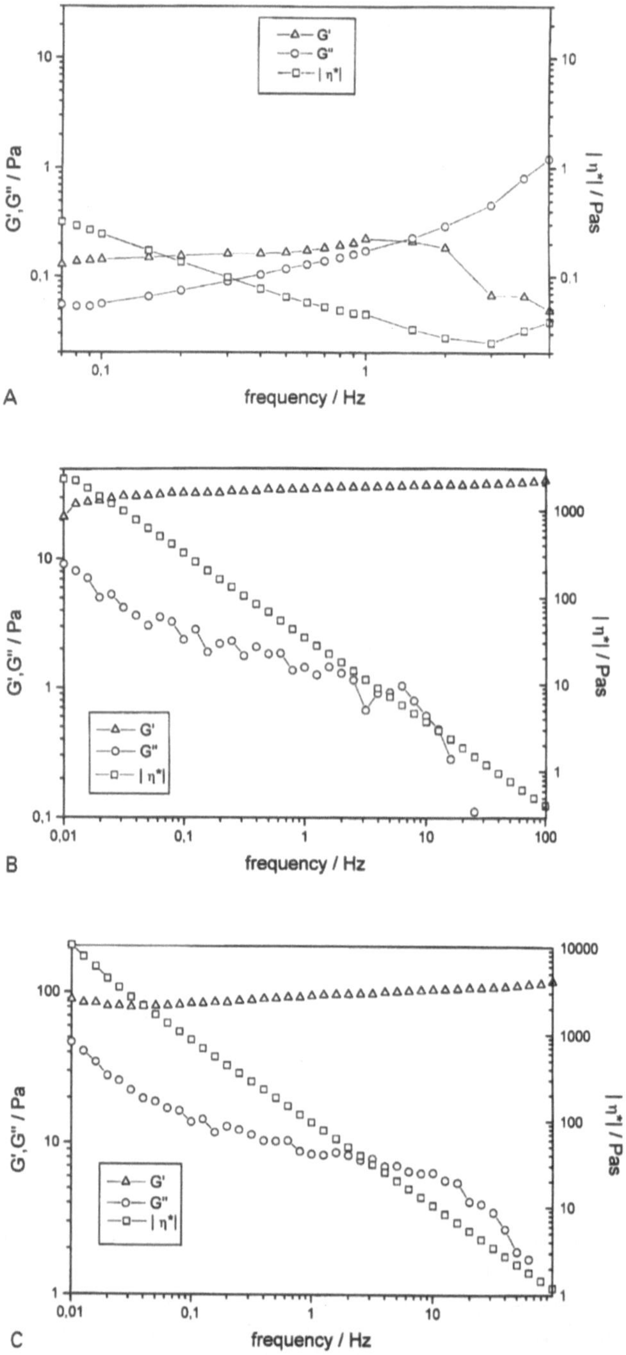

Fig. 2 (A) Electron micrograph of the sample immediately after the hydrolysis reaction (state I). The typical shape of stacked bilayers is displayed (B) Electron micrograph of the sample that was shaken by hand after completion of the hydrolysis reaction (state II). Large multilamellar vesicles with a high polydispersity in size are displayed (C) Electron micrograph of the multilamellar vesicle system (see Fig. 2B) that was sheared in a Couette system for 2 h at a shear rate of $4000\ s^{-1}$ (state III)

Fig. 3 (A) Rheogram of the sample immediately after completion of the hydrolysis reaction (state I). Plot of storage (G'), loss (G'') modulus and the complex viscosity versus frequency (B) Rheogram of the sample after hydrolysis reaction and after shaking by hand (state II). Plot of storage (G'), loss (G'') modulus and the complex viscosity versus frequency (C) Rheogram of the sample after hydrolysis reaction and having been sheared in a couette cell for 2 h at a shear rate of $4000\ s^{-1}$ (state III)

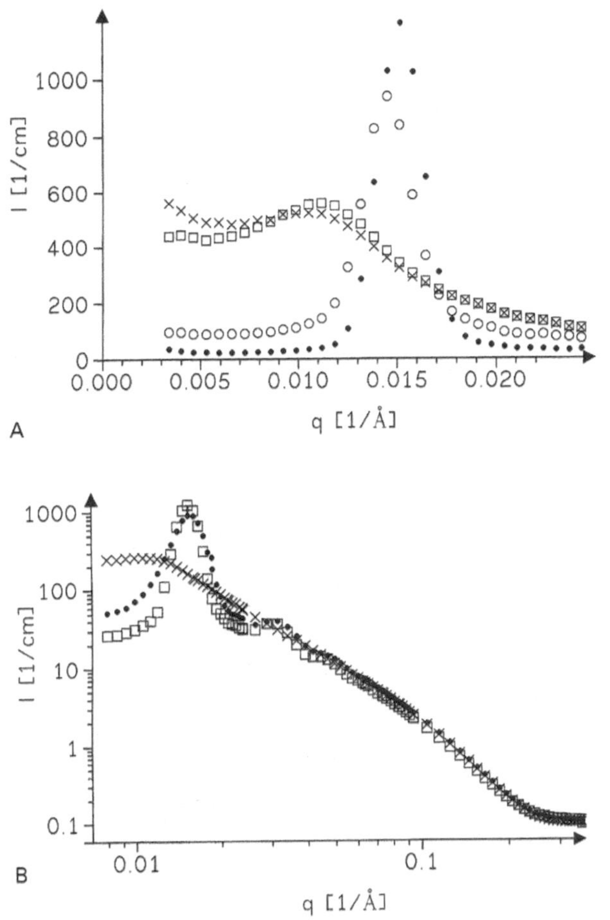

Fig. 4 (A) SANS intensity curves of the system 100 mM TDMAO/220 mM hexanol in D_2O. The original L_3-phase (□), directly after addition of 10 mM methyl formiate (×), after 80 mins (○), and after 28 h and having been tempered for 24 h at 40 °C (●) (B) SANS intensity curves of the system 100 mM TDMAO/220 mM hexanol in D_2O. The original L_3-phase (×), after 28 h and having been tempered for 24 h at 40 °C (□), and having been shaken vigorously (●)

intermediate q region (Fig. 4B), indicating a locally flat structure [17]. At very small q-values the scattering intensities are somewhat lower with the methylformiate. This could be the result of the increased repulsive interaction between the bilayers that is caused by a minute protonation of the bilayers, due to the onset of the hydrolysis reaction.

With passing time a new correlation peak appears that is shifted with respect to the L_3-peak to a higher q value. In relation to q_{max} for the L_3-phase the new peak appears at about 1.3 of q_{max}. This is exactly at the position where the peak is to be expected from previous scattering studies of L_3 and L_α-phases [18–20]. With evolving time the peak becomes sharper and shifts somewhat to even higher q-values. The shift and the narrowing of the peak position is

Table 1 Interlamellar spacing d and lamellae thickness D as obtained from the SANS curves for a system composed of 100 mM TDMAO and 220 mM 1-hexanol, where 10 mM methyl formiate were added. The values are given for the original L_3-phase, and the L_α-phase after the reaction, after being tempered, and finally after having been shaken vigorously

Sample	$d = 2\pi/q_{max}$ [Å]	D [Å]
L_3-phase	607	21.5
L_α after reaction	433	21.6
L_α tempered	415	21.7
L_α after shaking	409	21.7

a result of the increasing long-range order in the system that is due to the charge density of the bilayers. The increasing charge density may increase somewhat the bending constants of the bilayers and in particular the compression modulus of the whole phase. As a consequence the bilayers lose their wrinkles and become more flat and the interlamellar distance narrows somewhat. Similar effects have been observed when bilayers from nonionic surfactants were charged by adding small amounts of charged surfactants to the bilayers [7].

With time the scattering curves finally reach a stationary state that shows at least two higher-order correlation peaks. The thickness D of the bilayers was calculated from the decay of the scattering curve at high q and is given in Table 1. For this q-range the scattering curves for the L_α- and the L_3-phase are identical what shows that the thickness of the bilayers for the two phases remains the same. Finally, we compared the scattering curves of a sample that was shaken after the stationary state had been reached with the scattering curves for a sample that had always been at rest. Both curves are quite similar (Fig. 4B). After shaking the correlation peak is somewhat less pronounced which should be due to the fact that in the multilamellar vesicles only a limited number of shells is present, whereas in the case of the stacked bilayers the correlation of lamellae can be larger. In addition, for the vesicle state, one observes higher scattering intensity at low q. This should be due to the scattering of the individual vesicles that are so large that they give rise for strong scattering at very low q.

This shows that the scattering curves do not reflect clearly the change in the morphology of the L_α-phase that is brought about by the shear. This means that the classic stacked bilayer phase cannot straightforwardly be distinguished by SANS measurements from multi-lamellar vesicles (MLV-phase) that is produced by the shaking (as evidenced by the electron microscopy). The parameters that were evaluated from the scattering curves are given in Table 1. The interlamellar distance d that is determined from the SANS data and that is given in the table is in good agreement with the interlamellar distance as it can be determined from the FF-TEM micrograph.

We also determined a compression modulus K from the scattering data according to the equation

$$K = NkT/S(0)$$

where N is the number density of the vesicles and $S(0)$ the structure factor at zero scattering angle. The value of the compression modulus seems to control the viscoelastic behavior of the phase. According to the theory that was proposed by van der Linden and Dröge, the shear modulus depends both on the bending constants of the bilayers and the compression modulus [21]. The charge density seems to affect the bending constants much less than the compression modulus (as $S(0)$ will depend strongly on the effectiveness of the electrostatic interaction), since it has been found in other experiments that the charge density has only a very minute influence on the bending elasticity of amphiphilic films [22]. It is therefore likely that the shear modulus G_0 of the vesicle phase is mainly dependent on the compression modulus ($G_0 \sim K$) as has been found for other colloidal systems [23].

The scattering properties under shear

In order to monitor the structural changes that occur during the shear process in more detail we performed SANS experiments in a Couette shear cell. The details of the shear cell have been described before [24]. The q-range was chosen in such a way as to cover well the interaction peak that is observed in the unsheared sample and that is located at $q = 0.0112$ 1/Å. This interaction peak is due to the multilamellar structure of the vesicles and corresponds to a mean interlamellar spacing of 560 Å, a value that agrees well with the electronmicrographs (Fig. 2).

In Fig. 5A the radially averaged scattering intensity is given for various shear rates. In order to achieve steady-state condition an equilibration time of more than 1 h at a given shear rate was allowed for before the measurement. The application of shear leads to a marked decrease of the pronouncedness of the peak that is already largely diminished for a shear rate of 200 1/s and disappears completely for shear rates above 1000 1/s. This confirms nicely the result from electron microscopy, i.e., the shells of the multilamellar vesicles are stripped off and the interaction peak disappears when basically no more multilamellar vesicles are present. In addition, it is interesting to note that at still higher shear rates a broad and only little pronounced peak starts to appear at much lower q. For the highest shear rate its position is at $q = 0.00635$ 1/Å. Assuming this to be the interaction peak due to the interparticle interference of the newly formed unilamellar vesicles it corresponds to a mean distance of 1000 Å between neighboring vesicles. This should also be the mean particle diameter for a densely packed system and this is again in very good agreement with the electronmicrographs (Fig. 2C).

In a next step we looked at the relaxation process that takes place after switching off the shear. For this the shear was stopped abruptly after having sheared for more than 2 h at 6000 1/s and SANS curves were recorded in time intervals of 3 mins in the beginning and increasingly longer times after 1 h had passed. Some of the obtained intensity curves are given in Fig. 5B and from this it is clear that a relatively fast relaxation takes place in the first 30 min. Afterwards only fairly small further changes are observed. During this relaxation process the rather weak peak, that

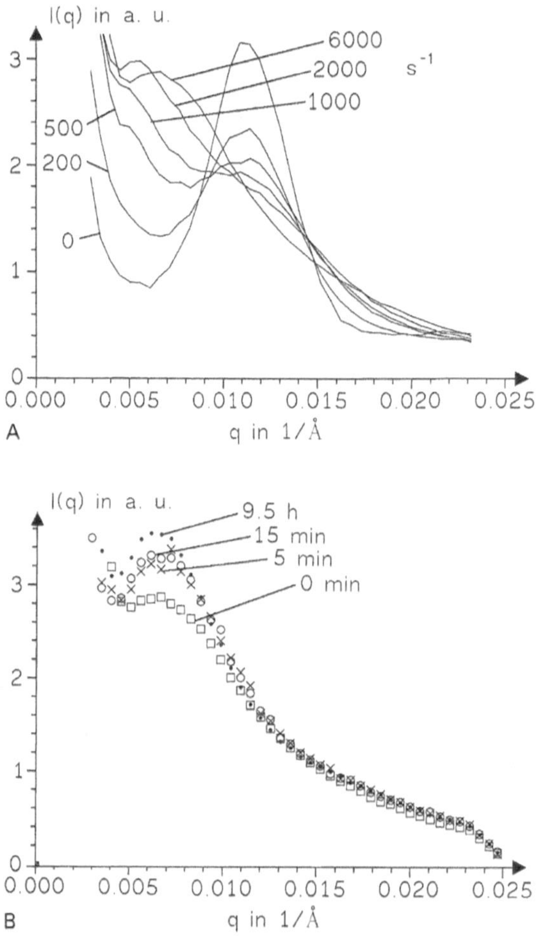

Fig. 5 (A) SANS intensity curves of the system 90 mM TDMAO, 220 mM hexanol and charged with 10 mM tetradecyltrimethyl ammonium bromide in D_2O. The system was sheared in a Couette cell at different shear rates. The system was allowed for at last 1 h to adjust to the various shear rates in order to obtain steady-state conditions (B) The same system as in Fig. 5A is allowed to relax after it had been sheared at 6000 s^{-1} for more than 2 h

is visible during shear, becomes markedly more prominent but it must be noted that its position remains unaltered. This can be interpreted in a way that during the relaxation process the size of the unilamellar vesicles remains constant but their degree of ordering increases. The system relaxes from a highly disordered state that is present under shear to one where the vesicles possess a higher degree of positional ordering. This increase in positional ordering may be accompanied also by a higher degree of monodispersity of the vesicles, since a fairly low degree of monodispersity has to be present in order to observe a more pronounced particle–particle interaction peak.

With all the evidence from the conductivity, rheological, FF-TEM, and SANS experiments [25], we have clearly demonstrated that the multilamellar vesicles can be transformed by shear to unilamellar vesicles. This occurs by consecutive stripping off of the outer shells and is the more pronounced the higher the shear rate. After turning off the shear the system remains permanently altered and the formed unilamellar vesicles are much more monodisperse than the original vesicles. This shows that the state of this vesicle system can be well-controlled by the application of a shear field.

Discussion

It is interesting to compare the results of this investigation with the results of the shear experiments on uncharged systems that were carried out by Roux [26]. These experiment were carried out on a classic L_α-phase that had been prepared in the conventional way. The L_α-phase was stabilized by undulation forces. It was observed that this L_α-phase is aligned by the shear forces and for shearing rates above a threshold value it is transformed into a multilamellar vesicle phase. At even higher shear rates the vesicles are transformed into a completely aligned lamellar phase that was free of defects.

So far we have not been able to observe the transition of the vesicle phase to the defect free L_α-phase. It is, however, conceivable that this transition may also occur in our system, but at shear rates that were not reached in this investigation. The threshold shear rate can depend on the interaction of the bilayers. It would thus be interesting to study the threshold value as a function of the charge density of the bilayers. In this context it is interesting to note that in a recent study on the system AOT in brine that forms a L_α-phase it was possible to form the vesicles under shear but not the defect free oriented bilayer state [27].

The experiments have shown that the charged L_α-phases from ternary systems can be prepared in three reproducible, different states: in a low viscous classic L_α-phase with domains of stacked bilayers, in a highly viscoelastic vesicle state. Both phases are stable for long times and it is impossible to conclude which state is the more stable one in the thermodynamic sense. Obviously one phase has some small excess energy with respect to the other one. However, there is a large activation energy barrier between the phases which can not be overcome by thermal energy. The low viscous phase can easily be transformed into the highly viscoelastic phase but the viscoelastic phase cannot be transformed back into the classic L_α-phase by a simple manipulation.

The L_α-phase can finally be transformed into a third reproducible state [25]. Under high shear rates of several thousand s^{-1} the multilamellar vesicles are transformed into unilamellar vesicles which have a long lifetime. It is yet not clear which of the three different states is the one which is thermodynamically stable.

At this point we would like to emphasize that we do not want to give the impression that vesicles and vesicle phases are only formed when bilayer phases are exposed to shear. For the vesicles that are present in the ternary phase diagrams of zwitterionic or nonionic surfactants and cosurfactants in the $L_{\alpha l}$-phase this is certainly not the case. These vesicles are probably stabilized by a different cosurfactant/surfactant ratio in the in- and out-side monolayer of the vesicles as has been theoretically shown to be expected for surfactant mixtures by Blankschtein [28].

In another experiment we produced the $L_{\alpha l}$-phases by kinetic reactions in which the cosurfactant is formed chemically. We started from a solution which contained 100 mM TDMAO and 140 mM hexylacetate. These solutions are perfectly clear and are in the L_1 state. The ester is solubilized by the surfactant in the interior of small microemulsion droplets. The hydrolysis of the ester under these conditions is very slow and the phases are stable for many weeks. The hydrolysis can however be accelerated by many orders of magnitude by the addition of sodium hydroxide. The reaction

$$C_6H_{13}\text{–}O\text{–}CO\text{–}CH_3 + NaOH \rightarrow CH_3\text{–}CO_2Na + C_6H_{13}OH$$

is complete in a few minutes. Preliminary results have shown that as the reaction proceeds, the microemulsion droplets are transformed to vesicles even though the samples were not stirred or shaken. We intend to study the reaction in a quantitative manner. The obtained result shows clearly that vesicles can indeed be formed without shear.

Conclusions

It can be concluded that the morphologies and the properties of apparently stable thermodynamic surfactant phases depend on the history of their preparation. We have shown this for a phase which can be prepared by mixing the different components or by having one component being produced by a chemical reaction. In this example a simple hydrolysis reaction for the production of an acid was used to ionically charge a system. There are many more possibilities available to change the composition of surfactant phases by chemical reactions and thus many surfactant phases can be produced by chemical reactions without shear forces. In the future we are likely to see more such experiments.

It is demonstrated for the first time that an ionically charged L_α-phase in which the large separation of the bilayers is maintained by electrostatic interaction can be prepared in the stacked bilayer state. This state is reached when shear during the preparation is avoided and the charging of the bilayers is accomplished in the solution at rest by a chemical reaction.

References

1. Dubois M, Zemb T (1991) Langmuir 7:1352
2. Rodgers J, Winsor PA (1969) J Colloid Interface Sci 30:247
3. Ekwall P, Mandell L, Fontell K (1996) Mol Cryst Liqu Cryst 8:157
4. Marignan J, Gauthier-Fournier F, Appell J, Akoum F, Lang J (1988) J Phys Chem 92(2):440
5. Illner JC, Hoffmann H (1995) Tenside Surf Det 32(318):4
6. Thunig C, Hoffmann H, Platz G (1989) Progr Colloid Polym Sci 79:297
7. Schomäcker R, Strey R (1994) J Phys Chem 98:3908
8. Bellocq AM, Roux D (1987) In: Friberg SE, Bothorel P (eds) "Microemulsions: Structure and Dynamics". CRC Press, Boca Raton, FL, USA, p 33
9. Helfrich, W (1994) J Phys Condens Mater 6A:79
10. Hoffmann H, Thunig C, Schmiedel P, Munkert U (1994) Il Nuovo Cimento 16D:1373
11. Oberdisse J, Couve C, Appell J, Berret JF, Ligoure C, Porte G (1996) Langmuir 12:1212
12. Haas S, Hoffmann H (1996) Progr Colloid Polym Sci 101:131
13. Oberdisse J, Porte G, to be published
14. Gradzielski M, Bergmeier M, Müller M, Hoffmann H (1997) J Phys Chem B 101:1719
15. Hoffmann H, Thunig C, Schmiedel P, Munkert U (1994) Langmuir 10:3972
16. Bergmeier M, Hoffmann H, Thunig C (1997) J Phys Chem B 101:5767
17. Kratky O, Porod G (1948) Acta Physica Austriaca 2:1331
18. Milner ,ST, Safran SA, Andelman D, Cates ME, Roux D (1988) J Phys France 49:1065
19. Porte G, Appell J, Bassereau P, Marignan J (1989) J Phys France 50:1335
20. Strey R, Schomäcker R, Roux D, Nallet F, Olsson U (1990) J Chem Soc Faraday Trans 86:2253
21. van der Linden E, Dröge JHM, Physica A 193:439
22. Gradzielski M, Farago B, to be published
23. Ottewill RH, Ber Bunsenges Phys Chem 89:517
24. Norden B, Elvingson C, Eriksson T, Kubista M, Sjöberg B, Takahashi M, Mortensen K (1990) J Mol Biol 216:223
25. Bergmeier M, Gradzielski M, Hoffmann H, Mortensen K, J Phys Chem B, accepted
26. Diat O, Roux D, Nallet F (1993) J Phys II France 9:1427
27. Bergenholtz J, Wagner NJ (1996) Langmuir 12:3122
28. Yuet PK, Blankschtein D (1996) Langmuir 12:3819

Progr Colloid Polym Sci (1998) 109:21–28

E. Brückner
H. Rehage

Solubilization of toluene in phospholipid vesicles studied by video-enhanced contrast microscopy

Received: 11 November 1997
Accepted: 18 November 1997

E. Brückner · Prof. Dr. H. Rehage (✉)
Universität GH Essen
Fachbereich 8
Institut für Umweltanalytik
D-45141 Essen
Germany

Abstract In the present study we employed toluene as a simple model liquid for detailed investigations of the solubilization process of toxic aromatic compounds as it may occur under environmental conditions. We focused our examinations on the interactions with lecithin vesicles because of the possibility that these phenomena enhance mobility of the toxic compounds and cause interactions with organisms.

From measurements of video-enhanced contrast microscopy (VEC), proton resonance, and fluorescence spectroscopy, three different structural regimes could be distinguished depending on the amount of toluene present. In the regime of low toluene concentrations, only spherical aggregates were formed. Increasing the amount of hydrocarbon caused the formation of non-spherical vesicles which exhibited characteristic shape fluctuations. At concentrations near the maximum solubilization capacity of the phospholipids, a separation of the toluene into clusters in the bilayer region occurred; this was followed by a rearrangement of the vesicle shape. Only globular particles were present at higher toluene concentrations. A model is proposed to describe the different solubilization mechanisms.

Key words Vesicles – phospholipids – solubilization – toluene – video-enhanced contrast microscopy

Introduction

Detailed studies of the interaction of hydrophobic compounds with phospholipid vesicles have been the subject of numerous investigations. A general purpose of these studies was to use vesicles as molecular containers for drug delivery systems [1, 2]. Furthermore, liposomes are known as simple model systems for biological cells, and are often used for studies on intra- and intercellular transport processes. In previous investigations, special emphasis was put on the interactions between phospholipid vesicles and toxic, aromatic compounds [3, 4]. These experimental studies were performed with hydrocarbons solubilized in the bilayer of lamellar phases, vesicles or black lipid membranes [5]. The surface active compounds typically used are natural lipids (e.g. egg lecithin) or synthetic, saturated lecithins like dimyristoylphosphatidylcholine (DMPC) and dipalmitoylphosphatidylcholine (DPPC).

The effect of the incorporation of hydrocarbons in the bilayer structure has been studied using experimental techniques like differential scanning calorimetry (DSC) [3], X-ray diffraction [6], ^{1}H-NMR-spectroscopy [7], fluorescence spectroscopy [8, 9] and freeze-fracture electron microscopy [10]. It was generally concluded that short chain alkanes like *n*-hexane are located in the centre of the bilayer and significantly thicken the membrane [5]. A decrease in the phase transition temperature was also observed as a result of the solubilization of small hydrocarbons and aromatic compounds. Long chain alkanes, which are situated between the lipid acyl chains, tend to increase the melting point of the paraffin chains [6, 7]. In general, it is found that the transition temperature depends on the concentration of the solubilized compound [4].

The purpose of the present investigations was primarily the evaluation of biological transport properties in environmental systems. As phospholipids play an important role in the composition of cell membranes, they are an integral component of each type of biological material. Phospholipids can act as surfactants and these "natural surfactants" may cause the solubilization and mobilization of hydrophobic and possibly toxic molecules in the soil and in water [11]. Furthermore, it is well known that vesicles can easily enter living cells. In situations where the membrane contains toxic chemical compounds, these transport processes might lead to previously unknown pathways for pollutant expansion. In order to get a better insight into the complicated mechanisms of surfactant-induced mobilization it is necessary to understand the influence of organic additives on the colloidal structure of the vesicles. It is therefore interesting to examine the interaction of phospholipids with toxic compounds at the microscopic and the molecular level.

In the most simple case, the interaction between a surfactant and an oil leads to dissolution or solubilization processes. On the other hand, the formation of microemulsions is also possible [12]. In these cases, the systems were observed with phospholipids without the addition of a special type of cosurfactant. Most of these microemulsions are only stable in a rather small temperature range.

In contrast to previous studies, we have focussed on large vesicles with bilayer-incorporated toluene as a hydrophobic, environmentally hazardous compound. For the optical measurements we have used video-enhanced contrast microscopy (VEC), a microscopic method that provides visual images with high contrast. This experimental technique can be applied as a powerful tool for the determination of different types of colloidal structures [13]. A special advantage of this instrument is its ability to allow investigations of dynamic shape transitions of amphiphilic structures in aqueous solutions.

Video-enhanced contrast microscopy was developed by Allen et al. [14]. It is based on a conventional light microscope, and includes contrast enhancing techniques like the Normarski differential interference contrast (VEC-DIC) or the polarization microscopy (VEC-POL) [15]. The picture is focussed onto the CCD-chip of a high-resolution black and white video-camera, and after electronically amplifying the contrast and simultaneously adjusting the offset level, the improved image can be visualized on a monitor in real time. This process allows the study of dynamic effects such as shape fluctuations or shape transitions, which cannot be observed with electron diffraction techniques. The observation of particles with a maximum resolution of 150–200 nm is possible with this technique; however smaller submicroscopic particles with diameters less than 20 nm are detectable (e.g. colloidal metal particles) [16]. Furthermore, structural details of colloidal particles are often emphasized by diffraction and hence also become visible.

Experimental part

Materials

Toluene, pyrene, dimyristoylphosphatidylcholine (DMPC) and dipalmitoylphosphatidylcholine (DPPC) were purchased from Fluka (Biochemika). All substances had purities of 99% or greater and were used without further treatment. Deuterium oxide (99.9 at%) was obtained from Aldrich Chemical Company. The water employed in this study was doubly distilled, degassed and saturated with argon prior to use.

Methods

Vesicle preparation

Giant vesicles were prepared according to the swelling method of Reeves and Dowben [17]. Dipalmitoylphosphatidylcholine was dissolved in chloroform and transferred to a round bottom flask; a thin film of the lipid was then obtained by evaporating the solvent under reduced pressure (20 hPa) at a temperature of 50 °C for about 3 h. Afterwards, oxygen-free water was added to obtain a lipid concentration of 1 mM. The dispersion was allowed to swell for several days. After this time, the dispersion was gently shaken under an argon atmosphere, leading to a mixture of multilamellar and unilamellar vesicles, most with diameters of about 10 μm.

Different samples of the dispersion were transferred to air-tight glass bottles, and toluene was added via a septum using a syringe. The samples were gently shaken for about one minute to avoid formation of emulsion droplets and afterwards the solubilization process was continued for several hours.

For the microscopic observations, a small amount of dispersed vesicles was placed in a microslide (0.2 mm height; Camlab, Cambridge), sealed and kept at constant temperature using a specially modified thermostat chamber.

Small unilamellar vesicles used for ^{1}H-NMR and fluorescence spectroscopy were prepared by dispersing dimyristoylphosphatidylcholine in deuterium oxide (^{1}H-NMR-study) or water (fluorescence spectroscopy). The samples were kept under an argon atmosphere with a surfactant concentration of 10 mM. Small unilamellar vesicles were formed by cavitation forces in an ultrasound bath reactor (Bandelin, Sonorex TK 52). The temperature was adjusted to 10° above the phase transition and the ultrasonic treatment was continued until the dispersion remained clear.

Video-enhanced contrast microscopy

Video-enhanced contrast microscopy was performed using a BH-2 microscope (Olympus Optical Company). The instrument was equipped with Normarski differential interference contrast and intensive illumination obtained by the use of a 250 W projector lamp. For the present investigations a 1.4 NA oil immersion condensor, a 1.4 NA oil immersion 60 × objective and a 6.5 × zoomocular were used. The pictures were visualized on a Sony, HR Trinitron PVM 1454 QM video screen. The device allowed magnifications of about 14 000 times. The video signal was contrast-enhanced by an image processor (Argus 20, Hamamatsu Photonics) including a frame memory to perform background subtraction, averaging and digital contrast manipulation. Real time observations were recorded on S-VHS video tape. Freeze frame images were digitally transferred to a personal computer to obtain better signal-to-noise ratios.

^{1}H-NMR spectroscopy

Proton magnetic resonance spectra of saturated phospholipids in small unilamellar vesicles in deuterium oxide were recorded at 200 MHz using a Varian Gemini XL 200 spectrometer. Dispersions of small unilamellar DMPC vesicles (10 mM) in deuterium oxide were investigated at temperatures above the phase transition. DPPC vesicles were only investigated in the gel state. Experiments with varying toluene/phospholipid ratios were carried out by titration with toluene using a syringe.

Fluorescence spectroscopy

Fluorescence measurements were carried out using a Jobin Yvon spectrofluorimeter JY 3. The excitation wavelength was set at 338 nm, with a bandwidth of 4 nm. A dispersion of small unilamellar DMPC vesicles (10 mM) and pyrene (2.5 μM) in degassed water was filled into a 1 cm septum-fitted cuvette under an argon atmosphere to avoid fluorescent quenching by oxygen. Measurements were performed by titration with argon saturated toluene using a syringe.

Results

Video-enhanced contrast microscopy

In a series of experiments we have systematically varied the amount of toluene which was dissolved in giant DPPC-vesicles. The studies were performed at temperatures at which vesicles of the pure phospholipid are in the gel state. Typical results of microscopic investigations with varying toluene content are summarized in Figs. 1–3.

At low toluene levels, with concentrations of up to 0.05% (v/v), spherical DPPC-vesicles were predominant (Fig. 1). At increasing toluene concentrations a phase transition to smaller vesicles and myelin-like structures occurred. Similar phenomena have been reported from the interaction of vesicles with benzene [18]. Up to toluene concentrations of 0.2% (v/v), non-spherical DPPC-vesicles with low symmetry were predominant (Fig. 2). These particles showed a wide range of shape fluctuations due to the interactions between toluene and the acyl groups of the phospholipids. At this point, the acyl–acyl van der Waals interaction of the phospholipids is reduced and the membrane exhibits a more fluid-like behaviour. The transition to the gel state of these aggregates occurs at much lower temperatures than of vesicles without toluene.

Adding more than 0.2% (v/v) of toluene to the vesicle phase an O/W emulsion in coexistence with vesicles could be directly observed. It is interesting to note that if the

Fig. 1 Giant DPPC-vesicle in the gel-state suspended in water after addition of 0.03% (v/v) toluene. This image gives a two-dimensional representation of the globular structure. The bilayer distance is enlarged due to the limited depth of focus

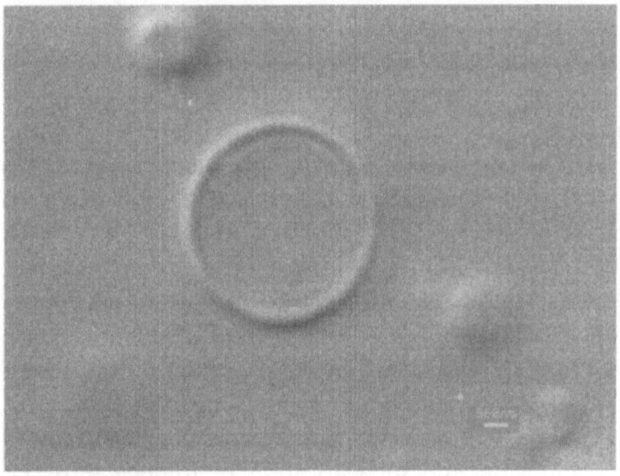

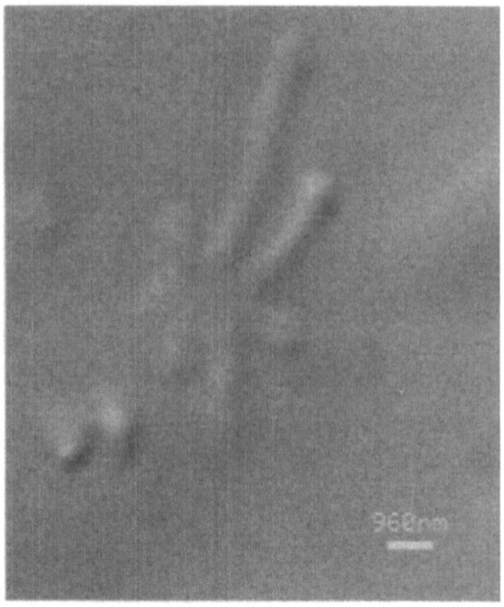

Fig. 2 Non-spherical vesicle structure of DPPC in pure water (1 mM) after addition of 0.15% of toluene. The structure shown is known as starfish vesicle [30]. The membrane shows tubular parts with very low curvatures. The end caps, however, have globular geometry with a small radius of curvature

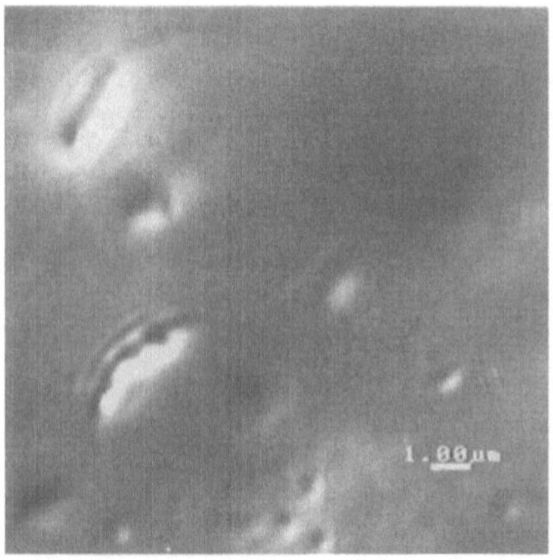

Fig. 3 Spherical vesicles with toluene droplets dissolved within the bilayer region (bright spots). Toluene can easily be identified by video-enhanced contrast microscopy because of the refraction index of this aromatic hydrocarbon ($n = 1.49$) is different to that of water ($n = 1.33$). The vesicles seem to be inflated and do not show undulations after incorporation of toluene

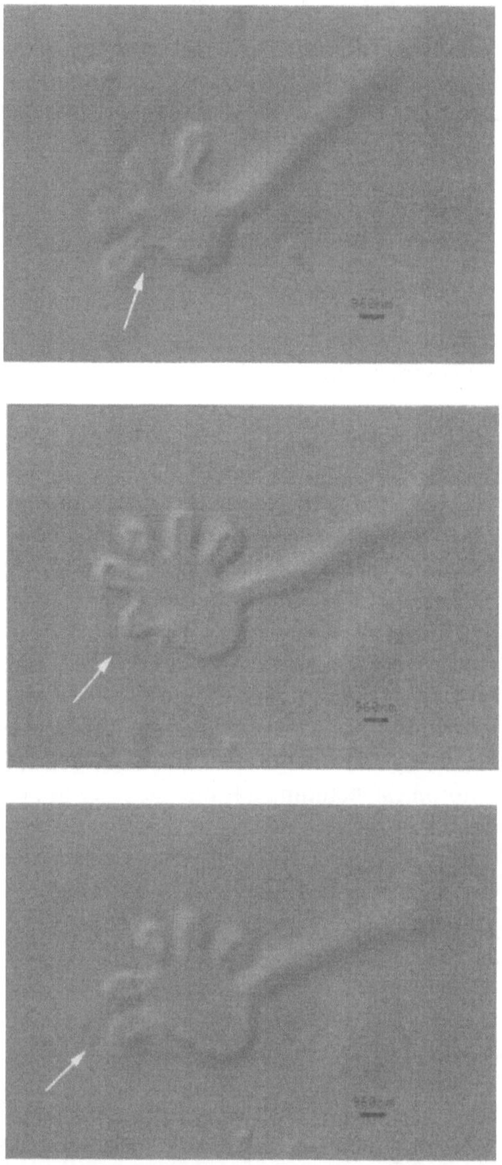

Fig. 4 Budding phenomena (marked by the arrow) observed with DPPC vesicles where toluene was added. The vesicles showed striking undulations motions indicating a low surface tension and bending elasticity of the fluid-like membrane. The bar denotes a distance of 960 nm

vesicle shapes were predominantly spherical, lens- or drop-like reservoirs of toluene were observed within the membrane (Fig. 3). At the present time, we cannot determine that the toluene is in fact located in the inner bilayer region or whether this phenomena is simply the result of an oil droplet adsorbed on the surface of the bilayer. In order to clarify this observation, the dispersions were also studied immediately after the addition of the aromatic hydrocarbon.

For toluene concentrations of 0.2% (v/v) various shape fluctuations were observable and budding transitions were frequent (Fig. 4). This shape modification can be explained by the diffusion and incorporation of toluene into the bilayer region, as follows. The swelling of the membrane changes the volume to area ratio of the vesicle. Assuming a constant vesicle volume (diffusion of water from the inner compartment into the bulk phase is slow [19]), the addition of excess hydrocarbon leads to a larger surface area; however, after some time this excess surface is again reduced by forming globular particles. The initial structures seem to be metastable and tend to rearrange to a spherical shape (Fig. 5). These globular aggregates are characterized by toluene droplets which are observed at different positions within the membrane.

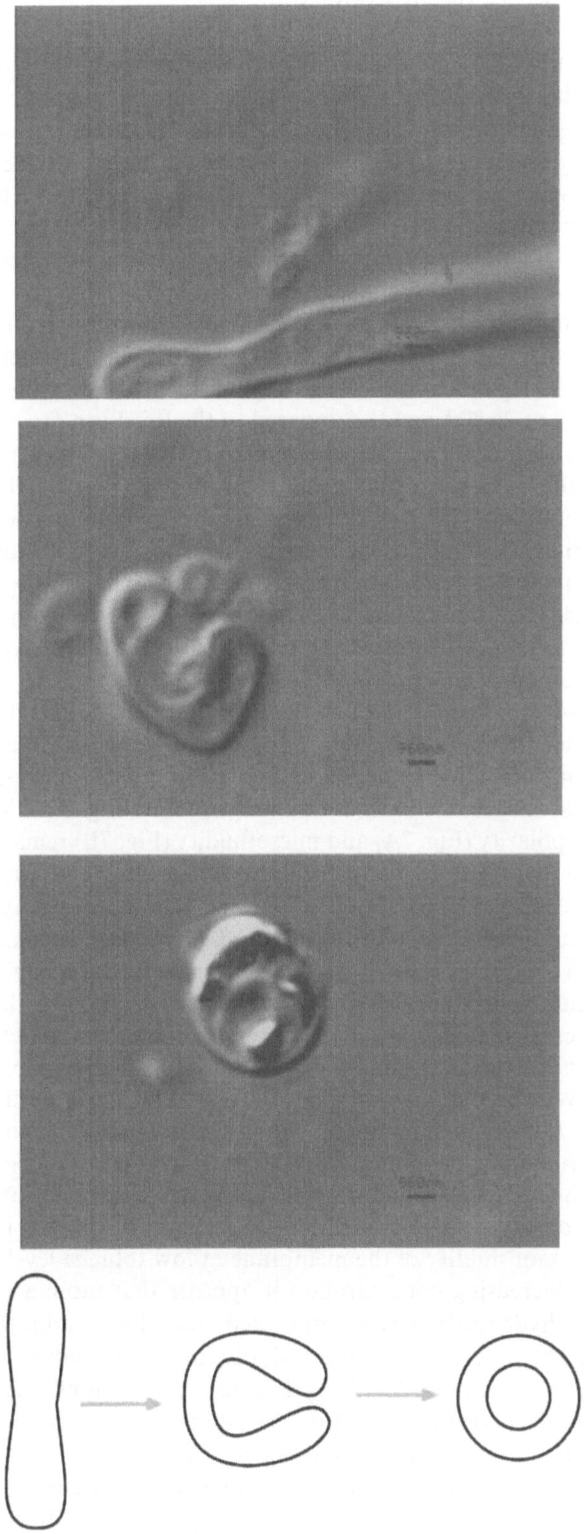

^{1}H-NMR Spectroscopy

Proton resonance spectroscopy was used to measure the influence of solubilized toluene at the molecular level. This experimental technique is especially useful to investigate small particles, which cannot be visualized by video-enhanced microscopy. The ^{1}H-NMR-spectra of small unilamellar vesicles of DMPC (fluid-state) and DPPC (gel-state) were recorded for different toluene concentrations, and these experiments were then compared to the microscopical investigations of larger aggregates. For ^{1}H-NMR-experiments, the phospholipid concentration was about 10 times larger than for microscopic investigations to obtain better intensities. The solvent was deuterium oxide instead of water. These restrictions limit the comparison with the microscopic studies to a qualitative level.

The proton resonance spectra of DMPC vesicles show a strong upfield shift of the methyl and methylene resonances of the lipid acyl chains. This holds up to a toluene/DMPC ratio of 4.7 and remains constant as the ratio is further increased (Fig. 6A). The shielding of the protons can be explained by the decrease of the van der Waals interaction of the acyl chains [20] and indicates that toluene is incorporated into the bilayer. Furthermore, it provides evidence of the interaction of the methylene and the terminal methyl groups of the lipid acylchains with the π-system of toluene. The protons of the acyl chains located above the aromatic ring are known to show an significant upfield shift caused by the ring current [21, 22].

The aromatic toluene-proton signals also show an upfield shift with a plateau regime starting at a toluene/DMPC ratio of 6.4 (Fig. 6B). This is a larger value than observed for the methyl resonance of the lipid acyl chains (4.7), and hence we can conclude that the toluene molecules are first interacting with paraffin chains of the surface active lipids of the vesicle. Upon the addition of excess aromatic hydrocarbons, toluene molecules are also in contact with each other, which leads to π-stacking interactions of the aromatic system [23].

The upfield shift of toluene solubilized in the membrane of small unilamellar DPPC-vesicles is summarized

Fig. 5 (a–c) Shape transition of a multilamellar vesicle. Following an invagination a smaller vesicle entrapped in the bigger aggregate can be observed. In this case the phase separation of toluene droplets could directly be observed (bright spots in the image). The sequence shows the same vesicle after time intervals of 10 min. It is interesting to note that after 20 min vesicles with phase separated toluene droplets were predominant in the whole sample. This indicates that the fluid reservoirs are formed by contraction and reorganisation processes within the membrane. (d) Schematic drawing of the shape transition shown in Fig. 5a–c. Vesicle shapes can be simplified as discocyte- (Fig. 5a), stomatocyte-like (Fig. 5b) and an entrapped vesicle (Fig. 5c)

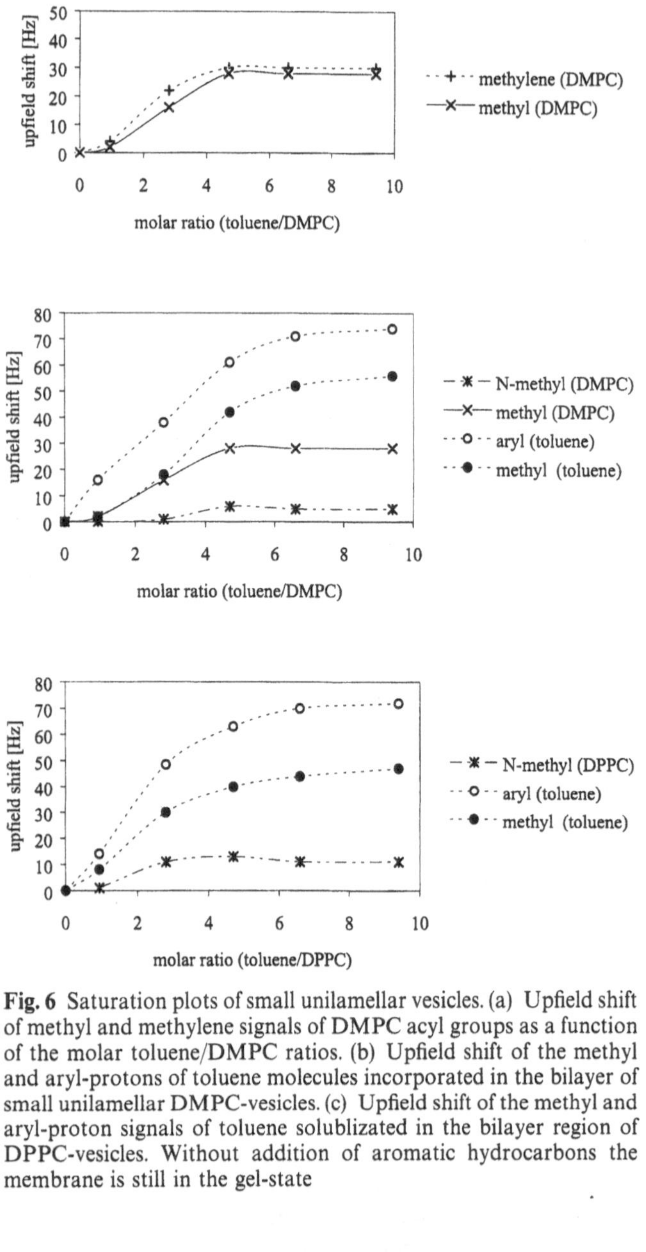

Fig. 6 Saturation plots of small unilamellar vesicles. (a) Upfield shift of methyl and methylene signals of DMPC acyl groups as a function of the molar toluene/DMPC ratios. (b) Upfield shift of the methyl and aryl-protons of toluene molecules incorporated in the bilayer of small unilamellar DMPC-vesicles. (c) Upfield shift of the methyl and aryl-proton signals of toluene solublizated in the bilayer region of DPPC-vesicles. Without addition of aromatic hydrocarbons the membrane is still in the gel-state

in Fig. 6C. These systems are still in the gel-state but become liquid-like upon the addition of toluene. A curve with a plateau, similar to that of DMPC vesicles, can be observed, but the saturation point occurs at a lower toluene/phospholipid ratio. The result that the short chain phospholipid shows a stronger solubilization capacity compared to the surface active compound with longer paraffin chains is at first glance surprising; however, this anomaly has been observed before on the addition of lindane [4]. The incorporation of lipophilic substances in the bilayer can be correlated with the degree of fluctuations in the membrane. Shorter acyl chains are known to sustain stronger fluctuations [3, 24].

Fluorescence spectroscopy

To study the microenvironment, especially the micropolarity and the microfluidity of the bilayer membrane as a function of solubilized toluene, we used pyrene as a fluorescence probe. This technique has been described in several publications [25, 26]. Pyrene is known to be incorporated deep into the lipophilic core of phospholipid bilayers. The fluorescence spectra of these molecules exhibit two important features: first, the monomer fluorescence intensities show five vibronic bands from the fine structure (peak *I–V*) which are strongly solvent dependent. The intensity of the 0–0 transition (peak I at 26 800 cm^{-1}) compared to the third peak (peak III) is significant enhanced if polar solvents are present, and the I/III ratio thus gives some information on the micropolarity [27]. Furthermore, the pyrene excited monomers can form excimers by face to face collision with pyrene molecules in the ground state. The emission intensity of the excimer as compared to the monomer (I_e/I_m) is an indicator of the microfluidity of the bilayer membrane.

Spectroscopic data of the intensity ratio I_1/I_3 and the excimer to monomer ratio I_e/I_m of pyrene in small unilamellar DMPC vesicles after addition of different toluene concentrations are summarized in Fig. 7. The micropolarity (Fig. 7A) and microfluidity (Fig. 7B) remain nearly constant at low toluene/DMPC ratios, and with increasing toluene concentrations the microviscosity increases. The micropolarity (Figure 7A) reaches the maximum at a molar ratio of about 7. This phenomenon might be due to the penetration of water molecules into the bilayer structure [28]. It is observed that this process occurs at concentrations of maximum solubilization capacity. We observed a threshold value of 0.95 for the intensity ratio I_1/I_3 of pyrene in toluene in the absence of a any phospholipid. This value describes the partition of pyrene in toluene droplets and water of an oil in water emulsion. The observed intensity ratio I_e/I_m (Fig. 7B) points to a constant fluidity of the membrane at low toluene levels. With increasing concentration it appears that more aromatic hydrocarbon is incorporated into the membrane until the saturation is reached (fluorescence studies of DPPC membranes in the gel state do not show such a distinct plateau value). Further addition of toluene results in a dramatic increase in the membrane's fluidity. At this concentration level the microviscosity decreases and with excess toluene a separation of droplets occurs. In the region of large toluene concentrations, we observe another plateau regime at an excimer to monomer ratio of 0.18 (Fig. 7B). This threshold value can also be measured in an oil-water emulsion of toluene without adding any surfactant material.

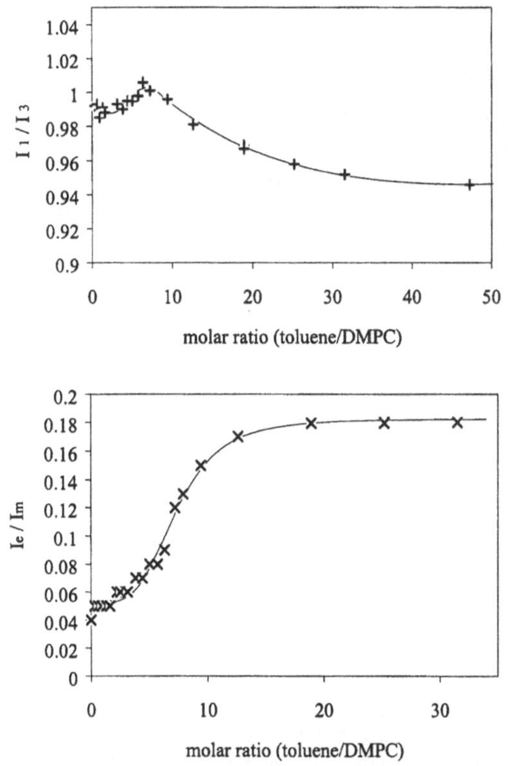

Fig. 7 (a) Variation of the I_1/I_3 ratio of pyrene fluorescence with the molar toluene/DMPC ratio in small unilamellar DMPC-vesicles (10 mM, $T = 25\,°C$, Pyrene concentration is 2.5 μM). (b) Determination of the microviscosity from fluorescence measurements. Plot of the excimer- to monomer-intensity ratio of pyrene in small unilamellar DMPC vesicles at different toluene concentrations

Discussion

In this publication we present our analysis of the shape and size of giant DPPC vesicles after addition of toluene by means of video-enhanced contrast microscopy further correlated to spectroscopic studies of the system. The experimental results show a strong influence of the shape of the vesicles on the amount of toluene incorporated into the membrane. Depending on the concentration of the aromatic hydrocarbon, three different structural regimes could be distinguished. Globular structures were mainly observed at low toluene concentrations. At increasing hydrocarbon content, non-spherical aggregates were formed and, at concentrations above the solubilization capacity, spherical vesicles seem to be in a stable coexistence with incorporated oil droplets. These results were compared with experiments performed with small unilamellar DMPC- and DPPC-vesicles. We used proton resonance and fluorescence spectroscopy in order to obtain information on the solubilization behaviour of toluene. We have found that the incorporation of aromatic compounds into the hydrophobic part of the bilayer has a significant influence on the lipid–lipid interaction. An upfield shift in the proton resonance spectra for the methyl and methylene groups of the acyl chains could be observed for DMPC vesicles with increasing toluene concentration. The *N*-methyl signals corresponding to the phospholipid head group showed only a slight shift at the same toluene concentrations, hence we can conclude that the average distance between the surfactant molecules is still unaffected under these experimental conditions. The addition of toluene into the membrane leads to the decoupling of the lipid–lipid acyl-chain interactions, a process with great potential influence on the physico-chemical features of amphiphilic systems. A change of the elastic membrane properties, and a strong decrease of the phase transition temperature is observed with microscopic techniques. This result is in excellent agreement with the fluorescence spectroscopic data. Our results show a strong fluidization and also the loss of the structural headgroup integrity of DMPC vesicles at high toluene concentrations. These experiments lead to the conclusion that the incorporation of toluene into the hydrophobic part of the bilayer occurs in separate steps. At low concentrations toluene is distributed uniformly between the paraffin chains of the surface active compounds. Further addition of excess toluene causes an inter-chain separation and swelling process in the lipids. At this point shielding of the acyl protons as apparent in the proton resonance spectrum reaches a maximum.

On a microscopic scale, the uptake of the solubilizate changed the spherical regime of DPPC vesicles to non-spherical aggregates with various shape and size fluctuations. This can be understood by considering that the strong aryl–acyl interaction affects the geometric packing $p = v/(a \cdot l)$ of the amphiphilic compounds. Here a denotes the optimum head group area (at the minimum energy), l is the critical chain length and v is the volume occupied by paraffin chains. The value of p changes, as expected, by increasing the effective volume v. As a further result of the increasing volume the truncated cone of the amphiphilic molecules attains a more cylindrical shape, which facilitates the formation of elongated, non-spherical structures [29].

It is interesting to note that at the maximum solubilization capacity the bilayer fluidization seems to reach a maximum. Under these conditions we have observed that the non-equilibrium vesicle structures undergo thermal shape fluctuations. These processes can be attributed to aryl–acyl interactions and a possible swelling process of the bilayer structure. For this concentration regime intermolecular toluene interactions become more and more pronounced and a small fluctuation may lead the system to adopt new conformations. This can occur if the energy gain given by the droplet formation of toluene overcomes the term of mixing entropy. This situation is observable with DPPC-vesicles in the gel-state.

Phase separation often accompanies with internal vesicle fusions and structural transitions. As result of such phase separations, spherical vesicles with microscopic amounts of solubilized toluene located in droplets or lens-like pockets within the bilayer region were observable (Fig. 3).

Such internal phase separations, however, could not be observed in the case of DMPC vesicles which were in the fluid state. These phospholipids exhibit a large amount of different conformations and they have a much stronger solubilization capacity for non-polar compounds. In these structures, we did not detect excess toluene droplets within the bilayer membrane.

Concluding remarks

Video-enhanced contrast microscopy in combination with spectroscopic techniques constitute a powerful tool for the investigation of the dynamic phenomena of vesicles in solution. The experimental method described here allows detailed examinations of the interaction of hydrophobic compounds with phospholipid vesicles.

Acknowledgement We gratefully acknowledge the financial support by grants of the Deutsche Forschungsgemeinschaft (DFG Re 681/6-1 and GRK 153/2-96: Verbesserung des Wasserkreislaufs urbaner Gebiete zum Schutz von Boden und Grundwasser) and of the Fonds der Deutschen Chemie.

References

1. Gregoriadis G (1977) Nature 263:407
2. Gregoriadis G (1983) In: Gregoriadis G (ed) Liposome Technology. CRC Press, Boca Raton, Florida
3. Sabra MC, Jorgensen K, Mouritsen OG (1995) Biochim Biophys Acta 1233:89
4. Antunes-Madeira MC, Madeira VMC (1989) Biochim Biophys Acta 982:161
5. Requena J, Billett DF, Haydon DA (1975) Proc R Soc London A 347:141
6. McIntosh TJ, Simon SA, MacDonald RC (1980) Biochim Biophys Acta 597:445
7. Hunt GR, Tipping LR (1978) Biochim Biophys Acta 507:242
8. Barenholz Y, Cohen T, Korenstein R, Ottolenghi M (1991) Biophys J 59:110
9. Wang S, Beechem JM, Gratton E, Glaser M (1991) Biochem 30:5565
10. McIntosh TJ, Costello MJ (1981) Biochim Biophys Acta 645:318
11. Eckelhoff A, Hirner AV (1997) Vom Wasser 89:297
12. Schurtenberger P, Peng Q, Leser ME, Luisi P-L (1993) J Colloid Interface Sci 156:43
13. Kachar B, Evans DF, Ninham BW (1984) J Colloid Interface Sci 100:287
14. Allen RD, Travis JF, Allen NS, Yilmaz H (1981) Cell Motil 1:291
15. Allen RD, Allen NS, Travis JF (1981) Cell Motil 1:275
16. Salmon ED (1992) Proc 50th Meeting of the Electron Microscopy Society of America, p 414
17. Reeves JP, Dowben RM (1969) J Cell Physiol 73:49
18. McDaniel RV, Simon SA, McIntosh TJ, Borovyagin V (1982) Biochem 21:4116
19. Ye R, Verkman AS (1989) Biochem 28:824
20. Schuh JR, Banerjee U, Müller L, Chan SI (1982) Biochim Biophys Acta 687:219
21. Miyagishi S, Nishida M (1980) J Colloid Interface Sci 78:195
22. Bunton CA, Cowell CP (1988) J Colloid Interface Sci 122:154
23. Zimmerman SC, Wu W, Zeng Z (1991) J Am Chem Soc 113:196
24. Ipsen JH, Jørgensen K, Mouritsen OG (1990) Biophys J 58:1099
25. Galla H-J, Sackmann E (1974) Biochim Biophys Acta 339:103
26. L'Heureux GP, Fragata M (1988) Biophys Chem 30:293
27. L'Heureux GP, Fragata M (1987) J Colloid Interface Sci 117:513
28. Kalyanasundaram K, Thomas JK (1977) J Am Chem Soc 99:2039
29. Israelachvili J (1992) In: Intermolecular & Surface Forces. Academic Press, London
30. Wintz W, Döbereiner H-G, Seifert U (1996) Europhys Lett 33:403

Progr Colloid Polym Sci (1998) 109:29–34

D. Exerowa
R. Ivanova
R. Sedev

On the origin of electrostatic interaction in foam films from ABA triblock copolymers

Received: 29 August 1997
Accepted: 5 September 1997

Prof. Dr. D. Exerowa (✉) · R. Ivanova
R. Sedev
Bulgarian Academy of Sciences
Institute of Physical Chemistry
Sofia 1113
Bulgaria

Abstract The effect of pH on the equilibrium thickness and the diffuse double layer potential of microscopic foam films from two non-ionic ABA triblock copolymers of ethylene oxide (A) and propylene oxide (B) – Synperonic P85 and Synperonic F108 – is studied by the microinterferometric method of Scheludko–Exerowa. A strong effect of pH (within the range 2.0–6.0) on the film thickness at constant ionic strength and capillary pressure is found. The film thickness decreases with decreasing pH. With Synperonic P85 at the critical value, pH_{CR} (=3.8), a transition to black films (thickness 15 nm) occurs. With Synperonic F108 dark gray films (thickness 40 nm) are observed below $pH_{CR,ST}$ (=3.2). Further decreasing of pH has no influence on the film thickness. Above the critical pH values the films are electrostatically stabilized. The values of the diffuse double layer potential, φ_0, and the surface charge density, σ, are estimated from the DLVO theory. Both φ_0 and σ diminish to zero when approaching pH_{CR}. These results corroborate our hypothesis that electrostatic interaction (i.e. φ_0 and σ) in foam films from non-ionic surfactants arises from preferential adsorption of OH^- ions at the solution/air interface. The films obtained below the critical pH values are sterically stabilized, i.e. decreasing the pH induces a transition from electrostatic to steric stabilization. The parameter $pH_{CR,ST}$ characterizes the transition to thicker than black, sterically stabilized films.

Key words Amphiphilic block copolymers – microscopic foam films – electrostatic interaction – DLVO

Introduction

In the last few years, foam films from aqueous non-ionic surfactant solutions have become an object of increasing interest [1–7]. These works continued the pioneering studies [8–11] and enlarged the scope of such investigations. Various non-ionic surface active substances, including zwitterionic phospholipids [2, 12] and amphiphilic block copolymers [3, 4] have been studied. All these studies convincingly showed that at low ionic strength electrostatic interaction is stabilizing the foam film.

The origin of the surface charge and potential at the solution/air interface in the case of nonionic surfactant solutions has been elucidated in foam film studies to determine the effect of pH on the diffuse double layer (DDL or φ_0) potential at low ionic strength [10, 11]. The surface charge and DDL potential in foam films from nonionic surfactants arise from preferential adsorption of OH^- ions [10].

It has been demonstrated that the φ_0-potential decreases with decreasing pH and drops to zero at a specific value pH*, termed pH "isoelectric" [10]. Decreasing the pH effectively suppresses the electrostatic repulsion and below pH* films do rupture [10, 11]. When surfactant concentration is high enough to obtain saturated adsorption layers a transition to Newton black films (bilayers) occurs at a specific value pH_{CR}, called pH "critical" [13, 14].

This hypothesis for the surface charge and potential formation is consistent with the observation of a stabilizing effect of Ca^{2+} ions on foam films from zwitterionic biosurfactant [2] and the charge reversal at the water/air interface due to adsorption of cationic surfactant [15]. It is also consonant with electrokinetic studies of microbubbles produced in aqueous solutions of electrolytes where the OH^- ions excess at the bubble interface lead to negative zeta potentials at the air/solution interface [16].

The determination of the φ_0-potential and/or the surface charge density from experimentally measured equilibrium foam film thickness via the DLVO theory is called "equilibrium foam film" method. A concise account of the scope, achievements and limitations of this method is given in Ref. [17].

Recently, we reported a significant effect of electrolyte concentration, C_{EL}, on the thickness of microscopic foam films from amphiphilic block copolymers of the Synperonic PE series [4]. This influence shows that electrostatic repulsion is operative. The film thickness which (at low electrolyte concentration) reflects the DDL thickness decreases with increasing C_{EL}. The neutral electrolyte (NaCl) shrinks the DDL thickness until a critical concentration $C_{EL,CR}$ is reached. Beyond $C_{EL,CR}$ steric repulsion becomes predominant and no further influence of C_{EL} on film thickness is detected. These results are similar to earlier findings for other non-ionic surfactants of the EO type [1].

This paper is a continuation of our previous reports on foam films from non-ionic PEO-PPO-PEO triblock copolymers [3, 4]. Our aim is to demonstrate the origin of surface charge and potential formation in foam films from these copolymers. To achieve this we investigate here the effect of pH on the equilibrium film thickness at constant temperature, surfactant concentration, ionic strength, and capillary pressure.

Materials and methods

The ABA triblock copolymers of polyethylene oxide (A blocks) and polypropylene oxide (B block) – Synperonic P85 and F108 – were received from ICI Surfactants, Witton, UK (Courtesy of Prof. Th.F. Tadros) and used without further purification. The molecular weight and PEO content are known from the manufacturer and the following chemical formulae are deduced: $EO_{27}PO_{39}EO_{27}$ for P85 ($M_w = 4600$) and $EO_{122}PO_{56}EO_{122}$ for F108 ($M_w =$ 14000). The two copolymers have hydrophobic PPO blocks of a similar size while the hydrophilic PEO blocks differ significantly.

NaCl (p.a.) and HCl were purchased from Merck. Prior to usage the NaCl was roasted at 500 °C for several hours to remove any organic (surface-active) contamination.

Aqueous solutions of the copolymers were prepared with triple distilled water (specific conductivity $10^{-6}\,\Omega^{-1}\,cm^{-1}$ and pH = 5.8 at 23 °C). The copolymer stock solutions were used for not more than a week. Copolymer solutions with various electrolyte concentrations and various pH (from 5.8 down to 2.0) were prepared by adding NaCl and HCl, keeping the ionic strength constant. The pH of the solutions was measured with a Radelkis OP-265 pH-Meter within ±0.05.

Synperonic PE copolymers are stable to and do not react with most acids or alkalies even at elevated temperatures [18] and therefore no changes in their chemical structure or degree of polymerization are expected under these experimental conditions.

The thickness of microscopic horizontal foam films (radius $r \sim 10^{-2}$ cm) was measured by the microinterferometric method of Scheludko–Exerowa [19–22]. The foam film is formed in the middle of a biconcave drop from the solution suspended in a vertical glass tube (radius $R \sim 0.2$ cm) enclosed in a vapor saturated environment. The film is observed in monochromatic light. From the changes of the intensity of the reflected light during film thinning an optical thickness can be deduced [23]. Assuming that the film is optically homogeneous with a refractive index equal to the refractive index of the solution the equivalent film thickness, h_w, is calculated. The accuracy of the method is estimated to 0.2 nm.

A vertical tube identical to that holding the biconcave drop was immersed in the same solution and the capillary pressure was directly measured by the capillary rise method [22, 24].

All experiments were carried out at temperature 23.0 ± 0.1 °C.

Results

The variation of the film thickness h_w with pH for P85 ($C_S = 3 \times 10^{-6}$ M) is shown in Fig. 1. The temperature is kept constant (23 °C). The ionic strength, I, is constant (10^{-3} M) and corresponds to a 1-1 electrolyte concentration lower than $C_{EL,CR}$ [4]. The capillary pressure counterbalancing the disjoining pressure inside the film is also constant (50 N/m^2). At higher pH the thickness h_w is constant and decreases when pH decreases below 5.0. In the range 3.8–5.6 either thicker silver films (○) or thinner black films (●) are obtained at one and the same pH, i.e. this is

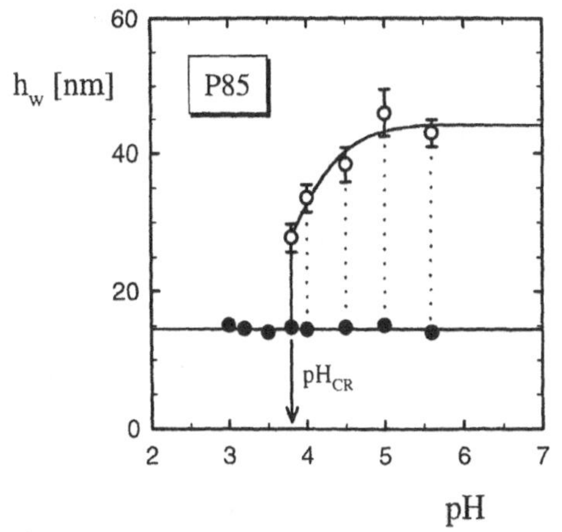

Fig. 1 Dependence of the equivalent film thickness, h_w, on pH of the aqueous solution of Synperonic P85 ($C_S = 3 \times 10^{-6}$ M P85, $I = 10^{-3}$ M, 23 °C). The dotted lines show the metastable zone where either silver (○) or black films (●) are formed at one and the same pH. Below $pH_{CR} = 3.8$ (the arrow) only black films are formed

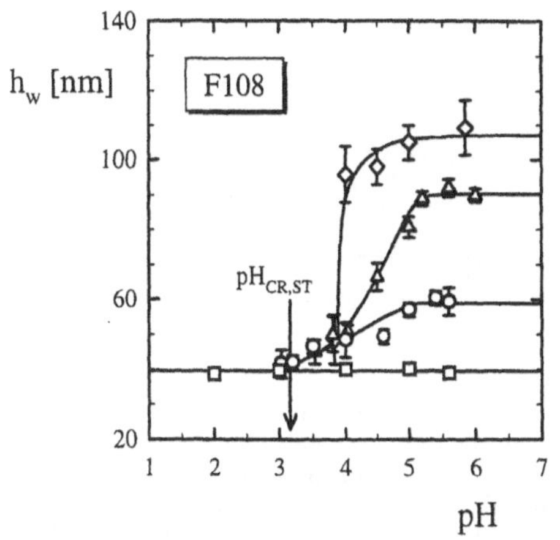

Fig. 2 Dependence of the equivalent film thickness, h_w, on pH of the aqueous solution of Synperonic F108 ($C_S = 7 \times 10^{-7}$ M F108; ◇: $I = 1.5 \times 10^{-4}$ M, △: $I = 3 \times 10^{-4}$ M, ○: $I = 10^{-3}$ M, and □: $I = 10^{-2}$ M; 23 °C). Below $pH_{CR,ST} = 3.2$ (the arrow) film thickness is pH independent

a metastable zone. Below $pH_{CR} = 3.8$ black films are formed only and their thickness (15 nm) is pH independent.

The effect of pH on h_w for F108 ($C_S = 7 \times 10^{-7}$ M F108) at four different values of the ionic strength is shown in Fig. 2. The overall trend is similar to that with P85: with decreasing pH the thickness decreases from a plateau to a common lower value of 40 nm. For all curves the plateau occurs above pH = 5.0 and its value decreases with increasing the ionic strength. The lower the ionic strength the steeper is the dependence. Below $pH_{CR,ST} = 3.2$ the thickness is independent of pH. At about pH = 4 all curves converge though they stop at different pH (the lowest possible pH at a given ionic strength). At the highest ionic strength (Fig. 2, □) corresponding to a 1–1 electrolyte concentration higher than $C_{EL,CR}$ the electrostatic repulsion is screened [4] and no influence of pH on the film thickness is observed.

Discussion

The behavior of the two copolymers is similar to that of low molecular non-ionic surfactants (e.g. decylmethylsulfoxide [13] or lysophosphatidylcholine [2]). In the case of P85 a metastable zone (pH = 3.8–5.6) is found. It is reminiscent of that observed with other non-ionic surfactants [2, 13, 14]. The parameter pH_{CR}, as defined earlier [13], characterize the transition to black films. The thinnest films obtained with F108 are sterically stabilized – they are neither black nor bilayers [4]. We introduce here the notation $pH_{CR,ST}$ (pH critical steric) to emphasize this difference.

The silver films obtained at pH higher than the critical values are electrostatically stabilized. For both copolymers the plateau values of the film thickness observed above pH = 5.0–5.2 are in excellent agreement with the values previously measured at pH = 5.6 and an ionic strength corresponding to a 1–1 electrolyte concentration lower than $C_{EL,CR}$ [4].

The φ_0-potential is estimated from the DLVO theory as follows. The total disjoining pressure, $\Pi (= \Pi_{EL} + \Pi_{vw})$, equals the capillary pressure, P_C, which is experimentally measured. Thus from the pressure balance at equilibrium

$$P_C = \Pi_{EL} + \Pi_{vw} ,$$

a value for the DDL potential at each film thickness can be obtained.

The evaluation of Π_{EL} is based on a three layer model of the foam film. The film is divided into an aqueous core and two adsorption layers (Fig. 3). The thickness of the adsorption layers, h_1, and their refractive index, n_1, can be reasonably estimated [3]. The aqueous core thickness, h_2, is calculated from the equivalent film thickness, h_w, as [25]

$$h_2 = h_w - 2\,\frac{n_1^2 - 1}{n_2^2 - 1}\,h_1 .$$

The values thus estimated are listed in Table 1. The φ_0-potential plane is located in the middle of the hydrophilic PEO brushes (see Fig. 3), i.e. the distance, d, between the two φ_0-potential plane is

$$d = h_2 + 2\,\frac{h_{PEO}}{2} = h_2 + h_{PEO} .$$

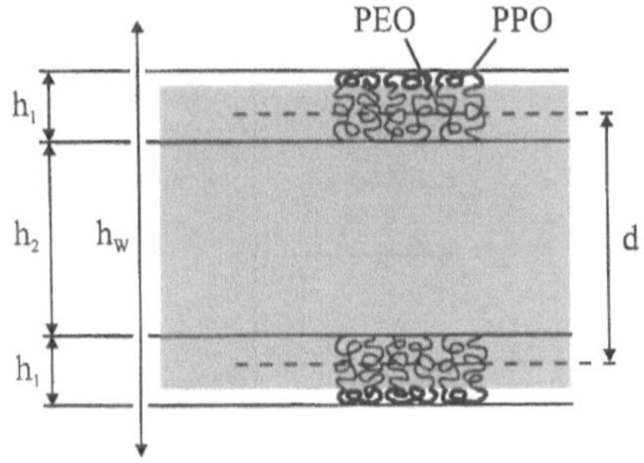

Fig. 3 Three-layer model. The film consists of two adsorption layers (thickness h_1 and refractive index n_1) and an aqueous core (thickness h_2 and refractive index n_2). The total film thickness $(2h_1 + h_2)$ is smaller than the equivalent thickness (h_w) because $n_1 > n_2$. The planes of the DDL potential (dotted lines) are located in the middle of the hydrophilic PEO brushes. The distance between the two φ_0-planes is d

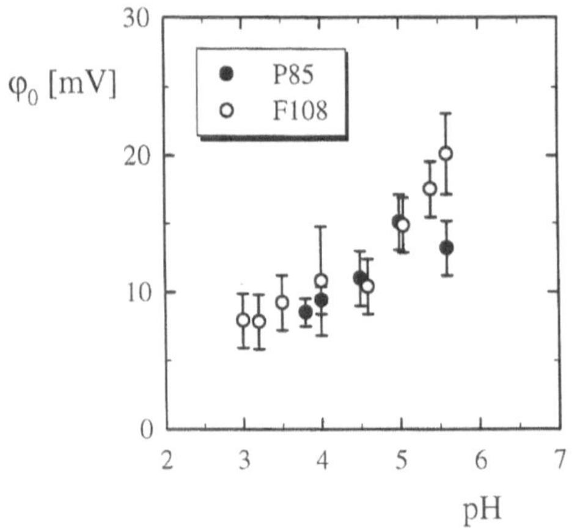

Fig. 4 φ_0-potential as a function of pH (●: 3×10^{-6} M P85; ○: 7×10^{-7} M F108; $I = 10^{-3}$ M)

Table 1 Parameters of the three-layer model of the foam film

Layer/Copolymer	n	*P*85 h [nm]	*F*108 h [nm]
Adsorption layers	1.42	3.2	10.6
Aqueous core	1.33	$h_2 = h_w - 8.5$	$h_2 = h_w - 28.0$

The van der Waals contribution, Π_{vw}, is estimated along the lines given in the paper of Donners et al. [26]. A more detailed discussion has been presented in Ref. [3]. The computational procedure yielding the values of the DDL potential has been previously described [1, 27].

The results for P85 and F108 at constant ionic strength are shown in Fig. 4. The DDL potential is quite low in agreement with previous findings for other non-ionic surfactants [1, 9, 10, 22]. It decreases with decreasing pH and reaches very low values when approaching pH_{CR}. We insist on this general trend rather than the absolute values of the DDL potential. The φ_0-potential calculated are model-dependent as exemplified by Cohen et al. [28] and cautionless camparison should be avoided.

The values for the two copolymers coincide and this is a manifestation of the peculiarity of the water/air interface. This result confirms our hypothesis that the effect is an interfacial one. When pH decreases the bulk concentration of H^+ ions increases. Consequently, their adsorption at the solution/air interface increases and they recombine with the excess of potential-forming OH^- ions. Thus the DDL potential decreases and eventually vanishes at the critical pH. Thus the critical values, pH_{CR} and $pH_{CR,ST}$, identify the isoelectric point of the solution/air interface for the two copolymers where electrostatic interaction drops to zero.

The results for the pH dependence of the DDL potential and the surface charge density for films stabilized with

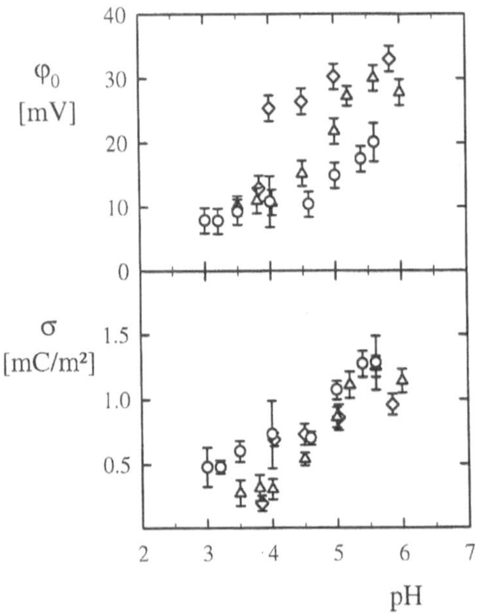

Fig. 5 φ_0-potential and surface charge density σ as a function of pH at different ionic strength (7×10^{-7} M F108; ◇: $I = 1.5 \times 10^{-4}$ M, △: $I = 3 \times 10^{-4}$ M, ○: $I = 10^{-3}$ M)

F108 are given in Fig. 5 for three different values of the ionic strength. Both φ_0-potential and surface charge decrease with decreasing the pH and decline to zero at $pH_{CR,ST}$. Above pH = 5 the surface charge appears to be constant as previously inferred from disjoining pressure isotherms [3].

The electrostatic interaction is mainly governed by the peculiar properties of the water/air interface rather than the dissolved non-ionic substances. This is demonstrated in Fig. 6 where the distance d between the φ_0-planes is shown as a function of pH at constant ionic strength

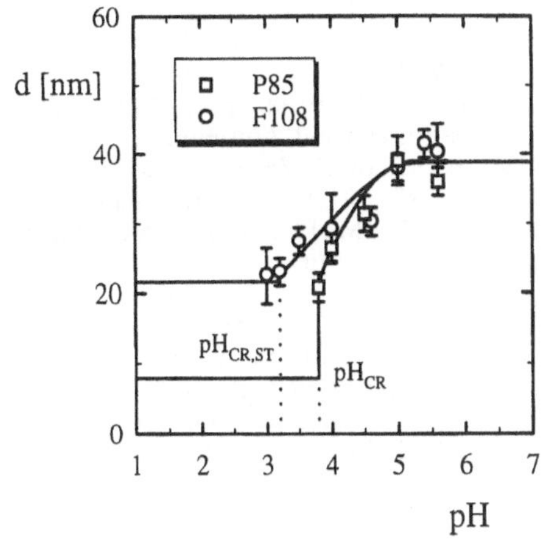

Fig. 6 Effect of pH on the distance d between the two φ_0-planes (○: F108, $P_C = 58\ N/m^2$; □: P85, $P_C = 50\ N/m^2$; $I = 10^{-3}$ M)

Table 2

Copolymer	h_w	h_2	R_F	h_2/R_F
P85	15.0	6.5	2.5	~2.6
F108	40.0	12.0	5.4	~2.2

(10^{-3} M). The capillary pressure is negligibly different for the two copolymers. The changes in the distance between the two φ_0-potential planes are interpreted only as changes in the value of Π_{EL} as the ionic strength and the capillary pressure are constant. Above the critical pH values the two curves coincide within the experimental scatter. On the contrary below the critical pH where the steric stabilization prevails the two sets of data diverge significantly.

Below the critical pH values Π_{EL} is entirely suppressed and the film thickness is determined by steric repulsion. The equivalent thickness of these sterically stabilized films are in agreement with the values found with films formed at high (above $C_{EL,CR}$) electrolyte concentration [4]. Their thickness is in close relation to the length of the PEO chains – the longer the polymer chain the thicker the film. Estimates show that these films are thicker than twice the adsorption layer thickness (Table 2). The thickness of the aqueous core, h_2, is of the order of the radius of gyration, R_F (Table 2). The dimensionless structure of the thickness of the thinnest films stabilized with the two polymeric surfactants is the same as described in Ref. [4].

The transition from electrostatic to steric stabilization occurs at different values of pH ($pH_{CR} = 3.8$ for P85 and $pH_{CR,ST} = 3.2$ for F108). This should be attributed to the different PEO chain length since this is the main difference between the two copolymers. A divergence between the two sets of data is already seen above pH_{CR} in Fig. 6 and therefore the charge creation mechanism appears to be more complex. The importance of the surfactant chain length as well as the effect of copolymer concentration on the isoelectric point of a solution/air interface with adsorbed block copolymers require further investigation.

Conclusions

The effect of pH on the equilibrium thickness of microscopic foam films is interpreted as a decrease of the DDL potential due to neutralization of the excess of OH^- ions adsorbed at the solution/air interface. This mechanism has been originally proposed by Exerowa et al. [10, 11].

Upon decreasing of pH the foam films become thinner but do not rupture, i.e. the DDL potential vanishes at the isoelectric point but the films are stable. This is attributed to steric repulsion between the PEO hydrophilic chains of the block copolymers.

A transition from electrostatic to steric stabilization can be induced at constant low ionic strength by decreasing the pH. Such a transition is observed for the first time with block copolymers and is characterized by the parameter $pH_{CR,ST}$.

Electrostatic stabilization is mainly related to the specific properties of the water/air interface. On the contrary steric stabilization is strongly dependent on the length of the polymeric surfactant molecules.

Acknowledgements The authors are indebted to Prof. Th.F. Tadros for supplying the Synperonic PE copolymers. Thanks are due to Dr. T. Kolarov for providing the program for the surface charge and potential calculation.

References

1. Kolarov T, Cohen R, Exerowa D (1989) Colloids Surfaces 42:49
2. Cohen R, Exerowa D (1994) Colloid Surfaces A 85:271
3. Sedev R, Ivanova R, Kolarov T, Exerowa D (1997) J Disp Sci Technol 18:751
4. Exerowa D, Sedev R, Ivanova R, Kolarov T, Tadros ThF (1997) Colloids Surfaces A 123:277
5. Manev E, Pugh R (1991) Langmuir 7:2253
6. Waltermo A, Manev E, Pugh R, Claesson P (1996) J Disp Sci Technol 15:273
7. Bergeron V, Waltermo A, Claesson P (1996) Langmuir 12:1336
8. Derjaguin B, Titijevskaya A, Vybornova V (1960) Kolloidn Zh 22:407
9. Exerowa D, Scheludko A (1964) In: Overbeek JThG (ed) Proc 4th Int Congr Surf Activity, Brussels. Gordon & Breach, London, p 1097
10. Exerowa D, (1969) Kolloid-Z 232: 703
11. Exerowa D, Zacharieva M (1972) In: Research in Surface Forces, Nauka, Moscow, Vol 4, p 234

12. Cohen R, Exerowa D, Kolarov T, Yamanaka T, Muller V (1992) Colloid Surfaces 65:201
13. Exerowa D, Kolarov T, Christov Chr (1971/1972) Ann Univ Sofia, Fac Chim 66:293
14. Exerowa D (1978) Comm Dept Chem Bulg Acad Sci 11:739
15. Kolarov T, Yankov R, Esipova NE, Exerowa D, Zorin ZM (1993) Colloid Polym Sci 271:519
16. Yoon R-H, Yordan JL (1986) J Colloid Interface Sci 113:430; Huddelston RW, Smith AL, In: Akers AJ (ed) Foams. Academic Press, London, p 163; Kelsall G, Tang S, Yurdakul S, Smith A (1996) J Chem Soc Faraday Trans 92:3887
17. Exerowa D, Kruglyakov PM (1998) Foam and Foam Films. Elsevier, Amsterdam
18. Schmolka I (1967) In: Schick M (ed) Nonionic Surfactants. Marcel Dekker, New York, p 324
19. Scheludko A (1967) Adv Colloid Interface Sci 1:392
20. Scheludko A, Exerowa D (1959) Comm Inst Chem Bulg Acad Sci 7:123
21. Kolarov T, Iliev L (1974/1975) Ann Sof Univ Fac Chim 69:107
22. Exerowa D, Zacharieva M, Cohen R, Platikanov D (1979) Colloid Polym Sci 257:1089
23. Vasicek CJ (1960) Optics of Thin Films, North-Holland, Amsterdam
24. Platikanov D, Zacharieva M, Exerowa D (1971/1972) Ann Univ Sofia 66:277
25. Duyvis EM (1962) PhD Thesis, University of Utrecht
26. Donners W, Rijnbout J, Vrij A (1977) J Colloid Interface Sci 60:540
27. Kolarov T, Exerowa D, Balinov B, Martinov G (1986) Kolloidn Zh 48:1076
28. Cohen R, Exerowa D, Kolarov T, Yamanaka T, Tano T (1977) Langmuir 13:3172

Progr Colloid Polym Sci (1998) 109:35–41

M.J. Rosen

Molecular interactions and the quantitative prediction of synergism in mixtures of surfactants

Received: 6 January 1997
Accepted: 14 April 1997

M.J. Rosen
Surfactant Research Institute
Brooklyn College City University of New York
Brooklyn, New York 11210
USA

Abstract The equations for determining the nature and strength of the interaction between two different surfactants in mixed monolayers at various interfaces (β^{σ} value)or in mixed micelles (β^{m} value), based upon nonideal or regular solution theory, and their utilization for quantitative prediction of synergism, are reviewed. From the data on numerous systems, where both the chemical structures of the two surfactants and their molecular environments were changed, it appears that electrostatic forces between the two hydrophilic groups dominate the interaction. The replacement of air by a liquid aliphatic hydrocarbon or by a nonpolar hydrophobic solid (Parafilm, Teflon) reduces the (attractive) interaction between the two surfactants slightly, but leaves unchanged other chemical structural effects.

The quantitative relationships between the β^{σ} and β^{m} values and the existence of synergism in surface or interfacial tension reduction efficiency, synergism in mixed micelle formation, and synergism in surface or interfacial reduction effectiveness are given, along with relationships for the mole fraction of the two surfactants at the point of maximum synergism and for the value of the relevant property at that point. Experimental and calculated values are compared. Future advances in the field may be in non-detergency industrial applications.

Key words Surfactant mixtures – synergism – molecular interactions – mixed micelle formation – surface tension reduction – interfacial tension reduction

Introduction

Interest in synergy in mixtures of surfactants stems from both academic and industrial sources. Academic, in that the existence of synergism between two different molecules must originate from interactions between them and consequently sheds light on these; industrial, in that it bears on the constant search of industrial research for improvement in the performance properties of materials, a search that has been accentuated in recent years by the difficulty of introducing entirely new chemical structures into the marketplace.

About fifteen years ago [1, 2], we derived in our laboratory a set of equations that allowed us to treat synergism in binary mixtures of surfactants in quantitative fashion. They revealed the exact conditions for the existence of synergy in the two fundamental properties of surfactants: mixed monolayer formation at an interface and mixed micelle formation in solution. In addition, they allowed us to calculate the ratio of the two different surfactants at the point of maximum synergism and the value of the relevant interfacial property at that point.

The treatment starts with the basic thermodynamic equations for mixed surfactant solutions [1]:

$$\alpha C_{12} = X_1 f_1 C_1^0 , \tag{1}$$

$$(1-\alpha)C_{12} = (1-X_1) f_2 C_2^0 , \tag{2}$$

where α is the mole fraction of surfactant 1 in the total surfactant in the solution phase, C_1^0, C_2^0, and C_{12} are the molar concentrations of individual surfactants 1 and 2 and their mixture at a given value of α, respectively, required to produce the same surface tension (γ) value of the solution at a given temperature. X_1 is the mole fraction of surfactant 1 in the total surfactant in the mixed monolayer and f_1, f_2 are the activity coefficients of individual surfactant 1 and 2, respectively, in the mixed monolayer.

Based upon the non-ideal or regular solution assumption, the relationships

$$f_1 = \exp[\beta^\sigma (1-X_1)^2] , \tag{3}$$

$$f_2 = \exp[\beta^\sigma X_1^2] \tag{4}$$

are used for the activity coefficients. β^σ is the interaction parameter for mixed monolayer formation.

Equations (1)–(4) yield

$$\frac{X_1^2 \ln(\alpha C_{12}/X_1 C_2^0)}{(1-X_1)^2 \ln[(1-\alpha)C_{12}/(1-X_1)C_2^0]} = 1 , \tag{5}$$

$$\beta^\sigma = \frac{\ln(\alpha C_{12}/X_1 C_1^0)}{(1-X_1)^2} . \tag{6}$$

Equation (5) is solved numerically for X_1, which is then substituted in Eq. (6) to yield β^σ.

Rubingh [3] derived the following analogous equations for evaluating β^m, the interaction parameter for mixed micelle formation:

$$\frac{(X_1^m)^2 \ln(\alpha_1 C_{12}^m/X_1^m C_1^m)}{(1-X_1^m)^2 \ln[(1-\alpha_1)C_{12}^m/(1-X_1^m)C_2^m]} = 1 , \tag{7}$$

$$\beta^m = \frac{\ln(\alpha_1 C_{12}^m/X_1^m C_1^m)}{(1-X_1^m)^2} , \tag{8}$$

where X_1^m is the mol fraction of surfactant 1 in the total surfactant in the mixed micelle; C_1^m, C_2^m, and C_{12}^m are critical micelle concentrations of surfactants 1, 2, and their mixture, respectively. In this case, Eq. (7) is solved for X_1^m, the mole fraction of surfactant 1 in the mixed micelle. The value of β^σ and β^m can readily be evaluated experimentally from surface tension–concentration curves for solutions of the two individual surfactant and at least one mixture of them at some value of α (Fig. 1).

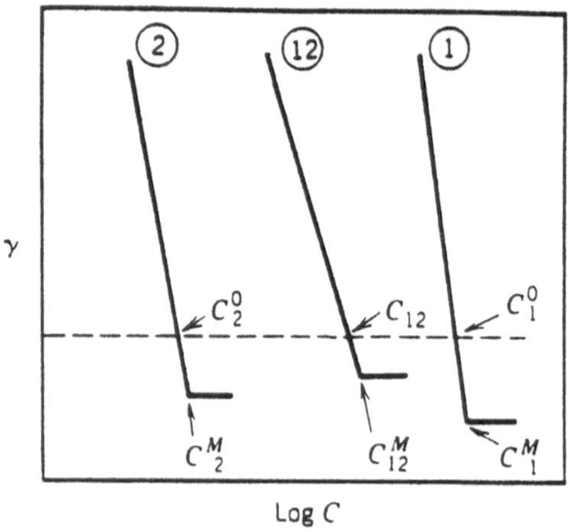

Fig. 1 Experimental evaluation of β^σ and β^m. ① Surfactant 1. ② Surfactant 2. ⑫ Mixture of surfactants 1 and 2 at mole fraction, α, in the solution phase

This treatment assumes that the effect of changes in the area per molecule of the surfactant at the interface upon mixing are insignificant. When interaction is strong ($\beta^\sigma > 5$), however, it has been shown [4] that it is necessary, in calculating β^σ and X_1, to take into consideration these changes. In that case, the following equations should be used:

$$\frac{X_1^2}{(1-X_1)^2} \times \frac{\ln \dfrac{\alpha C_{12}}{X_1 C_1^0} - \dfrac{\gamma A_1^0}{RT}\left[1 - \dfrac{A_{av}}{X_1 A_1^0 + (1-X_1)A_2^0}\right]}{\ln \dfrac{(1-\alpha)C_{12}}{(1-X_1)C_2^0} - \dfrac{\gamma A_2^0}{RT}\left[1 - \dfrac{A_{av}}{X_1 A_1^0 + (1-X_1)A_2^0}\right]} = 1 , \tag{9}$$

$$\beta^\sigma = \frac{\ln \dfrac{\alpha C_{12}}{X_1 C_1^0} - \dfrac{\gamma A_1^0}{RT}\left[1 - \dfrac{A_{av}}{X_1 A_1^0 + (1-X_1)A_2^0}\right]}{(1-X_1)^2} , \tag{10}$$

where A_1^0, A_2^0, A_{av} are the surface areas of surfactants 1, 2, and their mixture, respectively, at the air/water interface (from the slope of γ–log C plots).

Equations (5), (6), (9) and (10) have been shown to be valid for calculating β^σ by comparing the value of X_1, obtained from them with the value of X_1 obtained by an entirely different method, using the Gibbs adsorption equation [1, 4].

For determining the value of β^σ in a mixed monolayer at the liquid/liquid interface (β_{LL}^σ), plots of interfacial tension as a function of surfactant concentration are used; for determining its value at the liquid/solid interface (β_{LS}^σ), a plot of $\gamma \cos\theta$ as a function of surfactant concentration, where θ is the contact angle of the surfactant solution on

Table 1 Molecular interactions at the aqueous solution/air interface (25 °C)

System	β^{σ}	β^{m}	Ref.
$C_{12}H_{25}SO_3^-Na^+$–$[C_{10}H_{21}N^+(CH_3)_2CH_2CHOH]_2 \cdot 2Br^-$ (0.1 M NaCl)	−41.0	−19.0	[10]
$C_{12}H_{25}SO_3^-Na^+$–$[C_{10}H_{21}N^+(CH_3)_2CH_2CHOH]_2 \cdot 2Br^-$ (0.1 M NaBr)	−35.0	−20.0	[10]
$C_{10}H_{21}SO_3^-Na^+$–$[C_{10}H_{21}N^+(CH_3)_2CH_2CHOH]_2 \cdot 2Br^-$ (0.1 M NaBr)	−34.0	−14.0	[10]
$C_{12}H_{25}SO_3^-Na^+$–$[C_8H_{17}N^+(CH_3)_2CH_2CHOH]_2 \cdot 2Br^-$ (0.1 M NaBr)	−31.0	−15.0	[10]
$C_{10}H_{21}SO_3^-Na^+$–$[C_8H_{17}N^+(CH_3)_2CH_2CHOH]_2 \cdot 2Br^-$ (0.1 M NaBr)	−26.0	−12.0	[10]
$C_{10}H_{21}SO_3^-Na^+$–$C_{12}H_{25}N^+(CH_3)_3Br^-$ (H_2O)	−35.6	—	[11]
$C_{10}H_{21}SO_3^-Na^+$–$C_{12}H_{25}N^+(CH_3)_3Br^-$ (0.1 M NaBr)	−19.6	—	[4]
$C_{10}H_{21}SO_3^-Na^+$–$C_{12}H_{25}Pyr^+Br^-$ (0.1 M NaBr)	−19.7	—	[4]
$C_{12}H_{25}SO_3^-Na^+$–$C_{14}H_{29}N(CH_3)_2O$ (0.1 M NaCl, pH 5.8)	−10.3	−7.8	[9]
$C_{12}H_{25}SO_3^-Na^+$–$C_{12}H_{25}N^+(CH_3)(CH_2\phi)CH_2COO^-$ (0.1 M NaBr, pH 5.8)	−8.3	−5.0	[12]
$C_{12}H_{25}SO_3^-Na^+$–$C_{12}H_{25}(OC_2H_4)_8OH$ (0.1 M NaCl)	−2.6	−3.1	[13]
$C_{12}H_{25}SO_3^-Na^+$–$C_{12}H_{25}(OC_2H_4)_8OH$ (0.5 M NaCl)	−2.0	—	[13]
$C_{12}H_{25}SO_3^-Na^+$–$C_{12}H_{25}(OC_2H_4)_8OH$ (H_2O)	−2.7	−4.1	[13]
$C_{12}H_{25}SO_3^-Na^+$–$C_{12}H_{25}\phi SO_3^-Na^+$ (0.1 M NaCl)	−0.3	−0.3	[9]
$C_{12}H_{25}N^+(CH_3)(CH_2\phi)CH_2COO^-$–$C_{12}H_{25}(OC_2H_4)_8OH$ (H_2O)	−0.6	−0.9	[15]
$C_{12}H_{25}(OC_2H_4)_3OH$–$C_{12}H_{25}(OC_2H_4)_8OH$ (H_2O)	−0.2	—	[1]
$C_{15}H_{31}COO^-Na^+$–$C_{16}H_{33}SO_3^-Na^+$ (0.1 M NaCl, pH 10.6, 60 °C)	+0.7	+0.7	[14]
$C_{12}H_{25}SO_4^-Na^+$–$C_7F_{15}COO^-Na^+$ (0.1 M NaCl, 30 °C)	+2.0	—	[16]

$C_{12}H_{25}$Pyr–N-dodecylpyridinium bromide.

the solid surface, can be used in the case where the solid is a nonpolar, low surface energy solid.

When the value of β is negative, there is attractive interaction between the two different surfactants; when it is positive, then the interaction between them is repulsive. The greater the absolute value of β (whether positive or negative), the stronger the interaction between the two surfactants. Synergy occurs only when the molecular interaction is attractive (β is negative). However, as discussed below, this is a necessary, but not sufficient condition for synergy to exist. The existence of synergy depends also upon the difference in the values of the relevant property of the two different surfactants. When these values are close to each other, then the molecular interaction (β) value can be rather weak and still produce synergy; when the values are far apart, interaction between the two different surfactants must be strong for synergism to exist.

Published β values for many surfactant pairs [1–11] indicate that interactions between the two different surfactants are dominated by electrostatic forces between the two different hydrophilic head groups. Some examples are given in Table 1. Interactions are strongest when the surfactants each have a sign of opposite charge (e.g., mixtures of an anionic and a cationic surfactant). As expected, when one surfactant is doubly charged then its interaction with an oppositely charged surfactant is stronger than when it is singly charged. Interaction between oppositely charged surfactants is reduced by increase in the electrolyte content of medium, reflecting the resulting compression of the electrical double layers surrounding the ionic hydrophilic head groups. For mixtures containing quaternary ammonium halides, bromide ion reduces the interaction with anionic surfactants more than chloride ion, reflecting the tighter binding of the former than the latter to the positively charged hydrophilic head groups. When one surfactant carries a charge and the second surfactant, although carrying no charge, is capable of acquiring a charge of opposite sign by accepting or losing a proton (e.g., an ampholyte or amine oxide), then interaction between the two is weaker than between two oppositely charged materials, but still stronger than between ionic and nonionic surfactants. Attractive interactions are weakest when both surfactants carry no charges, or have ionic charges of the same sign. Mixtures of a long chain soap and a long chain alkyl sulfonate or alkylbenzenesulfonate show repulsive interactions, as do mixtures of an anionic hydrocarbon chain with an anionic fluorocarbon chain surfactant.

Interactions (β^{m} values) between two different surfactants in mixed micelles are generally weaker than their interactions in mixed monolayers at the aqueous solution/air interface. Noteworthy is the much weaker inter-

Table 2 Effect of the nature of the interface on molecular interactions at various interfaces (25 °C)

Systems	Interface	β^σ	Ref.
$C_{12}H_{25}SO_3^-Na^+$–N-octylpyrrolidinone	0.1 M NaCl(aq.)–air	−3.1	[7]
$C_{12}H_{25}SO_3^-Na^+$–N-octylpyrrolidinone	0.1 M NaCl(aq.)–Parafilm	−2.9	[7]
$C_{12}H_{25}SO_3^-Na^+$–N-octylpyrrolidinone	0.1 M NaCl(aq.)–Teflon	−2.5	[7]
$C_{12}H_{25}SO_3^-Na^+$–N-octylpyrrolidinone	0.1 M NaCl(aq.)—hexadecane	−1.7	[7]
$C_{12}H_{25}SO_3^-Na^+$–N-decylpyrrolidinone	0.1 M NaCl(aq.)–hexadecane	−2.3	[7]
$C_{10}H_{21}SO_3^-Na^+$–$C_{12}H_{25}N^+(CH_3)_3Br^-$	0.1 M NaBr(aq.)–air	−19.6	[4]
$C_{10}H_{21}SO_3^-Na^+$–$C_{12}H_{25}N^+(CH_3)_3Br^-$	0.1 M NaBr(aq.)–Parafilm	−15.3	[4]
$C_{10}H_{21}SO_3^-Na^+$–$C_{12}H_{25}N^+(CH_3)_3Br^-$	0.1 M NaBr(aq.)–Teflon	−14.1	[4]
$C_{10}H_{21}SO_3^-Na^+$–$C_{12}H_{25}Pyr^+Br^-$	0.1 M NaBr(aq.)–air	−19.7	[4]
$C_{10}H_{21}SO_3^-Na^+$–$C_{12}H_{25}Pyr^+Br^-$	0.1 M NaBr(aq.)–Parafilm	−15.5	[4]
$C_{10}H_{21}SO_3^-Na^+$–$C_{12}H_{25}Pyr^+Br^-$	0.1 M NaBr(aq.)–Teflon	−14.2	[4]
$C_{12}H_{25}SO_3^-Na^+$–$C_{12}H_{25}N^+(CH_3)(CH_2\phi)CH_2COO^-$	0.1 M NaBr(aq.)–air	−8.3	[12]
$C_{12}H_{25}SO_3^-Na^+$–$C_{12}H_{25}N(CH_3)(CH_2\phi)CH_2CH_2COO^-$	0.1 M NaBr(aq.)–Parafilm	−6.7	[12]
$C_{12}H_{25}SO_3^-Na^+$–$C_{12}H_{25}N^+(CH_3)(CH_2\phi)CH_2COO^-$	0.1 M NaBr(aq.)–Teflon	−6.2	[12]
$C_{12}H_{25}SO_3^-Na^+$–$C_{12}H_{25}(OC_2H_4)_8OH$	0.1 M NaCl(aq.)–air	−2.6	[13]
$C_{12}H_{25}SO_3^-Na^+$–$C_{12}H_{25}(OC_2H_4)_8OH$	0.1 M NaCl(aq.)–Teflon	−2.0	[4]
$C_{12}H_{25}SO_3^-Na^+$–$C_{12}H_{25}(OC_2H_4)_8OH$	0.5 M NaCl(aq.)–air	−2.0	[13]
$C_{12}H_{25}SO_3^-Na^+$–$C_{12}H_{25}(OC_2H_4)_8OH$	0.5 M NaCl(aq.)–Teflon	−1.7	[4]
$C_{12}H_{25}(OC_2H)_3OH$–$C_{12}H_{25}(OC_2H_4)_8OH$	H_2O–air	−0.2	[12]
$C_{12}H_{25}(OC_2H_4)_3OH$–$C_{12}H_{25}(OC_2H_4)_8OH$	H_2O–hexadecane	−0.7	[17]

$C_{12}H_{25}$PyrBr–N-dodecylpyridinium bromide.

action in a mixed micelle containing a surfactant with two hydrophilic and two hydrophobic groups (a "gemini" or "dimeric" surfactant) compared with the interaction in the mixed monolayer at the aqueous solution/air interface. This may be due to the greater difficulty of incorporating micelle hydrophobic groups into the interior of a convex machine than into a planar monolayer. An exception to this generally observed weaker interaction in mixed micelle than in mixed monolayers is the case where one of the surfactants is a polyoxyethylenated nonionic surfactant. Here, the interaction in the mixed micelle is somewhat stronger than in the mixed monolayer under the same conditions. At present, the exact reason for this is unknown. Perhaps, at the surface of a convex micelle, a large hydrophilic head group is more easily accommodated than at a planar interface.

The effect of the nature of the interface on molecular interactions between two different surfactants adsorbed there can be seen from the data listed in Table 2. The replacement of (nonpolar) air by a nonpolar solid or liquid (Parafilm, Teflon, hexadecane) results in a small reduction in the value of β^σ, with Teflon producing a somewhat greater reduction than Parafilm. Otherwise, the chemical structural effects seen from the data in Table 1 remain unchanged. A recent investigation [10] of the interaction in mixtures of gemini surfactants (containing two hydrophilic groups and two hydrophobic groups in the molecule) and surfactants containing a single hydrophilic and a single hydrophobic group in the molecule revealed that when attractive interaction between the two different surfactants becomes too large, they associate to form a complex with little or no surface activity. Thus, although mixtures of $[C_{10}H_{21}N^+(CH_3)_2CH_2CHOH]_2 \cdot 2Br^-$ with $C_{12}H_{25}SO_3^-Na^+$ in 0.1 M NaBr at 25 °C had β^σ and β^m values of −35.0 and −20.0, respectively, and showed strong synergism (see below) in surface tension reduction efficiency and effectiveness and in mixed micelle formation, mixtures of the slightly longer homolog, $[C_{12}H_{25}N^+(CH_3)_2CHOH]_2 \cdot 2Br^-$, with $C_{12}H_{25}SO_3^-$ Na^+ under the same conditions showed no synergism in any of these respects. In the latter case, a soluble 1:1 molar complex of the two surfactants, with little or no surface activity, appears to have formed.

Synergism

Three different types of synergism, dependent upon interaction in the mixed monolayer at an interface, on interaction in a mixed micelle, or on both, have been identified and investigated in quantitative fashion [2, 18], based upon the thermodynamic treatment described above.

Synergism in surface tension reduction efficiency

Synergism of this type exists in a mixture of the two surfactants in solution when a given surface tension (reduction) can be attained at a total mixed surfactant concentration less than that required of either surfactant by itself (Fig. 2). The conditions for the existence of this type of synergism are [2]:

1. β^σ must be negative,
2. $|\beta^\sigma| > |\ln C_1^0/C_2^0|$.

Thus, the two surfactants must attract each other in the mixed monolayer at the interface, and in addition, this attraction must be larger than the difference between the natural logarithms of the two concentrations needed to produce the same reduction in the surface tension of the solvent.

The mole fraction, α^*, of surfactant 1 in the total surfactant, which results in the maximum synergism in this respect, i.e., minimum total surfactant concentration (in molar units) required to produce any given surface tension (reduction), is given by the equation:

$$\alpha^* = \frac{\ln(C_1^0/C_2^0) + \beta^\sigma}{2\beta^\sigma} . \quad (11)$$

The minimum total surfactant concentration C_{12}, min, in molar units, at this point of maximum synergism, is given by the relationship:

$$C_{12,\min} = C_1^0 \exp\left\{\beta^\sigma\left[\frac{\beta^\sigma - \ln(C_1^0/C_2^0)}{2\beta^\sigma}\right]^2\right\} . \quad (12)$$

Analogous equations have been derived [7, 12] for synergism in interfacial tension reduction efficiency at the aqueous solution/aliphatic hydrocarbon and aqueous solution/hydrophobic solid interfaces. Some data at various interfaces [7] are shown in Table 3. Although the experimental values (α_1) are not exactly at the mole fraction for maximum synergism (α_1^*), the data show very good agreement between experimental and calculated values for $C_{12,\min}$.

The degree of synergism in surface or interfacial tension reduction efficiency possible in a mixture is measured by the maximum decrease in the molar concentration of (mixed) surfactant needed to produce a given surface tension value relative to that required of the more efficient of the two individual surfactants by itself; i.e., $(C_1^0 - C_{12,\min})/C_1^0$ or $1 - (C_{12,\min}/C_1^0)$, where $C_{12,\min}$ is the minimum concentration of mixed surfactant that can produce this surface tension and C_1^0 is the concentration required of the more efficient surfactant. The ratio, $C_{12,\min}/C_1^0$ is given by the relationship

$$C_{12,\min}/C_1^0 = \exp\left\{\frac{[\beta^\sigma - \ln(C_1^0/C_2^0)]^2}{4\beta^\sigma}\right\} . \quad (13)$$

The greater the value of $1 - (C_{12,\min}/C_1^0)$, the greater the degree of synergism.

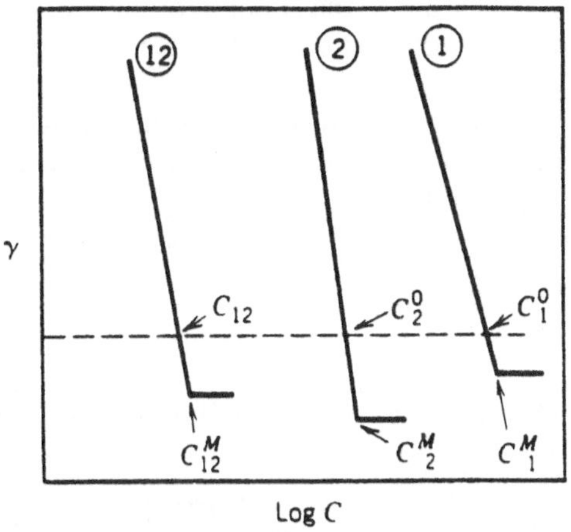

Fig. 2 Synergism in surface tension reduction efficiency or in mixed micelle formation. ① Surfactant 1. ② Surfactant 2. ⑫ Mixture of surfactants 1 and 2 at mole fraction, α in the solution phase. Synergism in surface tension reduction efficiency: $C_{12} < C_1^0, C_2^0$. Synergism in mixed micelle formation: $C_{12}^m < C_1^m, C_2^m$

Synergism in mixed micelle formation [2]

When the critical micelle concentration (cmc) of the mixture is less than the critical micelle concentration of either component surfactant under the same conditions, then synergism in mixed micelle formation is present. By use of a thermodynamic treatment similar to that used for synergism in surface tension reduction efficiency, the conditions for the existence of this type of synergy are [2]:

1. β^m must be negative and
2. $|\beta^m| > |\ln(C_1^m/C_2^m)|$.

Table 3 Synergism in surface or interfacial tension reduction efficiency; N-octylpyrrolidone/$C_{12}H_{25}SO_3Na$ mixtures in 0.1 M NaCl (aq) at 25 °C, $\pi = 32$ mN m^{-1} [7]

Interface	C_1^0(M)	C_2^0(M)	C_{12} exptl (M)	C_{12}, calcd (M)
H_2O–air	1.91×10^{-3}	1.51×10^{-3}	7.8×10^{-4}	7.7×10^{-4}
H_2O–Parafilm	1.93×10^{-3}	1.41×10^{-3}	8.1×10^{-4}	7.9×10^{-4}
H_2O–Teflon	3.31×10^{-3}	2.59×10^{-3}	1.58×10^{-3}	1.55×10^{-3}
H_2O–hexadecane	1.32×10^{-3}	5.48×10^{-4}	4.56×10^{-4}	4.69×10^{-4}

Here, β^m is the measure of the strength of the interaction between the two surfactants in the mixed micelle in the solution phase.

The mole fraction α_1^{m*} of surfactant 1 in the total surfactant that produces maximum synergy in this respect, i.e., the minimum cmc for the system, is given by the expression:

$$\alpha_1^{m*} = \frac{\ln(C_1^m/C_2^m) + \beta^m}{2\beta^m} \tag{14}$$

and the minimum cmc value that the mixture can attain is given by the expression:

$$C_{12,\min}^m = C_1^m \exp\left\{\beta^m \left[\frac{\beta^m - \ln(C_1^m/C_2^m)}{2\beta^m}\right]^2\right\}. \tag{15}$$

Some data in aqueous solution, in the absence or presence of a second hydrocarbon phase [20] are shown in Table 4. Again, although the experimental values (α_1) are not at the mole fraction for maximum synergism (α_1^{m*}), the data show good agreement between calculated and experimental values for $C_{12,\min}^m$.

The degree of synergism possible in mixed micelle formation in a mixture is measured by the maximum decrease in the cmc produced by the surfactant mixture relative to the lower cmc of the two individual surfactants comprising the mixture; i.e., $(C_1^m - C_{12,\min}^m)/C_1^m$ or $1 - (C_{12,\min}^m/C_1^m)$, where $C_{12,\min}^m$ is the lowest possible cmc of the mixed surfactant system and C_1^m is the lower cmc of the two surfactants constituting the mixture. The ratio, $C_{12,\min}^m/C_1^m$, is given by the relationship

$$C_{12,\min}^m/C_1^m = \exp\left\{\frac{[\beta - \ln(C_1^m/C_2^m)]^2}{4\beta^m}\right\}. \tag{16}$$

The greater the value of $1 - (C_{12,\min}^m/C_1^m)$, the greater the degree of synergism.

Synergism in surface tension reduction effectiveness [18]

The surface tension at the cmc (γ_{cmc}) of a surfactant solution is a measure of the surface tension reduction effectiveness of the surfactant [5b]. Synergism in this respect exists when the mixture of surfactants at its cmc reaches a lower surface tension value than that obtainable with either surfactant of the mixture at its cmc. Figure 3 illustrates this for a mixture of $[C_{10}H_{21}N^+(CH_3)_2\ CH_2\ CHOH]_2 \cdot 2\beta r^-$ and $C_{12}H_{25}SO_3^-Na_2^+$ in 0.1 M NaBr at 25 °C [10].

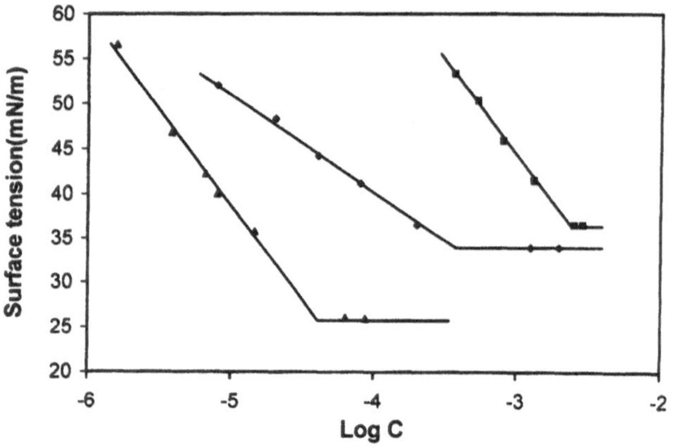

Fig. 3 Synergism in surface tension reduction effectiveness for a mixture of $[C_{10}H_{21}N^+(CH_3)_2CH_2CHOH]_2 \cdot 2\ Br^-(C_{10})$ and $C_{12}H_{25}SO_3^-Na^+(C_{12})$ in 0.1 M NaBr at 25 °C ($\alpha_{C10} = 0.67$). ■, C_{10}; ◆, C_{12}; ▲, mixture. From [10]

The conditions for this to occur are:

1. $\beta^\sigma - \beta^m$ must be negative.

2. $|\beta^\sigma - \beta^m| > \left|\ln\left[\frac{C_1^{0,\mathrm{cmc}} \cdot C_2^m}{C_2^{0,\mathrm{cmc}} \cdot C_1^m}\right]\right|$,

where $C_1^{0,\mathrm{cmc}}$, $C_2^{0,\mathrm{cmc}}$ are the molar concentrations of individual surfactants 1 and 2, respectively, required to yield a surface tension value equal to that of any mixture of the two surfactants at its cmc.

The mole fraction, $\alpha_1^{*,e}$, at which the mixed system shows maximum synergism in surface tension reduction effectiveness, i.e., produces the lowest value of the surface tension, is given by the Eq. [18]

$$\alpha_1^{*,e} = \frac{[\mathrm{cmc}_1^0 X_1^*/\mathrm{cmc}_2^0(1 - X_1^*)] \exp[\beta^m(1 - 2X_1^*)]}{1 + [\mathrm{cmc}_1^0 X_1^*/\mathrm{cmc}_2^0(1 - X_1^*)] \exp[\beta^m(1 - 2X_1^*)]}. \tag{17}$$

The value of X_1^*, the mole fraction of surfactant in the total surfactant at the interface, is calculated iteratively from the relationship

$$\frac{\gamma^0\mathrm{cmc}_1 - K_1(\beta^\sigma - \beta^m)(1 - X_1^*)^2}{\gamma^0\mathrm{cmc}_2 - K_2(\beta^\sigma - \beta^m)(X_1^*)^2} = 1, \tag{18}$$

where γ^0 cmc$_1$ and γ^0cmc$_2$ are the surface tension values of surfactants 1 and 2, respectively, at their cmc. K_1 and K_2 are the slopes of the γ–ln C plots of individual surfactants 1 and 2, respectively.

Table 4 Synergism in mixed micelle formation for N-dodecyl-N-methyl-N-benzylglycine/$C_{12}H_{25}SO_3Na$ aqueous mixtures in the presence and absence of a second liquid phase at 25 °C [20]

Interface	α_1 (α_1^*)	C_{12}^m, exptl (M)	C_{12}^m, calcd (M)
Air–H_2O	0.69 (0.80)	4.0×10^{-4}	4.5×10^{-4}
Cyclohexane–H_2O	0.95 (0.89)	3.0×10^{-4}	3.4×10^{-4}
Hexadecane–H_2O	0.95 (0.87)	4.3×10^{-4}	4.6×10^{-4}

Table 5 Synergism in surface of interfacial tension reduction effectivenes for N-dodecyl-N-methyl-N-benzylglycine/ $C_{12}H_{25}SO_3Na$ aqueous mixtures at various interfaces, 25 °C [20]

Interface	$\gamma^\circ cmc_1$ [$mN\,m^{-1}$]	$\gamma^\circ cmc_2$ [$mN\,m^{-1}$]	γ^* cmc, exptl [$mN\,m^{-1}$]	γ^*cmc, calcd [$mN\,m^{-1}$]
H_2O–air	32.8	39.0	28.0	29.9
H_2O–*n*-heptane	1.8	7.7	1.1	0.6
H_2O–isooctane	2.0	8.3	1.0	1.3
H_2O–hexadecane	3.5	9.9	1.4	1.4

The value of the surface tension at the point of maximum synergism (γ^*cmc), i.e., the lowest surface tension attainable in the mixed system, is given by the equations:

$$\gamma^*\text{cmc} = \gamma^\circ\text{cmc}_1 - K_1(\beta^\sigma - \beta^\text{m})(1 - X_1^*)^2 , \quad (19)$$

or

$$\gamma^*\text{cmc} = \gamma^\circ\text{cmc}_2 - K_2(\beta^\sigma - \beta^\text{m})(X_1^*)^2 . \quad (20)$$

From Eq. (19), the degree of synergism in this respect $1 - (\gamma^*\text{cmc}/\gamma^\circ\text{cmc})$ is $K(\beta^\sigma - \beta^\text{m})(1 - X^*)^2/\gamma^\circ\text{cmc}$, where K, X^*, and γ°cmc refer to the component having the lower value of γ°cmc.

Analogous equations can be used for predicting synergism in interfacial tension reduction effectiveness at various interfaces [19]. Some data are shown in Table 5.

Future prospects

It is expected that the current difficulties involved with the introduction of chemicals with novel structures into the marketplace will continue for the forseeable future. As a result, in the search for surfactants with new or improved properties, the investigation of mixtures of existing structural types of surfactants for possible synergy will continue to be an area of scientific exploration. Of particular interest will be:

(1) surfactant mixtures with greatly enhanced surface activity that can be used at lower concentrations, with consequent reduced environmental impact.

(2) surfactants, for use in industrial processes, as replacements for organic solvent in cleaning operations.

(3) surfactants in consumer and industrial products, as replacements for organic solvents as solubilizers of water-insoluble material, e.g., fragrances, inks and dyes, coatings.

(4) surfactants with both minimum skin irritation and maximum antipathogenic (antimicrobial, antiviral or antifungal) activity.

In general, it is felt that advances in the field will be greatest in the area of new, non-detergency industrial applications of surfactants.

References

1. Rosen MJ, Hua XY (1982) J Colloid Interface Sci 86:164
2. Hua XY, Rosen MJ (1982) J Colloid Interface Sci 90:212
3. Rubingh DN (1979) Mittal KL (ed) Solution Chemistry of Surfactants, Vol 1. Plenum, New York, pp 337–354
4. Gu B, Rosen MJ (1989) J Colloid Interface Sci 129:537
5. Rosen MJ (1989) Surfactants and Interfacial Phenomena, 2nd ed. Wiley, New York, (a) pp 398–402; (b) p 404
6. Rosen MJ (1991) Langmuir 7:885
7. Rosen MJ, Gu B, Murphy DS, Zhu ZH (1989) J Colloid Interface Sci 129:468
8. Rosen MJ, Zhu ZH, Gao T (1993) J Colloid Interface Sci 157:254
9. Rosen MJ, Gao T, Nakatsuji Y, Masuyama A (1994) Colloids and Surfaces A 88:1
10. Liu L, Rosen MJ (1996) J Colloid Interface Sci 179:454
11. Rodakiewicz-Nowak (1982) J Colloid Interface Sci 84:532
12. Rosen MJ, Gu B (1987) Colloids Surfaces 23:119
13. Rosen MJ, Zhao F (1983) J Colloid Interface Sci 95:443
14. Rosen MJ, Zhu ZH (1989) J Colloid Interface Sci 133:473
15. Rosen MJ, Zhu BY (1984) J Colloid Interface Sci 99:427
16. Zhao G-X, Zhu BY (1986) In: Scamehorn JF (ed) Phenomena in Mixed Surfactant Systems. ACS Symp Series, Vol 311. Amer Chem Soc, Washington DC, pp 184–198
17. Rosen MJ, Murphy DS (1991) Langmuir 7:2630
18. Hua XY, Rosen MJ (1988) J Colloid Interface Sci 125:730
19. Murphy DS, Zhu ZH, Yuan XY, Rosen MJ (1990) J Amer Oil Chem Soc 67:197
20. Rosen MJ, Murphy DS (1989) J Colloid Interface Sci 129:208

Progr Colloid Polym Sci (1998) 109:42–48

O.A. El Seoud
L.P. Novaki

Water solubilization by surfactant aggregates in organic solvents: Limitations of the multi-state water model

Received: 3 September 1997
Accepted: 15 September 1997

Prof. Dr. Omar A. El Seoud (✉)
Luzia P. Novaki
Instituto de Química
Universidade de São Paulo
C.P. 26.077
05599-970 São Paulo, S.P.
Brazil

Abstract Models for water solubilization by surfactant aggregates in organic solvents are discussed. In the multi-state model, water is present as interfacial-, bound-, intermediate-, and bulk-like water, respectively. The following evidence shows that this picture is an oversimplification: (i) interpretation of the dependence of a property of solubilized water, e.g., its chemical shift, on the ratio [water]/[surfactant], *W/S*, in terms of bound-, and bulk-like water is not unequivocal because the interpretation is model-dependent; (ii) values of the so-called deuterium/protium "fractionation factor, φ" for water solubilized by anionic-, cationic-, and non-ionic reverse aggregates indicate that it is not present in layers of different structures, as implied by the multi-state model; (iii) the same conclusion is deduced from IR studies of solubilized HOD. Therefore, available data indicate that water appears to be present as one layer whose properties vary continuously as a function of increasing *W/S*. The following forces preclude any long range order within the aqueous nanodroplet: (i) the surface potential of the micellar interface; (ii) water structure perturbation due to hydration of species present in the "water pool"; and (iii) dependence of surfactant–water interactions on *W/S*. Calculation of the decay of the electric potential within the water pool for two opposite surfactant molecules as a function of *W/S* shows that these potentials either cancel out, or their values reach zero mV within a short distance of each other, e.g., ca. 8 Å at *W/S* of 50.

Key words Water solubilization by reverse micelles – water-in-oil microemulsions – model for reverse aggregate-solubilized water – NMR of reverse aggregate-solubilized water – IR of reverse aggregate-solubilized water

Introduction

Several classes of surfactants aggregate in organic solvents of low polarity and dielectric constant, e.g., hydrocarbons and chlorinated hydrocarbons to form reverse micelles. One of the most important properties of these aggregates is their ability to solubilize water and aqueous solutions, leading to the formation of an aqueous nanodroplet surrounded by a monolayer of surfactant molecules. The peculiar properties of reverse micellar "cores" have been exploited, e.g., in catalysis of chemical and enzymatic reactions, and in the preparation of polymers and inorganic particles of controlled size [1–8].

It is important to understand fully the properties of solubilized water because of its participation in solvation

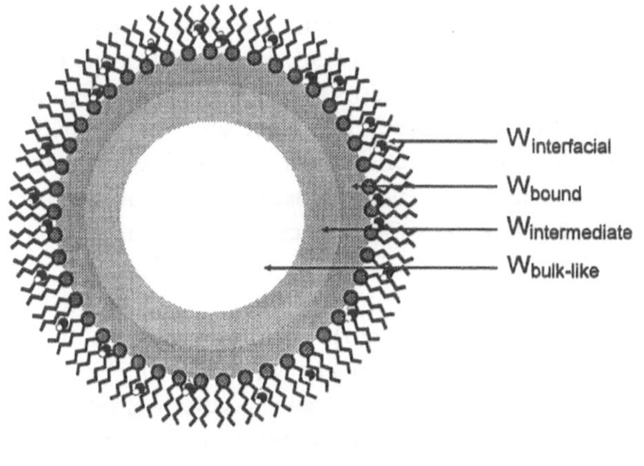

Fig. 1 Schematic representation of a W/O μE, showing the different types of water present, namely interfacial, $W_{interfacial}$; bound, W_{bound}; intermediate, $W_{intermediate}$; and bulk-like, $W_{bulk\text{-}like}$. See text for explanation

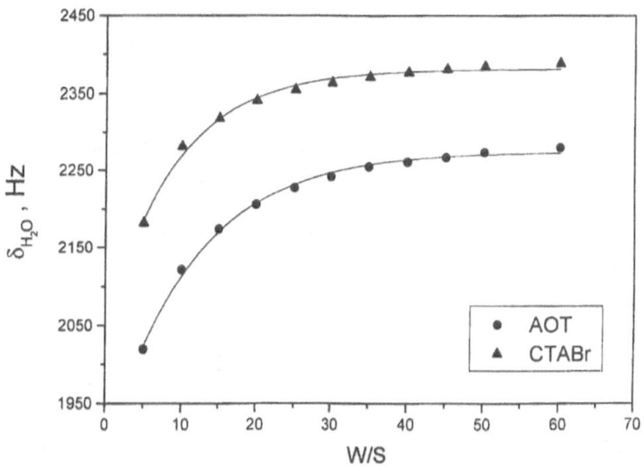

Fig. 2 Dependence of the chemical shift of water, δ_{H_2O}, solubilized by reverse aggregates of AOT (isooctane, at 36 °C), and CTABr ($CDCl_3$-*n*-heptane, 1:1, v/v, at 25 °C) on *W/S*, at 500.13 MHz (for protons). Data are re-plotted from ref. 14b. The points are experimental and the solid curves were calculated from Eq. (1), by using 2275.6, −389.5 and −0.088 for A, B, and C, respectively (AOT), and 2381.4, −341.6, and −0.113 for A, B, and C, respectively (CTABr)

of the surfactant headgroup and, when reverse aggregates are used as catalysts, solvation of reactants, transition states, and also in proton transfers. The physico-chemical properties of the system are usually discussed in terms of the molar ratio [water]/[surfactant], *W/S*. In *wet* reverse micelles, RMs, the amount of solubilized water is smaller than, or equal to, the amount necessary to hydrate the surfactant headgroup. Solubilization of more water results in the formation of a water-in-oil microemulsion, W/O μE, which is isotropic and thermodynamically stable [8].

An important aspect of RMs and W/O μEs is the physico-chemical model that describes the state of solubilized water. Figure 1 is a schematic representation of a W/O μE, showing the different types of water that have been claimed to be present in the micellar system. In principle, there can be up to three "layers" within the micellar water "pool" [9–15]. The first one, W_{bound}, is at the periphery of the pool, and is made of water molecules tightly bound to the surfactant headgroup. Properties of this type of water deviate appreciably from those of bulk water. The second, intermediate "layer", $W_{intermediate}$, refers to distorted H-bonded water species, e.g., cyclic dimers or higher aggregates with unfavorable H-bonds [10, 11]. The third, central "layer" contains bulk-like water, $W_{bulk\text{-}like}$, and its formation coincides with formation of the W/O μE [8]. According to some authors, there is one more type of water which lies at the oil side of the interface. This interfacial water, $W_{interfacial}$, refers to "monomeric water molecules which are not bound to any other molecules or groups but are trapped between the polar headgroups of the surfactant at the interface" [12].

This model is largely based on the observation that several properties of reverse aggregate-solubilized water show a pronounced change in the wet reverse micelle regime, followed by a smaller one in the W/O microemulsion domain [8]. Figure 2 is a typical example of such behavior. It shows the dependence of chemical shift of solubilized water, δ_{H_2O}, on *W/S* for reverse aggregates of two surfactants, viz., sodium bis(2-ethylhexyl) sulfosuccinate, AOT, and cetyltrimethyl-ammonium bromide, CTABr. The *W/S* range shown extends up to 60, the phase separation boundary. For each surfactant, the dependence of δ_{H2O}, on *W/S* can be fitted either to three straight lines, each covering a *W/S* range, or to a single curve. In the first case, the lines cover *W/S* ranges of 5–15, 20–35 and 40–60, respectively. The following are the linear regression results for AOT (slope of the line, corresponding correlation coefficient, cc, and corresponding sum of the squares of the residues, ΣQ): 15.50, 0.983, 401.9; 3.18, 0.993, 18.9; and 0.98, 0.986, 6.0, for the above mentioned *W/S* ranges, respectively. Results for CTABr are (slope, corresponding cc, and corresponding ΣQ): 13.61, 0.966, 668.0; 1.93, 0.987, 12.0; and 0.58, 0.987, 1.9, for the above mentioned *W/S* ranges, respectively. The solid curves shown in Fig. 2 were calculated from the expression

$$\delta_{H_2O} = A + B\exp(C\,W/S)\,, \tag{1}$$

where *A*, *B*, and *C* are regression coefficients. The good fit is clear from the figure, and from values of ΣQ, 28.3 and 38.9 for AOT, and CTABr, respectively. That is, the same set of experimental data can be fitted according to different models! Whereas the first approach implies three types or "states" of water within the pool [10–12, 16d] the latter one requires no such assumption.

Thus the problem of the model that describes water structure within reverse aggregates is a complex one whose solution requires use of a variety of techniques, each sensitive to a different property of the system, and has its own time-scale. We limit our discussion to two *noninvasive* techniques, viz., NMR and IR, these are discussed separately below.

Use of NMR to probe the structure of reverse aggregate-solubilized water

Chemical shifts and relaxation times (T_1 and T_2) of both surfactant headgroup, and solubilized water have been used to study the dependence of water-surfactant interactions on W/S [16]. Because diffusion of water molecules within the pool is faster than the NMR time-scale, average water signals are observed, and the dependence of, e.g., δ_{H_2O} on W/S has been treated in terms of the fraction (f) and the corresponding chemical shift of each type of water, e.g., Eq. (2) [16]:

$$\delta_{H_2O} = f_{\text{bound}}(\delta_{H_2O})_{\text{bound}} + f_{\text{bulk-like}}(\delta_{H_2O})_{\text{bulk-like}} \,. \quad (2)$$

That is, a *preconceived model* is required in order to fit NMR data, the information obtained, e.g., estimates of f_{bound} and $f_{\text{bulk-like}}$ is, therefore, *model-dependent.*

Using NMR, we have measured the so-called D/H "fractionation factor, φ" of solubilized water and used its value to get insight into the structure of aggregate-solubilized water. Details of the determination of φ by NMR are given elsewhere [17, 18], so that only essentials will be covered here: consider a species (i) dissolved in H_2O–D_2O. For simplicity, (i) carries no exchangeable hydrogen, and is hydrated by one solvent molecule. The equilibrium constant for the isotopic exchange depicted by Eq. (3) is shown by Eq. (4):

$$\text{i(LOH)} + \text{LOD} \rightleftharpoons \text{i(LOD)} + \text{LOH} \,, \quad (3)$$

$$K = [\text{i(LOD)}]\,[\text{LOH}]/[\text{i(LOH)}]\,[\text{LOD}] \,, \quad (4)$$

that is

$$K = \{[\text{i(LOD)}]/[\text{i(LOH)}]\}/\{[\text{LOD}]/[\text{LOH}]\} \,, \quad (5)$$

or simply

$$K = \varphi_i = [D/H]_i/[D/H]_{\text{bulk solvent}} \,, \quad (6)$$

where L is an unspecified isotope of hydrogen. The deuterium/protium fractionation factor at (*i*), φ_i, expresses the deuterium preference for the site in question (i.e., the hydration shell of (*i*) relative to an average solvent site [17]). The next point is to show how φ_i can be rationalized in terms of the degree of structure of water of hydration of (*i*), relative to bulk water. Note that for bulk solvent $\varphi = 1$, by definition. In a system where D and H equilibrate among a number of different sites, e.g., hydration shell of (*i*) and bulk solvent, deuteriums accumulate, relative to protiums, at sites where they are most closely confined by potential barriers. Fractionation factors *less than unity* imply a greater preference for deuterium in bulk solvent than in the site in question. The converse of this argument shows that fractionation factors *greater than unity* are associated with stronger binding potentials in the site than those in bulk solvent [19, 20]. This interpretation of φ_i in terms of the degree of solvent structure at the site (*i*) relative to an average solvent site has been employed to explain fractionation factors in solution [21–25], in the vapor phase in equilibrium with electrolyte solutions [26], for water in salt hydrates [27], and that embedded in ion exchange resins [28].

The structure of reverse aggregate-solubilized water can be inferred from its fractionation factor, as follows. In the case of W/O μEs, it is usually assumed that the central water layer is akin to bulk water. Therefore D/H fractionation, if it occurs, should be between water molecules in the peripheral and central layers, i.e., between W_{bound}, and $W_{\text{bulk-like}}$, *provided that the binding potential (i.e., water structure) in both layers is different.* Because the two-state water solubilization model implies that water in the first layer is more structured that in the second layer, *the fractionation factor for the aggregate-solubilized water is expected to be greater than unity.*

Fractionation factor for water solubilized by RMs and W/O μEs can be most conveniently calculated from the dependence of the NMR chemical shift of a surfactant headgroup nucleus, e.g., ^{1}H or ^{13}C on the isotopic composition of solubilized H_2O–D_2O mixtures, at constant [surfactant] and W/S. Alternatively, provided that the surfactant monomer is a strong electrolyte, i.e., does not hydrolyze in the water pool, φ can be determined from the dependence of the chemical shift of solubilized H_2O–D_2O mixtures on their isotopic composition, at constant [surfactant] and W/S. Both approaches have been verified experimentally [25, 29, 30].

In case of a W/O μE, applying the two-state water solubilization model, Eq. (6) is given by

$$\varphi_M = (D/H)W_{\text{bound}}/(D/H)W_{\text{bulk-like}} \,, \quad (7)$$

where φ_M is the fractionation factor for aggregate-solubilized water. The equation that is used to determine φ_M from NMR data is given by [29, 30]

$$(\delta_{HD} - \delta_H)/(\delta_D - \delta_H) = \{\varphi_M/[(1 - \chi_D) + \varphi_M\chi_D]\}\chi_D \,, \quad (8)$$

where δ_{HD}, δ_H, and δ_D refer to observed chemical shifts of *either* the surfactant headgroup nuclei (^{1}H and ^{13}C) in the presence of solubilized H_2O–D_2O mixtures, solubilized pure H_2O, and solubilized pure D_2O, respectively, *or* to chemical shift of solubilized H_2O–D_2O mixtures, solubilized pure H_2O, and solubilized pure D_2O, respectively, and χ_D is the atom fraction of deuterium in the aggregate-solubilized water. A plot of the left-hand term of Eq. (8) versus χ_D should be *linear* for $\varphi_M = 1$, *curve down* for $\varphi_M < 1$, or *curve up* for $\varphi_M > 1$.

We have determined φ_M for water solubilized by RMs and W/O μEs of the following anionic, cationic and nonionic surfactant systems: AOT; magnesium bis(2-ethylhexyl) sulfosuccinate; CTABr; cetyldimethylbenzylammonium chloride; cetyldimethyl-3-phenylpropylammonium chloride; and polyoxyethylene (4) dodecyl ether. *In all cases* a fractionation factor of *unity* was obtained, indicating, therefore, no measurable *D/H* fractionation between "layers" in the water pool, i.e., our results are not in agreement with the coexistence of two or more structurally different states of solubilized water [29, 30].

Regarding this approach, two important points should be kept in mind:

(i) because φ_M is an equilibrium constant, fast diffusion of water molecules between different loci in the water pool (e.g., between the first and second water layers) *has no bearing* on its calculation, i.e., φ_M is not an average fractionation factor for the different "types" of water within the pool, if present; (ii) as shown by Eq. (8), *no preconceived model* is required for the calculation of φ_M. Consequently, this approach provides a *model-independent* picture of the structure of reverse aggregate-solubilized water.

Use of IR to probe the structure of reverse aggregate-solubilized water

In IR and Raman spectroscopy studies of water solubilization by reverse aggregates, data treatment involves deconvolution of ν_{OH} band of H_2O, or ν_{OD} band of D_2O into two, or three peaks [10–12]. The deconvoluted peaks are attributed to water with different structures within the micellar system, i.e., W_{bound}, $W_{intermediate}$, $W_{bulk\text{-}like}$, and $W_{interfacial}$. Alternatively, deconvolution of ν_{OD} (or ν_{OH}) band of HOD has been interpreted in terms of the presence of *one type* of aggregate-solubilized water whose properties change continuously as a function of increasing *W/S* [13, 15].

Before addressing the reason for these conflicting interpretations (in terms of a water structure model), it is important to discuss the following point which is crucial for any curve fitting of data to a particular chemical model: quantitative treatment of IR and Raman experimental data requires some "a priori" hypothesis on the origin of the vibrational dynamics of the system under analysis. The suggested model should fit the data accurately (i.e., with the least possible error), and *agrees with chemistry*. With this proviso, some conclusions drawn from curve fitting of bands of reverse aggregate-solubilized water become untenable. For example, it is not obvious how to rationalize the following IR and Raman spectroscopy results for AOT RMs and W/O μEs: (a) discrepancy in the number of water types (one, two or three) present within the aggregate [10–13, 15]; (b) presence of 3 types of water at $W/S \leq 3$ [10a, b]; (c) large differences in the dependence of W_{bound} on *W/S*. This dependence is quadratic in one case [10a], and complex (higher than fifth power dependence!) in other cases [11c, 12a]; (iv) discrepancy in the curve fitting-based threshold of formation of $W_{bulk\text{-}like}$, 3.5 [10], 6.7 [11e], and 12 [12a]; (v) unexpected water distribution among the layers under supercritical conditions (reverse aggregates in ethane) where most of the solubilized water is present as W_{bound}, not $W_{bulk\text{-}like}$ even in the μE domain [31].

As an example, we consider IR results of RMs and W/O μEs of CTABr in a chloroform/*n*-dodecane mixture (6:4, v/v), *W/S* from 2.1 to 40.2 [15a]. Curve fitting of the ν_{OD} band of solubilized water (4% D_2O in H_2O, v/v) showed the presence of a main peak at $2518 \pm 7\ cm^{-1}$ and two small ones at $2365 \pm 42\ cm^{-1}$ and $2662 \pm 20\ cm^{-1}$. *At any W/S*, the main peak corresponds to $90 \pm 2\%$ of the total peak area, and it obeys Beer's law. As a function of increasing *W/S*, the frequency of the main peak decreases, whereas its full width at half-height increases, values of both properties at *W/S* of 40.2 are close to the corresponding ones of HOD in bulk aqueous phase [32].

There are two alternative interpretations of these results: (a) the three peaks obtained by curve fitting at ca. $2662\ cm^{-1}$, ca. $2518\ cm^{-1}$, and ca. $2365\ cm^{-1}$ correspond to different types of water in the pool, namely, W_{bound}, $W_{intermediate}$, and $W_{bulk\text{-}like}$, respectively; (b) there is one type of water present which gives rise to the observed main peak at ca. $2518\ cm^{-1}$, additional peaks at ca. $2662\ cm^{-1}$ and ca. $2365\ cm^{-1}$ need not be associated with HOD molecules present in layers of different structures, as implied by the multi-state water solubilization model. If the former model were correct then one expects a correlation – *that agrees with chemistry* – between areas of the three peaks and *W/S*. Considering that the fraction of W_{bound} levels off at *W/S* of ca. 6 [8], and that the fraction of $W_{intermediate}$ should level off at a higher *W/S* [10a, 11a], then it is *expected* that areas of W_{bound}, at ca. $2662\ cm^{-1}$, and of $W_{intermediate}$ at ca. $2518\ cm^{-1}$ should reach *limiting values* at certain *W/S*, whereas that of $W_{bulk\text{-}like}$ at ca. $2365\ cm^{-1}$ should increase continuously as more water is solubilized. Our results show, however, that this is not the case because: (a) all three peaks increase in area as a function of *W/S*; (b) the ratio between the area of each peak and the total peak area is *practically independent* of *W/S*;

(c) even well within the W/O microemulsion domain, the main peak is that at ca. 2518 cm^{-1}, i.e., which *presumably corresponds to* $W_{intermediate}$, *not* $W_{bulk\text{-}like}$! Therefore, the two small Gaussian peaks which were introduced to get a better fit need not be associated with two additional types of water in the pool, namely W_{bound} and $W_{bulk\text{-}like}$. Their use was necessary because the ν_{OD} peak of HOD is asymmetric, as given elsewhere [32]. That is, our data are best explained without resorting to the coexistence of layers of water of different structure within the pool. We arrived at the same conclusion from our IR studies on water (4% D_2O in H_2O, v/v) solubilization by inverse aggregates of AOT in heptane and magnesium bis(2-ethylhexyl) sulfosuccinate in toluene [15b].

Therefore, the basic premise involved in assuming that each of the D_2O or H_2O bands obtained by curve fitting may be attributed to a different type of water seems possibly suspect because these bands may originate from coupled water molecule vibrations, and from a bending overtone often reported in the spectrum of liquid water [13, 33]. On the other hand, deconvolution of ν_{OH} or ν_{OD} vibrations of HOD is straightforward because both frequencies are essentially decoupled, provided that $[D_2O] \leq 10\%$ [32]. This advantage has been recognized both for bulk aqueous phase [32], and for reverse aggregates [10c, 13b].

The presence of one type of water within the reverse aggregate also agrees with: fluorescence measurements in RMs [34], NMR studies of concentrated salt solutions [35], IR results of HOD in bulk aqueous phase [32], theoretical calculations on molecular dynamics of water [36], dielectric relaxation of water in hydrated phospholipid bilayers [37], and measurement of water chemical potential in the presence of phospholipid bilayer membranes [38].

Proposed model for reverse aggregate-solubilized water

In the preceding sections we have discussed results, including those of our IR and NMR work, that indicate that treatment of experimental data in terms of the coexistence of structurally different water "layers" within the pool is probably an oversimplification. The change of the slope of graphs of certain physical properties as a function of increasing *W/S*, which is observed at the threshold of formation of W/O μEs, need not be attributed to the formation of a second, bulk-like water. It may well reflect the expected decrease in water–surfactant interactions after completion of the hydration of its headgroup. Thus properties of solubilized water change continuously as a function of increasing *W/S*, akin to the dilution of a concentrated electrolyte solution.

The principle factor which contributes to an averaging of water structure over the *whole volume* of the aqueous nanodroplet is that *W/S* inside reverse aggregates are much smaller than the corresponding ratios in aqueous surfactant solutions (≤ 60 and 535 for W/O μEs, and a 0.1 M CTABr aqueous solution, respectively [14]). Consequently, it is highly likely that water molecules within the pool are simultaneously affected by several forces, this precludes any long range order within the aqueous nanodroplet. These forces include: (i) the surface potential of the micellar interface; (ii) water structure perturbation due to hydration of species present in the pool; and (iii) dependence of surfactant–water interactions on *W/S*. Points i–iii are best understood from Fig. 3 which is a schematic representation of an AOT W/O μE, at *W/S* of 50, i.e., close to the phase separation boundary. The bases which we used to draw of Fig. 3 are explained in the Calculations section, detailed results are reported in Table 1. Briefly, the following points were taken into account: the W/O interface is not rigid, but fluctuates [39]; the *whole* sulfosuccinate moiety of the surfactant (i.e., NaO_3S–CH(COO)CH_2–COO) interacts with solubilized water [12, 40]; and the water pool contains sodium ions due to the dissociation of AOT, and also surfactant monomers due to their migration from the W/O interface [41]. Regarding point (i), focusing on the two depicted AOT monomers (*at the interface*), Fig. 3 and Table 1 show that the *opposite* surface potential decay curves either cancel out, or their values reach zero mV within a short distance of each other, e.g., ca. 8 Å at *W/S* of 50. Note that these electrostatic interactions (e.g., repulsion) cancel out due fluctuation of the interface, so that the electrical neutrality within

Fig. 3 Schematic representation of a W/O μE, at *W/S* of 50 showing fluctuation of the W/O interface, exponential decay of the surface potential in the water pool, and presence within the pool of dissociated sodium ions (●), and free surfactant molecules (▲) that migrated from the W/O interface. The dashed circle represents the rigid, or static W/O interface

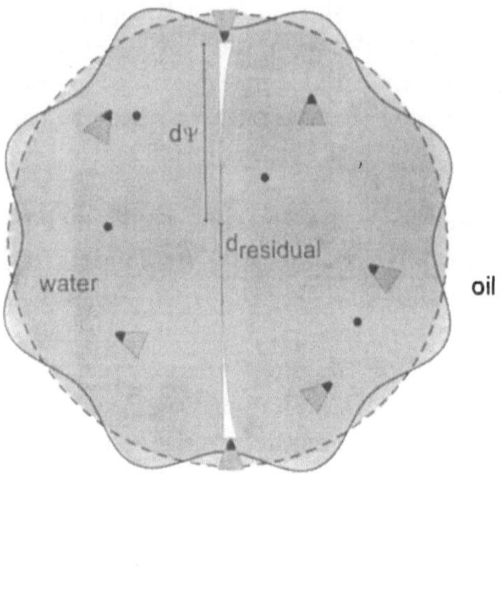

Table 1 Parameters used to represent RMs and W/O μEs of AOT, Fig. 3

W/S [a)]	R_w [b)] [Å]	d_1 [c)] [Å]	d_2 [d)] [Å]	d_ψ [e)] [Å]	$d_{residual}$ [f)] [Å]
4	10	10	1	3.7	g
8	16	22	13	9.1	g
12	22	34	25	14.5	g
16	29	48	39	19.9	g
20	35	60	51	25.3	0.5
25	43	76	67	32.0	3.0
30	51	92	83	38.7	5.6
35	58	106	97	45.4	6.1
40	65	120	111	52.2	6.7
50	79	148	139	65.6	7.8

a) W/S = [Water]/[surfactant].
b) R_w = Radius of the water pool, calculated from data of Maitra [42].
c) d_1 = Water pool diameter considering a surface fluctuation of ±5 Å from the rigid, or static W/O interface, $d_1 = 2R_w - 10$.
d) d_2 = Distance between headgroups of the two depicted AOT molecules (at the interface), Fig. 3. $d_2 = d_1 - (2 \times 4.5$ Å). The length of the headgroup is taken as 4.5 Å, see text.
e) d_ψ = Distance from the W/O interface at which the surface potential reaches zero mV [43].
f) $d_{residual}$ = Residual distance between d_ψ of the two depicted AOT molecules (at the interface), Fig. 3. $d_{residual} = d_2 - (2 \times d_\psi)$.
g) Calculations resulted in (physically meaningless) negative values.

the water pool is maintained [39, 43, 44]. The effect of point (ii) can be appreciated from our calculation that the structure of ca. 8% of the water present is perturbed due to hydration of the dissociated sodium ions, and free surfactant monomers. Regarding point (iii), water solubilization changes the relative stability, i.e., the population of different conformers of the sulfonate and the ester headgroups, these conformers bind water differently [12a, 13b, 40].

Calculations

A minimum-energy conformation for AOT was calculated by the MM2 method. Energy of this structure was further minimized by the PM3 semi-empirical method (version 6.0, MOPAC program package). The structure resulted is a truncated cone, with the following characteristics: dihydral angle between planes of the two carbonyl groups, 68.1°; angle between the *two* terminal methyl groups ($\underline{C}H_3$–$C_4H_9(C_2H_5)$) and carbon atom of the sulfosuccinate moiety (NaO_3S–$\underline{C}H(COO)CH_2$–COO), 65.2°; and distance between ($\underline{C}H_3$–$C_4H_9(C_2H_5)$) and the oxygen atom of the sulfonate group, *ca* 11.8 Å. The structure of minimum energy shows that the two carbonyl groups are *gauche* with respect to each other, as suggested by IR studies [13b, 16b] and the monomer length is similar to that determined by X-ray diffraction, 11.95 Å [45]. The headgroup is considered to be composed of the *whole* sulfosuccinate moiety, vide supra. Consequently, the surfactant molecule is divided into the hydrocarbon tail, 7.3 Å, and the headgroup, 4.5 Å. The radii of the water pool at different W/S, R_w, were taken from data of Maitra [42]. The interface was assumed to fluctuate within ±5 Å from its rigid, or static position (i.e., from the dashed circle of Fig. 3 [39]). Regarding the species present in the water pool, for [AOT] = 0.1 M, we took 0.2; 0.008 moles/liter; 4; and 39 as the degree of dissociation of the surfactant [44], the concentration of free monomers in the water pool [46], the hydration number of Na^+ ion [48], and the hydration number of surfactant monomers in the water pool [49], respectively. Using these figures, we calculated that 0.08 mole of water (1.6% of solubilized water) is attached to dissociated sodium ions, and 0.31 mole of water (6.2% of solubilized water) is included in the hydrophobic hydration shell [50] of surfactant molecules inside the pool. The dependence of the exponential decay of surface potential within the pool on W/S was calculated by interpolation of the data of Caselli and Mangone. These show an excellent linear relationship (cc 0.9998) between W/S and the distance at which value of the surface potential reaches zero mV [43].

Acknowledgments L.P. Novaki thanks FAPESP for a postdoctoral fellowship, O.A. El Seoud thanks FAPESP and FINEP for financial support, and CNPq for a research productivity fellowship. We would like to thank Guilherme A. Marson, Paulo A.R. Pires, and Reinaldo C. Bazito for their help during the preparation of this manuscript.

References

1. Fendler JH (1982) Membrane Mimetic Chemistry. Wiley, New York
2. Attwood D, Florence AT (1984) Surfactant Systems: Their Chemistry, Pharmacy, and Biology. Chapman and Hall, London
3. Eicke H-F, Kvita P (1984) In: Luisi LP, Straub BE (eds) Reverse Micelles: Biological and Technological Relevance of Amphiphilic Structures in Apolar Media. Plenum Press, New York, p 21
4. Langevin D (1984) In: Luisi LP, Straub BE (eds) Reverse Micelles: Biological and Technological Relevance of Amphiphilic Structures in Apolar Media. Plenum Press, New York, p 287
5. a) Candau F, Leong YS, Poyet G, Candau SJ (1984) J Colloid Interface Sci 101:167; b) Candau F, Leong YS, Fitch RM (1985) J Polymer Sci 23:195; c) Candau F, Zekhini Z, Durand JP (1986) J Colloid Interface Sci 114:398
6. Kon-no K, Kitatahara A, El Seoud OA (1987) In: Schick M (ed) Nonionic Surfactants: Physical Chemistry. Marcel Dekker, New York, p 185

7. a) Goba RD, Kon-no K, Kandori K, Kitahara A (1983) J Colloid Interface Sci 93:293. b) Lianos P, Thomas JK (1986) J Colloid Interface Sci 117:505
8. El Seoud OA (1994) In: Hinze WL (ed) Organized Assemblies in Chemical Analysis, Vol 1. JAI Press, Greenwich, p 1
9. a) Yoshioka H (1983) J Colloid Interface Sci 95:81; b) Yoshioka H, Kazama S (1983) J Colloid Interface Sci 95:240; c) Goto A, Yoshioka H, Kishimoto H, Fujita T (1992) Langmuir 8:441
10. a) Onori G, Santucci A (1993) J Phys Chem 97:5340; b) D'Angelo M, Onori G, Santucci A (1994) J Phys Chem 98:3189; c) Amico P, D'Angelo M, Onori G, Santucci A (1995) Nuovo Cimento 17D:1053
11. a) Boicelli CA, Giomini M, Giuliani, AM (1984) Appl Spectroscopy 38:537; b) MacDonald H, Bedwell B, Gulari E (1986) Langmuir 2:704; c) D'Aprano A, Lizzio A, Liveri VT, Aliotta F, Vasi C, Migliardo P (1988) J Phys Chem 92: 4436; d) Aliotta F, Migliardo P, Donato DI, Liveri VT, Bardez E, Larry B (1992) Prog Colloid Polym Sci 89:258; e) Giammona G, Goffredi F, Liveri VT, Vassallo G (1992) J Colloid Interface Sci 152:465
12. a) Jain TK, Varshney M, Maitra A (1989) J Phys Chem 93:7409; b) Maitra A, Jain TK, Shervani Z (1990) Colloids Surfaces 47:255
13. a) Pacynko WF, Yarwood J, Tiddy GJT (1987) Liquid Crystals 2:201; b) Christopher DJ, Yarwood J, Belton PS, Hills B (1992) J Colloid Interface Sci 152:465
14. a) El Seoud OA, El Seoud MI, Mickiewicz JA (1994) J Colloid Interface Sci 163:87; b) El Seoud OA, Okano LT, Novaki LP, Barlow GK (1996) Ber Bunsenges Phys Chem 100:1147
15. a) Novaki LP, El Seoud OA (1997) Ber Bunsenges Phys Chem 101:1928; b) Novaki LP, El Seoud OA (1998) J Colloid Interface Sci, in press
16. a) Chachaty C (1987) Prog NMR Spectroscopy 19:183, and references cited therein; b) Heatley F (1988) J Chem Soc Faraday Trans 1 84:343; c) Hauser H, Haering G, Pande A, Luisi PL (1989) J Phys Chem 93:7869; d) Kawai T, Hamada K, Kon-No K (1993) Bull Chem Soc Jpn 66:2804; e) Yoshino A, Okabayashi H, Yoshida T (1994) J Phys Chem 98:7036; f) Waysbort D, Ezrahi S, Aserin A, Givati R, Garti N (1997) J Colloid Interface Sci 188:282
17. Schowen KB (1978) In: Gandour RD, Schowen RL (eds) Transition States for Biochemical Processes, Plenum Press, New York, p 225
18. Albery J (1975) In: Caldin E, Gold V (eds) Proton-Transfer Reactions, Chapman & Hall, London, p 263
19. Kreevoy MM (1964) J Chem Educ 41:636
20. El Seoud OA (1997) J Mol Liq 72:85
21. a) Gold V, Grist S (1971) J Chem Soc B 1665; b) Gold V, Morris KP, Wilcox CF (1982) J Chem Soc Perkin Trans 2 1615
22. a) Kresge AJ, Tang YC (1979) J Phys Chem 83:2156; b) Keeffe JR, Kresge AJ (1989) Can J Chem 67:792
23. a) Al-Rawi JMA, Bloxsidge JP, Elvidge JA, Jones JR, More RA, O'Ferrall R (1979) J Chem Soc Perkin Trans 2 1593; b) Balzer L, Bergman NA (1982) J Chem Soc Perkin Trans 2 313
24. a) El Seoud OA, El Seoud MI, Pires PAR, Farah JPS (1984) J Phys Chem 88:2669; b) El Seoud MI, El Seoud OA (1984) Ber Bunsenges Phys Chem 88:742; c) El Seoud OA, Kiyan NZ, Vieira PC (1987) Ber Bunsenges Phys Chem 91:825
25. a) Jarret RM, Saunders M (1985) J Am Chem Soc 107:2648; b) ibid (1986) 108:7549
26. Pepezin J, Jalki G, Jansco G, Van Hook WA (1972) J Phys Chem 76:743
27. Pradhananga TM, Matsuo S (1985) J Phys Chem 89:1869
28. a) Gupta AR, Nandan D, Sarpal SK (1982) J Phys Chem 86:3257; b) Gupta AR, Nandan D (1985) ibid 89:4329
29. El Seoud OA, El Seoud MI, Mickiewicz JA (1994) J Colloid Interface Sci 163:87
30. El Seoud OA, Okano LT, Novaki LP, Barlow GK (1996) Ber Bunsenges Phys Chem 100:1147
31. Ikushima Y, Saito N, Arai M (1997) J Colloid Interface Sci 186:254
32. a) Senior WA, Verrall RE (1969) J Phys Chem 73:4242; b) Mundy WC, Gutierrez L, Spedding FH (1973) J Chem Phys 59:2173; c) Wall TT, Hornig DF (1965) J Chem Phys 43:2079; d) Wall TT, Hornig DF (1967) J Chem Phys 47:784; e) Walrafen GE (1968) J Chem Phys 48:244; f) Schiffer J, Hornig DF (1968) J Chem Phys 49:4150; g) Lucas M, De Trobriand A, Ceccaldi M (1975) J Phys Chem 79:913; h) Kristiansson O, Eriksson A, Lindberg J (1984) Acta Chem Scand A38:609
33. Scherer JR (1980) In: Adv Clark RJ, Hester RE (eds) Infrared Raman Spectroscopy, Vol 5. Wiley, New York, p 149
34. a) Belletête M, Droucher GJ (1989) J Colloid Interface Sci 134:289; b) Belletête M, Lachapelle M, Droucher G (1990) J Phys Chem 94:5337
35. a) Boden N, Mortimer M (1978) J Chem Soc Faraday Trans 2 74:353; b) Baianu IC, Boden M, Lightowlers D, Mortimer M (1978) Chem Phys Lett 54:169
36. Teleman O, Joensson B, Engstroem S (1987) Mol Phys 60:193
37. Enders H, Nimtz G (1984) Ber Bunseges Phys Chem 88:512
38. Lis LJ, McAlister M, Fuller N, Rand RP, Parsegian VA (1982) Biophys J 37:657
39. Leodidis EB, Hatton TA (1989) Langmuir 5:741
40. a) Okabayashi H, Taga K, Tsukamoto K, Matsushita K, Kamo O, Yoshikawa K (1987) Colloids and Surfaces 24:337; b) Moran PD, Bowmaker GA, Cooney RP, Bartlett JR, Woolfrey JL (1995) Langmuir 11:738; c) Moran PD, Bowmaker GA, Cooney RP, Bartlett JR, Woolfrey JL (1995) J Mater Chem 5:295
41. Zulauf M, Eicke H-F (1979) J Phys Chem 83:480
42. Maitra A (1984) J Phys Chem 88:5122
43. Caselli M, Mangone A (1992) Ann Chim 82:303
44. Karpe P, Ruckenstein E (1990) J Colloid Interface Sci 137:408
45. Ekwall P, Mandell L, Fontell K (1970) J Colloid Interface Sci 33:215
46. The concentration of surfactant monomers which migrate from the interface into the water pool was taken as equal to the critical micelle concentration of SDS, 0.008 M [47]
47. Mukerjee P, Mysels KJ (1971) In: Critical Micelle Concentrations of Aqueous Surfactant Systems, NSRDS-NBS 36. US Government Printing Office, Washington, DC
48. Marcus Y (1994) Biophys Chem 51:111
49. The hydration number of the AOT monomer, 39, was taken as equal to the hydration number of tetra-n-butylammonium ethanesulfonate [50]
50. Nakayama H, Yamanobe M, Baba K (1991) Bull Chem Soc Jpn 64:3023

Progr Colloid Polym Sci (1998) 109:49–59

T.P. Lockhart
E. Borgarello

Colloid science in oil and gas production and transportation

Received: 17 July 1997
Accepted: 21 July 1997

T.P. Lockhart (✉) · E. Borgarello
Eniricerche SpA
20097 San Donato Milanese
Italy

Abstract The importance of colloidal and surface phenomena in the upstream end of the oil and gas business is illustrated by examining several technological problems that presently attract strong interest within the industry. Examples from the authors' own laboratory of recent technical developments grounded in colloid science are briefly discussed. These include the development of more effective clay dispersants for environmentally-friendly drilling fluids, the use of carbon black dispersions to block gas migration in well cements, the evaluation of polymer gels to control water production by means of experimental tests within porous media, and the definition of the phase behavior of polymeric scale inhibitors. Other areas discussed include organic solids deposition (asphaltenes, wax, and gas hydrate) and fluid flow in porous media. Finally, the challenges of studying in the laboratory the physical and chemical colloidal phenomena underlying field problems are considered, and a typical porous media flow apparatus illustrated.

Key words Clay dispersions – fluid-rock interactions – polymer gels – carbon black – organic deposits – inorganic scale – porous media

Introduction

For most colloid scientists, whatever familiarity they might have with the oil industry begins and ends with the efforts of the 1970s and 1980s to develop "enhanced oil recovery" (EOR) technology. With the high oil prices of that period and the fear of rapidly depleting the world's reserves of fossil fuels, EOR research aimed to develop methods capable of producing a greater portion of the oil from petroleum reservoirs, oil which otherwise remains abandoned in the reservoir at the end of a field's economic life. Many of the chemical EOR processes studied intensively in the oil companies and universities in that period have a strong "colloidal" element: aqueous foams, micellar and polymer solutions, and microemulsions, *inter alia* [1].

With the collapse of the price of oil in the second half of the 1980s, interest in these chemical EOR processes declined sharply on account of their unfavorable economics. This in turn led to a progressive, and by now essentially total, withdrawal from basic and applied R&D in this area, and to the removal of petroleum-related technology issues from the awareness of most colloid scientists.

Colloidal and surface phenomena are, however, pervasive in the production and transportation of oil and gas. They are present as the source of operational problems and are important to improving the ability to forecast reservoir productivity. They are also central to the technologies used to drill wells, to optimize well productivity, and to assure the flow of oil and gas through pipelines. The "colloidal dimension", in fact, is characteristic of the porous reservoir rock itself, which is comprised of mineral phases that may have an extension of tens of microns or less and whose microscopic pores are occupied by two or more immiscible phases (water and oil and/or gas). Furthermore, the wettability of the rock, which determines the rate and ultimate level of oil recovery, is a complex function of the type and distribution of mineral phases, the nature of the brine and crude oil, and the physical conditions and geological and production history of the reservoir. The oils and brines produced from the reservoir themselves may be so composed that the change in thermodynamic conditions they experience during production provokes the nucleation and separation of solid organic or mineral phases within the well or surface facilities or during transport through undersea pipelines.

The importance of these issues for the near- and medium-term profitability of the oil industry creates a continuing demand for technological development and drives research in the oil and service companies, universities, and various research institutes. In fact, the industry's success in developing more effective and lower cost technologies for finding, producing, and transporting oil is one of the reasons that the bleak forecasts on the future availability and cost of fossil fuels, which had driven EOR research, failed to come about. Colloid science has played a part in this process, and colloidal issues continue to furnish new challenges to the efforts of the industry to maintain this remarkable pace of technological evolution.

The visibility of the colloidal dimension of the oil industry within the community of colloid scientists could and, indeed, should be greater. On the one hand, the oilfield offers a number of remarkable colloidal problems and phenomena that require further study. On the other, a more systematic and profound application of the concepts and methods of colloid science could facilitate the resolution of these problems and accelerate technological progress in the industry. In this article, therefore, we will highlight some of the important processes and problems in the field of oil and gas production and transport in which colloidal phenomena play a leading role. Though our survey is anchored in the technological requirements of the oil industry, we will indicate some areas where there is a strong need for improved understanding of basic collodial phenomena.

Table 1 reports, within the broadly identifiable stages of oil and gas production and transportation, some of the areas in which colloidal and surface phenomena and

Table 1 Some colloidal issues in oil and gas production and transportation

Technological area	Colloidal aspects
Exploration and appraisal	
Petrophysical evaluation of reservoir cores	Wettability; capillary pressure; relative permeability
Crude oil, brine deposition tendency	Physical state of components in the oil; thermodynamics, kinetics of nucleation and growth of solid phases
Basin modelling	Diagenetic phenomena, including nucleation events
Production	
Drilling fluids	Clay dispersions in water, oil Mechanism of destabilization of shales by aqueous drilling fluids Fluid invasion/filtration properties in porous media
Cements	Additives to reduce gas migration in un-cured cement
Well productivity	
• acidizing	Dissolution, nucleation/precipitation of mineral phases
• corrosion inhibition	Surface adsorption characteristics; retention/release in reservoir rock
• fracturing	Rheology of concentrated polymer solutions and gels and filtration behavior in porous media
• sand control	Chemical consolidation of clay-rich sands without loss of permeability Sand filtration on gravel bed
• scale inhibition	Thermodynamics, kinetics of nucleation; threshhold inhibition mechanisms; inhibitor retention/release in reservoir rock
Produced fluids management	
• oil–water separations	Centrifugation technology; emulsion-breaking additives
• water re-injection	Water quality, filtration properties
• downhole water shut-off	Polymer solution and gel rheology, cross-linking; propagation, injectivity, gel stability in porous media
• organic deposit formation	Thermodynamics, kinetics of nucleation/precipitation of asphaltenes, wax, hydrates; inhibition/remediation technologies
Transportation	
Heavy oil transportation as aqueous emulsion	Emulsion formation and rheology; stabilization and destabilization
Multiphase flow	Emulsion formation, rheology

technologies based on colloidal systems are important. This will be a useful reference for the succeeding portions of this text, in which several technological areas rich in colloidal issues are discussed.

Colloid science in oil and gas production and transportation

Drilling fluids

The so-called drilling fluids or "muds" are pumped down the hollow center of the drill stem to the drill bit, which they serve to cool and lubricate. Most importantly, the drilling fluid conveys the rock cuttings to the surface, where they are removed by filtration or sedimentation devices so that the drilling fluid can be recirculated back downhole. While the drilling fluid must satisfactorily carry out these and a number of other "tasks", the most essential physical property of the fluid is its rheology (most drilling muds are Bingham-like fluids) which must assure adequate solids-carrying properties even under static conditions and non-vertical well orientations [2].

Initially, drilling was carried out using water-based fluids. Beginning in the 1950s, however, hydrocarbon-based fluids (oil-based muds) formulated as water-in-oil emulsions came to be preferred for the more demanding drilling situations. Though far more expensive per unit volume of fluid, oil-based muds are often economically convenient because fewer problems are encountered during drilling. More recently, the use of oil-based muds has been increasingly constrained by strong environmental pressures, which has led to a return to water-based fluids and to research efforts to overcome their technical limitations.

Water-based muds are complex chemical mixtures that can be divided into several distinct categories [2]. The lowest-cost and most widely used aqueous fluids are formulated with dispersed bentonite clay solids which impart the basic rheological properties. Other aqueous drilling fluids employ bio- or synthetic polymers for rheology control, the latter being preferred for high temperature wells, but are sensitive to the clay solids incorporated during drilling. With respect to oil-based muds, water-based muds all suffer from three principal weaknesses: (1) Their rheological properties decay rapidly at the high temperatures encountered in deeper wells. (2) Their interaction with shale (clay rich) intervals can provoke caving of solids into the wellbore or swelling and closure of the hole, resulting in costly delays or loss of the well. (3) The invasion of the oil or gas reservoir by the drilling fluid can severely reduce the permeability of the reservoir rock, compromising well productivity.

In order to reduce damage to well productivity, drilling fluids are formulated with solids (clay, sized salt or carbonates) that filter-out on the rock face, forming a low permeability cake. Increasing demands placed on well productivity, reduced possibilities to drill into the hydrocarbon reservoir using non-damaging oil-based muds, and new well construction requirements are responsible for strong re-focusing of attention on this area. More insight is needed into the basic filtration properties of the mud solids in order to clarify the factors that influence the rate of filtercake build-up and the depth of solids invasion into the reservoir rock [3]. Also, achieving clean lift-off of the filtercake is necessary to recover the permeability of the formation. Both filtercake formation and removal are likely to be strongly influenced by the pore structure of the reservoir rock and by interactions between the filtered solids and other additives (polymers and thinners) present in the mud, variables that have not yet been examined adequately.

One route to improving the resistance of water-based muds to high temperature is to achieve better control over the flocculation behavior of the clay solids, a problem centered on the activity and thermal stability of the clay dispersants ("thinners"). We have pursued this approach in our laboratory, focusing our attention on the low-cost chromium-lignosulfonate and chromium-lignite thinners. These "workhorse" thinners suffer from two principle limitations: poor control of the bentonite based fluid rheology above 120 °C, and environmental restrictions on the use of chromium, which is present in these thinners both as Cr(III) and Cr(VI) (the latter is cancer suspect and is the more strictly regulated of the two).

The industrial goals of our program were to remove Cr(VI) and, if possible, also Cr(III) from these thinners, while simultaneously increasing their resistance to high temperature. Because little insight into the properties and function of these thinners had been developed since their introduction in the early 1950s, we felt that it was essential first to better define the chemical and functional role played by chromium. By means of a series of rheological, adsorption, and surface potential measurements on bentonite clay dispersions of chromium-containing and chromium-free lignite and lignosulfonate thinners, it was shown that Cr(III) strongly enhances the level of thinner adsorption and clay dispersion activity [4, 5]. These results were interpreted in terms of bridging of Cr(III) between the surface of the clay particles and the thinner (cf. upper right-hand corner of Fig. 1).

Other studies showed that Cr(VI) does not contribute to the thinning power of lignosulfonate [6]. However, thermal aging tests on bentonite suspensions with Cr(III) lignosulfonate showed that rheological stability at high temperature is strongly enhanced in the presence of Cr(VI).

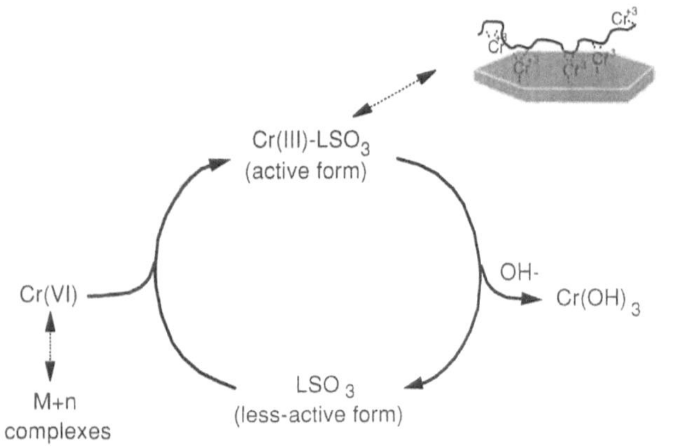

Fig. 1 Postulated cycle of reactions explaining the thermal resistance of betonite drilling fluids thinned with chromium-lignosulfonate (LSO_3). In the active form of the thinner, Cr(III) is chemically bound, and enhances adsorption of the thinner onto the surface of the clay particles (as illustrated at the upper right corner). Deactivation of the thinner at high temperature is attributed to hydrolysis of the Cr(III). Cr(VI), or appropriate complexes of Cr(III) and other metal ions (see text), regenerate the active form of the thinner, maintaining the drilling fluid rheology

An important clue to the role of Cr(VI) came from tests on suspensions of bentonite with chromium-free lignosulfonate. While the addition of Cr(VI) had no impact on the (poor) thinning power of this additive, the rheology of the suspension was improved markedly after aging at high temperature. On the basis of these and other observations [6, 7], a hypothesis consistent with the known aqueous chemistry of chromium and capable of explaining the observed behavior was formulated (Fig. 1).

According to this picture, the active form of the lignosulfonate thinner is chemically combined with Cr(III). In the alkaline drilling fluid (pH > 9) at high temperature, Cr(III) bound to the thinner is prone to hydrolysis to $Cr(OH)_3$, which constitutes a thermodynamic sink for the system. The inactive, chromium-free lignosulfonate is "renewed" by slow Cr(VI) reduction at elevated temperature which regenerates the Cr(III)-containing, active form of the thinner. Thus, Cr(VI) serves as a "reserve" of Cr(III) for the thinner, a role made possible by its stability to alkaline hydrolysis.

The ultimate test of a scientific hypothesis lies in its predictive ability. In this sense, the cycle of reactions hypothesized in Fig. 1 has proved to be a rich source of new ideas. First, we thought that, in analogy to Cr(VI), more hydrolytically stable, complexed forms of Cr(III) might also be able to function as a Cr(III) reservoir and extend the useful lifetime of the bentonite drilling fluid. This prediction was verified with Cr(III) oxalate, which imparts thermal stability to lignosulfonate–bentonite suspensions identical (for equimolar amounts of chromium) to that of Cr(VI) [6]. With this result we reached our first objective, namely eliminating the use of Cr(VI) salts without compromising the thermal stability of these water-based fluids.

Next, we explored the possibility that other, more environmentally-benign, metals might be made to play the same role as the Cr(III) complexes. This led to the discovery that certain Zr(IV) complexes used in combination with chromium-free lignosulfonates provide excellent rheological control at temperatures well beyond those that can be achieved with chromium-containing formulations [7, 8]. These novel chromium-free aqueous drilling fluids have proven their value in a number of field applications [9].

The third issue limiting the broader use of water-based drilling fluids is the instability of the wellbore in proximity to clay-rich shale intervals. This complex and economically important issue has commanded considerable attention over the past decade. While some progress has been made in finding new additives capable of reducing wellbore stability problems, a general solution to the problem is still far from hand [10]. Recent studies have called seriously into question the basic physical mechanisms long-assumed to be responsible for shale destabilization [11]. Better understanding of the problem appears to be a condition for making significant technological progress in this area. Among the questions to resolve are the following: What is the role of fluid invasion into the low-permeability clay matrix? What is the role of fluid-clay particle interactions, including intercalation of charged species between the clay lattices, charge dispersion, etc.? How is the problem influenced by the composition of the shale matrix? How do the chemical additives employed by the industry function? Colloidal and surface phenomena are clearly central to these issues.

Gas impermeable cements

Cements are employed to support the steel casing of the well and to provide a hydraulic seal to the formation. A problem encountered when cementing against pressurized gas formations is migration of the gas into the uncured cement slurry. This can lead to the formation of channels open to gas and fluid flow, with negative consequences for safety, the environment, and well productivity. The colloidal additives silica fume and polymer latex are employed to block gas migration in the cement slurry. The function of these additives has been attributed to two factors. First, by virtue of their small size, they are believed to produce a better elastic response of the cement column against the back pressure of the gas from the formation. Secondly, these additives each undergo chemical or physical transformations within the cement: silica fume hydrates and participates in cement hardening, while the polymer latex can fuse to form an impermeable film within the cement.

Examining the available evidence, we speculated that perhaps the most important characteristic of these additives was their small size relative to the cement matrix, which should greatly increase the tortuosity of the path taken by the gas molecules though the uncured cement and result in a reduced rate of gas invasion. If this factor is indeed sufficient, then additive function does not depend on a combination of small particle size and a perfectly timed hydration or fusion process. Rather, any well-dispersed colloidal solid of sufficiently small particle size should impart gas impermeability. This hypothesis was verified for carbon black, dispersed with anionic dispersants, which renders the usual well cements impermeable to gas migration, without modifying the mechanical properties of the hardened cement [12]. This chemically inert additive satisfies the cement performance requirements at a fraction of the cost of silica fume and polymer latex and is now used routinely in Italy.

Polymer gels in oil and gas production

Aqueous polymer gels, composed of low concentrations of synthetic or biopolymers and a chemical cross-linking agent, are used in impressive volumes (up to several thousand m^3 per treatment) in several industrially important processes [13]. The most important and well-established application of these lyophilic colloids is for the hydraulic fracturing of formations. These fracturing operations are carried out routinely in order to bypass near-wellbore formation damage that restricts the inflow of fluids to the wellbore, or to increase the drainage radius of a well in the case where well productivity is limited by low formation permeability. In such applications, the cross-linked fluid is pumped into the well at a pressure exceeding the mechanical resistance of the formation, which fractures in a direction and to an extent that depends in part on the rheology of the gel and its rate of leak-off into the formation along the faces of the fracture. While it is now possible to predict to some degree the extension and direction of these fractures, the resulting productivity of the well is often far below the theoretical value. The source of this problem is the invasion of the gel into the porous rock, which can irreversibly block the pore throats and provoke unfavorable changes in the local water saturation and the oil and water relative permeability in the rock along the fracture face. Further development of this technology, therefore, depends on better controlling gel leak-off, and on achieving more complete breakdown and removal of the gel after treatment.

More recently, polymer gels have been used to reduce water production from gas and oil wells. These treatments involve injecting a fluid, gel-forming composition ("gelant") into the reservoir to saturate the zone (e.g., a high permeability layer or fracture) through which water enters the well; after gelation, fluid flow through the treated zone is eliminated. Provided the offending zone has been correctly identified and the gelant correctly placed, it is possible to improve the economic performance of the well significantly [14].

Ideally, gel-forming compositions for water shut-off applications should satisfy four requirements:

1. Programmable gelation delay. This aids treatment design and assures both against premature gelation in the well and unacceptably long closure of the well while waiting for the gel to form.
2. Good injectivity. This allows the entire gelant volume to be placed in the formation without exceeding pressure limitations.
3. Good propagation of the gelant components (i.e., limited retention within the porous medium). This assures that gel will form throughout the entire volume of rock invaded by the gelant.
4. Durability of the permeability reduction. For maximum economic return, the water problem should be eliminated for the remaining life of the well.

Although polymer gels formulated with biopolymers (Xanthan gum, Schleroglucan) have been employed for this purpose, most attention presently is focused on polyacrylamide and its copolymers which can be cross-linked with Cr(III) ions, or with a mixture of phenol and formaldehyde or their derivatives. While extensive studies have been carried out on these complex colloidal systems in bulk, the focus has increasingly moved to understanding gel performance *within porous rock*, where a number of fluid-rock interactions emerge as important. For example, studies on Cr(III) cross-linked gelants show that mineralogy dependent adsorption and precipitation reactions involving Cr(III) ions can strip the cross-linker from a large fraction of the gelant injected into the formation [15]. By way of mechanisms not yet fully understood, the retained Cr(III) also contributes to strongly reduce gelant injectivity. Gelants employing phenol and formaldehyde for cross-linking, on the other hand, are insensitive to the mineral phases present in the rock, but the phenol may be strongly retained by partitioning into the drops of oil present in the pores of the rock [16]. The injectivity of these gelants is reduced significantly compared to their bulk gelation time owing to filtration within the porous medium of micron-sized aggregates of cross-linked polymer [17].

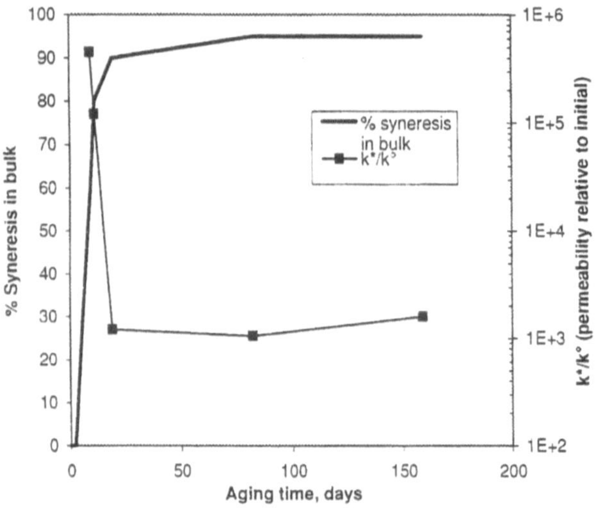

Fig. 2 Evolution of gel syneresis in bulk samples of a polyacrylamide gel as a function of aging at 100 °C, compared with the decreasing permeability reduction (compared to the untreated core) to brine of gel-treated sandstone cores aged at the same temperature

Another striking example showing the relevance of the "physical context" for understanding gel performance is provided by studies of gel stability. As polyacrylamide ages at high temperature (>80 °C) acrylate units are formed by hydrolysis along the polymer backbone. In the presence of Mg^{2+} and Ca^{2+} ions these gels undergo spontaneous and extensive syneresis (the volume occupied by gel shrinks to as little as 5% of the original volume) [18]. It has long been assumed that gels displaying this behavior under reservoir conditions are unsuitable for application. Recently, we examined this point directly by studying the persistance of the permeability reduction of porous rock cores treated with polymer gels [19]. Surprisingly, the permeability reduction was found to persist even where massive syneresis of the gel occurs in bulk (Fig. 2). Chemical tracer studies established that extensive syneresis had also taken place within the porous medium. In the light of these and other considerations we have proposed that the strong attraction of the gel for the rock surface leads to the formation of gel lenses in the pore throats of the rock, thereby maintaining to a large degree the reduced permeability of the porous medium (Fig. 3).

It is evident from this discussion that traditional experiments on bulk samples do not capture important processes that operate in the reservoir. It is important in such cases to employ experimental methods that allow fluid flow in porous rock, and physical and chemical fluid-rock interactions to be studied. The final section of this paper will briefly describe the facilities employed for porous media studies.

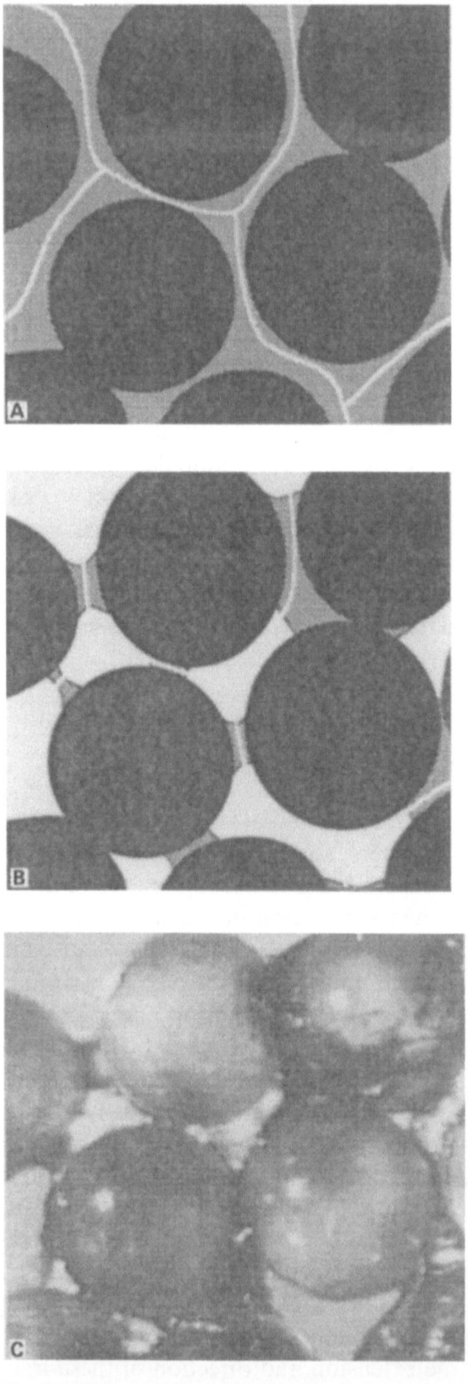

Fig. 3 Diagram illustrating hypothesis for how polymer gel syneresis can progress without compromising permeability reduction of porous media. (a) prior to syneresis, the gel fills the entire pore space, and fluid flow occurs along channel (white line) through the gel. Syneresis initiates in the center of the pores and, (b) at a high degree of syneresis, the water pockets expand to fill the pore bodies leaving a thin gel film on the pore walls and gel lenses in the pore throats. (c) Photograph of actual syneresed gel in glass bead pack (1 mm spheres) displaying the predicted behavior

Phase separation from produced fluids

As noted in the Introduction, predicting and controlling the phase stability of fluids during production and transport in pipelines are important technological challenges. Though it will be convenient to consider the specific issues of hydrocarbon and brine phase stability separately, at the most general level they present many similarities. For both, it is important to develop models capable of predicting the thermodynamic conditions under which phase separation is favored, and the amount and rate of deposition as a function of the fluid dynamic conditions in the well and pipeline. Next, where it is not possible or economically convenient to operate inside the phase stability envelope, it is important to find additives capable of inhibiting phase separation. Though such additives may operate by shifting the thermodynamics for stability, more interesting are "threshhold" (kinetic) inhibitors that delay the onset of precipitation by inhibiting the nucleation and growth of the solid phase. Below, some of the issues facing attempts to predict and control the stability of the hydrocarbon and aqueous phases are discussed.

Hydrocarbon phase stability. Three solid phases, asphaltenes, waxes, and gas hydrates, may separate from hydrocarbon streams under unfavorable operating conditions. Asphaltene phase stability is largely governed by pressure, hence deposits generally form within the reservoir surrounding the well, in the wellbore, or in the surface facilities. Phase separation of waxes, instead, is favored by low temperature, and hence tends to be a problem in surface facilities and especially in pipelines (the temperature on the seafloor may be 5 °C or less). Gas hydrates, spontaneously formed clathrate structures of water and gas molecules, can form deposits at low temperature and moderate pressures, and constitute a particularly important issue for undersea pipelines.

Predicting the conditions for the onset of phase separation presents challenges for all three of these systems. While reliable thermodynamic models for hydrate formation exist for simple gas–water mixtures, the situation is much more complex for hydrocarbon mixtures and for hard brines. For waxes and asphaltenes, the existing models are for the most part highly empirical and, in the case of the asphaltenes, many require use of expensive samples of crude oil recovered at the well bottom and studied at reservoir temperature and pressure. Although the flow conditions in the well/pipeline are important for the formation of these deposits, efforts to describe their influence on deposit build-up are still primitive.

The asphaltenes represent a particularly fascinating challenge for colloid scientists. The asphaltene fraction of the crude oil, defined operationally by solubility criteria, is comprised of a vast number of polar compounds of moderately high molecular weight (roughly 500–2000 g/mol). To date there remains considerable confusion regarding their chemical structure and physical disposition within crude oil [20]. While it is well-established that asphaltenes spontaneously aggregate in the crude oil to form colloidal aggregates, referred to as "micelles," there is an open debate on the factors that govern their stability in the crude oil. The widely invoked steric-stabilization model, represented in Fig. 4, hypothesizes that the asphaltene aggregates are peptized in the crude oil by resin molecules (other, equally ill-defined polar components of the oil). However, the success of recent attempts to model asphaltenes as lyophilic colloids, and their phase behavior in terms of polymer solution theory, together with serious inconsistencies of the steric-stabilization model, strongly suggest the opportunity of discarding this paradigm [20].

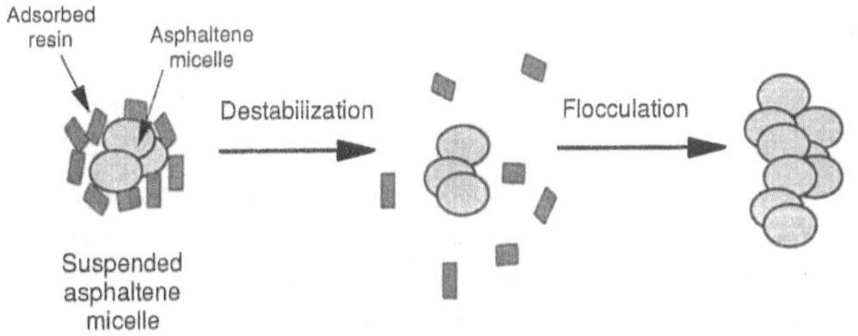

Fig. 4 Schematic representation of common interpretative model of asphaltene colloidal structure and phase behavior: Asphaltene aggregates ("micelles") are assumed to be suspended in the oil through the peptizing action of resin molecules, which are thought to adsorb onto the micelle surface. Changes in temperature or pressure, or addition of non-solvent, is believed to provoke phase separation through desorption of the peptizing resin molecules. This model is criticized in the text and in ref. [20]

For all three of these organic solids, effort is underway to develop more effective additives capable of inhibiting their nucleation and deposition [21]. In the case of hydrate inhibitors, pipelines currently are kept free from deposits by adding organic co-solvents (e.g., glycol) to the humid gas mixtures pumped through undersea pipelines. The trend toward pipelines of increasing length and depth, however, requires the use of larger volumes of solvent and increasing costs of solvent separation and recycling. Thus there is considerable interest within the industry for threshold inhibitors capable of working at levels below 1%; recent results suggest that such inhibitors may reduce the cost of controlling hydrate formation by as much as 50% [22, 23].

Inorganic scale. Thermodynamic conditions favoring the separation of inorganic scales, especially $CaCO_3$ and $BaSO_4$, from produced brine are encountered relatively frequently in oil production. Their formation in the wellbore and perforations (through which fluids enter the well) can lead to a severe loss of productivity. Though it is often possible to remediate $CaCO_3$ deposits by chemical washes of the well and adjacent formation, $BaSO_4$ deposits are very difficult to remove. For this reason, operators prefer to "manage" scaling by preventing the formation of deposits. This is accomplished with threshold inhibitors, chemical additives capable, even at the level of 1–20 ppm, of inhibiting scale nucleation and/or growth. The scale inhibitors presently in widest use are polyphosphonates and acrylate (co)polymers.

Interestingly, the colloidal issues here, which begin with the thermodynamics and kinetics of scale formation and the mechanisms of inhibition, extend to the technology for applying the inhibitors in the field, which perhaps presents the greatest challenges. In order to protect the near-well formation and performations from scaling, the inhibitors must be present in the produced water *before* it reaches the well. To accomplish this, aqueous solutions of inhibitor are periodically pumped into the formation. The success of these treatments depends on the retention of the inhibitor by the formation and its prolonged release at an effective level into subsequently produced water. The high cost of inhibitor treatments has driven a continuing search for ways to prolong the life of the treatments.

There are two predominant strategies employed for inhibitor retention/release. The most common relies on the *adsorption* of inhibitor onto the reservoir rock; subsequent release of the inhibitor is governed by desorption processes. Inhibitor adsorption and desorption behavior has been modelled mathematically, and experimental methods developed which allow empirical isotherms of inhibitors to be determined for different brines and rock mineralogies [24, 25]. Such mathematical models incorporated into two- and three-dimensional reservoir simulators have become important tools for treatment design and optimization.

The second principle mechanism employed for inhibitor retention is based on precipitation of the inhibitor within the formation rock and slow dissolution of the separated phase in the subsequently produced brine. Interest in this technology derives from the possibility to achieve long treatment lifetimes independent of the reservoir mineralogy. Successful application, however, requires knowledge of the phase behavior of the inhibitor, and the solubility of the precipitated phase in the reservoir brine. The theoretical understanding of these treatments is still relatively primitive. Only recently, in fact, were the parameters (brine salinity, temperature, and pH) controlling the extent of phase separation clarified, and the role of polymer fractionation by molecular weight established [26]. Figure 5 illustrates the pore-level precipitation events and fractionation of the inhibitor, together with a typical return curve (Fig. 6). The continuing need for better inhibitor treatment performance and greater reliability provides a strong impetus for further work in this area.

Fluid flow in porous rock

Petrophysical evaluation of reservoir cores recovered during drilling is important for estimating a reservoir's production potential and for evaluating different recovery schemes. Reservoir performance depends strongly on the capillary forces and water and oil relative permeabilities of the reservoir, which ultimately are determined by the wettability state of the rock. The issue of reservoir wettability has provided a rich and complex area for study for several decades [27]. The wettability state of the reservoir depends on the minerology, pore shape and size of the porous medium, and on the brine and oil composition. Current work aims at rationalizing and elucidating the interdependence of these phenomena, with the ultimate goal of connecting wettability to capillary forces and relative permeability.

Independently, work is underway to develop chemical treatments capable of modifying the oil and water relative permeability of the reservoir rock in order to selectively reduce inflow of water to the well [28]. Presently, it is known that both adsorbed water soluble polymers and water-based gels have the property of reducing the water permeability of porous rock to a greater extent than the oil permeability, although insight into the basic mechanisms operating is still lacking. Such knowledge is likely to be crucial to develop the full potential of these treatments [29].

The laboratory study of oilfield phenomena

It is characteristic of many industrial processes that their study in the laboratory necessarily requires a reductive approach. This is dramatically true of the chemical and physical processes involved in the production and transport of hydrocarbons. Reservoirs, wells, and pipelines often present conditions that are difficult to simulate singly, much less simultaneously, in the laboratory. These include high temperature and pressure, complex flow conditions (radial geometry in the reservoir outside the welbore), immense throughput volumes of fluids, and oil compositions (which strongly effect the oil phase physical properties) unstable at ambient conditions. The reservoir rock contributes further complications, being markedly heterogeneous from the scale of microns to meters, and not lending itself to direct investigation of pore-level phenomena. Thus careful attention to the design of

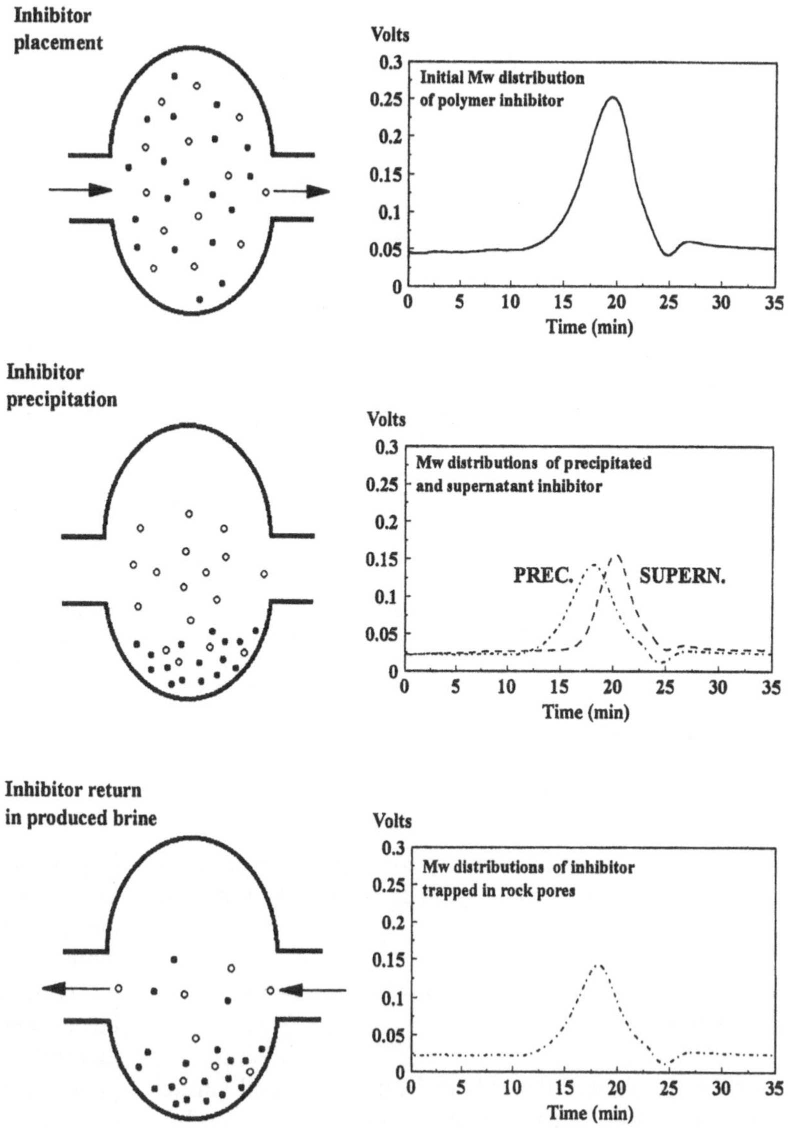

Fig. 5 Pore scale scenario and inhibitor molecular weight distributions for precipitation-based scale inhibitor treatments. Top figures show arrival of inhibitor slug with original M_w distribution of the inhibitor. Middle section shows preferential phase separation of higher molecular weight end of distribution during shut-in (filled circles). Bottom figures show precipitated inhibitor (higher molecular weight fraction) in equilibrium with reservoir brine after the well is put back on production

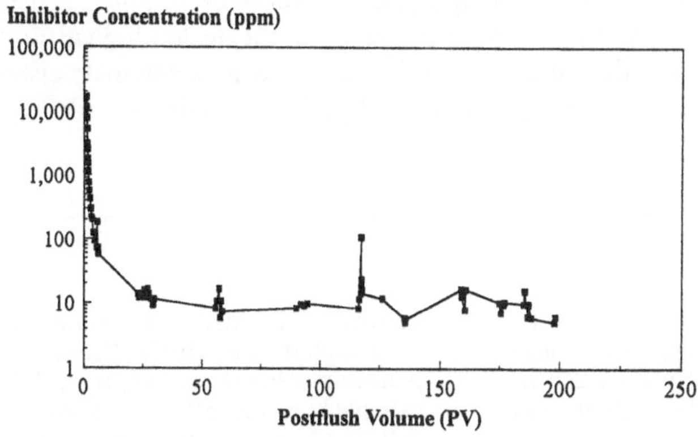

Fig. 6 Inhibitor return concentration in brine during post-flush of porous medium after inhibitor precipitation treatment (PV = pore volumes of brine put through the sandpack). Long flat inhibitor concentration (about 10 ppm) assures effective inhibition against scaling

laboratory equipment and procedures and to the interpretation and scaling of experimental results is especially important.

The study of reservoir-level processes, in particular, requires use of porous media flow cells that can be likened in several respects to chromatographic columns [30]. Figure 7 represents schematically a computer-driven coreflood apparatus and indicates some of the experimental features that can be adopted as required for the investigation of specific problems. This includes measurement of the pressure drop across the entire core (or separate core sections) in order to determine the permeability, and to monitor for changes in permeability with changing fluid saturations or wettability state, filtration phenomena, chemical reaction, etc. Propagation of chemical species can be monitored by effluent analysis with reference to chemical tracers, and can be used to

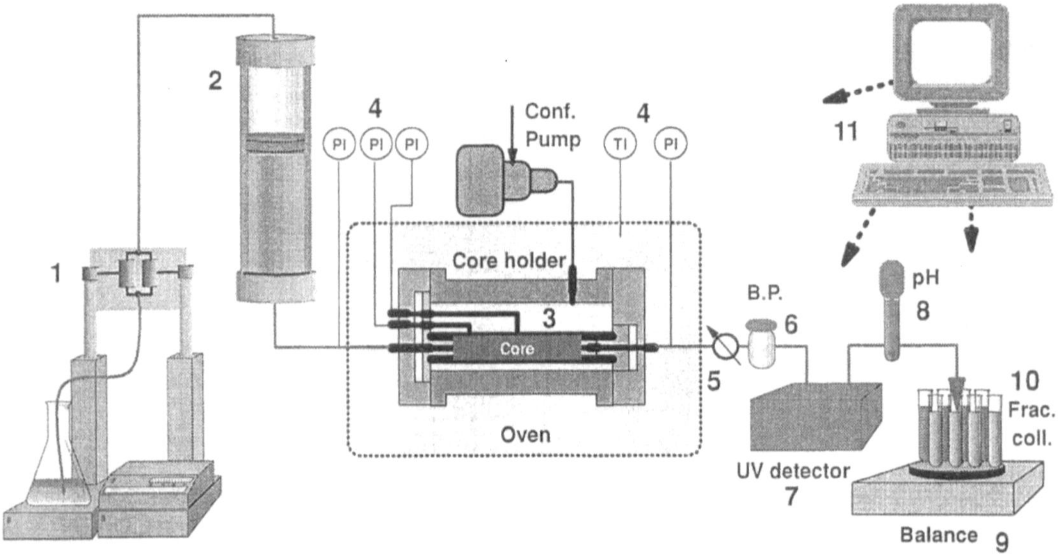

Fig. 7 Schematic showing porous media experimental facility. Pumps (1) drive de-aerated liquids from transfer vessel (2) through rock cores contained in Hassler cells (3) (pressurized rubber sleeve avoids fluid by-pass of the core). By means of pressure taps (4) core permeability and changes in permeability can be monitored along the length of the core. A capillary fitted to the end of the core holder (not shown) permits measurement of effluent viscosity. Cooling coil (5) and back-pressure regulator (6) reduce effluent to ambient T and P. On-line effluent analysis may include UV (7) and pH (8) measurement, while a balance (9) and fraction collectors (10) are employed to quantify fluid throughput and characterize the effluent chemical composition. The entire instrumentation and data collection is computer driven (11)

characterize adsorption/desorption phenomena and chemical reactions, including their thermodynamic and kinetic aspects.

Conclusions

As we have shown, the colloidal dimension and surface phenomena are present in many problems in the production and transportation of petroleum, and underlie many of the chemical technologies that have been developed to assure high well and pipeline productivity. Many of these problems are fascinating for their complexity and for the range and type of colloidal issues they present. Importantly from an industrial perspective, many of these could also benefit greatly from a more systematic application of colloid and surface science principles and methodologies.

Though the industry tends to focus on shorter-term targets and often favors the application of trial and error methods in the search for incremental technology improvements, many important problems facing the industry require a discontinuous advance in understanding and technology. Our experience suggests that a knowledge-based approach, employing the tools of colloid science, combined with an awareness for the technological and economic constraints of the oilfield, can be extemely effective for resolving operational problems in the oil industry.

References

1. Lake LW (1989) Enhanced Oil Recovery. Prentice-Hall, Englewood Cliffs, NJ
2. Gray GR, Darley HCJ (1980) Compositions and Properties of Oil Well Drilling Fluids, 4th ed. Gulf, Houston
3. Jiao D, Sharma MM (1992) Society of Petroleum Engineers paper 23823, presented at the SPE Int Symp on Formation Damage Control, Lafayette, February 26–27
4. Rabaioli MR, Miano F, Lockhart TP, Burrafato G (1993) Society of Petroleum Engineers paper 25179, presented at the SPE Int Symp on Oilfield Chem, New Orleans, March 2–5
5. Rabaioli MR, Miano F (1994) Colloids Surf A: Physicochem Eng Aspects 84:229
6. Miano F, Carminati S, Burrafato G, Lockhart TP (1994) Royal Soc Chem Spec Publ 159:71
7. Burrafato G, Miano F, Carminati S, Lockhart TP (1995) Society of Petroleum Engineers paper 28962, presented at the SPE Int Symp on Oilfield Chem, San Antonio, February 14–17
8. Miano F, Carminati S, Lockhart TP, Burrafato G (1996) SPE Drilling & Completion, p 147

9. Burrafato G, Guarneri A, Lockhart TP, Nicora L (1997) Society of Petroleum Engineers paper 37288, presented at the SPE Int Symp on Oilfield Chem, Houston, February 18–21
10. van Oort E (1997) Society of Petroleum Engineers paper 37263, presented at the SPE Int Symp on Oilfield Chem, Houston, February 18–21
11. Santarelli FJ, Carminati S (1995) Society of Petroleum Engineers paper SPE/IADC 29421, presented at the SPE/IADC Drilling Conf, Amsterdam, 28 February–2 March
12. Calloni G, Moroni N, Miano F (1995) Society of Petroleum Engineers paper 28959, presented at the SPE Int Symp on Oilfield Chem, San Antonio, February 14–17
13. Borchardt JK (1989) ACS Symp Ser, 196:3
14. Sydansk RD, Moore PE (1992) Oil & Gas J, January 20 1992 40
15. Bryant SL, Bartosek M, Lockhart TP, (1996) J Petrol Sci Eng 16:1
16. Bryant SL, Bartosek M, Lockhart TP (1997) J Petrol Sci Eng 17:197
17. Bryant SL, Bartosek M, Borghi G, Lockhart TP (1997) Society of Petroleum Engineers paper 37244, presented at the SPE Int Symp on Oilfield Chem, Houston, February 18–21
18. Albonico P, Lockhart TP (1997) J Petrol Sci Eng 18:61
19. Bryant SL, Rabaioli MR, Lockhart TP (1996) SPE Production & Facilities 11:209
20. Cimino R, Correra S, Del Bianco A, Lockhart TP Plenum, New York (1996) In: Sheu EY (ed) Asphaltenes – Fundamentals and Applications
21. Forsdyke IN (1997) Society of Petroleum Engineers paper 37237, presented at the SPE Int Symp on Oilfield Chem, Houston, February 18–21
22. Agro CB, Blain RA, Osborne CG, Priestley ID (1997) Society of Petroleum Engineers paper 37255, presented at the SPE Int Symp on Oilfield Chem, Houston, February 18–21
23. Pakulski M (1997) Society of Petroleum Engineers paper 37285, presented at the SPE Int Symp on Oilfield Chem, Houston, February 18–21
24. Sorbie KS, Wat RMS, Todd AC (1992) SPE Production Eng 7:307
25. Jordan MM, Sorbie KS, Jiang P, Yuan MD, Todd AC, Thiery L (1994) NACE Annu Conf and Corrosion Show, Baltimore, February 27–March 3
26. Rabaioli MR, Lockhart TP (1996) J Petrol Sci Eng 15:115
27. Morrow NR (1990) J Petrol Technol 1476
28. Zaitoun A, Kohler N, Guerrini Y (1991) J Petroleum Technol 862
29. Liang J, Sun H, Seright RS (1995) SPE Reservoir Eng 10:282
30. Marle CM (1981) Multiphase Flow in Porous Media. Editions Technip, Paris

Progr Colloid Polym Sci (1998) 109:60–73

B. Wu
S.D. Christian
J.F. Scamehorn

Recovery of surfactant from micellar-enhanced ultrafiltration using a precipitation process

Received: 16 July 1997
Accepted: 21 July 1997

B. Wu · S.D. Christian (✉)
J.F. Scamehorn
Institute for Applied Surfactant Research
University of Oklahoma
73019 Norman, Oklahoma
USA

Abstract A precipitation method is proposed to recover anionic surfactants from a micellar-enhanced ultrafiltration (MEUF) process. Monovalent potassium ion is used to precipitate the dodecylsulfate anion from sodium dodecylsulfate (SDS). The recovery process is quick and clean and requires adding smaller amounts of the electrolyte than if the sodium cation is used to precipitate dodecylsulfate. No substantial loss of organic solutes (phenol and *tert*-butylphenol) occurs during the precipitation, as was observed when sodium ion was used as the precipitant. It has been shown that potassium dodecylsulfate (KDS) is similar to SDS in its cmc, water solubility, and ability to solubilize organic solutes in micellar solutions. However, KDS has a significantly higher Krafft point than SDS (35 °C). An MEUF experiment using KDS as the model surfactant gave large rejection of *tert*-butylphenol (96%). Based on these findings, KDS is proposed as a potential surfactant for use in MEUF, with recovery of the surfactant to be accomplished by lowering the system temperature below the Krafft point.

Key words Micellar-enhanced ultrafiltration (MEUF) – surfactant recovery – precipitation – Krafft point

Introduction

Traditional separation methods for removing organic solutes from water may involve such processes as solvent extraction followed by distillation, or adsorption on activated carbon followed by regeneration of the carbon. These techniques generally involve a phase change and they can be quite energy-intensive. Therefore, it is desirable to develop low-energy separation processes that may be of general use in the environmental remediation area as well as in the treatment of industrial process streams.

One of the most promising alternative separation processes for removing organic solutes from aqueous streams is ultrafiltration. Conventional ultrafiltration (UF) is unable to effectively remove organic solutes having molecular weights less than 300 Da from wastewater (1). Flux through the membrane decreases as pore diameters and molecular weight cut-off values decrease, so that the direct ultrafiltration of low molecular weight compounds from water is usually not attractive economically, even when it is technically feasible. Our research group has developed the technology of adding water-soluble colloids to aqueous streams and subsequently using ultrafiltration to remove target ions and molecules that cannot be removed by conventional ultrafiltration [1–9]. These types of separations are generically termed colloid-enhanced ultrafiltration (CEUF) methods. One of these techniques, micellar-enhanced ultrafiltration (MEUF), has been shown to be able to achieve >99% rejections for metal ions and organic solutes (1). In order for MEUF to be economical in many applications, it is necessary to recover most of the surfactant following the ultrafiltration so that it can be reused in the process.

Background

Removal of organic solutes via micellar-enhanced ultrafiltration (MEUF)

In the MEUF process, the added surfactant concentration is maintained well above the critical micelle concentration (cmc), so that most of the surfactant molecules are in micelles. Under favorable circumstances, the unwanted organic solutes are also largely solubilized in the micelles, owing to the tendency of the solute to dissolve in the hydrophobic interior of the micelle [33]. When the solution is forced through an ultrafiltration membrane with pore diameters smaller than the micelle diameters, the micelles, along with the solubilized solute molecules, are rejected by the membrane (the retentate stream). Moreover, if ionic surfactant is used, oppositely charged ionic species absorb or bind onto the micelles and can be removed together with the organic impurities [1]. Figure 1 is a schematic diagram of MEUF to remove metal ions and organic solutes from water. The removal efficiency (rejection) of an organic solute (A) in the MEUF process is defined as

$$\text{rejection}\ (\%) = (1 - [A \text{ in permeate}]/[A \text{ in retentate}]) * 100\% \quad (1)$$

where [A in permeate] is the concentration of the organic solute A in the permeate form and [A in retentate] is the concentration of A in the retentate. Various solubilization equilibrium constants (K) have been defined to indicate the solubilization capacity of a micellar solution for an organic solute [12]. The definition used here is

$$K = X/C , \quad (2)$$

where X is the mole fraction of organic solute in the micellar phase, and C is the concentration of unsolubilized or monomeric organic solute not solubilized in the surfactant micelles.

Review of existing surfactant recovery methods for the MEUF process

Vacuum, air, or steam stripping, liquid–liquid extraction, and precipitation methods have been proposed as possible procedures for recovering the surfactant from the retentate solution [20]. Stripping is only effective in removing volatile organics from the retentate. In the use of liquid–liquid extraction to remove either solute or surfactant from the retentate, a downstream separation of the solvent from the extracted material is necessary. Also, solubilization of the solvent into the retentate must be avoided.

The concept of precipitating the surfactant from the retentate, leaving the organic solute dissolved in solution, is attractive because the solute need not be volatile [14, 35]. The surfactant can be precipitated by the

Fig. 1 Scheme of micellar-enhanced ultrafiltration (MEUF) to remove dissolved heavy metal cations and organic solutes from water

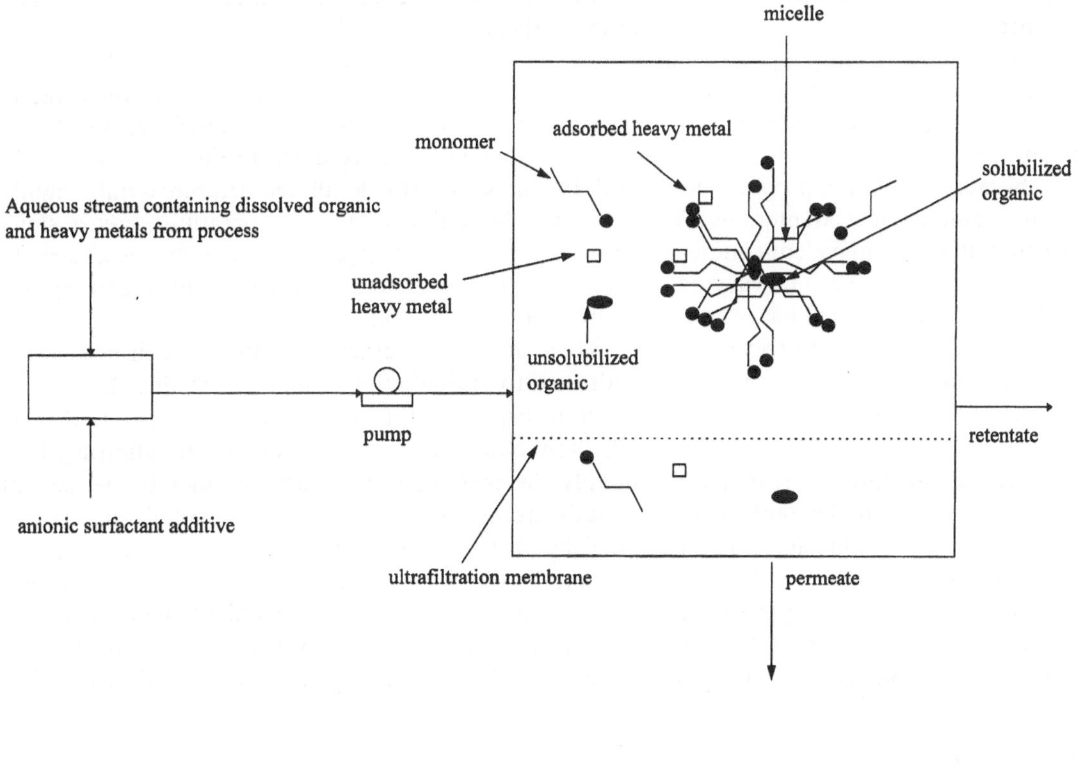

addition of an ion having a charge opposite that of the surfactant or by decreasing the solution temperature below the Krafft point. Presumably, the precipitate can be separated from the solution by gravity settling, ordinary filtration, or centrifuging. The resulting surfactant filter cake can then be recycled back to the process or treated further if necessary. If the pollutants that have been removed from water in the surfactant-based separation are organic species that remain dissolved in the water after the surfactant has been removed, those materials will be expected to remain in the filtrate from the surfactant precipitation/recovery step. On the other hand, if the materials are liquid organics having little water solubility and a density less than that of water, they will tend to float to the top of the solution as the surfactant is precipitated, so that these compounds can be recovered in relatively pure form. But if the organic is slightly soluble in water and denser than the filtrate solution, it will form a precipitate or liquid phase that will fall to the bottom and mix with the surfactant that has separated as a solid precipitate [14].

Previously, studies have been made of the effectiveness of a monovalent ion (sodium) in precipitating a model surfactant, sodium dodecylsulfate (SDS) [14, 16]. The major advantage of the use of sodium ion as a precipitating agent is that large crystals tend to form [17], making crystal recovery relatively inexpensive by gravity settling. However, because surfactants like SDS have relatively large solubility products, unacceptably large concentrations of salt may be required to precipitate most of the surfactant; thus ref. [15] shows that in order to precipitate 99% of the SDS in a 0.05 M SDS solution at 30 °C, for example, 1.5 M NaCl is needed.

Attempts have also been made to develop a process in which a multivalent counterion is added to the surfactant, at concentrations slightly in excess of stoichiometric conditions, thereby causing precipitation of a large fraction of the surfactant [15, 16], because solubility products for salts of an ionic surfactant and multivalent counterions are in general much smaller than those for univalent counterions. However, the resulting precipitate (barium, calcium, or aluminum dodecylsulfate, for example) will have a very low solubility in water, making it impossible to recycle the surfactant to the process without further treatment. One way to precipitate and recover the surfactant may be to mix the precipitate with an aqueous solution of a co-ion (ion having the same charge as the surfactant) that can form a very insoluble compound with the multivalent counterion [15]. Thus, a solution of sodium sulfate might be used to redissolve $Ba(DS)_2$ as soluble sodium dodecylsulfate (SDS), leaving solid $BaSO_4$ as the only precipitate. This procedure would of course increase the cost, both by introducing another unit operation and because expensive chemicals may be required, first to precipitate the surfactant and then to redissolve it [15].

Polyelectrolytes may also be considered as potential precipitants for dodecylsulfate and other ionic surfactants. A quaternary ammonium polyelectrolyte (polyquat) will form a 1:1 insoluble complex with SDS [18]. Polyalkylamine (PAA) and polyethyleneimine (PEI) will also precipitate dodecylsulfate anions at moderate to low pH values [19]. But this technique would still require an additional unit operation to recover the surfactant from the precipitate.

Use of temperature cycling to recover surfactant in MEUF

Instead of using sodium as the counterion to precipitate the anionic surfactant dodecylsulfate, in this work potassium is used. The hydrated radius of potassium ion is smaller than that of sodium ion; thus the resulting potassium dodecylsulfate (KDS) has a higher Krafft temperature (lower solubility product) than SDS. At surfactant concentrations above the cmc (e.g. in the retentate), the Krafft temperature is the temperature below which the surfactant precipitates [34]. Above the Krafft temperature, the surfactant is entirely soluble in water. The Krafft temperature of the KDS (ca. 35 °C – to be discussed) is closer to room temperature than that of SDS (16 °C). This is an advantage in surfactant recovery because MEUF is often used near this temperature.

The process investigated in this study involves operating MEUF above the Krafft temperature, then cooling the retentate below the Krafft temperature, causing almost all of the surfactant to precipitate from solution. After separating the solid surfactant from the solution (by gravity settling, centrifugation, or filtration), it can be dissolved in the hotter feed stream to the MEUF unit for reuse. Generally, the retentate stream is much smaller than the original feed stream. Because the Krafft temperature is only slightly below the MEUF operating temperature, only a small amount of heat exchanger duty may be necessary to achieve the required cooling of the retentate, making this a low-energy separation.

There are several crucial questions which need to be addressed to evaluate this technique: whether the organic solute incorporates into the precipitated surfactant crystals (how clean the solute/surfactant separation is); how rapidly the precipitation occurs; whether the surfactant crystals are dense enough and large enough to be separated by gravity settling; what fraction of the surfactant precipitates (both at equilibrium and under actual conditions). This paper addresses several of these questions. Recovery of surfactant for reuse is important to the economic viability of many surfactant-based separation processes [34]. Therefore, the results of the present work could have broad application in industry.

Materials and methods

Materials

Electrophoresis grade sodium dodecylsulfate (SDS) was purchased from Fisher Chemical and used without further purification. Potassium dodecylsulfate (KDS) was prepared by recrystallization; 200 ml of 0.1 M SDS aqueous solution was added into 300 ml 0.1 M KCl aqueous solution. The precipitate was separated by filtration and washed first with 0.1 M KCl aqueous solution and then with deionized water. The product was recrystallized from 0.1 M KCl aqueous solution and then with deionized water. The product was recrystallized from 0.1 M KCl aqueous solution and again washed with 0.1 M KCl solution and deionized water, respectively. A silver nitrate solution was used to test for the presence of Cl^-. After the Cl^- had been quantitatively removed, the resulting product, potassium dodecylsulfate, was oven-dried at approximately 80 °C and stored in a sealed container. The purity of the product KDS was checked by mass spectrometry.

Reagent-grade ACS phenol and 4-*tert*-butylphenol (TBP) were obtained from Fisher Chemical and used without further treatment. The acetic acid buffer used in ultraviolet analysis of dodecylsulfate was prepared by mixing 12 ml anhydrous acetic acid, 142 g sodium sulfate, 7.4 g disodium ethylenediaminetetraacetrate (EDTA), and 800 ml of deionized water; then 2 M NaOH aqueous solution was used to adjust the pH of the resulting solution to 5.0.

Redistilled and deionized water was used throughout.

Methods

The concentrations of surfactants (SDS and KDS) were determined from the ultraviolet (UV) spectra, using a Hewlett-Packard Diode Array Spectrophotometer, Model 8452A. Six ml of the surfactant solution was mixed with 0.375 ml acetic acid buffer, 0.15 ml ethyl violet, and 3 ml toluene. The mixture was thoroughly shaken and centrifuged for 4 min at full speed to obtain a good separation. The absorbance of the toluene phase was determined by UV analysis at a wavelength of 612 nm.

TBP and phenol solutions were also analyzed by UV at wavelengths of 276 and 270 nm, respectively.

Surface tension measurements were made with a Sensa-Dyne Model 6000 maximum bubble pressure Surface Tensiometer.

The solubility experiments were conducted in the thermostatted water bath. An excess of the surfactant and/or the organic compound was thoroughly shaken with water and the mixture was allowed to stand for 24 h before analysis. Aqueous solutions were filtered through filter paper before any analysis was made on the homogeneous solutions.

The apparatus used in studies of micellar-enhanced ultrafiltration (MEUF) is the same as that described in ref. [5]. The volume capacity of the retentate cell was 300 ml and the pressure applied was 4 atm. The C-type Molecular/Pore ultrafiltration membrane was purchased from Spectrum Medical Industries. It was made of cellulose with an effective diameter of 76 mm and a 5000 Da molecular weight cut-off (MWCO). The approximate permeate flow rate in MEUF operations was about one drop per second. The experimental run was terminated when the volume of the retentate solution equalled 100 ml or less.

Results and discussions

Using KCl to precipitate SDS from aqueous solution

The first test involved the addition of KCl into SDS aqueous solution at 25 °C. This had two goals: first, if precipitation occurs initially at a certain mole ratio of potassium to sodium and if the precipitate is KDS as expected, the preliminary experiments could be expected to provide a qualitative indication of the feasibility of precipitating KDS – namely, the speed of precipitation, the clarity of the solution above the KDS precipitate, and the approximate Krafft temperature of KDS: second, these tests should indicate whether a simple way could be found to prepare pure KDS crystals.

Figure 2 shows results of the tests. The initial concentration of the SDS solution was 0.05 M, which is more than five times the cmc at 25 °C [32]. Solid KCl was added to the SDS solution and the volume change was ignored. In the experiment, precipitation of a white crystalline solid was observed to occur when the mole ratio of KCl to SDS exceeded 0.5. Near-quantitative conversion of SDS to KDS was achieved at higher mole ratio of KCl to SDS; for example, about 99% of the SDS is converted to KDS at a mole ratio 4 to 1. The kinetics of precipitation is relatively rapid and at a mole ratio of 1 to 1 or greater, the precipitate can be seen within seconds after KCl is added. When the mixture was allowed to stand for several hours without disturbing, the supernatant became quite clear indicating little presence of suspended solids. In general, the precipitate was totally dissolved by the time the temperature had been raised to approximately 35 °C and the

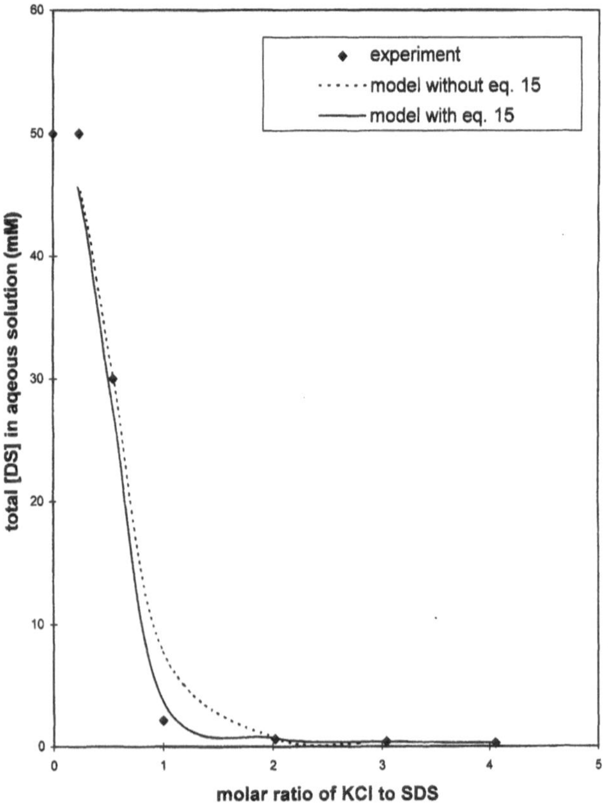

Fig. 2 Use of KCl to precipitate sodium dodecylsulfate. (temperature: 25 °C, initial [SDS] = 50 mM)

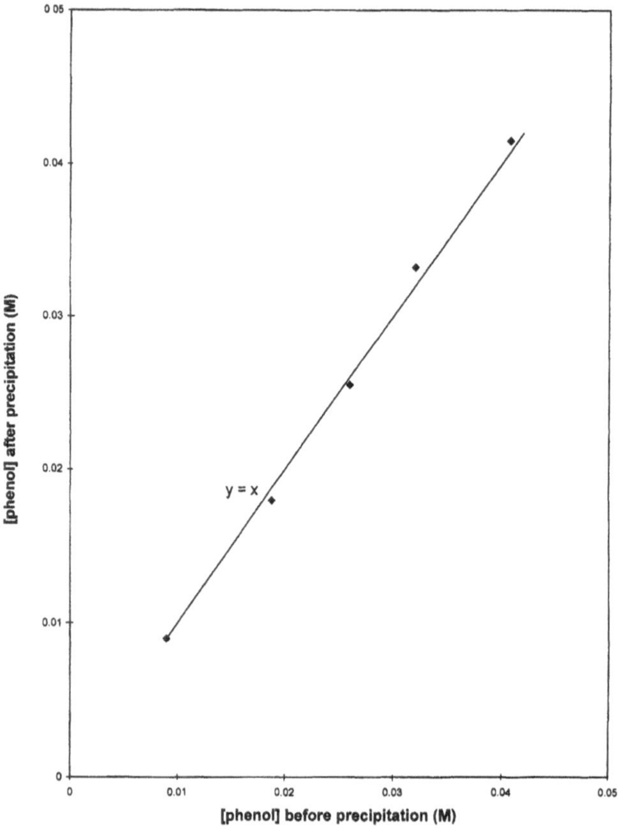

Fig. 3 Effect of adding KCl to SDS aqueous solution on dissolved phenol at constant KCl to SDS ratio. (temperature: 25 °C, initial [SDS] = 50 mM, [KCl]/[SDS] = 1)

solution was observed to be clear. Therefore, 35 °C is approximately the Krafft temperature of KDS.

The purity of recrystallized KDS was checked by mass spectrometry, as indicated previously. Mass spectra of the KDS precipitate and KDS recrystallized from water show the presence of significant levels of sodium. When 0.1 M KCl solution, rather than pure water, was used as the medium for recrystallizing KDS, with SDS as the starting surfactant material, the mass spectra indicated that the sodium impurities were almost gone. This method of preparing KDS is much faster and less costly than a previous method, which used neutralization of dodecyl hydrogen sulfate with potassium hydroxide, where the organic acid had been prepared by ion exchanging SDS with resin [10].

Tests were also conducted to examine the effectiveness of precipitation in separating the surfactant from the dissolved organic solutes. Phenol and TBP were selected as organic solutes, representing phenolic pollutants with varying water solubility and hydrophobicity. Solute concentrations of TBP were chosen to be somewhat less than the maximum solubility of TBP in water (4.8 mM at 25 °C).

In order for the proposed process to be effective in a single stage, the precipitated surfactant must not contain a substantial amount of the organic solute. To test this, the surfactant was precipitated in the presence of either organic solute and the composition of the supernatant analyzed for the change in solute concentration 24 h later. The phenol and TBP final concentrations are shown in Figs. 3 and 4, respectively, at a KCl/SDS ratio of 1:1. The final phenol concentration is shown in Fig. 5 as a function of the KCl/SDS ratio. In all cases, the final and initial organic solute concentrations are equal, indicating that an insignificant amount of the solute coprecipitates with the surfactant or adsorbs on the surfactant crystals.

Modeling the equilibrium precipitation

In mathematically describing the precipitation, the equilibrium between monomer and precipitate and the monomer–micelle equilibrium must be considered. Equation (3) (below) describes the solubility product as the product of the thermodynamic activity (activity coefficient multiplied

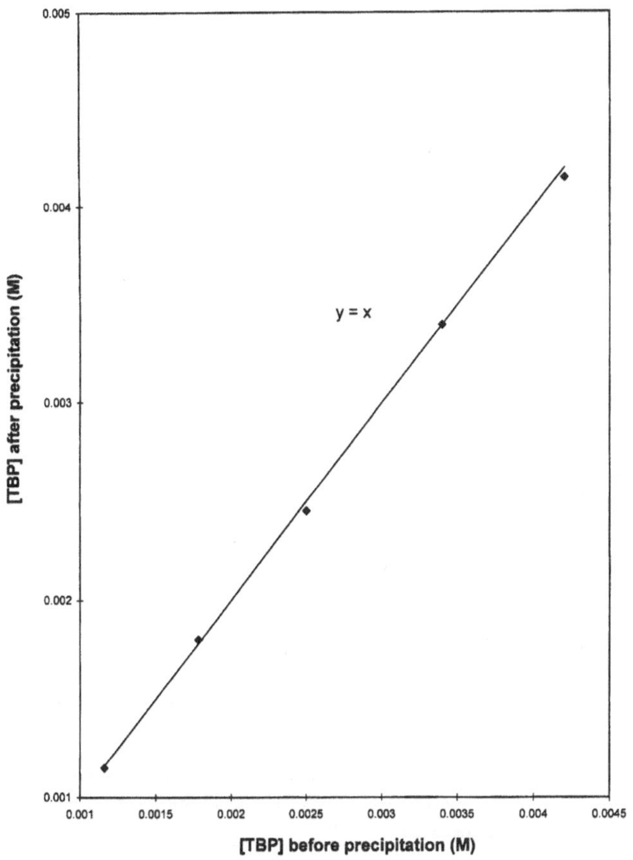

Fig. 4 Effect of adding KCl to SDS aqueous solution on dissolved TBP at constant KCl to SDS ratio. (temperature: 25 °C, initial [SDS] = 50 mM, [KCl]/[SDS] = 1)

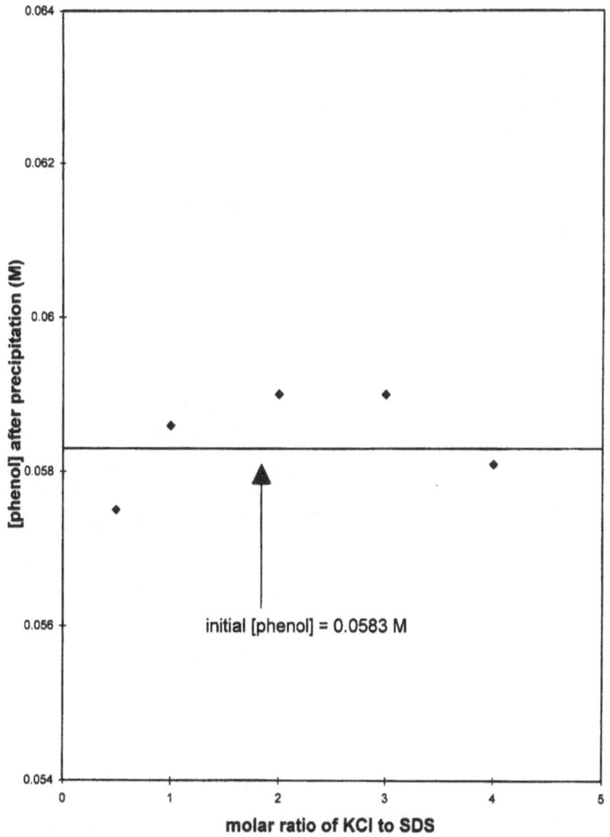

Fig. 5 Effect of adding KCl to SDS aqueous solution on dissolved phenol at varying KCl to SDS ratios. (temperature: 25 °C, initial [SDS] = 50 mM)

by concentration) of the surfactant monomer anion and the unbound counterion (K^+). The relationship between the concentration of micelles and monomeric surfactant and unbound counterions (Na^+ and K^+) is described by the mass action model (Eq. (4)) Equations (5) and (6) apply the Debye–Hückel theory to calculate the activity coefficients of the ionic surfactant and counterions in solution. Equations (7–14) are material balances for the various species present. The final equation (Eq. (15)) has been included to account for the amount of SDS that is co-precipitated with KDS:

$$K_{sp} = \gamma_m^2 * cK^+ * cDS^-_{mono} , \quad (3)$$

$$(cK^+cNa^+)^q(cDS^-_{mono})^n = (K_{mic})^{(n+q)} C_{micelle} , \quad (4)$$

$$\log \gamma_m = -0.511 I^{1/2}/(I^{1/2}+1) , \quad (5)$$

$$I = 0.5(cK^+ + cNa^+ + cCl^- + cDS^-_{mono}) , \quad (6)$$

$$\text{Ratio} = cK^+/(cK^+ + cNa^+) , \quad (7)$$

$$cK_{tot} = cK^+ + KDS_{precpt} + c_{micelle} q/n \text{ Ratio} , \quad (8)$$

$$SDS_{tot} = KDS_{precpt} + c_{DS^-mono} + c_{micelle} , \quad (9)$$

$$cK^+ + cNa^+ = cCl^- + c_{DS^-mono} + nc_{micelle}(1 - q/n) , \quad (10)$$

$$cK_{tot} = R_{K:Na} cNa_{tot} , \quad (11)$$

$$cNa_{tot} = SDS_{tot} , \quad (12)$$

$$cCl^- = cK_{tot} , \quad (13)$$

$$DS_{tot} = SDS_{tot} - KDS_{prcpt} , \quad (14)$$

$$SDS_{prcpt} = K\, KDS_{prcpt} \gamma_m^2 cNa^+ cDS^-_{mono} . \quad (15)$$

K_{sp} is the solubility product of KDS, which is held constant at 1.98×10^{-5} M^2 (the value obtained from the solubility experiment described later), γ_m is the mean activity coefficient calculated by the extended Debye–Hückel theory, cM is the concentration of any ion M dissolved in solution, cDS^-_{mono} is the concentration of monomeric dodecylsulfate anion, $c_{micelle}$ is the surfactant concentration in micellar form, n and q are aggregation numbers which are held constant (62 and 40 for the dodecylsulfate ion and counterion respectively), K_{mic} is the micellization constant, I is the ionic strength, Ratio is the mole fraction of K^+ ion in the total aqueous dissolved cation concentration, cM_{tot} is the total amount of an ion M, $R_{K:Na}$ is the concentration ratio of KCl to SDS, SDS_{tot} is the total amount of SDS before adding KCl, KDS_{prcpt} is the amount of KDS (in mol) precipitated, and DS_{tot} is the total amount of dodecylsulfate remaining in solution after precipitation. As the volume of the system is considered constant (as solid KCl is added), all quantities of materials are expressed as concentrations. An assumption made here is that the ratio of K^+ to Na^+ is the same in both the monomeric and the micellar phase. SDS_{prcpt} is the amount of SDS incorporated into the KDS solid. Equation (15) is based on the assumption that there is a direct proportionality (with proportionality constant K) between the amount of SDS incorporated in the crystal and the product of the total amount of solid KDS formed and the activity of unprecipitated SDS in the supernatant liquid.

The equations were solved by using the computer program SEQS [13, 23], obtained from the copyright owner. Experimental results are well correlated by the model (without using Eq. (15)) except at KCl to SDS ratios near 1 (see the dotted line in Fig. 2). The overall fit of data is improved by including a nonzero (positive) value of K in Eq. (15).

Figure 2 displays the predicted values from both approaches (with or without Eq. (15)) in comparison with the experimental data. Values calculated from both approaches fit the experimental data well at high mole ratios of KCl to SDS (mole ratio ≥ 2), and both predict that precipitation will occur at a mole ratio of 0.24, which was not observed in the experiments. The excellent agreement between experimental and calculated results at high mole ratios shows the utility of the solubility product as a parameter in estimating activities of the surfactant anions and also verifies the accuracy of the K_{sp} value determined from solubility measurements. The differences between the model and experiment at low mole ratios could be attributed to supersaturation. The approach including Eq. (15) is in better agreement with experiments at moderate mole ratios (around 1) than the approach without Eq. (15), showing the importance of accounting for the loss of sodium ions from solution in this region due to the coprecipitation. From the model estimation, the amount of sodium precipitated becomes a maximum at a mole ratio of about 1, which accounts for the relatively poor predictions obtained with the approach excluding Eq. (15) in this region.

If Eq. (15) is included, the model also predicts that at a mole ratio of KCl to SDS of about 1:1 or greater the total amount of dodecylsulfate (in both monomeric and micellar forms) in solution equals the amount of monomer dodecylsulfate anion. This is the point at which dodecylsulfate micelles completely disappear and a break in the surface tension curve is expected. Such breaks do appear in plots of the measured surface tension (see Fig. 6). When

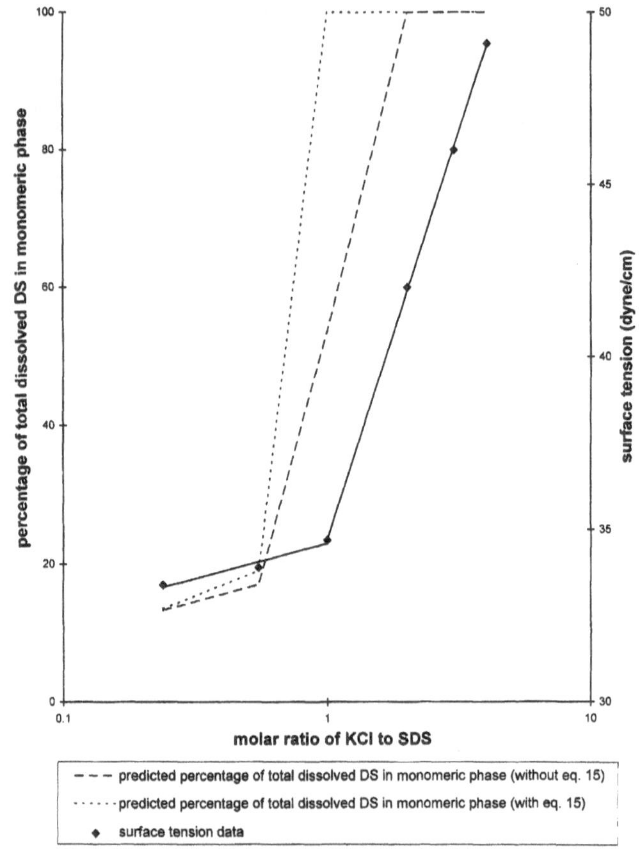

Fig. 6 Experimental indication and model predictions of disappearance of the micellar phase as a result of precipitation of dodecylsulfate salts

KCl is added, the surface tension increases slowly until the KCl to SDS ratio reaches 1, where a sudden break is evidenced. This break occurs at the point where the monomer–micelle–solid (pseudo-three-phase system) converts into a two-phase monomer–solid system. Beyond the break, the surface tension increases with increasing KCl to SDS ratio, reflecting the influence of added K^+, which continues to precipitate out dodecylsulfate anions. This surface tension plot of precipitating out the surfactant is an exact reverse curve of that of accumulating a surfactant in aqueous solution. The KCl to SDS ratio for the surface tension break would be 2:1 if Eq. (15) is excluded from the model, which further justifies the importance of considering the loss of sodium ions from the aqueous phase during the precipitation process.

Properties of KDS

The potential use of KDS in MEUF separation processes is evaluated by conducting experiments to determine its

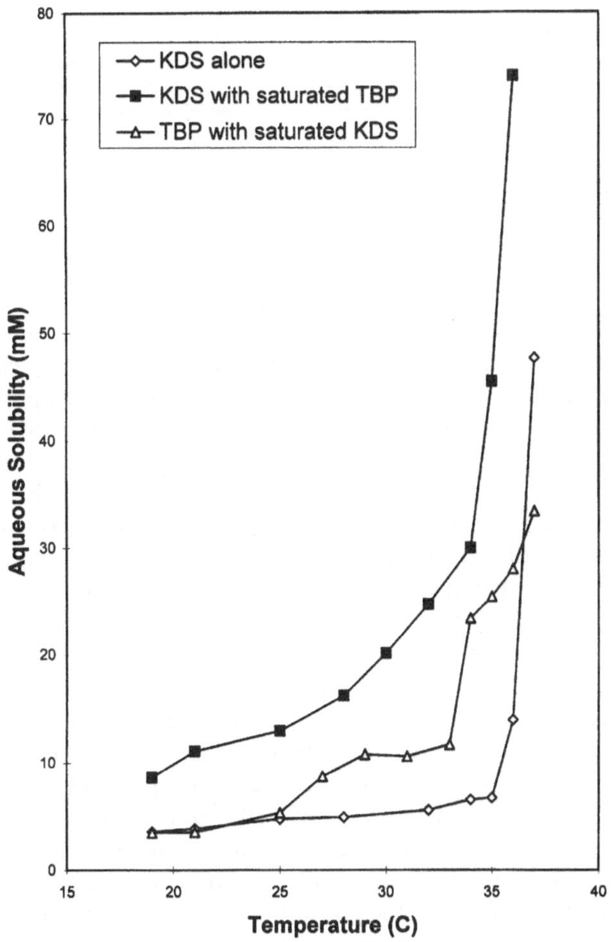

Fig. 7 Aqueous solubility data for KDS and TBP

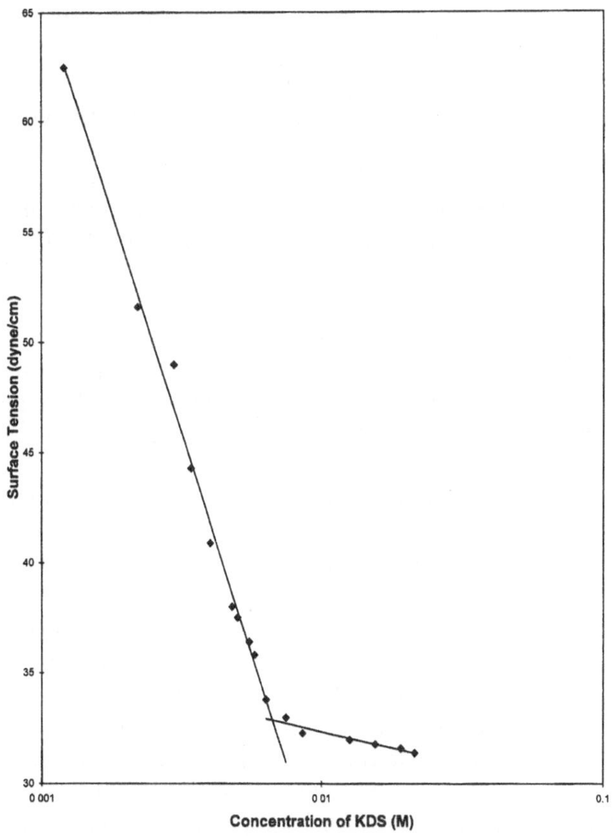

Fig. 8 Surface tension measurements for KDS aqueous solution to determine its cmc value (temperature: 39 °C)

ability to solubilize organic solutes and to be recycled by varying the operating temperature. The following properties of KDS were determined for the purpose.

Krafft temperature: The Krafft temperature of KDS was measured by gradually raising the temperature of the aqueous solution, in the presence of excess KDS solid, from room temperature until all of the KDS dissolved. By this method, the Krafft temperature was found to be 35 °C, which is identical to that found in the precipitation test. It can be also be determined from the plot of KDS solubility in water as a function of temperature (Fig. 7), giving a value of 35 °C.

cmc: Surface tensions of aqueous solutions of KDS were determined at 39 °C (several degrees above the Krafft point) to make forming micelles possible. The cmc value of KDS at 39 °C was determined to be 6.9 mM (Fig. 8). From the plot, using the Gibbs equation, the surface excess quantity of KDS at 39 °C was calculated to be approximately 3.1×10^{-10} mol/cm^2, corresponding to an area per molecule of 53 A^2. Nearly identical values of these quantities have been reported for SDS at 25 °C [25]. From the solubility experiment, the cmc of KDS at 35 °C is 6.7 mM, which corresponds to the water solubility of KDS at the Krafft temperature. Although the cmc from surface tension measurement was determined at 39 °C, there is expected to be very little difference between the cmc value at 39 °C and at the Krafft temperature (35 °C) [10].

Water solubility: The solubility data for KDS in water are plotted in Fig. 7. A dramatic increase in the solubility of KDS occurs at temperatures near 35 °C, which is regarded as the Krafft temperature. The solubility of KDS at the Krafft temperature is about 7 mM which is considered to be the cmc at that temperature. These two results are consistent with those from the Krafft point and surface tension experiments.

Solubility data for KDS and TBP in solutions saturated with respect to both compounds (useful in simulating MEUF operating conditions), have also been obtained (Fig. 7). The solubility of KDS greatly increases between 34 °C and 35 °C, corresponding to the Krafft temperature of KDS in the presence of saturated TBP. As expected, the Krafft temperature for the solutions containing TBP is somewhat smaller than that of KDS in pure water. Similarly, the solubility curve for TBP in saturated KDS aqueous solution shows a sharp upward break at a temperature between 33 °C and 34 °C.

Theoretically, below the Krafft temperature, the presence of KDS should not significantly increase the solubility of TBP above the solubility of the compound in pure water at the same temperature. In fact, at temperatures well below the Krafft temperature, the solubility of TBP in KDS solutions only slightly exceeds the solubility of TBP in pure water; for example, the solubility of TBP in water at 25 °C is 4.8 mM and that in water saturated by KDS at the same temperature is 5.4 mM. But at temperatures closer to the Krafft temperature, the solubilities obtained in the present study are apparently considerably larger than those determined in pure water at the same temperatures. Although solubility data for TBP in water at temperatures other than 25 °C have not been found, the fact that the solubility of TBP determined with saturated TBP at 29 °C is approximately twice as great as that at room temperature suggests that some experimental problem exists. It is possible that formation of a macroemulsion or miniemulsion of TBP at temperatures approaching the Krafft temperature causes excessive concentrations of TBP and KDS to penetrate through the filter paper. Thus, sets of mutual solubility values for both TBP and KDS at temperatures below the Krafft temperature are suspect.

Solubility product (K_{sp}) values for KDS: Table 1 shows K_{sp} value of KDS at 25 °C calculated from different sources. The most direct way is to use the solubility data of KDS in water (4.79 mM from Fig. 7) at 25 °C. An alternative for K_{sp} estimation is to use data points at high mole ratios of KCl to SDS in Fig. 2. At ratios where no micelles are present, the system results shown in Fig. 2 reflect only the precipitation equilibrium. The solubility product of KDS at mole ratios of KCl to SDS of 3.04 and 4.06 can easily be calculated from experimental concentrations for dodecylsulfate ion and calculated concentrations from mass balance for potassium ion. Activity coefficients calculated from the extended Debye–Hückel equation were used in the estimations of K_{sp} values. The K_{sp} values from different sources are in good agreement.

Table 2 is a comparison of selected properties of KDS obtained from various sources. The cmc values from ref. [21] (from conductivity measurements) and ref. [10] (from surface tension data) are in good agreement with the value from the present work (from surface tension data). Ref. [21] does not report an exact value of the Krafft temperature for KDS, although a value between 30 °C and 40 °C is indicated. The Krafft temperature of KDS reported in ref. [10] is 23 °C, much lower than the value 35 °C reported here and not consistent with that from ref. [21]. Another property of interest that can be compared is the water solubility. Although the solubility of KDS in water at 30 °C is reported in ref. [21] to be 3 mM, the solubility of KDS at 30 °C obtained here is approximately 5 mM, while that from ref. [10] is much higher (7 mM), presumably because that temperature is well above the Krafft temperature reported in ref. [10]. In general, values from the present work agree well with those from ref. [21], but differ considerably from those in ref. [10], except for the cmc value. The Krafft point and the solubility data in this work are quite consistent with other experimental values: the solubility of KDS at the Krafft point agrees well with the cmc value obtained from surface tension measurements; the K_{sp} value derived from solubility data at 25 °C is consistent with that calculated from the experiments of adding KCl to the SDS solutions.

Table 1 Comparison of K_{sp} values for KDS at 25 °C from different experimental methods

	γ_m	$[K^+]$ [mM]	$[DS^-]$ [mM]	K_{sp} $[10^{-5} M^2]$
[KCl]:[SDS] = 3.04	0.76	102	0.391	2.30
[KCl]:[SDS] = 4.06	0.75	153	0.273	2.35
Solubility experiment	0.93	4.79	4.79	1.98

Table 2 Comparison of KDS properties from different sources

	cmc [mM]	Krafft point [°C]	Aqueous solubility at 30 °C [mM]
Ref. [10]	6.9	23	7
Ref. [21]	7.8	$30 < K_f < 40$	3
This work	6.9	35	5

Using KDS as the surfactant in micellar-enhanced ultrafiltration (MEUF)

Utilizing property data for KDS, an MEUF process using KDS as the surfactant was designed and operated. The temperature of the experiment was maintained at 38 °C. The initial concentrations of TBP and KDS were 6.51 and 20 mM, respectively. The initial concentration of TBP was chosen to be approximately equal to or just greater than the solubility of TBP in water, corresponding to conditions that might be expected in polluted water. The initial concentration of KDS was chosen to be well above the cmc.

The initial volume of solution was 300 ml. The TBP remaining in the retentate is concentrated from 6.5 mM for the feed to 28 mM at the end of the run, but the TBP concentration in permeate is determined to be approximately 1 mM (lower than its solubility in water) during the

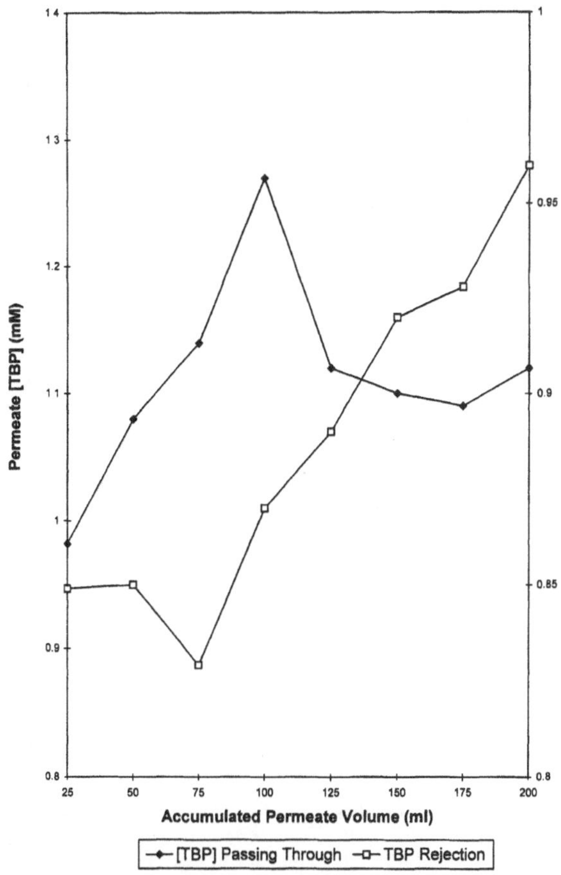

Fig. 9 Effectiveness of using KDS in MEUF to remove TBP from water stream (temperature: 38 °C, initial retentate [TBP] = 6.5 mM)

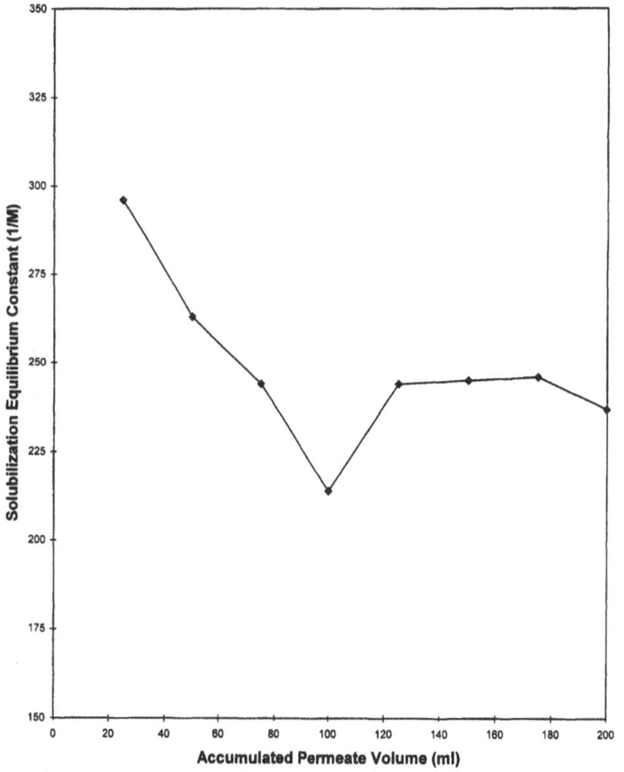

Fig. 10 Solubilization equilibrium constant values for TBP determined from the MEUF run (temperature: 38 °C)

whole process (Fig. 9). The highest rejection of organic obtained in the run is 96%. The organic rejection could be made even greater by increasing the KDS concentration.

The solubilization equilibrium constants, defined by Eq. (2), were estimated for TBP in micellar KDS from the MEUF data. During the MEUF process, small concentrations of both KDS and TBP transfer from the retentate into the permeate. The concentration of surfactant in the permeate is about the same as the cmc value in the retentate [7, 24]. Since the concentrations of KDS and TBP in the permeate are much lower than those in the retentate, the actual concentrations of both KDS and TBP in the retentate increase continuously as the volume of the retentate solution decreases. At larger concentrations of KDS, the relative concentration of surfactant monomers (compared to micellar KDS) decreases, according to the relation [11]

$$\log \mathrm{cmc} = -a \log[K^+] + b\,, \qquad (16)$$

where $[K^+]$ is the concentration of free potassium counterions. Thus, increasing the concentration of KDS in the retentate as the ultrafiltration proceeds lowers the concentration of free surfactant monomer, with a resulting increase in the fraction of the organic solute that is rejected. An additional influence on the cmc is the presence of the organic solute, but this factor was not taken into account in the present study. Another assumption made here is that the binding constant of K^+ remains at 0.6 throughout whole process. The details of the calculation of solubilization equilibrium constant are given in the Appendix. The calculated results are plotted in Fig. 10.

Examining separation of organic solutes and KDS precipitates in recycling KDS in MEUF

The conclusion from a previous section that organic solutes do not precipitate and adsorb to any significant extent along with the solid KDS is supported by the results of experiments in which KDS was converted from aqueous solution (at temperatures above the Krafft temperature) to a solid phase (below its Krafft temperature). Once again, two possible causes of loss of organic solutes (incorporation in the precipitate and adsorption) were examined. KDS aqueous solutions were prepared at KDS concentrations well above the cmc, with dissolved organic solutes

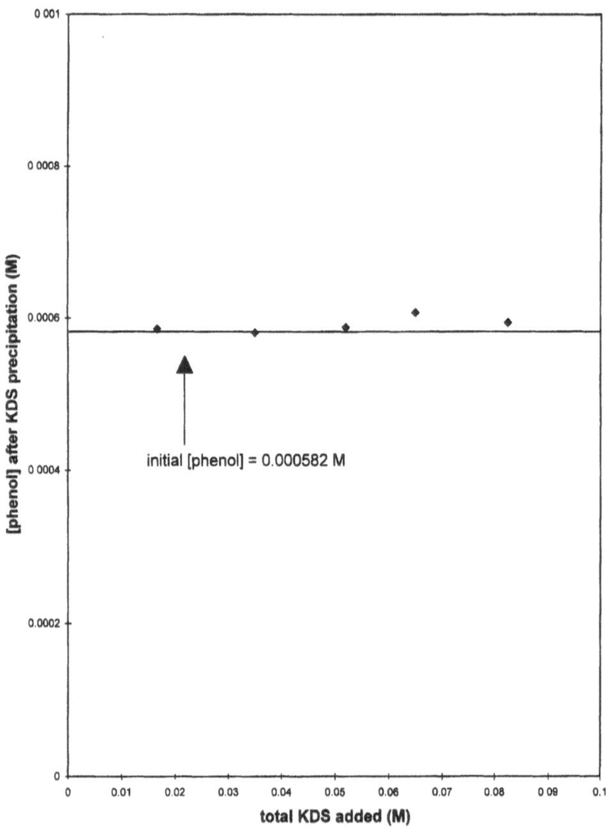

Fig. 11 Effect of precipitation of KDS resulting from cooling the solution below the Krafft temperature on dissolved phenol

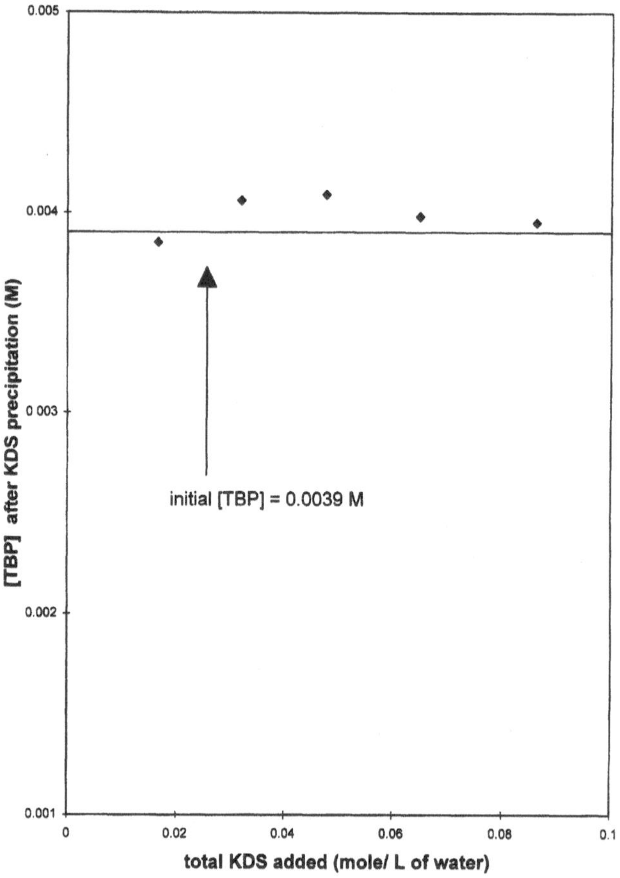

Fig. 12 Effect of precipitation of KDS resulting from cooling the solution below the Krafft temperature on dissolved TBP

present, at a temperature above the Krafft temperature. These solutions were cooled at 25 °C (below the Krafft temperature) and rapid precipitation of KDS occurred. The precipitated KDS crystals were allowed to equilibrate with aqueous solution for 24 h at 25 °C prior to analysis of the supernatant liquid solution for organic solutes. As expected, experimental results (Figs. 11 and 12) indicate that the precipitation of solid KDS has no significant effect on the concentrations of organics in water.

Conclusions and recommendations

Effectiveness of KDS as the surfactant in MEUF

The surfactant potassium dodecylsulfate (KDS) was studied as a possible candidate for use as a micellar surfactant in MEUF separation processes. KDS may be expected to resemble the surfactant sodium dodecylsulfate (SDS) in many of its physical properties, including surface adsorption, critical micelle concentration, solubility in water, and solubilities of organic compounds in the micellar solution. If these expectations are correct, KDS should be potentially as useful as SDS in the MEUF process. An important difference between KDS and SDS is that KDS is known to have a much higher Krafft temperature than SDS, making it feasible to recover KDS by the new precipitation method.

The surface excess quantity for KDS is about 3.1×10^{-10} mol/cm^2 at 39 °C, identical to that reported previously for SDS (3.1×10^{-10} mol/cm^2 at 25 °C) [25]. The cmc of KDS at 39 °C is 6.9 mM, not greatly different from that of SDS at 40 °C, which is about 8 mM. Considering the cost of the surfactant, it is usually beneficial in MEUF separations to choose a surfactant with a low cmc, because there will then be less loss of surfactant into the permeate and less make-up surfactant will be needed to attain the required micellar concentration in the feed stream.

No quantitative comparison has been made of the solubilities of KDS and SDS in water because of the difference in Krafft temperatures of these compounds. As is characteristic of ionic surfactants, the aqueous solubilities are much smaller below the Krafft temperature than above. At the Krafft temperature, the solubility may be regarded as a good approximation to the cmc value. Thus, the cmc of SDS near the Krafft temperature (at about 16 °C) is about 8 mM and that of KDS at the Krafft temperature (at 35 °C) is 6.7 mM. The general behavior and extent of solubilization of organics in KDS and SDS micellar aqueous solutions are also very similar.

The similarity in properties of KDS and SDS implies that the use of these surfactants in removing organics and heavy metals by MEUF should also be similar. Although experiments on the use of KDS in removing metal ions have not yet been performed, the large rejections of organic solutes achieved with KDS in the MEUF process indicate that KDS should be approximately as effective as SDS in MEUF separations.

The Krafft temperature of KDS (35 °C), which is much higher than that of SDS (16 °C), is still in a usable temperature range for many MEUF separations. The higher Krafft temperature of KDS makes this compound an attractive candidate for use in MEUF, utilizing a scheme proposed here for regenerating the surfactant by utilizing temperature variations.

Recovering anionic surfactants: precipitation by adding electrolytes versus precipitation through varying system temperature

Two precipitation methods were examined in this study to recover and recycle surfactants in MEUF processes. Adding counterions to anionic surfactant solutions is a more conventional way, and one key to its success is selecting appropriate surfactants and counterions. Surfactants suitable for counterion precipitation should have a Krafft temperature lower than the operating temperature of MEUF, be sufficiently soluble so that the MEUF processes can operate without the formation of precipitants, and be capable of precipitating in the presence of selected counterions. Counterions chosen for precipitating surfactants should be effective at small concentrations. Sodium salts are not good candidates for precipitating SDS because the Krafft temperature of SDS is significantly lower than room temperature; therefore, large quantities of sodium salts are needed to cause the precipitation of SDS. In addition to the extra material cost associated with using sodium salts, a disadvantage noted previously was that the use of NaCl to precipitate SDS from micellar solutions containing organic solutes caused significant quantities of the organics to be included in the precipitate mass [26]. Fortunately, potassium ion, which has a larger hydrated radius than sodium ion, is a more effective precipitant. The resulting KDS precipitate has a Krafft temperature higher than room temperature and a low solubility at room temperature, just like salts of dodecylsulfate with multivalent counterions. KDS retains the tendency of sodium dodecylsulfate to precipitate from aqueous solution as large, well-formed crystals.

The process of recycling surfactants in MEUF through varying the system temperature (from above to below the Krafft temperature) should be economically attractive. It avoids the problems associated with adding electrolytes, such as NaCl, to precipitate SDS. With possible candidate surfactants like KDS, the separation of surfactants from streams by temperature cycling could be simple, rapid, and clean. However, this method may be somewhat difficult to scale up for treating large volume streams.

A major finding of the present work is the observation that KDS can be precipitated cleanly from aqueous solutions without incorporating water-soluble organics in the precipitate, thus solving a potentially serious surfactant regeneration problem. A possible explanation for the difference between the present results and those obtained previously [26], where large concentrations of NaCl were used to precipitate SDS, may be that the presence of large salt concentrations can cause some salting out of the organic solutes. In any case, loss of organic solutes from aqueous solution did not occur in precipitating KDS under the conditions used here.

Finally, the method described in the present work to prepare KDS crystals is quite simple, yielding the solid material in highly purified form, without generating concentrated electrolyte streams that require further treatment. The attractiveness of employing KDS in MEUF separations is enhanced by the fact that the compound can be readily prepared in the form required for use in membrane separation processes.

Work needs to be done to develop methods for determining the solubility of organic compounds in KDS solutions, particularly near the Krafft temperature, where emulsion formation may make it difficult to utilize ordinary solubility measurements. To obtain thermodynamic solubilities, the semi-equilibrium dialysis (SED) method [27–31] might be used to prevent excess concentrations of organic solutes and KDS from passing through the membrane, as apparently occurred in the conventional filtration experiments performed with phenol and TBP. It will be necessary to estimate the cost of using KDS (as the added colloid in various MEUF separations) compared with the costs of other membrane separation methods, including MEUF utilizing various surfactants and alternative recovery methods.

Acknowledgments Financial support for this work was provided by National Science Foundation Grant CBT 8814147, an Applied Research Grant from the Oklahoma Center for the Advancement of Science and Technology and the TAPPI foundation. In addition, support was received from the industrial sponsors of the Institute for Applied Surfactant Research including Akzo Nobel, Amway, Amoco,

Aqualon, Colgate-Palmolive, Dial, Dow, Dowelanco, DuPont, Henkel, ICI, Lever, Kerr-Mcgee, Reckitt and Colman, Lubrizol, Philliips Petroleum, Pilot Chemical, Schlumberger, Shell, Sun, and Witco. Dr. Scamehorn holds the Asahi Glass Chair in chemical engineering at the University of Oklahoma.

Appendix

The following equations were utilized to calculate the solubilization equilibrium constant K of TBP in KDS micellar aqueous solution from MEUF results at 38 °C:

$$\mathrm{Log\,cmc}(I) = -a \log c\mathrm{K}^+(I) + b\,, \tag{A1}$$

$$c\mathrm{K}^+(I) = c\mathrm{KDS}(I)_{\mathrm{ret}} + c\mathrm{K}(I)_{\mathrm{bind}}\,, \tag{A2}$$

$$c\mathrm{K}(I)_{\mathrm{bind}} = K_{\mathrm{bind}}\, c\mathrm{KDS}(I)_{\mathrm{ret}}\,, \tag{A3}$$

$$n\mathrm{KDS}_{\mathrm{tot}} = c\mathrm{KDS}(I)_{\mathrm{ret}}(V_0 - I\,\Delta V) + n\mathrm{KDS}(I)_{\mathrm{per}}\,, \tag{A4}$$

$$c\mathrm{KDS}(I)_{\mathrm{ret}} = \mathrm{cmc}(I) + c\mathrm{KDS}(I)_{\mathrm{mic}}\,, \tag{A5}$$

$$n\mathrm{KDS}(I)_{\mathrm{per}} = n\mathrm{KDS}(I)_{\mathrm{lost}} + n\mathrm{KDS}(I-1)_{\mathrm{per}}\,, \tag{A6}$$

$$n\mathrm{KDS}(I)_{\mathrm{lost}} = \mathrm{cmc}(I)_{\mathrm{avg}}\,\Delta V\,, \tag{A7}$$

$$\mathrm{cmc}(I)_{\mathrm{avg}} = 0.5(\mathrm{cmc}(I-1) + \mathrm{cmc}(I))\,, \tag{A8}$$

$$n\mathrm{TBP}(I)_{\mathrm{tot}} = c\mathrm{TBP}(I)_{\mathrm{ret}}(V_0 - I\,\Delta V) + n\mathrm{TBP}(I)_{\mathrm{per}}\,, \tag{A9}$$

$$c\mathrm{TBP}(I)_{\mathrm{ret}} = c\mathrm{TBP}(I)_{\mathrm{mic}} + c\mathrm{TBP}(I)_{\mathrm{mono}}\,, \tag{A10}$$

$$c\mathrm{TBP}(I)_{\mathrm{mono}} = c\mathrm{TBP}(I)_{\mathrm{expt}}\,, \tag{A11}$$

$$n\mathrm{TBP}(I)_{\mathrm{per}} = n\mathrm{TBP}(I)_{\mathrm{lost}} + n\mathrm{TBP}(I-1)_{\mathrm{per}}\,, \tag{A12}$$

$$n\mathrm{TBP}(I)_{\mathrm{lost}} = 0.5(c\mathrm{TBP}(I-1)_{\mathrm{expt}} + c\mathrm{TBP}(I)_{\mathrm{expt}})\,\Delta V\,, \tag{A13}$$

$$K(I)_{\mathrm{sol\,eq}} = c\mathrm{TBP}(I)_{\mathrm{mic}}/(c\mathrm{TBP}(I)_{\mathrm{mic}} + c\mathrm{KDS}(I)_{\mathrm{mic}})/c\mathrm{TBP}(I)_{\mathrm{mono}}\,. \tag{A14}$$

Equation (A14) is the expression of the solubilization equilibrium constant. Two assumptions are made here: the concentration of TBP outside micelles in the retentate solution equals the concentration of TBP penetrating the membrane during ultrafiltration (Eq. (A11)), and the concentration of monomeric KDS can be represented by the Corrin–Harkins relationship (Eq. (A1)). Equation (A2) and (A3) are mass balance equations for potassium ions. Equations (A4)–(A7) are mass balance for KDS. Equations (A9)–(A13) are mass balance for TBP. Constant a in the Corrin–Harkins equation was set equal to 0.6, and constant b was calculated as -3.458 by letting both K^+ and cmc in the equation be 6.9 mM which was obtained from the surface tension measurement. $c\mathrm{K}^+(I)$ is the concentration of free K^+. I is the sequence number of variables corresponding to volume of the permeate solution $25I$ ml. $c\mathrm{K}(I)_{\mathrm{bind}}$ is the concentration of potassium binding to the micelles. K_{bind} is the counterion binding constant which is set to be 0.6 in this work. $c\mathrm{KDS}(I)_{\mathrm{ret}}$ is the concentration of KDS in the retentate and $n\mathrm{KDS}_{\mathrm{tot}}$ is total amount of KDS in the retentate which is 0.06 mol in the experiment. V_0 is the initial volume of the retentate solution which is 300 ml in the experiment $n\mathrm{KDS}(I)_{\mathrm{per}}$ is the amount of KDS in the permeate. ΔV is 25 ml. $c\mathrm{KDS}(I)_{\mathrm{mic}}$ is the concentration of the KDS existing as micelles. $n\mathrm{KDS}(I)_{\mathrm{lost}}$ is the amount of KDS lost from the retentate to the permeate in the region I. cmc(I) is the critical micellar concentration at certain permeate volume $25\,I$ ml. $\mathrm{cmc}(I)_{\mathrm{avg}}$ is the arithmetic mean value of cmc(I) in the region I. $n\mathrm{TBP}_{\mathrm{tot}}$ is total amount of TBP in the system which is 0.002 mole in the experiment. $c\mathrm{TBP}(I)_{\mathrm{ret}}$ is the concentration of TBP in the retentate. $n\mathrm{TBP}(I)_{\mathrm{lost}}$ is the amount of TBP lost from the retentate to the permeate in the region I. $n\mathrm{TBP}(I)_{\mathrm{per}}$ is the amount of TBP in the permeate. $c\mathrm{TBP}(I)_{\mathrm{mic}}$ is the concentration of TBP solubilized in the micellar KDS aqueous solution. $c\mathrm{TBP}(I)_{\mathrm{mono}}$ is the concentration of TBP in the monomer phase in the retentate. $K(I)_{\mathrm{sol\,eq}}$ is the solubilization equilibrium constant of TBP in the micellar KDS aqueous solution in the region I.

Those equations were solved by using the SEQS program. The resulting solubilization equilibration constants of TBP are plotted in Fig. 12.

Glossary of terms

K_{sp}	solubility product of KDS (M^2)
γ_m	mean activity coefficient calculated by the extended Debye–Hückel equation
$c\mathrm{M}$	concentration of unbounded ion M (M)
$c\mathrm{DS}^-_{\mathrm{mono}}$	concentration of monomeric dodecylsulfate anion (M)
K_{mic}	micellization constant, in molar units
n	aggregation number for dodecylsulfate anion
q	aggregation number for counterion
c_{micelle}	surfactant concentration in micellar form (M)
I	ionic strength (M)
Ratio	mole fraction of K^+ ion in the total aqueous dissolved cation concentration
$c\mathrm{M}_{\mathrm{tot}}$	total amount of ion M (mol)
$\mathrm{KDS}_{\mathrm{prcpt}}$	amount of KDS precipitated (mol)
$\mathrm{SDS}_{\mathrm{tot}}$	amount of SDS before adding KCl (mol)
$R_{\mathrm{K:Na}}$	concentration ratio of KCl to SDS
Ds_{tot}	total amount of dodecylsulfate remaining in solution after precipitation (mol)
$\mathrm{SDS}_{\mathrm{prcpt}}$	amount of SDS incorporated into the KDS solid (mol)
K	proportionality constant for SDS incorporation

References

1. Scamehorn JF, Christian SD, Ellington RT (1989) In: Scamehorn JF, Harwell JH (eds) Surfactant-Based Separation Processes, Chaps 1 and 2, Marcel Dekker, New York
2. Sasaki KJ, Burnet SL, Christian SD, Tucker EE, Scamehorn JF (1989) Langmuir 5:363
3. Christian SD, Bhat SN, Tucker EE, Scamehorn JF, El-Sayed DA (1988) AIChE J 34:189
4. Scamehorn JF, Christian SD (1988) Proc 9th AESF/EPA Conf on Environmental Control for the Metal Finishing Industry
5. Dunn RO, Scamehorn JF, Christian SD (1989) Colloid Surf 35:49
6. Scamehorn JF, Ellington RT, Christian SD, Penney BW (1986) AIChE Symp Ser No. 250, 82:48
7. Christian SD, Tucker EE, Scamehorn JF (1992) Am Environmental Lab 2:13
8. Krehbiel DK, Scamehorn JF, Ritter R, Christian SD, Tucker EE (1992) Sep Sci Tech 27:1775
9. Simmons DL, Schovanec AL, Scamehorn JF, Christian SD, Taylor RW (1992) In: Vandergridt GF et al. (eds) ACS Symposium Series, Vol 509. Washington DC, 180
10. Lu JR, Marrocco A, Su TJ, Thomas RK, Penfold J (1993) J Colloid Interface Sci 158:303
11. Corrin ML, Harkins WD (1947) J Am Chem Soc 69:684
12. Nguyen CM, Christian SD, Scamehorn JF (1988) Tenside Surfactants Deterg 25:328
13. Morgan ME (1992) M.S. thesis, Department of Chemistry, University of Oklahoma
14. Brant LL, Stellner KL, Scamehorn JF (1989) In: Scamehorn JF, Harwell JH (eds) Surfactant-Based Separation Processes, Chap 12, Marcel Dekker, New York
15. Uchiyama H (1993) Private correspondence, Department of Chemistry, University of Oklahoma
16. Stellner KL, Scamehorn JF (1986) J Am Oil Chem Soc 63:566
17. Stellner KL, Scamehorn JF (1989) Langmuir 5:77
18. Tucker EE, Christian SD, Scamehorn JF, Uchiyama H, Guo W (1992) In: Sabatini DA, Knox RC (eds) ACS Symposium Series, Vol 491. Washington DC, p 86
19. Dharmawardana U (1993) Private correspondence, Department of Chemistry, University of Oklahoma
20. Roberts B (1992) PhD dissertation, Department of Chemical Engineering, University of Oklahoma
21. Meguro K, Kondo T (1956) J Chem Soc Japan Pure Chem Sec 77:156
22. Tipton RJ (1989) M.S. thesis, Department of Chemistry, University of Oklahoma
23. SEQS computer program (1990) CET Research Group Ltd
24. Uchiyama H, Christian SD, Tucker EE, Scamehorn JF (1993) J Phys Chem 97:10868
25. Dahanayake M, Cohen AW, Rosen MJ (1986) J Phys Chem 90:2413
26. Floyd R (1992) Private correspondence, Department of Chemical Engineering, University of Oklahoma
27. Lee BH (1990) PhD dissertation, Department of Chemistry, University of Oklahoma
28. Lee BH, Christian SD, Tucker EE, Scamehorn JF (1990) Langmuir 6:230
29. Higazy WS, Mahmoud FZ, Taha AA, Christian SD (1988) J Solution Chem 17:191
30. Mahmoud FZ, Christian SD, Tucker EE, Taha AA, Scamehorn JF (1989) J Phys Chem 93:5903
31. Christian SD, Tucker EE, Scamehorn JF, Lee BH, Sasaki KJ (1989) Langmuir 5:876
32. Weil JK, Smith FS, Stirton AJ, Bistline RG (1963) J Am Oil Soc 40:538
33. Dunaway CS, Christian SD, Scamehorn JF (1995) In: Christian SD, Scamehorn JF (eds) Solubilization in Surfactant Aggregates, Surfactant Science Series, Vol 55, Marcel Dekker, New York, p 3
34. Scamehorn JF, Harwell JH (1993) In: Ogino K, Abe M (eds) Mixed Surfactant Systems, Surfactant Science Series, Vol 46, Marcel Dekker, New York, p 283
35. Yin Y, Scamehorn JF, Christian SD (1995) In: Sabatini DA, Knox RC, Harwell JH (eds) Surfactant-enhanced Subsurface remediation, ACS Symposium Series, Vol 594. American Chemical Society, Washington DC, p 231

Progr Colloid Polym Sci (1998) 109:74–84
© Steinkopff Verlag 1998

J.R. Baran Jr
G.A. Pope
W.H. Wade
V. Weerasooriya

An overview of surfactant enhanced aquifer remediation

Received: 3 July 1997
Accepted: 16 July 1997

J.R. Baran Jr* · W.H. Wade (✉)
V. Weerasooriya
Department of Chemistry
University of Texas
Austin, TX 78712
USA

G.A. Pope
Department of Petroleum Engineering
University of Texas
Austin, TX 78712
USA

**Present address*
3M, Industrial and Consumer
Sector Laboratory
St. Paul, MN 55144-1000
USA

Abstract A systematic study of the phase behavior of chlorinated hydrocarbons with various surfactant species in aqueous solution is discussed. Chlorinated hydrocarbons are more polar and better solvents than hydrocarbons and therefore require more hydrophilic surfactants in order to exhibit classical Winsor type phase behavior. These chlorocarbons and surfactants still obey the mathematical relationships for mixing that we previously established for hydrocarbons. Mass-transport calculations show that surfactants should drastically decrease the time required to clean up toxic spills involving chlorocarbons.

Key words Chlorocarbons – surfactants – microemulsions

Introduction

In the early 1970s, with the oil crisis brought on by the price increases of oil exports from OPEC, researchers began investigating ways of enhancing the efficiency of oil recovery from wells. Our research in that area was to devise a systematic approach for selecting and optimizing surfactant systems for Enhanced Oil Recovery (EOR) [1–9]. We were able to establish a set of mathematical relationships to determine the best surfactant system to optimize recovery at various temperatures, salinities, and oil compositions.

With the subsequent lowering of prices of imported oil into the United States, funding for this type of research essentially disappeared. Researchers began looking for other applications of the knowledge gained from these studies. Recent interest in environmental contaminants afflicting the world's potable groundwater supplies [10] led us to investigate the applicability of the knowledge garnered from EOR to Surfactant Enhanced Aquifer Remediation (SEAR). These contaminants enter the subsurface as a separate organic phase or nonaqueous phase liquid (NAPL). Many contaminants such as chlorocarbons are more dense than water (therefore, are colloquially known as dense nonaqueous phase liquids, DNAPLs) and migrate to the bottom of the aquifer. A portion of the liquid phase is retained in the soil pores as immobile ganglia held by capillary forces at high interfacial tension [11, 12]. This trapped phase slowly dissolves into the water over long periods of time, thus representing a long-term health hazard.

One method to remove NAPL from an aquifer is to pump the contaminated water out of the aquifer and treat it at the surface. This pump and treat method has not been very effective, since the NAPL is not very soluble in water and a large portion of the residual NAPL cannot be pumped out. The addition of a surfactant to the system would allow more NAPL to be solubilized by incorporation into a microemulsion phase that would also lower the interfacial tension. This was the most common approach for EOR [13]. The use of surfactants for subsurface remediation was recently reviewed by West and Harwell [14].

A systematic approach to selecting and optimizing surfactants for SEAR, would be to study the classical Winsor type phase behavior of the systems [15]. In order to obtain meaningful results from this type of investigation, all system variables, except one, should be held constant. In our studies we chose to vary only the salinity within the systems of interest. At low salinity, Winsor type I (oil-in-water, O/W) microemulsions are formed. As the salinity is increased, both the extent of solubilization and the opacity of the microemulsion increases. At a certain salinity, there is a transition to a Winsor type III (middle phase) system, which begins with the middle phase having a water/oil volume ratio (WOR) near infinity. As the salinity is increased, the system gradually passes through the *optimum state* where the middle phase WOR = 1, and ultimately, to a system where the middle phase WOR approaches 0. Further increases in salinity generate Winsor type II (water-in-oil, W/O) systems with decreasing opacity and water solubilized in oil. For optimum system composition, the salinity required is called the *optimum salinity*, S^*, where equal volumes of oil and water that are solubilized per unit weight of surfactant plus cosurfactant (if present) are designated as the *optimum solubilization parameter*, σ^*, and simultaneously the I/III and III/II interfacial tensions, γ^*, are equal. The *range of salinities*, ΔS, over which one has a Winsor type III system has also been described in these studies since it is an easy method to assess the sensitivity to system variation in that large values of σ^* give small values for ΔS [16]. A generic diagram of such a system is shown in Fig. 1.

We choose to exclusively identify system conditions required for Winsor type III optimum middle-phase systems. These systems have the surfactant equipartitioned into both the excess aqueous and oleic phases.

Another useful property of type III systems is the inverse relationship between efficiency of solubilization and interfacial tension, i.e. the better the solubilization, the lower the interfacial tension (see Fig. 1) [17, 18]. If one wants to mobilize a chlorocarbon, a Winsor type III system should be designed, because of their low interfacial tensions. Type II systems should be avoided since the chlorocarbons would be a chemical potential sink for the surfactant. If one wants only to solubilize the chlorocarbon, type I systems far from the I/III boundary should be designed so that the interfacial tensions are not low enough to mobilize [19].

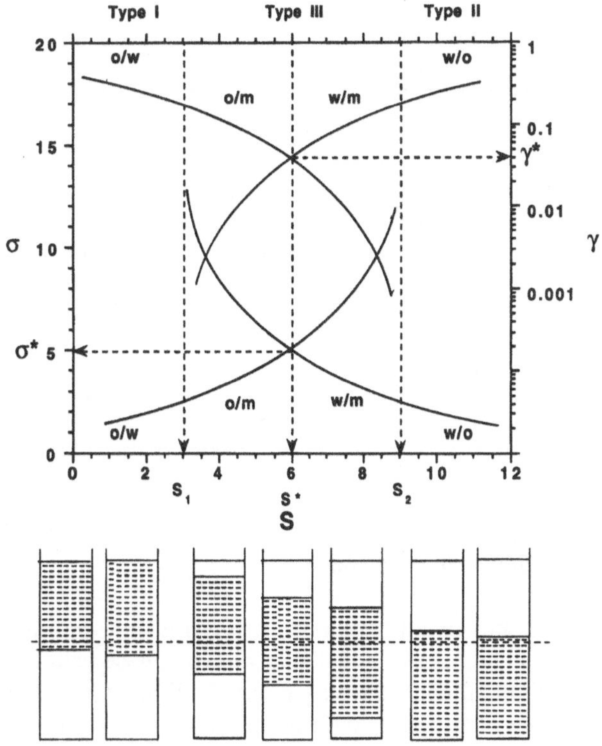

Fig. 1 Classical phase behavior for a variation in electrolyte concentration with accompaning interfacial tensions and solubilization parameters for a properly designed chlorocarbon/water/surfactant system

Surfactants

While many surfactants had proven successful to various degrees in producing middle-phase microemulsions for EOR, we found early on that most were clearly not useful for establishing classical phase behavior with chlorocarbons. Chlorocarbons are far better solvents than most hydrocarbons, thereby reducing the surface activity of the surfactant. Those surfactants that did demonstrate surface activity either formed complex phase behavior (i.e. liquid crystal formation) or Winsor type II systems, severely aggravated by ground temperatures in the range of 10–20 °C. Calcium tolerance is a necessary requirement for any usable surfactant. All of those mentioned here exhibit this characteristic. Most of these experiments were conducted at 25 °C. But the effect of temperature was studied on some systems at 40 °C and at 15 °C.

Table 1 Parameters for phase behavior of 2 wt% MA

	S^* [wt%]	σ^* [mL/g]	ΔS [wt%]
PCE			
NaCl	4.10	2.40	2.80
$CaCl_2$	2.40	1.00	5.60
CCl_4			
NaCl	2.48	3.30	0.77
$CaCl_2$	0.16	2.65	0.10
TCA			
NaCl	1.62	4.80	—
$CaCl_2$	—	—	—
TCE			
NaCl	1.31	4.65	0.42
$CaCl_2$	—	—	—

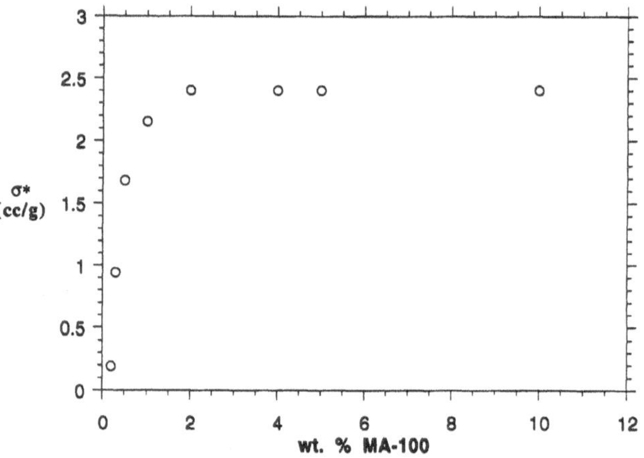

Fig. 2 Effects of varying MA-100 concentration on the optimum solubilization parameter of PCE at 4.1 wt% NaCl

Sodium dialkylsulfosuccinates

The first commercially available surfactants that were successful in exhibiting classical Winsor type phase behavior were sodium dialkylsulfosuccinates. Sodium dihexylsulfosuccinate, obtained from Cytec and sold under the brand name Aerosol MA-80I, was successful with only a limited number of DNAPLs. In all studies in this report, the MA-80I was purified via drying in a vacuum oven, resulting in a white solid that is 100% active, and this is referred to this report as just MA. It produced classical phase behavior with tetrachloroethylene, trichloroethylene, 1,1,1-trichloroethane and carbon tetrachloride (PCE, TCE, TCA and CCl_4, respectively) [20–22]. The optimum salinities (S^*), optimum solubilization parameters (σ^*) and salinity window (ΔS) for these systems are listed in Table 1.

As expected, as S^* decreased, σ^* increased and ΔS decreased. We believe that the decrease in salinity in going from PCE to CCl_4 to TCA to TCE is directly due to the increase in solvency along this trend. PCE and CCl_4, when emulsified with MA, also exhibited classical phase behavior with $CaCl_2$, although at lower salinities. TCA and TCE do not show classical phase behavior with pure $CaCl_2$. Instead, one obtains what we defined as the *switching salinity*, $S^{\ddagger}$ [21]. We defined this as the point at which the system changes from a Winsor type I system to a Winsor type II system without going through a Winsor type III system. This is due to the width of the salinity window going to zero.

Attempts to use other sodium dialkylsulfosuccinates produced mixed results. It was possible to produce Winsor type III systems with sodium diamylsulfosuccinate (AY) with these oils [23]. Since the amyl group is less lipophilic than the hexyl group, the optimum salinities for these oils increase. Sodium dioctylsulfosuccinate (OT) only produced Winsor type II systems with all chlorocarbons studied. In effect, these systems have a "negative optimum salinity", due to the increased solubility of OT in these chlorocarbons over that of MA.

Table 2 Critical micelle concentrations in the presence of oil at optimum salinity

Surfactant/oil	CMC [wt%]	S^* [wt% NaCl]
MA/PCE	0.19	4.1
MA/hexane	0.12	6.9
AY/PCE	0.16	7.2
AY/hexane	0.07	12.1

Since these surfactants have particularly high critical micelle concentrations (CMCs), it is possible to determine the CMC in an unusual manner for aqueous solutions of MA and AY in the presence of both PCE and *n*-hexane [22]. The values were obtained by plotting the solubilization parameter vs. wt% surfactant, followed by extrapolation. As the surfactant concentration is lowered, the optimum solubilization parameter becomes smaller. At a certain concentration, the middle phase disappears. Figure 2 demonstrates this relationship for MA with PCE. The concentration of surfactant below which it will no longer aggregate to form microemulsions is the CMC in the presence of oil. This value can be obtained directly from the graph.

As the results show, for a given surfactant the higher the S^*, the lower the CMC. The CMCs for MA and AY in both *n*-hexane and PCE are given in Table 2. The addition of electrolyte leads to a lower CMC and since microemulsions form spontaneously, the presence of oil must also lower the CMC. Presently, we cannot distinguish the relative importance of each.

Ethylene oxide and propylene oxide containing surfactants

While searching for other surfactants that might produce classical phase behavior with chlorocarbons and at the same time have sufficient tolerance of Ca^{2+}, we studied the effects of surfactants containing ethylene oxide (EO) and propylene oxide (PO). Previously our laboratory had used C_{14} and C_{16} Guerbet alcohol hydrophobes with various amounts of EO or PO and end-capped with sodium sulfate for EOR studies [24]. Unfortunately, none of these surfactants were capable of exhibiting classical phase behavior with chlorocarbons themselves, but required MA as a cosurfactant [20, 21].

We found that for all species, S^* increased as the amount of EO in the molecule increased [20, 21]. It was found that by going from PCE to CCl_4 to TCE, the respective optimum salinities decreased. But on a plot of log S^* vs. EON the rate of change was the same for all three systems. We believe this is indicative of the differences in polarity of the molecules [21]. The lower the optimum salinity, the more polar the chlorocarbon.

In mixtures of Guerbet/MA (50/50 and 70/30) [20], the S^* increased with the addition of EO exponentially, to the point that the ratio of the slopes of the surfactant mixtures on a plot of ln S^* vs. EON was directly related to the level of Guerbet species in the formulation. The 50/50 and 70/30 curves crossed near the EO_6 value, indicating that surfactants with fewer EOs are more lipohilic than MA. Those with more EOs were more hydrophilic than MA and that EO_6 must have an hydrophilic/lipophilic balance (HLB) nearly identical to MA.

The optimum solubilization parameters demonstrated a much more complex relationship [20, 21]. In general, as more EO was added, the solubilization parameter increased. However, all systems showed a local minimum in solubilization parameter at EO_8 and a local maximum at EO_4. We believe this to be due to the complex interplay of increasing salinity causing reduction in the solubilization parameter, while the increased structural length leads to increases in the solubilization parameter. Increasing the Guerbet/MA ratio in the system from 50/50 to 70/30 causes an increase in the solubilization parameter. The simple explanation is that the average structural length is greater. The amount of increase is anywhere from 10% with CCl_4 to 2–3 times with PCE.

In general, ΔS decreased as more EO was added to the molecule [20, 21]. This was expected since the solubilization parameter and the salinity window have an inverse relationship. The 70/30 data lie slightly below the 50/50 data. Increasing the overall hydrophobe structural length has been found to decrease ΔS [5, 6]. But, increasing the structural length by adding EO also increases S^*, with the net result being that the ratio $\Delta S/S^*$ is essentially constant.

As noted these systems are calcium tolerant [20, 21]. When all of the NaCl was replaced by $CaCl_2$, the optimum salinity decreased, the optimum solubilization parameter increased and the salinity window decreased.

Substituting propylene oxide for ethylene oxide in these surfactants was found to have very little effect on the optimum salinity, but a major effect on the solubilization parameters [20, 21]. Optimum solublization parameters were much larger than that of the non-PO containing surfactants. This is due to the majority of the PO containing surfactant residing in the oleic phase.

Increasing the temperature to 40 °C caused very little change in the optimum salinities (<1%), but caused a 15–25% reduction in the solubilization parameter [20, 21]. The salinity window generally increased as the temperature was increased.

Decreasing the temperature to 15 °C led to extremely stable macroemulsions (usually six months or longer). The effect on optimum salinity, optimum solubilization parameter and salinity window was quite small. For systems containing lower EO species, the optimum salinity was lower at lower temperatures due to the dominance of the $-SO_4^-$ temperature coefficient [25]. With higher EO species, the optimum salinity increases with the lowering of the temperature, since the temperature coefficient of the EO dominates.

We also synthesized several propoxylated surfactants end-capped with sodium sulfonate that successfully produced middle-phase systems by themselves. These contained branched hydrophobes with various amounts of PO added. Larger hydrophobes and more PO in the molecule resulted in lower optimum salinities. They all exhibited larger optimum solubilization parameters than the sulfosuccinates. We reached a point in the use of these compounds where only complex phase behavior (i.e. liquid crystal formation) was exhibited. This occurred when the surfactant molecule became too large. In the case of the octyl hydrophobe, the onset of liquid crystal formation occurred with nine moles of PO, with the dodecyl hydrophobe it occurred with six moles of PO.

Alkylbenzene sulfonates [23]

None of the various monoisomeric alkylbenzene sulfonated surfactants that were studied, produced Winsor type III phase behavior by themselves. All were combined in various ratios with MA to form Winsor type III systems with PCE, TCE and CCl_4. Although this class of surfactants proved to be very useful in EOR, they only formed middle-phase systems when present as a minor part of the surfactant mixture. In general, they had prolonged equilibration times, increasing as the amount of alkylbenzene sulfonate was increased. The solubilization parameters were quite small in these mixtures.

Even with these problems, some generalizations could be ascertained about the surfactants. As expected, the larger the hydrophobic part of the molecule, or the more branching within isomers of the same hydrophobe, led to a reduction in the optimum salinity.

TWEEN®s and SPAN®s [26]

These were the first nonionic surfactants that we investigated to produce Winsor type phase behavior. In general, to observe classical phase behavior, we found it necessary to use TWEEN® 21 (sorbitan monolaurate with 4 moles of EO) as the major component of the surfactant system. The only oils that exhibited this behavior were PCE and CCl_4. These systems produced small solubilization at rather high salinities. For example, TWEEN® 21 with PCE produced an optimum system at $S^* = 8.2$ wt% NaCl with $\sigma^* = 2.8$ mL/g and $\Delta S = 8.55$ wt% NaCl. These parameters are actually worse than those for the anionic MA.

Alkylglucamide surfactants [27, 28]

These surfactants were synthesized by a method patented by Proctor and Gamble [29] and are glucamine based

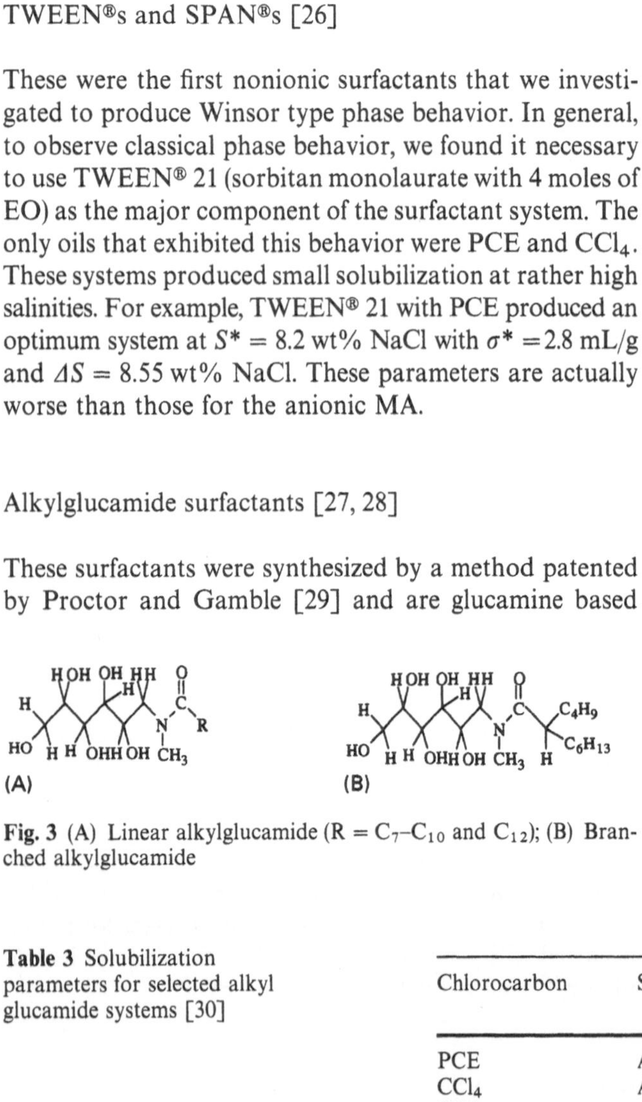

Fig. 3 (A) Linear alkylglucamide (R = C_7–C_{10} and C_{12}); (B) Branched alkylglucamide

with an amide linkage formed between methylglucamine and a variety of carboxylic acids (see Fig. 3).

These surfactants are extremely hydrophilic and proved to be excellent candidates for the formation of chlorocarbon-in-water microemulsions. Various analogues from the patent, as well as a branched C_{12} species, derived from the Guerbet carboxylic acid starting material, were synthesized. Mixtures of these surfactants proved successful in producing classical Winsor type phase behavior at room temperature, with little or no added electrolyte, with PCE, TCE, TCA and CCl_4. The solubilization parameters (ranging from 1.2 to 3.75 ml/g), were smaller than those obtained with MA (see Table 1).

In addition to these DNAPLs, it was possible to produce classical Winsor type phase behavior with chloroform ($CHCl_3$) and 1,1,2,2,-tetrachloroethane (1,1,2,2-TCE), using various mixtures of these surfactants. When 1,2-dichlorobenzene (1,2-DCB), dichloromethane (CH_2Cl_2) and 1,2-dichloroethane (1,2-DCE) were the oil phases, various degrees of complex phase behavior were detected.

Table 3 lists the solubilization parameters for mixtures of alkylglucamides that produced a Winsor type I system at 0 and 0.1 wt% NaCl, and a Winsor type III system at 0.5 wt% of NaCl. The solubilization parameter for a type I system is defined as the volume of oil dissolved in water divided by the weight of the surfactant (mL/g). For a type III system, there are two solubilization parameters, σ_o and σ_w. σ_o is defined as the volume of oil dissolved in the middle-phase divided by the weight of surfactant. σ_w is defined as the volume of water dissolved in the middle-phase divided by the weight of surfactant.

Generally, the solubilization parameter increases as one nears the type III region of a system [21]. Here, this also holds true, except when proceeding from 0 to 0.1 wt% NaCl. In comparing these two values, one notes that the solubilization parameter shows no definite trend. There is no apparent explanation for this anomaly, except that operating at 0% NaCl is new territory and may not be

Table 3 Solubilization parameters for selected alkyl glucamide systems [30]

Chlorocarbon	Surfactant mixture[a)]	$\sigma_{0\%}$ [mL/g]	$\sigma_{0.1\%}$ [mL/g]	$\sigma_{o,0.5\%}$ [mL/g]	$\sigma_{w,0.5\%}$ [mL/g]
PCE	A12G/A9 (3/2)	1.25	1.00	1.75	0.75
CCl_4	A12G/A9 (42/58)	2.00	0.75	1.50	4.25
TCA	A12G/A9 (34/66)	2.25	1.00	1.50	4.25
TCE	A12G/A9 (1/3)	3.50	3.50	3.00	4.00
1,2-DCB	A12G/A9 (27/73)	0.50	1.00	2.25	3.00
1,2-DCE	A12G/A9 (1/4) (4%)	0.25	0.25	0.50	1.38
$CHCl_3$	A9/A7 (1/4) (4%)	1.00	1.13	1.25	1.75
CH_2Cl_2	A9	2.05	2.50	2.75	2.50
1,1,2,2-TCE	A9/A7 (1/9) (8%)	0.63	0.63	0.63	0.75

a) Surfactant concentration is 2 wt% of the aqueous phase, unless otherwise noted. The numbers in parenthesis are weight ratios.

governed by the same rules that apply to anionic systems containing electrolyte.

These surfactants were found to be basically temperature insensitive [28]. For example, at 25 °C, TCE with a mixture of A12G/A9 (25/75) formed an optimum system at $S^* = 9.60$ wt% NaCl and a $\sigma^* = 3.75$ ml/g. At 16 °C, the same combination of oil and surfactants produced an optimum system at $S^* = 9.75$ wt% and a $\sigma^* = 3.70$ ml/g. Finally, as opposed to anionic surfactants, these surfactants demonstrate rapid equilibration times at 0% NaCl.

Alkyl polyglucoside surfactants

Since we were unable to license the Proctor & Gamble patent, we instituted a search for similar types of surfactants that are, or could be, commercially available. We were approached by AKZO Nobel offering experimental samples of alkyl polyglucosides, and have begun a joint research project with them to explore three surfactant parameters: (a) hydrophobe molecular weight, (b) hydrophobe branching, and (c) degree of polymerization (DP) of glucosides. We do not have the studies completed at this time, but it is safe to say their properties are in every way similar to the alkylglucamides in that they make Winsor III systems at low temperatures, require no added electrolyte, and their phase behavior is rather insensitive to temperature.

EACN determination for chlorocarbons

Early studies on microemulsion formation with crude oils provided mixing rules for a wide variety of aromatic, aliphatic and cyclic aliphatic components that comprise normal crude oils. Mixtures of hydrocarbons were found to obey simple mole fraction averaging rules for microemulsion stability [31]. These rules could be expanded to any number of components and general relationships such as,

$$\mathrm{EACN}^*_{\mathrm{MIX}} = \sum_i X_i \mathrm{ACN}^*_i \tag{1}$$

could be developed. EACN (Equivalent Alkane Carbon Number) is a numerical representation for the hydrocarbon(s) being studied that does not represent any physical characteristic of the oil. EACNs can be integral and/or fractional. For instance an equimolar mixture of pentane (EACN = 5) and decane (EACN = 10) would have an EACN = 7.5. This equation has been successfully applied to a 26 component oil phase, which mimics a crude oil [32].

The Salager equation was developed to describe the partitioning of the surfactant in Winsor type III systems [1].

$$\ln \frac{S^*}{S^{*\mathrm{o}}} = k\{\mathrm{EACN}^*_{\mathrm{MIX}} - N_{\mathrm{MIN}}\} + f(A) + a(T-25)\,. \tag{2}$$

The above equation relates the system's variables: optimum salinity, S^*, oleic phase EACN*, alcohol cosolvent concentration (A), and temperature, T, where k, f and a are constants. At 25 °C, with no alcohol and with a defined standard state optimum salinity, $S^{*\mathrm{o}} = 1$ wt% Eq. (2) reduces to

$$\ln S^*\,(\mathrm{wt\%}) = k\{\mathrm{EACN}^*_{\mathrm{MIX}} - N_{\mathrm{MIN}}\}\,. \tag{3}$$

Note that at $S^* = 1$ wt%,

$$N_{\mathrm{MIN}} = \mathrm{EACN}^*_{\mathrm{MIX}}\,. \tag{4}$$

Therefore, N_{MIN} is the EACN for an optimum Winsor type III system with no alcohol, $T = 25$ °C and $S^* = 1$ wt%, and is considered to be an invariant property of the surfactant. This allows one to rank surfactant partitioning characteristics in an alternative way to the HLB approach [33].

Mathematical manipulations of these equations show that with a given surfactant one expects a linear relationship between ln S^* (wt%) and the mole fraction composition of the oleic phase. This can be written generally as,

$$\ln S^*_{\mathrm{MIX}} = \sum_{i=1\ldots} X^*_i \ln S^*_i\,. \tag{5}$$

The ability to determine the EACN relies completely on the experimental observation that, with all other system variables held constant, including remaining at the optimum state, there is a linear dependence of preferred EACN on mole fraction composition of the surfactant mixture, i.e.

$$\mathrm{EACN}^*_{\mathrm{MIX}} = \sum_j X^*_j \mathrm{EACN}^*_j \tag{6}$$

where j now indexes the surfactant species. By virtue of the validity of Eq. (4), there are relationships identical to Eqs. (2) and (5), except with the mole fractions being those of the surfactant species with a constant oil phase.

These mathematical relationships allow a thorough investigation of the classical Winsor type phase behavior of pure NAPLs and DNAPLs, mixtures of either NAPLs or DNAPLs, or both.

At this point, the validity of the EOR mole fraction averaging equations was tested for chlorocarbons. These studies were conducted with MA as the sole surfactant species and with NaCl as the electrolyte. Initially, we applied these equations to binary mixtures of PCE/CCl_4, PCE/TCE and CCl_4/TCE to test their validity. A plot of

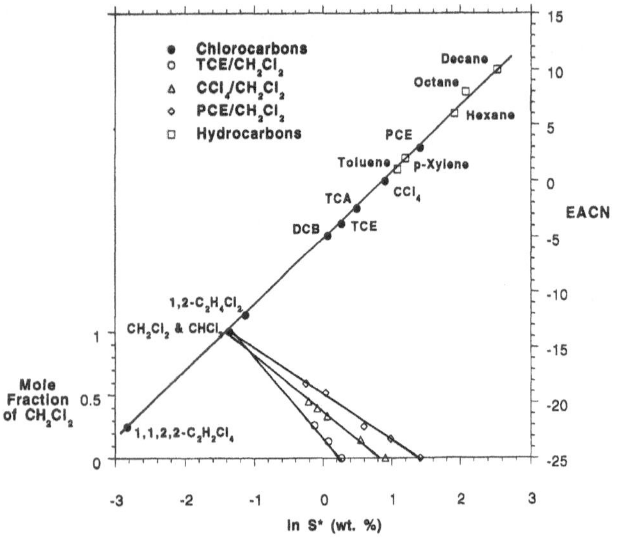

Fig. 4 ln S^* vs. EACN for species as noted

Table 4 S^* and EACN values of various chlorinated hydrocarbons [34]

Oil	S^* [wt% NaCl]	EACN
PCE	4.1	2.9
CCl_4	2.48	−0.06
TCA	1.64	−2.49
TCE	1.31	−3.81
DCB	1.09	−4.89
1,2-$C_2H_4Cl_2$	0.32	−12.10
$CHCl_3$	0.245	−13.67
CH_2Cl_2	0.24	−13.79
1,1,2,2-$C_2H_2Cl_4$	0.058	−22.15

ln S^*_{MIX} vs. mole fraction was made for these systems and found to obey Eq. (5) [34].

A ln S^* vs. EACN graph for MA using a wide range of hydrocarbons was constructed. The slope of this plot was found to be 0.17, which is in good agreement with the value obtained for hydrocarbons in previous studies (0.16) [2]. This is the value of k in Eq. (3). Solving this equation for N_{MIN}, we obtained −5.4 for MA, thereby permitting us to rewrite Eq. (3) as

$$\ln S^* (\text{wt\%}) = 0.17\ [\text{EACN} + 5.4] \tag{7}$$

for systems containing MA. By knowing the optimum salinities (Table 1) of the pure oils, we were able to find the EACN values of PCE, CCl_4, TCA TCE directly from this graph (Fig. 4 and Table 4).

The "negative" EACN values are due to factors other than molecular weight governing a DNAPL's relationship between EACN and ln S^*. Again, we believe that the dominating factor for chlorocarbons is often their polarities. The more polar the oil, the more negative the EACN. Although there is no physical meaning to a negative EACN value, we propose that these negative values can be introduced into the various equations for the purpose of calculating mixing properties.

We extended the mixing rules to binary mixtures of PCE, CCl_4, or TCE with one of the intractable oils (CH_2Cl_2, $CHCl_3$, 1,2-$C_2H_4Cl_2$, 1,1,2,2-$C_2H_2Cl_4$ or DCB). We found that we could make middle-phase systems out of these binary mixtures up to a certain amount of the intractable oil. When the concentration of the intractable oil reached a limiting stage, the salinity window went to zero [21]. Earlier we introduced the term *switching salinity* to describe the salinity at which a system went directly from Winsor type I to Winsor type II. Similarly, here we introduced the term *switching mole fraction*, $X^‡$ [34], to describe the mole fraction of the intractable species at which the salinity window went to zero.

When these binary systems were plotted on a ln S^* vs. mole fraction graph, an extrapolation of the data permitted one to obtain an optimum salinity for the pure, intractable species (see Fig. 4). The oil's EACN could be obtained from this graph. These values are listed in Table 4.

A set of ternary mixtures of chlorocarbons were made from PCE, TCE and 1,2-$C_2H_4Cl_2$ and were experimentally shown to have optimum salinities only slightly higher than those predicted by the mathematical calculations, though all were within 0.15 wt% NaCl [32].

A more complex mixture of hydrocarbons and chlorocarbons was created using PCE, DCB, 1,1,2,2-$C_2H_2Cl_4$, decane and toluene [34]. This mixture covered a range of 32 EACN units and an optimum salinity range of over two orders of magnitude, and therefore was a good test for the calculations. The EACN_{MIX} calculated from mole fractions and pure oil EACNs was found to be −3.86, corresponding to an optimum salinity of 1.30 wt% NaCl. Experimentally, these values were determined to be −2.78 and 1.56 wt% NaCl, respectively. Considering the range covered, we believe that the agreement was quite good.

N_{MIN} determination for anionic surfactants

Since the oil phase mixing rules proved to be successful in predicting the EACN of a mixture of chlorocarbons and/or hydrocarbons, we sought to verify the set of equations that dealt with surfactants.

If one knows the EACN of the oleic phase, the N_{MIN} of the surfactant can be calculated. For a mixture of surfactants,

$$N_{MINmix} = \sum X_i N_{MINi} \quad (8)$$

where X is the mole fraction of the particular surfactant. Rearranging Eq. (8) for a two surfactant system,

$$N_{MINmix} = (N_{MINa} - N_{MINb})X_a + N_{MINb} \; . \quad (9)$$

Therefore, if N_{MINa} is known, the y-intercept of a graph of X_a vs. N_{MINmix} will be the N_{MIN} of the unknown surfactant.

A table of N_{MIN}s permits a systematic selection of surfactants. For a typical salinity scan to encompass the middle-phase region of a system at greater than 1 wt% NaCl, one must choose a surfactant (or surfactant mixture) with an N_{MIN} less than the EACN of the oil phase. Just because this criterion is met does not guarantee Winsor type phase behavior due to complex phase behavior (e.g. liquid crystal formation). Our study on the N_{MIN}s of various surfactants concentrated on using PCE, TCE and CCl_4 as the oil phase. Single surfactants that formed middle-phase systems were used by themselves to determine the N_{MIN} value. Various combinations of MA and AY were used to validate the surfactant mixing rules (eqs. (8) and (9)). Surfactants that had to be combined with MA in various ratios to form middle phase systems had the N_{MIN} of the species determined from the N_{MIN}s of the mixes using Eq. (9). A comprehensive list of surfactants discussed here are presented in Tables 5–8.

As stated earlier, OT does not exhibit classical phase behavior with any of the DNAPLs. It only produced Winsor type II systems since it is too hydrophobic. As expected AY would be the most hydrophilic (most negative N_{MIN}) and OT the most hydrophobic (most positive N_{MIN}).

These values can be used in collaboration with the EACNs of oils to determine if a surfactant will exhibit classical phase behavior with a particular oil. For instance, OT has an N_{MIN} greater (more positive) than any of the EACNs of the chlorocarbons, therefore OT would not be expected to form classical phase behavior with any of the chlorocarbons. MA and AY have N_{MIN}s more negative than PCE, TCE, TCA, CCl_4 and 1,2-DCB, and therefore these surfactants would be expected to produce classical phase behavior with these oils at the appropriate salinity. They all do, except for 1,2-DCB, which exhibits complex phase behavior with these surfactants.

The nomenclature used to describe the alkylbenzene sulfonates in Table 6 is such that, $1\Phi C_8$ is an octane hydrophobe bound to the benzene ring at the first carbon in the chain and sulfonated in the *para* position. As expected, the phase behavior of the commercially available sodium dodecylbenzene sulfonate falls between the 1,4 and 6 isomers of dodecylbenzene sulfonate. Two expected trends are seen in the data. First, as the length of the hydrophobic part of the molecule is increased, the more positive the N_{MIN}. Second, as the branching within isomers of the same hydrophobic increases, the N_{MIN} also becomes more positive.

As stated earlier, the alkylbenzene sulfonates required prolonged equilibration times and had small solubilization parameters. These surfactants typically were minor components in the surfactant system and thus, required long extrapolations to determine N_{MIN}s. Considering all of these

Table 5 N_{MIN} values for sodium dialkylsulfosuccinates [23]

Surfactant	N_{MIN}
OT-100	16.1 ± 1.8
MA-100	−5.40 ± 0.01
AY-100	−8.66 ± 0.10

Table 6 Alkylbenzene sulfonate surfactants [23]

Surfactant	N_{MIN}
$7\Phi C_{14}$	4.07 ± 0.16
$4\Phi C_{14}$	−4.29 ± 1.34
$4\Phi C_{12}$	−5.81 ± 0.77
$6\Phi C_{12}$	−6.85 ± 0.02
$1\Phi C_{14}$	−8.66 ± 0.99
$2\Phi C_{14}$	−8.87*
DDBS†	−9.14 ± 0.40
$3\Phi C_{10}$	−9.47*
$1\Phi C_{10}$	−10.86 ± 0.48
$5\Phi C_{10}$	−11.75 ± 0.93
$1\Phi C_{12}$	−12.08 ± 0.60
$1\Phi C_8$	−16.80 ± 0.14

* Only one experimentally determined value was obtained.
† Commercially available sodium dodecylbenzensulfonate.

Table 7 EO and PO guerbet alcohol surfactants [23]

Surfactant	N_{MIN}
$C_{16}PO_{6.5}SO_4Na$	9.08 ± 1.85
$C_{16}PO_4SO_4Na$	6.83 ± 2.54
$C_{16}PO_{2.6}SO_4Na$	4.49 ± 2.16
$C_{16}EO_2SO_4Na$	2.99 ± 1.27
$C_{16}EO_0SO_4Na$	2.51 ± 0.43
$C_{16}EO_4SO_4Na$	0.32 ± 1.26
$C_{14}EO_{1.6}SO_4Na$	−2.08 ± 0.78
$C_{14}EO_{2.9}SO_4Na$	−2.96 ± 1.41
$C_{16}EO_6SO_4Na$	−3.63 ± 0.74
$C_{14}EO_4SO_4Na$	−4.37 ± 0.51
$C_{16}EO_8SO_4Na$	−6.45 ± 0.95
$C_{16}EO_{10}SO_4Na$	−11.01 ± 1.08

Table 8 New generation surfactants [23]

Surfactant	N_{MIN}
$C_{12}PO_3SO_4Na$	−3.03*
$C_8PO_6SO_4Na$	-4.80 ± 0.61
$C_8PO_3SO_4Na$	-11.07 ± 0.36

* Only one experimentally determined value.

Table 9 HLB of alkyl glucamide surfactants [28]

Surfactant	HLB
A7	13.2 ± 0.6
A9	12.6
A12G	9.3 ± 0.4

facts, we do not feel extremely confident about the values in Table 6.

Table 7 lists the EO and PO Guerbet alcohol-based surfactants discussed earlier. In general, the larger the hydrophobic part of the molecule the more positive the N_{MIN}. One notices that as more EO is added to the surfactant molecule the surfactant becomes more hydrophilic and the N_{MIN} gets smaller (more negative). Conversely, if PO is added to the molecule the surfactants becomes still more hydrophobic and the N_{MIN} gets larger (more positive).

The surfactants in Table 8 were synthesized in-house, but should be commercially viable. Larger hydrophobes and more PO lead to more positive N_{MIN}s. C_8PO_9 and $C_{12}PO_6$ were also synthesized, but produced only complex phase behavior (i.e. liquid crystal formation) in all systems studied. Therefore, no larger molecules of this type were prepared.

HLB determinations for nonionic surfactants

Since the alkylglucamide surfactants are nonionic, they are not expected to obey the above equations describing ionic surfactants. Similar to anionic systems, an equation used to predict the behavior of nonionic systems has been elucidated for alkylphenol ethoxylates [5, 6, 35]. With HLB_{MIX} as the dependent variable,

$$HLB_{MIX} = a - k(EACN_{MIX}) + f[A] + b[S] + c(T - 28\,°C) \quad (10)$$

where a, k, f, b and c are constants whose values differ depending on whether one is interested in the type I/III boundary, the optimal system, or the type III/II boundary [5].

Previous results with four different nonionic surfactants have shown that the constants a and k are independent of surfactant type as long as one is interested in only optimal phase behavior [6]. Since all systems were prepared and studied at 25 °C, the temperature term is essentially negligible and can be eliminated. Since there is no alcohol in the systems, the $f[A]$ term is zero.

The effect of salinity on the phase behavior of nonionic surfactants is much less than that with anionic surfactants, as demonstrated by the large Winsor III ranges found here as well as in earlier results [5]. Thus the salinity term can essentially be eliminated. Inserting the previously determined constants. Eq. (10) reduces to

$$HLB_{MIX} = 11.0 - 0.115\,(EACN)\,. \quad (11)$$

Only CH_2Cl_2 produced an optimum system with a single surfactant (see Table 3). Using the EACN value in Table 4 for CH_2Cl_2, one finds that the HLB for A9 is 12.6. The HLB_{MIX} of the surfactant mixes in Table 3 were then determined using Eq. (11). Once those were determined, the HLB of the unknown surfactant could be obtained using Eq. (12).

$$HLB_{MIX} = \sum_{i=1\ldots} X_i^* HLB_i^* \quad (12)$$

The HLB of a mixture of surfactants had previously been determined to behave on a mole fraction basis [6, 35]. The average values of each surfactant are listed in Table 9.

This is what one would qualitative expect. The shorter the hydrophobic part of the molecule, the larger the HLB. Previous results in a series of straight-chained hydrophobes have shown that for every additional carbon added to the chain, one expects the HLB to decrease by approximately 0.475 units [6]. The value here is 0.3 ± 0.3 units per carbon. If the A12G surfactant was a straight-chained species, one would expect it to have a HLB value of approximately 11.5 (based on 0.3 HLB units per carbon). But, previous results have shown that going from a straight chain to branching at the fourth carbon, results in the HLB being lowered by another 2 units [36]. Therefore, the A12G is expected to have an HLB of approximately 9.5, which is within the experimental error of the obtained value.

Applications

We have completed a series of calculations to see if the addition of surfactants to the pump and treat method will indeed lessen the time required for remediation of a chlorocarbon spill [22]. Johnson and Pankow [37] found that it took 134.4 yr to clean up a 10 m spill of TCE with a groundwater velocity of 0.5 m/d. If a 2 wt% solution of MA was added to this spill, it would take 0.97 yr to clean up the same spill! Of course this time can be shortened even further by increasing the concentration (a 5 wt% solution would require 0.38 yr for remediation) or using a better surfactant system (a 2 wt% solution of a 50/50 mix of $MA/C_{16}GA(EO)_{10}SO_4Na$ would require 0.20 yr for remediation).

Recently a field test using the dihexylsulfosuccinate, MA, was used incorporating the principles earlier discussed in this manuscript to a contaminated site at Hill Air Force Base, Utah. The contaminant was primarily TCE, and $2\frac{1}{2}$ pore volumes of a 7 wt% surfactant solution was injected. When this solution was extracted, 99% of the contaminant was removed. When an additional 10 pore volumes of water was injected, 96% of the surfactant was removed. This represents by far the best recovery efficiency ever reported to the present time.

Conclusions

The results presented here have shown that we can successfully predict chlorocarbon behavior with the mathematical expressions first introduced to describe the interaction of surfactants with hydrocarbons. The surfactants must be somewhat more hydrophilic, due to the increased polarity of the chlorocarbons, compared to hydrocarbons, used as the oil phase.

These results show that the use of surfactants as part of an aquifer remediation program has great promise. Further work is being conducted to find better and cheaper surfactants to accomplish remediation of toxic spills.

Acknowledgments The authors express their appreciation to Dr. Larry Britton of Condea Vista Chemical Company for his help in biodegradation studies, and INTERA Inc. for conducting the field test mentioned above. We further express our thanks to Ingegärd Johannson of AKZO Nobel for providing a series of polyglucoside surfactants.

This work was partially financed by the Texas Higher Education Coordinating Board of the State of Texas.

References

1. Cash RL, Cayias, JL, Fournier Okasis, Jacobson JK, LeGear CA, Schares T, Schechter RS, Wade WH (1979) Soc Pet Eng J 17:22
2. Salager JL, Vasquez E, Morgan JC, Schechter RS, Wade WH (1979) Soc Pet Eng J 19:107
3. Vasquez E, Salager JL, El-Emary M, Koukounis C, Schechter RS, Wade WH (1979) Solution Chem Surfactants 2:801
4. Salager JL, Bourrel M, Schechter RS, Wade WH (1979) Soc Pet Eng J 19:271
5. Graciaa A, Schechter RS, Wade WH, Yiv SH, Barakat Y (1982) J Colloid Interface Sci 89:217
6. Graciaa A, Fortney L, Schechter RS, Wade WH, Yiv SH (1982) Soc Pet Eng J 22:743
7. Barakat Y, Fortney LN, Schechter RS, Wade WH, Yiv SH (1983) J Colloid Interface Sci 92:561
8. Abe M, Schechter D, Schechter RS, Wade WH, Weerasooriya U, Yiv SH (1986) J Colloid Inteface Sci 114:343
9. Lalanne-Cassou C, Carmona I, Fortney L, Samii A, Schechter RS, Wade WH, Weerasooriya U, Weerasooriya V, Yiv SH (1987) J Disp Sci Technol 8:137
10. MacKay DM, Cherry J (1989) Environ Sci Technol 23:630
11. Melrose JC, Brandner CF (1974) J Can Pet Technol 13:54
12. Morrow NR, Songkran B (1981) In: Shah DO (Ed) Surface Phenomena in Enhanced Oil Recovery. Plenum Press, New York, p 387
13. A comprehensive review studies as applied to hydrocarbon systems can be found in: Bourrell M, Schechter RS (1988) Microemulsions and Related Systems. Marcel Dekker, New York
14. West CC, Harwell JH (1992) Environ Sci Technol 26:2324
15. Winsor PA (1954) Solvent Properties of Amphiphillic Compounds. Butterworths, London
16. Barakat Y, Fortney LN, LaLanne-Cassou C, Schechter RS, Wade WH, Weerasooriya U, Yiv SH (1983) Soc Pet Eng J 913
17. Huh C (1979) J Colloid Interface Sci 71:408
18. Israelachvili J (1986) In: Mittal KL, Bothorel P (Eds) Surfactants in Solution. Plenum Press, New York, p 3
19. Taber JJ (1981) In: Dinesh Shah (Ed) Surface Phenomena in Enhanced Oil Recovery. Plenum Press, New York, p 13
20. Baran Jr JR, Pope GA, Wade WH, Weerasooriya V (1994) Langmuir 10:1146
21. Baran Jr JR, Pope GA, Wade WH, Weerasooriya V, Yapa A (1994) Environ Sci Technol 28:1361
22. Baran Jr JR, Pope GA, Schultz C, Wade WH, Weerasooriya V, Yapa A (1996) In: Chattopadhyay AK, Mittal KL (Eds) Surfactants in Solution. Marcel Dekker, New York, p 393
23. Baran Jr JR, Pope GA, Schultz C, Wade WH, Weerasooriya V (1996) In: Tedder DW, Pohland FG (Eds) Emerging Technologies in Hazardous Waste Management VI. American Chemical Society, Washington DC, p 111
24 Sunwoo C, Wade WH (1992) J Disp Sci Technol 13:491
25. Anton RE, Graciaa A, Lachaise J, Salager JL (1995) J Disp Sci Technol 13:565
26. Baran Jr JR, Pope GA, Wade WH, unpublished results
27. Arenas E, Baran Jr JR, Pope GA, Wade WH, Weerasooriya V (1996) Langmuir 12:588
28. Baran Jr JR, Pope GA, Wade WH, Weerasooriya V (1996) Environ Sci Technol 30:2143
29. Connor DS, Scheibel JJ, Kao JN. US Patent #5,338,487
30. The notation used in this table has the A referring to the surfactant being a glucamide followed by a number, which represents the number of carbons in the acid that is combined with the glucamine to prepare the glucamide surfactant. The G denotes the Guerbet species.

31. Cash RL, Cayias JL, Fournier GR, McAllister D, Schares T, Schechter RS, Wade WH (1977) J Colloid Interface Sci 59:39
32. Cayias JL, Schechter RS, Wade WH (1976) Soc Pet Eng J 16:351
33. Griffin WC (1949) J Soc Cosmet Chem 1:311
34. Baran Jr JR, Pope GA, Wade WH, Weerasooriya V, Yapa A (1994) J Colloid Interface Sci 168:67
35. Bourrel M, Salager JL, Schechter RS, Wade WH (1980) J Colloid Interface Sci 75:451
36. Graciaa A, Barakat Y, El-Emary M, Fortney L, Schechter RS, Yiv SH, Wade WH (1982) J Colloid Interface Sci 89:209
37. Johnson RL, Pankow JF (1992) Environ Sci Technol 26:896

Progr Colloid Polym Sci (1998) 109:85–92

K. Ohbu
M. Fujiwara
Y. Abe

Physicochemical properties of α-sulfonated fatty acid esters

Received: 26 August 1997
Accepted: 5 September 1997

Kazuo Ohbu (✉) · M. Fujiwara · Y. Abe
Material Science Research Center
Lion Corporation
Hirai 7-13-12
Edogawa-ku, Tokyo 132
Japan

Abstract The physicochemical properties and surface activities of α-sulfonated fatty acid esters (SFE) were reviewed mainly based on the recent investigation of the authors. The relationship between the behavior in solution and molecular structures were examined. Four solid phases with different hydration were detected in the phase diagram of C_{16} SF methyl ester/water system. The favorable characteristics of SFE, such as hard water tolerance and low Krafft point, were discussed with regard to their crystal structures.

Key words α-Sulfo fatty acid esters – surface activity – water hardness tolerance – Kraft point – phase diagram – crystal structure

Introduction

The application of α-sulfonated fatty acid esters, especially the methyl ester, has attracted considerable attention, mainly due to the recent demand for renewable sources for the hydrophobic part of surfactants in addition to their characteristic behavior in hard water. General formula of α sulfonated fatty acid esters, also called α-ester sulfonates, abbreviated SFE, is R–CH(SO_3M)–COO–R′ (where R = long alkyl, M = alkali counter ion, R′ = alkyl, polyoxyalkylene or polyol). It is noteworthy that SFEs obtained by the ordinary sulfonation reaction with SO_3 are regarded as racemic mixtures since their structures have chirality on the α-carbon atom.

Fatty acid esters used for producing α-sulfonated fatty acid esters are readily obtained by transesterification of the corresponding triglycerides with alcohols. Although investigations of the sulfonation reaction of fatty acids and their esters have been intensively carried out since the 1930s, the large scale production and application of α-sulfonated fatty acid esters began very recently, as technological advances concerning the reduction of unfavorable byproducts, e.g., disalts of the α-sulfonated fatty acids, which show poor surface activities.

An extensive review covering a wide range of α-sulfonated fatty acid esters, such as their synthesis, physicochemical properties, and applications, was recently made by Schwuger and Lewandowski [1]. In this article, characteristic properties and fundamental surface activities of α-sulfonated fatty acid methyl and longer alkyl esters and some other esters in aqueous solutions and solid forms will be reviewed mainly on the basis of our recent investigations from the viewpoint of their molecular aggregation states including crystal structures.

Chemical structures of the sulfonation products of fatty acid methyl esters

The sulfonation of sufficiently hydrogenated fatty acid methyl esters with sulfur trioxide proceeds through several intermediate reactions [2–4]. The neutralized products of the sulfonation reaction are racemic mixtures since they have chirality on their α-carbon as shown in Fig. 1. In this article, each SFE is abbreviated as shown in Fig. 1, so that

$$C_{m-2}H_{2(m-2)+1}CH{-}COOC_nH_{2n+1}$$
$$|$$
$$SO_3Na$$

Fig. 1 Molecular structures of α-sulfonated fatty acid esters (SFE) abbriviated as C_m–C_n SF

C_m and C_n mean a fatty acid residue and alkyl residue of the ester substituent, respectively. The α-sulfonated fatty acid ester salt abbreviated as $C_{m=14}$–$C_{n=1}$ SF Na, for instance, means an SFE sodium salt derived from the myristic acid methyl ester.

Solubilities, c.m.c. and Krafft points

Effects of hydrophobic chain length and structures of ester substituents in SFE Na salts

The temperature dependence of solubilities for a series of C_m–SFE sodium salts is similar to that of other conventional anionic surfactants, such as alkyl sulfates. Solubility curves of SFE with various carbon chain lengths both in the fatty acid residue and in the alkyl ester residue are shown in Fig. 2 [5]. The longer the acyl chain length, m, of the SFE-methyl ester, the solubility curves move to a higher temperature region. On the contrary, equilibrium solubilities of SFE with $m = 14$ (derived from myristic acid) show a maximum in $C_{n=3}$ (there is no plot for $C_{m=14}$ $C_{n=3}$ in Fig. 2 since they are readily soluble in water even in the vicinity of 0 °C). When C_m exceeds 3, however, the solubility curves move again to a higher temperature region. Krafft points and the c.m.c. of C_m–C_n SF are listed in Table 1. The existence of an optimum solubility may be due to the difference in molecular alignments in the hydrated solid states of SFE. The n-propyl group as the ester substituent of SFE seems to most significantly contribute to the excellent solubility by its sterically restraining effect.

The logarithmic plots of the c.m.c. of sodium salts of C_m–$C_{n=1}$ SF and C_{14}–C_n SF decrease linearly with the increase in C_m as shown in Fig. 3. The cohesive energy change to transfer a methylene group of a surfactant molecule from a hydrophobic environment to an aqueous medium, ω, is calculated from the slope of the plots of ln c.m.c. ~ C_m and the degree of counter ion dissociation from micelles. For alkyl sulfates, ω is reported to be $1.1\kappa T$ [6]. As shown in Fig. 3, the cohesive energy change for C_m–C_1 SF is substantially equal to ω for alkyl sulfates. However, ω for C_{14}–C_n SF is half the value of C_m–C_1 SF.

It is well known that the increase in the unit number of the polyoxyethylene group inserted between the alkyl and hydrophilic groups in alkylpolyoxyethylene ether sulfate would cause a decrease in the c.m.c., while it increases the surface tension and the dissociation degree of the micelle [7]. SFE with polyethyleneglycol esters, however, increase the c.m.c. regardless of their hydrophobic chain length, while keeping the Krafft points below 0 °C [8]. A typical relationship between the c.m.c. and the oxyethylene unit number of C_{14}–$(C_2H_4O)_n$ SF is shown in Fig. 4. The role of the polyoxyethylene group in SFE-polyethyleneglycol esters is different from that in alkylpolyoxyethylene ether sulfates.

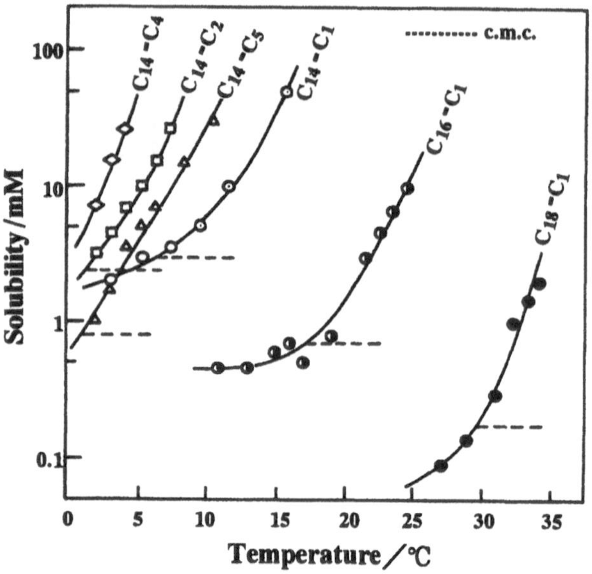

Fig. 2 Solubility curves for various C_m–C_n SF as a function of temperature

Table 1 Krafft points and c.m.c. of C_m–C_n SF

C_m–C_n SF	Krafft point [°C]	c.m.c. [mM]
C_{18}–C_1	30	0.18[a]
C_{16}–C_1	17	0.73[b]
C_{14}–C_1	6	2.80[c]
C_{14}–C_2	1	2.25[d]
C_{14}–C_3	<0	1.35[d]
C_{14}–C_4	<0	0.90[d]
C_{14}–C_5	1	0.75[d]

[a] 33 °C.
[b] 23 °C.
[c] 13 °C.
[d] 25 °C.

Surface activities and properties of micelles

Some results of the surface activity measurements are listed in Table 2. C_{16}–C_1 SF shows a favorable depression

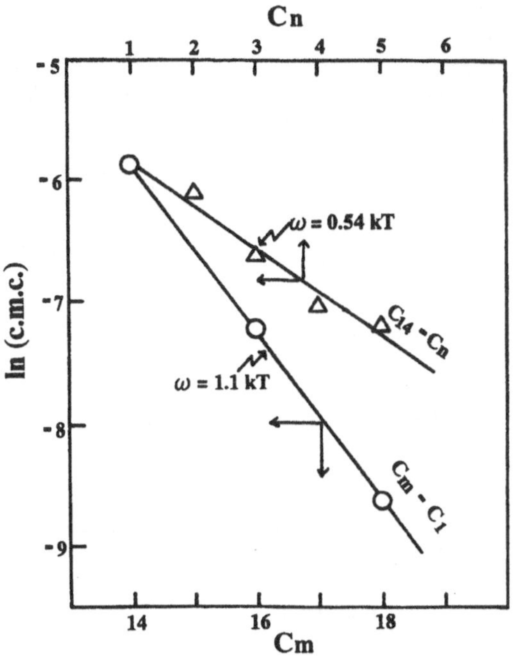

Fig. 3 Logarithmic plots of c.m.c. vs. number of carbon atoms for C_m–C_1 SF and C_{14}–C_n SF

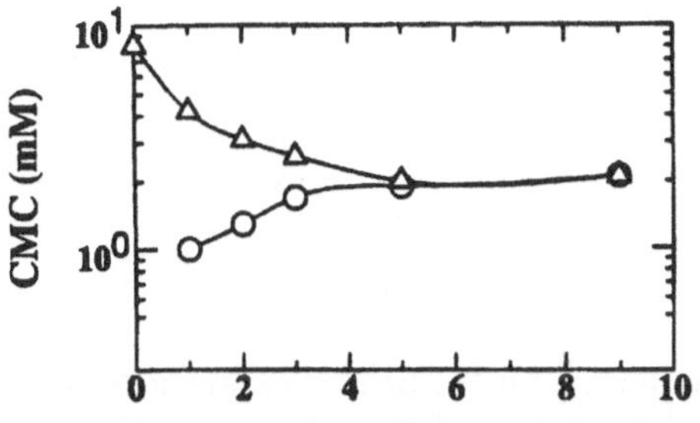

Fig. 4 Comparison of c.m.c. value vs. oxyethylene unit number relationships in C_{14}–SF-polyethyleneglycol monoester and dodecylpolyoxyethyleneether sulfate [8] ○; C_{14}–$(C_2H_4O)_n$ SF Na, △: $C_{12}H_{25}O(C_2H_4O)_nSO_3Na$

of the interfacial tension to less than 10 mN/m and solubilizes a large quantity of oil-soluble dye. Regarding the penetration capacity, surfactants with the hydrophilic group in the middle part of their molecular structure have a favolable performance. Wetting time decreases as the alkyl chain length of the ester substituents become longer. The relationship of various combinations of C_m and C_n in SFE with several surface activities were examined in detail [9]. The molecular structure of SFE with a similar carbon chain for both a hydrophobic group and an ester substituent, such as C_{10}–C_8 SF and C_{12}–C_{10} SF, showed a very short wetting time of less than 2 s.

The solubilization capacity of SFE for oleic acid is much higher than LAS [10], as shown in Table 3. Comparing the solubilization limits of $C_{14/16}$ SF Na with commercially available LAS Na, the value of SFE is larger by 1.6–1.9 times than LAS. Static light scattering measurements revealed that the solubilized amount of oleic acid in SFE micelles is more than twice that in LAS micelles. This result would be interpreted by a temporary decrease in the micellar aggregation number and the subsequent occurrence of intermicellar aggregation due to the incorporation of the bulky oleic acid molecules.

Solubilities of SFE Ca salts

The equilibrium solubility curves of the sodium and calcium salts of C_m–C_1 SFE are shown in Fig. 5 [5]. The Krafft point and c.m.c. of each SFE are listed in Table 4. The Krafft points of the SFE calcium salts are considerably lower than those of the alkyl sulfate calcium salts with a same carbon number. It can be postulated that low Krafft points are attributable to the relatively labile structures of the hydrated solid phase.

Since there is a good linearity for both SFE Na and SFE1/2Ca as shown in Fig. 6, the micelle formation of SFE Na and SFE1/2Ca is considered to obey the following

Table 2 Surface activities of C_m–C_n SF Na

α-SF	Wetting[a) [s]	Interfacial tension[b) [mN/m]	Solubilizing[c) [g Y.OB/mole-SF]
No surfactant	∞	27	0
C_{18}–C_1	—	9	—
C_{16}–C_1	35	7	4.88
C_{14}–C_1	34	16	3.60
C_{14}–C_2	21	15	3.53
C_{14}–C_3	12	13	3.50
C_{14}–C_4	7.3	10	4.03
C_{14}–C_5	5.4	—	4.26

[a)] Conc. of SFE: 10 mM, 10 × 10 × 2 mm wool felt, 25 °C.
[b)] Conc. of SFE: 300 ppm, edible oil, pendant drop method, 25 °C.
[c)] Solubilizate: Yellow–OB, 25 °C.

Table 3 Solubilization limit of oleic acid in $C_{14/16}$–C_1 SF Na and LAS Na solutions at 30 °C

Surfactant	Concentration [mM]	Solubilization limit	
		Max. concentration of oleic acid [mM]	[oleic acid]/[micellar surfactant][c] [mole/mole]
$C_{14/16}$ SF[a]	10	2.80	0.31
	15	4.26	0.30
	20	5.32	0.28
LASA[b]	10	1.60	0.19
	15	2.30	0.17
	20	2.84	0.15

[a] C_{14}/C_{16}; 3/7, c.m.c.: 0.86 mM.
[b] c.m.c.: 1.35 mM.
[c] [micellar surfactant] = [total surfactant] – [surfactat at c.m.c.]

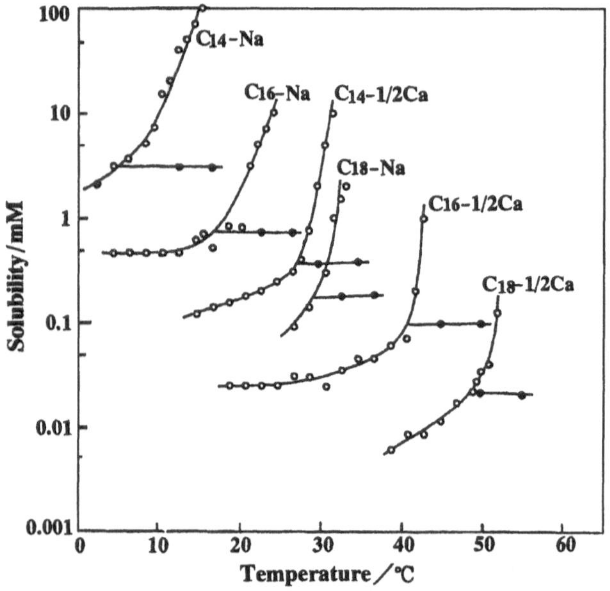

Fig. 5 Solubility curves for C_m–C_1 SF Na and C_m–C_1 SF 1/2Ca as a function of temperature [5]

equation [5, 13]:

$$\ln(\text{c.mc.}) = m\,\omega/kT + K_g/Z_i\{\ln(2000\,\pi\sigma^2/DNkT) - \ln C_g\} + \text{const}\,. \quad (1)$$

In Eq. (1) K_g signifies the degree of counter ion binding to the micelle. The K_g values of C_{14}–C_1 SF Na and C_{14}–C_1 SF1/2Ca were estimated to be 0.61 and 0.70, respectively [5]. Meanwhile, the counterion binding of the C_{14}–C_1 SF Na micelle was calculated to be 0.70–0.72 in the temperature range of 20–50 °C from the thermodynamic data of micelle formation [14]. These values are very close to that of the C_{12} alkyl sulfate Na, i.e., 0.66. The dissociation

Table 4 Krafft points and c.m.c. of SFE sodium and calcium salts as compared with alkyl sulfates

Surfactant and counter ion	Krafft point [°C]	c.m.c. [mM]	Ref.
C_mC_1 SF			
C_{14} 1/2Ca	28	0.66	
C_{16} 1/2Ca	41	0.19	
C_{18} 1/2Ca	49	0.042	
C_m AS			
C_{12} Na	9	8.1	[11]
C_{14} Na	30	2.1	[11]
C_{12} 1/2Ca	50	1.2	[12]
C_{14} 1/2Ca	71	0.68	[12]

Refonding C_m–C_1 SF Na, see Table 1.

degrees of the counter ion for the SFE Na and Ca salts are substantially equal to that for the alkyl sulfate.

Since the plots of ln(c.m.c.) vs. C_m for SFE Na and AS Na overlap each other as shown in Fig. 6, the effective hydrophobic chain length of SFE corresponds to that of the alkyl sulfates.

Precipitation phase boundary of SFE Na in the presence of calcium ion

Solubility in hard water is an important indicator of each surfactant for various practical applications. As clarified in Fig. 5 and Table 4, SFE1/2Ca has a poor solubility in water under equilibrium condition. In practice, however, SFE Na is generally regarded as a surfactant having a very low sensitivity to hard water, and suited for detergent formulations applicable to cold water washing. Time dependent changes of precipitation phase boundary diagrams were examined in order to combine the above two results [5].

Figure 7 is the precipitation phase boundary diagrams of SFE Na and alkyl sulfate Na with time. The precipitation boundary for the alkyl sulfate at 10 min is very close to the boundary at equilibrium. The greater part of the alkyl sulfate precipitates within 10 min in the water hardness region of more than 3° DH. On the other hand, C_{14}–C_1 SF do no actually precipitate for a few days in the same water hardness region, although the equilibrium precipitation region spreads over the range of 3–10° DH. Even C_{16}–C_1 SF can be kept in a dissolved state for some 10 min in the ordinary water hardness region.

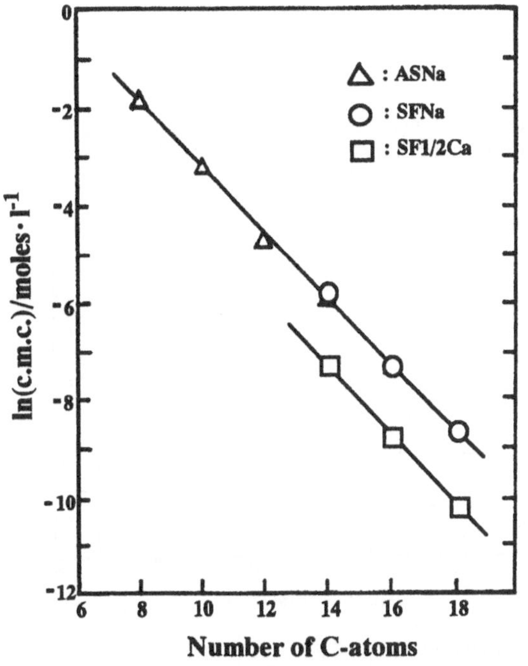

Fig. 6 Logarithmic plots of c.m.c. vs. number of carbon atom (m) of C_m–C_1SF Na, C_m–C_1 SF 1/2Ca, and C_m alkyl sulfate Na [5]. ○; $C_{m-2}H_{2(m-2)+1}CH(SO_3Na)$–$COOCH_3$, □; C_{m-2}–$CH_{2(m-2)+1}CH(SO_31/2Ca)$–$COOCH_3$, △: $C_mH_{2m+1}SO_4Na$

Phase diagram of a SFE sodium salt/water system

Phase diagrams and related data, particularly temperature and water content ranges where each crystalline or liquid crystalline phase exists and physicochemical and mechanical properties of each phase, are the fundamental information indispensable in handling every kind of surfactants appropriately. Especially, the properties of surfactants in crystal phases, such as hydration, phase transition, hygroscopy, are important factors for the application to granular detergents. The behaviour of C_{16}–C_1 SF Na derived from the palmitic acid were investigated by means of powder X-ray diffraction and a DSC [15].

The DSC curves for the C_{14}–C_1 SF Na with water contents of 0–8.9% are shown in Fig. 8. Two separated peaks were detected simultaneously in the long spacing region of the X-ray diffraction in the water content range from 1.6 to 5.6%. A single peak was obtained in the powder with a water content of 8.9% (corresponds to C_{16}–C_1 SF Na·$2H_2O$). Similarly, C_{16}–C_1 SF Na·$5H_2O$ (water content: 19.5%) and C_{16}–C_1 SF Na·$10H_2O$ (32.6%) were detected. According to the results from the DSC and X-ray diffractometry, SFE Na with an intermediate water content between every two crystal phases seems to be an eutectic mixture. In an eutectic mixture consisting of two components, the phase transition

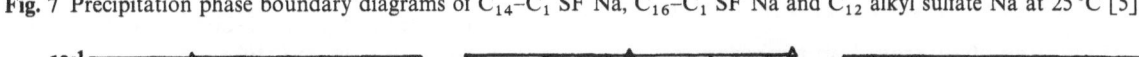

Fig. 7 Precipitation phase boundary diagrams of C_{14}–C_1 SF Na, C_{16}–C_1 SF Na and C_{12} alkyl sulfate Na at 25 °C [5]

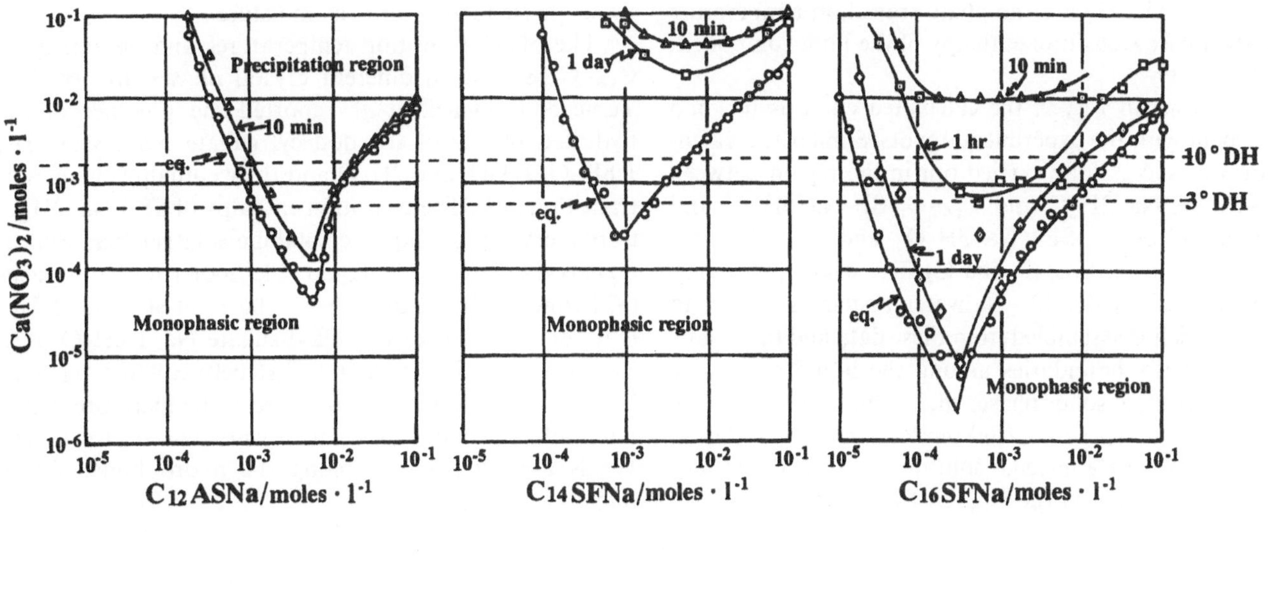

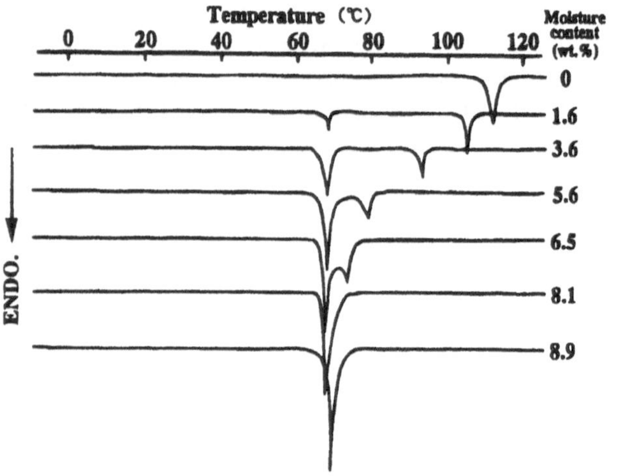

Fig. 8 Differential scanning calorimetry curves for C_{16}–C_1 SF Na with moisture contents of 0–8.9% [15]

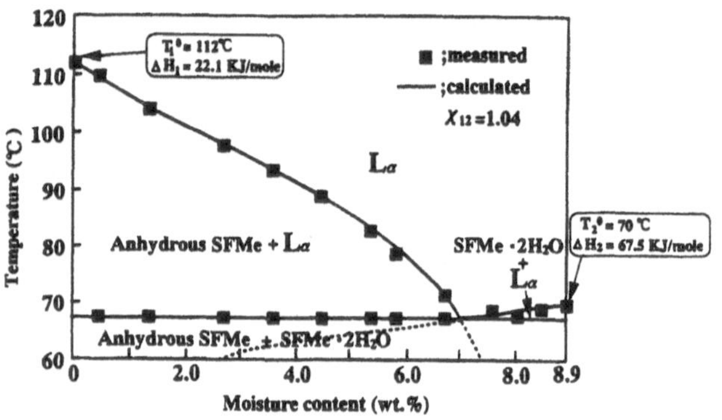

Fig. 9 Crystal–liquid crystalline transition temperature (T_c) for C_{16}–C_1 SF Na with moisture contents of 0–8.9%. Theoretical curves were calculated by equations (2) and (3) [15]

temperature of component 1 (T_1) is expressed by the Flory–Huggins–Scott equation [16] as follows:

$$1/T_1 - 1/T_1^0 = -(R/\Delta H_1)(V_1/V_2)(\ln\Phi_1 + \chi_{12}\Phi_2^2)\,, \qquad (2)$$

where T_1^0 and ΔH_1 are the phase transition temperature and the phase transition enthalpy of the pure component 1, respectively, V_1, V_2, Φ_1, and Φ_2 are the partial molar volumes and the volume fractions of components 1 and 2, respectively. A calculation was carried out on the assumptions that V_1 approximately equals V_2, and that Φ_1 and Φ_2 are approximately equal to the weight fractions of component 1 and 2. χ_{12} is the interaction parameter between components 1 and 2, and is successively adjusted so that the calculated curve has a good fit with the experimentally obtained plots. R is the gas constant. Similarly, the phase transition temperature and the phase transition enthalpy of component 2 are expressed by

$$1/T_2 - 1/T_2^0 = -(R/\Delta H_2)(V_2/V_1)(\ln\Phi_2 + \chi_{12}\Phi_1^2)\,, \qquad (3)$$

where T_2 and ΔH_2 are the phase transition temperature and the phase transition enthalpy of the pure component 2, respectively.

As shown in Fig. 9, the calculated curve is in good agreement with the experimental plots. Similarly, a calculation was successfully carried out in the region between C_{16}–C_1 SF Na·$2H_2O$ and C_{16}–C_1 SF Na·$5H_2O$, and between C_{16}–C_1 SF Na·$5H_2O$ and C_{16}–C_1 SF Na·$10H_2O$ using Eqs. (2) and (3). The whole phase diagram of the C_{16}–C_1 SF Na/water system is shown in Fig. 10 which is assembled from these data and the results relating to the boundaries among the liquid crystalline phase, hydrated solid phase, and dilute solutions [5]. There are four types of crystal phases, two types of liquid crystalline phases, a micellar solution phase, and a monomer solution phase in the C_{16}–C_1 SF Na/water system.

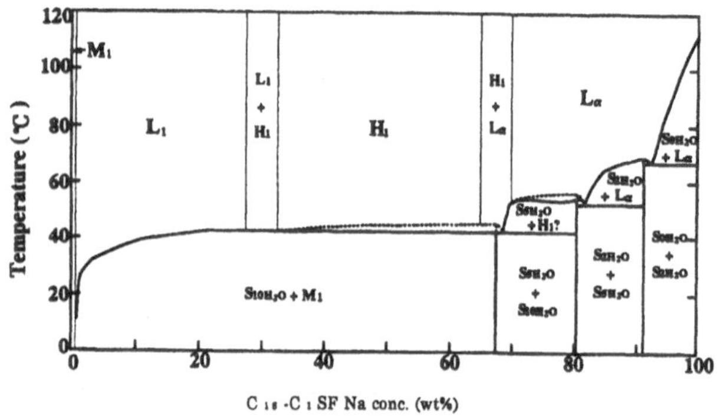

Fig. 10 Phase diagram of C_{16}–C_1 SF Na–water system [15]. S_{0H_2O}; anhydrous SFE crystal, S_{2H_2O}; SFE·$2H_2O$ crystal S_{5H_2O}; SFE·$5H_2O$ crystal, S_{10H_2O}; SFE·$10H_2O$ crystal, L_α; lamellar liquid crystalline, H_1; hexagonal liquid crystal, L_1; micellar solution, M_1; monomer solution. The c.m.c. line is schematic

The most stable solid phase is C_{16}–C_1 SF Na·$2H_2O$, which retains crystalline water under the condition of up to 50 °C and humidity range of 10–80% RH. The anhydrous phase readily absorbs moisture.

The phase transition temperatures and enthalpies of C_{16}–C_1 SF Na in different crystal phases are listed in Table 5. Kékicheff et al. reported the presence of the hydrated phases of the dodecyl sulfate Na crystal, i.e., $1/8H_2O$, $1H_2O$, and $2H_2O$ and their enthalpy changes of transition [17]. The transition enthalpy of SFE Na·$2H_2O$ from a crystal to a liquid crystalline is larger than that of C_{12} alkyl sulfate Na·$1/8H_2O$ (12 kJ/mole) or palmitic acid (53 kJ/mole). On the other hand, the T_c of SFE Na·$2H_2O$ is lower than that of C_{12} alkyl sulfate Na·$1/8H_2O$. It is assumed that SFE has a crystal structure, which tends to become unstable with an increase in temperature, while intermolecular interactions mainly attributable to van der Waals attractive force among the hydrophobic chains remain rather firmly.

Crystal structures of SFE connected with the physicochemical properties

In order to understand the fundamental nature of SFE and the characteristic behavior, such as the low Krafft points, the hard water tolerance, and the large heat of transition, the molecular arrangement and the intermolecular bonding in the crystal structures of surfactants determined by X-ray analysis are some of the most indispensable information that are required.

Arrangements of surfactant molecules are strictly influenced by the effective size of the hydrophilic group including bound water. Surfactant molecules with a bulky hydrophilic group tend to form crystal structures in which their hydrophobic chains interdigitate with each other. On the other hand, surfactant molecules, in which the effective cross section of the hydrophobic chain equals the hydrophilic group, tend to form a stacked double layer structure [18]. According to previously reported crystal

Table 5 Phase transition temperature (T_c) and enthalpy changes (ΔH) of C_{16}–C_1 SF Na solids

Phase	T_c [°C]	ΔH [kJ/mole]
C_{16}–C_1 SF Na (anhydrous)	112	22.1
C_{16}–C_1 SF Na·$2H_2O$	70	67.5
C_{16}–C_1 SF Na·$5H_2O$	53	39.4
C_{16}–C_1 SF Na·$10H_2O$	42	46.1

Fig. 11 Perspective views of the SFE crystal structures

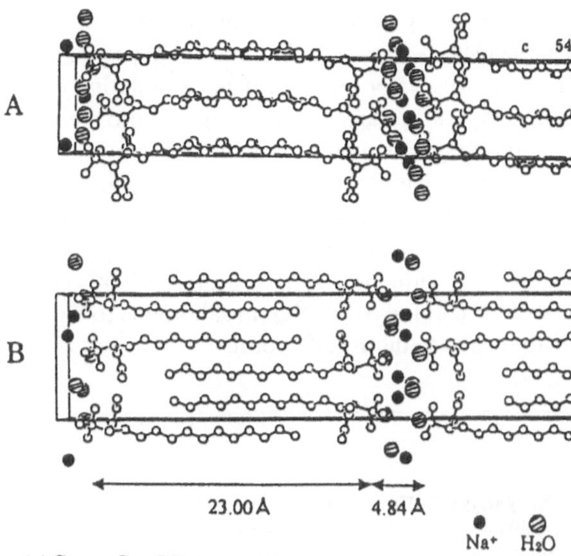

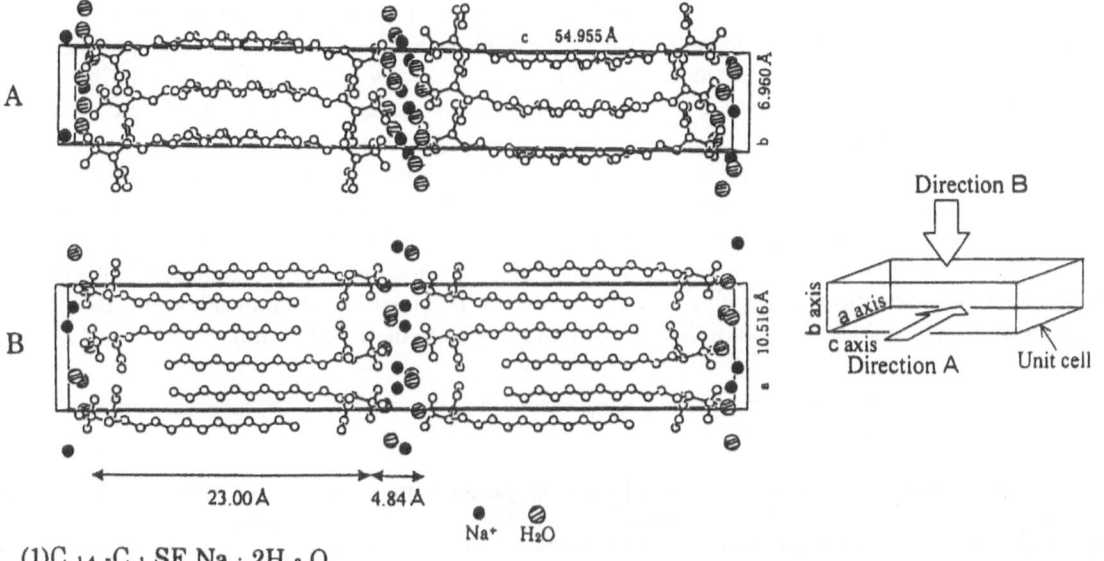

(1)C 14 -C 1 SF Na · 2H 2 O

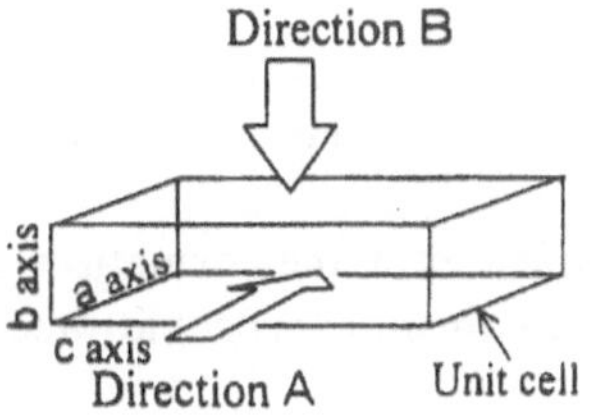

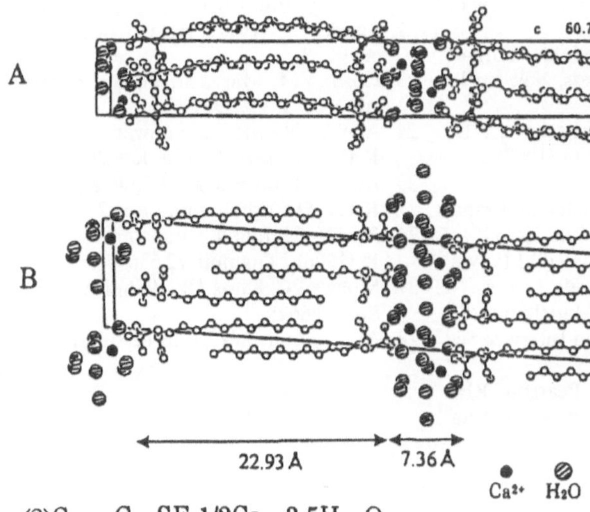

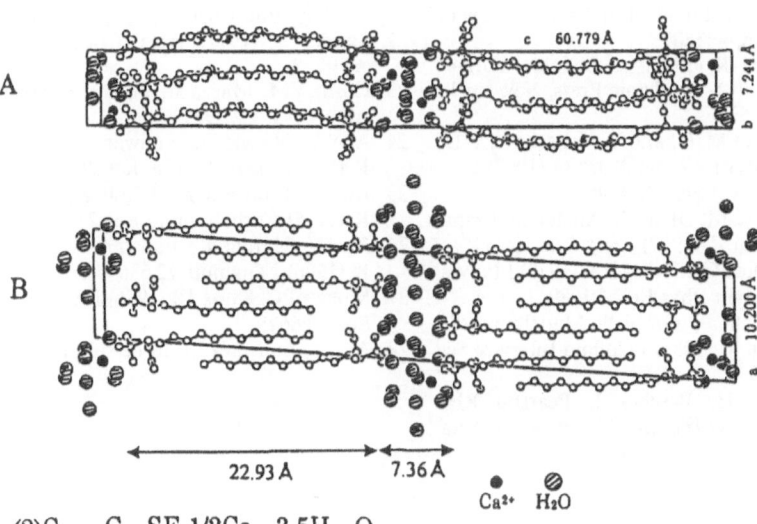

(2)C 14 -C 1 SF 1/2Ca · 3.5H 2 O

analysis data, ionic surfactants, such as fatty acid [19], fatty potassium salt [20], distearyldimethylammonium chloride [21] and alkyl sulfate [22, 23], form a stacked double layer structure, in which the surfactant molecule ends face each other. Sugar based surfactants, such as methylacylglucosides and alkyl glucosides, form interdigitated molecular arrangement [24–27].

The crystal structures of C_{14}–C_1 SF Na · $2H_2O$, C_{14}–C_1 SF 1/2Ca · $3.5H_2O$ and some other homologs were ascertained by a single crystal X-ray analysis [28]. The perspective views of the molecular arrangement are shown in Fig. 11 (crystal system: orthorhombic and monoclinic, respectively).

The hydrocarbon chains in both the sodium salt and calcium salt are in the interdigitated form, where a bimolecular pair consisting of R- and S-SFEs becomes a fundamental unit.

A cross section of a surfactant molecule can be estimated from the crystallographic data. The cross sections of the hydrocarbon chains in SFE Na salt · $2H_2O$ and SFE 1/2Ca · $3.5H_2O$ are 18.9 and 18.5 $Å^2$, respectively. The cross section of dodecyl sulfate Na · $1/8H_2O$ is known to be 19.8 $Å^2$. The large difference in the heat of transition mentioned above may be attributable to these distances.

It can be presumed that the formation of the fairly stable crystal including the indispensable integration of the R-, S-molecular pairs acts as a kinetic barrier to retard the nucleation and nuclear growth of SFE. The extremely slow precipitation of SFE calcium salts shown in Fig. 7 would be explained by the incorporation of the above concept.

Table 6 Coordination numbers of counter ions in SFE crystals

Counter ion	SO_3	H_2O	Total
Na^+	2	4	6
Ca^{2+}	1	6	7

The coordination numbers of counter ions were elucidated from the crystallographic data. The coordination numbers of sodium and calcium ion in each SFE crystal are shown in Table 6. The electrostatic potential of an ionic bond generally decreases in the presence of bound water molecules. As for the Na ion, the predominant coordination to water may cause a relatively low transition temperature as shown in Fig. 10 and Table 5.

From the same viewpoint, calcium ion in the SFE crystal is supposed to be labile, since the calcium ion is surrounded by six coordination bonds to water molecules and has only one coordination bond with the oxygen atom of the sulfonate residue in the SFE molecule. This coordination structure of the SFE calcium salt seems to work as a factor that lowers the Krafft point.

References

1. Schwuger MJ, Lewandowski H (1995) In: Stache HW (ed) Anionic surfactants: Organic Chemistry, Surfactant Science Series, Vol 56, Ch 8. Marcel Dekker, Basle, pp 461–499
2. Nagayama M, Okumura O, Sakatani T, Hashimoto S, Noda S (1975) Yukagaku J Jpn Oil Chem Soc 24:395
3. Schmid K, Bauman H, Stein W, Dolhaine H (1984) Proc 1st World Surfactants Congr (CESIO), Vol II, Gelnhausen K, p 105
4. Inagaki T (1991) Proc World Conference on Oil chemicals, p 269, Kualalumpur, Am Oil Chem Soc
5. Fujiwara M, Miyake M, Abe Y (1993) Colloid Polym Sci 271:780
6. Moroi Y, Sugii R, Akine C, Matsuura R (1985) J Colloid Interface Sci 108:180
7. Lange H, Schwuger MJ (1980) Colloid Polym Sci 258:1264
8. Okano T, Egawa N, Fujiwara M, Fukuda M (1996) J Am Oil Chem Soc 73:31
9. Okano T, Tanabe J, Fukuda M, Tanaka M (1992) J Am Oil Chem Soc 69:44
10. Fujiwara M, Kaneko Y, Ohbu K (1995) Colloid Polym Sci 273:1055
11. Shinoda K, Maekawa M, Shibata Y (1986) J Phys Chem 90:1228
12. Hato M, Shinoda K (1973) Bull Chem Soc Jpn 46:3889
13. Shinoda K (1963) In: Colloidal Surfactants, Ch 1, Academic Press, New York, pp 55–57
14. Fujiwara M, Okano T, Nakashima T-H, Nakamura AA, Sugihara G (1997) Colloid Polym Sci 275:474
15. Fujiwara M, Okano T, Amano H, Asano H, Ohbu K (1997) Langmuir 13:3345
16. Wittman JC, Manley RSt (1977) J Polym Sci, Polym Phys Ed 15:1089
17. Kékicheff P, Grabielle-Madelmont C, Ollivon M (1989) J Colloid Interface Sci 131:112
18. Hauser H, Pascher I, Pearson RH, Sundel S (1981) Biochim Biophys Acta 650:21
19. Vand V, Morley WM, Lomer TR (1951) Acta Crystallogr 4:324
20. Lewis ELV, Lomer TR (1969) Acta Crystallogr B27:702
21. Okuyama K, Soboi Y, Iijima N, Hirabayashi K, Kunitake T, Kajiyama T (1988) Bull Chem Soc Japan 61:1485
22. Sundel S (1977) Acta Chem Scand A31:799
23. Coiro VM, Mazza F, Pochtti G (1986) Acta Cryst C42:991
24. Abe Y, Harata K, Fujiwara M, Ohbu K (1995) Carbohydrate Res 269:43
25. Abe Y, Fujiwara M, Ohbu K, Harata K (1995) Carbohydrate Res 275:9
26. Abe Y, Harata K, Fujiwara M, Ohbu K (1996) Langmuir 12:636
27. Moews PC, Knox JR (1976) J Am Chem Soc 98:6628
28. Abe Y, Fujiwara M, to be submitted

Progr Colloid Polym Sci (1998) 109:93–100

S.E. Friberg
L. Fei

Vapor pressures of phenethyl alcohol and phenethyl acetate in aqueous solutions of sodium xylene sulfonate and polyvinylpyrrolidone

Received: 2 July 1997
Accepted: 3 July 1997

Prof. Dr. S.E. Friberg (✉) · L. Fei
Department of Chemistry
Clarkson University
Potsdam, New York 13699-5810
USA

Abstract Vapor pressures of two fragrance materials in aqueous solutions of polyvinylpyrrolidone and sodium xylenesulfonate were determined using gas chromatography of head space samples. The association between polymer and hydrotrope was evaluated from the values of surface tension and electrical conductance. The result showed an association of the hydrotrope and the polymer leading to enhanced surface tension after addition of polymer to hydrotope solutions in a certain hydrotope concentration range. The polymer hydrotope association structure resulted in a reduction of fragrance vapor pressure due to solubilization of the fragrance into the association structure.

Key words Polymers – hydrotropes – polymer surfactant association – detergents – microemulsions

Introduction

The vapor pressure is a useful fundamental property; the most direct manner to determine the activity of a solute and if sufficient information is available, of the solvent as well.

The vapor pressure of fragrance molecules is an essential property also from an applied point of view, because its relation to the structural changes during evaporation from a formulation and to its perception limit is a decisive economic factor.

The recent trend of changing fragrance formulations from solvent based to solubilized ones [1, 2] has ignited an interest in the phase equilibria and vapor pressures in colloidal fragrance dispersions. Abe et al. [3–7] and Kayali et al. [8] have published phase diagrams of fragrance systems and Friberg et al. [9, 10] have demonstrated the vapor pressure dependence on the different colloidal structures encountered.

The second important area is the influence of polymers on the vapor pressure of fragrance and flavor substance in aqueous systems. This phenomenon is especially important in foods [11–13], but has also relevance for the personal care industry, because the presence of a polymer modifies the association structures of surfactants in a most decisive manner [14–19].

We found the problem of the interaction between a negatively charged hydrotrope [20] and a polymer to be intriguing and with this article we present the results investigating this problem using the vapor pressure of two solubilized fragrances phenethyl alcohol and phenethyl acetate.

Experimental

Materials

Sodium xylenesulfonate (SXS), 90%; phenethyl alcohol (PEA), 99%; phenethyl acetate (PEAc), 99%; Aldrich Chemical Co., Milwaukee, WI. Polyvinylpyrrolidone USP, GAF Co., Wayne, NJ. Water, deionized. SXS was purified by washing with hexane to remove organic compounds, followed by the hexane evaporation in a vacuum until constant sample weight. The purity was checked by surface tension measurement, using the disappearance of the minimum in the surface tension curve versus the logarithm of SXS concentration in aqueous solution as a sign that hydrophobic impurities were removed.

Phase diagram

The isotropic liquid solubility region of the phase diagram was determined by visual and microscopic observation of samples during titration of one liquid component.

Vapor pressure

The equilibrium vapor pressure of phenethyl alcohol or phenethyl acetate in each sample was measured by Headspace Gas Chromatography, which has been described in detail in previous articles [9, 10]. The only difference in current investigation was the following column temperature program: initial temperature 40 °C; temperature program rate, 20 °C/min 5 min; which followed by another program rate, 30 °C, until the final temperature, 200 °C, 0.5 min. A 3 min column equilibrium time at initial temperature before each sample injection was also added into the program. The retention time of PEA and PEAc was 4.12 and 5.25 min, respectively.

Electrical conductivity

Thirty ml solutions with different SXS concentration were prepared and stored at room temperature (22 ± 0.1 °C). The electrical conductivities of these solutions were measured with the electrical conductivity meter (ATI Orion, Model 170) which had been calibrated with 0.01 M KCl at the same temperature before the experiments.

Surface tension

Surface tension of sodium xylenesulfonate aqueous solutions with and without polymer was measured with the Surface Tensiomat (model 21 from Fisher Scientific). Calibration was carried out with doubly distilled water.

Results

The isotropic aqueous solution regions in the systems phenethyl alcohol, sodium xylenesulfonate and water (solid line), and phenethyl alcohol, sodium xylenesulfonate, and 5% PVD aqueous solution (dashed line) are given in Fig. 1. The regions are similar with minor difference in the solubility of PEA and SXS.

Figure 2 shows isotropic aqueous solution regions in the system phenethyl acetate–sodium xylenesulfonate–water, and the system phenethyl acetate–sodium xylenesulfonate–5% PVD aqueous solution. Phenethyl acetate is virtually insoluble in water or in the polymer aqueous solution. The solubility of PEAc is increased by adding

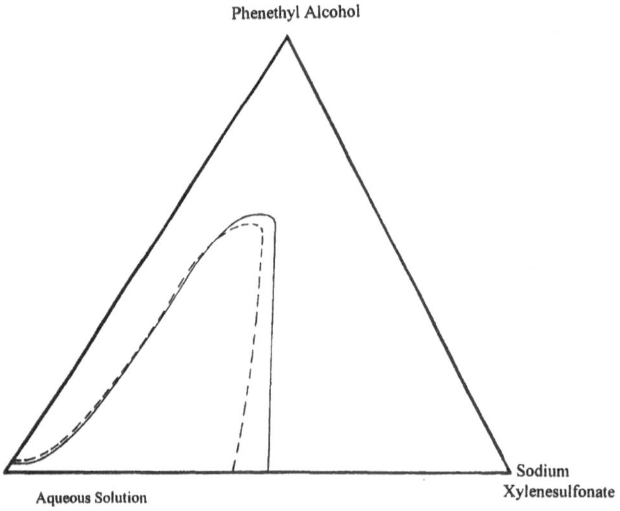

Fig. 1 The isotropic solution region in the system sodium xylenesulfonate/phenethyl alcohol with water (—) and with a water/polyvinylpyrrolidone (95/5) solution (- - - -)

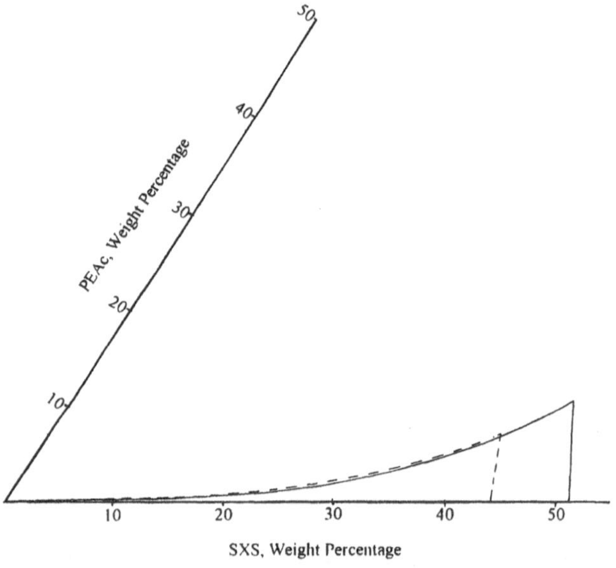

Fig. 2 The isotropic solution region in the system sodium xylenesulfonate/phenethyl acetate with water (—) and with a water/polyvinylpyrrolidone (95/5) solution (- - - -)

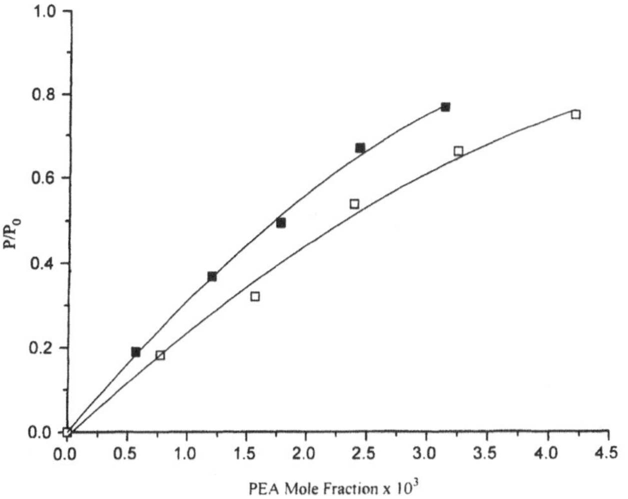

Fig. 3 Relative vapor pressure of phenethyl alcohol in water (■) and in a 5% by weight solution of polyvinylpyrrolidone in water (□). p = measured vapor pressure, p_o = vapor pressure of pure phenethyl alcohol

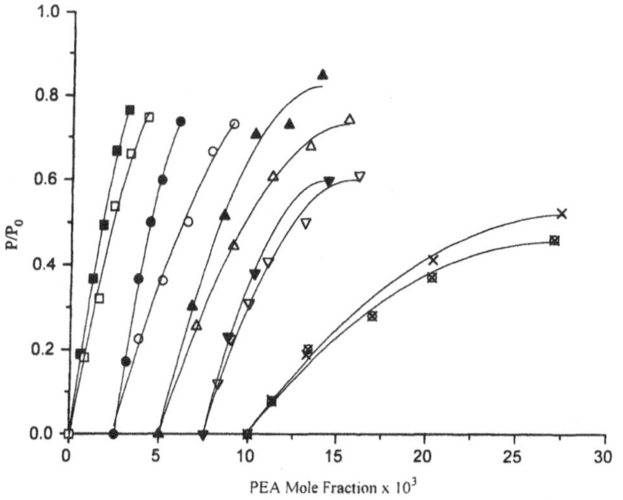

Fig. 5 Relative vapor pressure of phenethyl alcohol in aqueous solutions of sodium xylene sulfonate (SXS) and polyvinylpyrrolidone (PVP)

	Wt. percent SXS	PVP	
(■)	0	0	
(□)	0	5	
(●)	5	0	abscissa 0 at 2.5×10^{-3}
(○)	5	4.75	abscissa 0 at 2.5×10^{-3}
(▲)	7.5	0	abscissa 0 at 5×10^{-3}
(△)	7.5	4.63	abscissa 0 at 5×10^{-3}
(▼)	10	0	abscissa 0 at 7.5×10^{-3}
(▽)	10	4.5	abscissa 0 at 7.5×10^{-3}
(×)	30	0	abscissa 0 at 1×10^{-2}
(⊗)	30	3.5	abscissa 0 at 1×10^{-2}

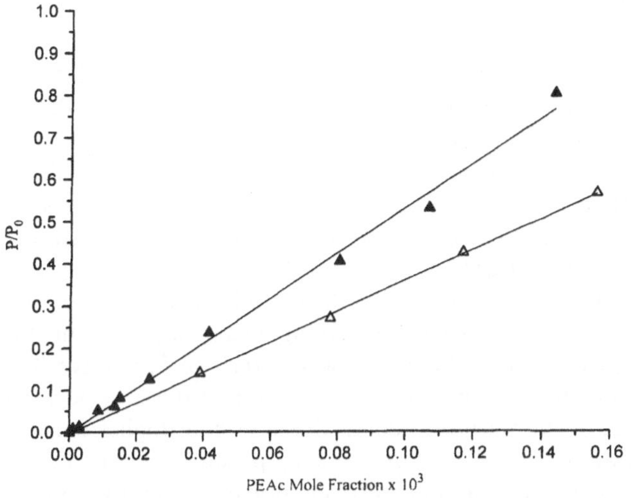

Fig. 4 Relative vapor pressure of phenethyl acetate in water (▲) and in a 5% by weight solution of polyvinylpyrrolidone in water (△)

SXS in both cases. The maximum values in both systems appear at the saturated SXS aqueous solution, 10% PEAc in the former, 7% PEAc in the latter.

The vapor pressure of PEA in water and in the 5% PVP solution is given in Fig. 3 and the corresponding values for PEAc in Fig. 4. The vapor pressures are extremely high with that of PEA two magnitudes grater than that of an ideal solution and that of PEAc three magnitudes higher. Figure 5 combines this information with the vapor pressures of PEA in aqueous solutions of SXS and with PVP. The results reveal the reduction in the vapor pressure increase with enhanced amounts of SXS and, in all cases, show a lower vapor pressure for the samples containing PVP. The corresponding values for PEAc are given in Fig. 6 demonstrating a most pronounced reduction in the PEAc vapor pressure increase with SXS concentration as well as less of a difference between aqueous solutions with and without polymer.

The surface tension of the sodium xylenesulfonate aqueous solutions with and without polymer versus the logarithm of the SXS concentration is given in Fig. 7. In the solutions without polymer, the reduction of surface tension becomes less at an approximate weight fraction of 0.1 in agreement with earlier results [21]. The surface tension values for concentrations in excess of this level were not constant but continued to be reduced with less of a slope reflecting the stepwise association of the hydrotrope. When 5% polymer is added to water, the surface tension values were lower than those of solutions without polymer at SXS concentrations less than 1.5%. For concentrations of SXS in excess of this value the surface tension was greater than that of the solutions with

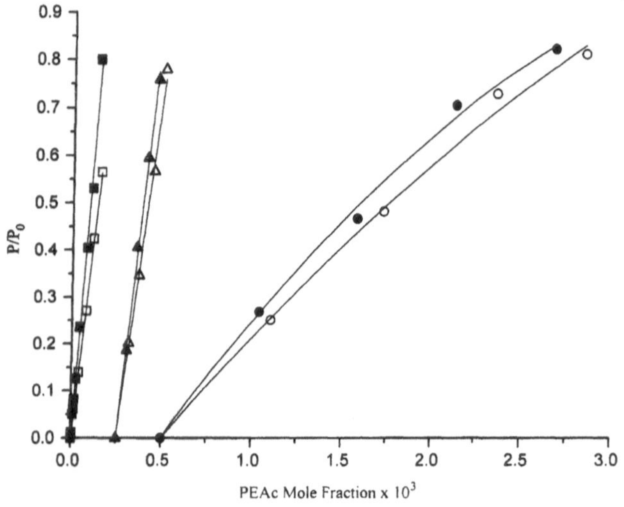

Fig. 6 Relative vapor pressure of phenethyl acetate in aqueous solutions of sodium xylene sulfonate (SXS) and polyvinylpyrrolidone (PVP)

	Wt. percent SXS	PVP	
(■)	0	0	
(□)	0	5	
(▲)	10	0	abscissa 0 at 2.5×10^{-4}
(△)	10	4.5	abscissa 0 at 2.5×10^{-4}
(●)	27	0	abscissa 0 at 5×10^{-4}
(○)	27	3.65	abscissa 0 at 5×10^{-4}

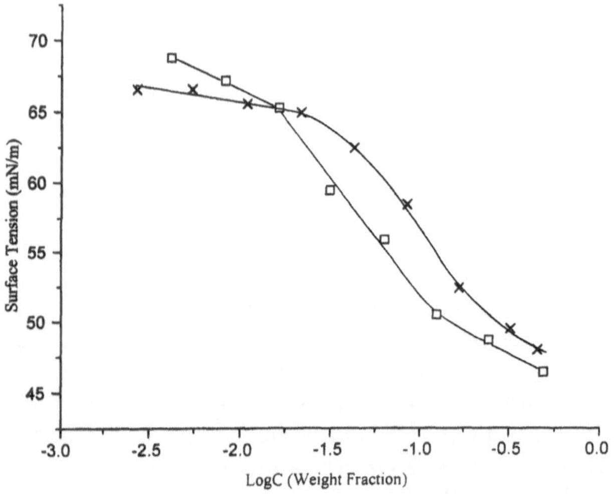

Fig. 7 Surface tension of aqueous solutions of sodium xylenesulfonate (SXS) and polyvinylpyrrolidone (PVP) versus concentration of SXS. PVP%: (□) 0; (×) 5

hydrotrope only; a relation that prevailed to the highest concentration of hydrotrope; its limit of solubility in water.

Electric conductivity, Fig. 8, shows lower conductivity with 5% polymer than for the SXS solutions without polymer with aqueous solutions.

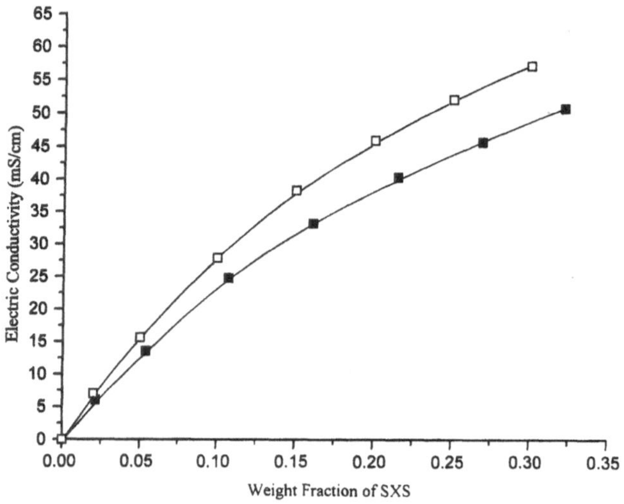

Fig. 8 Electric conductivity of aqueous solutions of sodium xylenesulfonate (SXS) and polyvinylpyrrolidone (PVP) versus concentration of SXS. PVP%: (□) 0; (■) 5

Discussion

The results are relevant for several phenomena such as the influence by the polymer on the solubilization of the fragrance, the interaction between polymer and hydrotrope and the influence of structures resulting from such an interaction on the vapor pressure of the fragrances. In addition, the hydrotrope/fragrance interaction turned out to be significant.

Figures 1 and 2 clarify the first phenomenon; the presence of the polymer had no significant influence on the maximum solubilization of any of the two fragrances. The solubility of them in pure water and in water plus polymer was for all purposes identical and even at higher concentrations of hydrotrope, where, presumably, there was a strong interaction between polymer and hydrotrope (vide infra) no significant difference in *maximum* solubilization was found, except the one dependent on the reduction in maximum solubility of the hydrotrope.

The interaction between hydrotrope and polymer will first be examined from the results in Figs. 7 and 8. In the interpretation of these results, it is illustrative to realize both the fundamental difference from and the similarities with polymer/surfactant interactions. The difference lies with the fact that hydrotropes do not form micelles, but show stepwise association [20]. The surface tension measurements agree with this kind of association [22, 23]; the knick point at 10 wt% in the surface tension curve versus the logarithm of concentration is not followed by a horizontal part of the curve. However, the surface tension curve for the hydrotrope with added polymer shows distinctly higher values over a wide range of concentrations, Fig. 7, indicating a polymer/hydrotrope association. For these pre-self-association phenomena the rich information about surfactant–polymer interactions [24] is useful. By now the original contributions by Lindman et al. [25, 26] have been followed by a huge output treating the more fundamental aspect [27–31] properties of materials [32, 33] and influence on phase equilibria [34–38].

Following these extensive investigations the interpretation is straightforward of the values given by Fig. 7. The hydrotrope associates with the polymer in the concentration range 2.5×10^{-3}–2.8×10^{-2} weight fraction as demonstrated by the fact that the surface tension in this hydrotrope concentration range is virtually independent of its concentration. For concentrations in excess of this range, the concentration of free hydrotope is increased initiating a significant reduction in surface tension. The difference from the behavior of surfactant/polymer systems is noticeable. In a surfactant system [39] the saturation of polymer associated micelles and the consequent initiation of the reduction of surface tension takes place at concentrations of the surfactant greater than its critical micellization concentration for polymer concentrations at the level of 5%. In the present case, with polymer–hydrotrope association, the saturation of polymer is established at hydrotrope concentrations far less, approximately four times, than that at the onset of its eigenassociation. This result is not unexpected; the short hydrocarbon chain of the hydrotrope results in an extremely high value of its eigenassociation concentration. With this condition in mind, it appears reasonable that the polar interaction between the pyrrolidone groups and the negative charge of hydrotrope plays a significant role in the polymer/hydrotrope association. Accepting this conclusion, an estimation shows an approximate upper limit of one-third of the available pyrrolidone groups of the polymer to be occupied before surface tension begins to be reduced by an increase of the free hydrotrope concentration. In the range in which the surface tension is reduced the amount of SXS bound to the polymer may be calculated from the concentration difference between the two curves. In the surface tension range 62.5–50.0 mN/m the fraction bound is constant at 0.47 ± 0.01.

Hence, it appears reasonable to describe the hydrotrope/polymer interaction in the following manner. For low concentrations of hydrotrope the binding of it to the polymer is strong, probably with the electrostatic component a major factor, and the increase in the free hydrotrope (monomer) concentration is small. Above a certain hydrotrope concentration ($\approx 2.8 \times 10^{-2}$ wt fraction), a simple equilibrium is established

$$C_A^H = kC_{tot}^H , \tag{1}$$

where C_A^H is amount adsorbed and C_{tot}^H is total concentration of hydrotrope.

A corresponding calculation may be made based on the results of electric conductance, Fig. 8. Taking the values for conductivity at 50, 40, 30, 20 and 10 mS/cm, Fig. 8, one now arrives at a bound fraction 0.31 ± 0.03; a lower value than that found from surface tension results. The values are different because they represent different phenomena. The values for surface tension reflect the reduction of anionic hydrotrope species in the solution due to association with the polymer, while the conductance measurements illustrate the corresponding total loss of ions.

The combined results may be used to estimate the fraction of sodium ions adhering to the polymer/hydrotrope association structure. The specific conductance of the sodium and the xylenesulfonate ions is, of course, different; a reasonable estimate may be made using available data on sodium benzoate [40]. For the present estimate a ratio of specific conductivity of xylenesulfonate to sodium ion of 0.6 is used. With the knowledge of the reduction in conductance equal to 0.31 and assuming a fraction σ of sodium ions adhering to the polymer the expression becomes

$$0.31 = (0.47 \cdot 311 + \sigma \cdot 518)/829 ,$$

$$\sigma = 0.21 . \tag{2}$$

This value is far less than those found for counterion adsorption to micelles in aqueous solutions [41]. This result is not unexpected; the micellar interface is well defined while there is reason to expect the one from the polymer/hydrotrope association to be irregular, because of the binding conditions discussed earlier.

With these conditions discussed the vapor pressure dependence may be analyzed beginning at the values for low content of PEA and PEAc. For fragrance concentrations approaching zero, Henry's law is valid:

$$P = Kx , \tag{3}$$

where P is the fragrance vapor pressure, x is its mole fraction and K is Henry's constant. It is of advantage to use our modified version [9] of Henry's law

$$P = FxP_0 , \tag{4}$$

where the P_0 is the vapor pressure of the pure compound, because the F constant becomes equal to one for an ideal solution. In this manner, the constant illustrates the deviation from ideality. $RT \ln F$ is the energy to transfer 1 mol of solubilizate from an ideal solution to the solution in question and F is the activity coefficient in the well-known expression for the chemical potential

$$\mu = \mu^0 + RT \ln Fx . \tag{5}$$

Table 1 The value of the F constant

Composition, solvent				F value, fragrance	
W	PVP	SXS (tot)	SXS (free)	PEA	PEAc
100	—	—	—	$(3.4 \pm 0.1) \times 10^2$	$(5.3 \pm 0.1) \times 10^3$
95	5	—	—	$(2.7 \pm 0.3) \times 10^2$	$(3.6 \pm 0.05) \times 10^3$
95	—	5	—	$(3.3 \pm 0.2) \times 10^2$	—
90.25	4.75	5	2.7	$(1.6 \pm 0.1) \times 10^2$	—
92.5	—	7.5	—	$(1.8 \pm 0.1) \times 10^2$	—
87.875	4.625	7.5	4.0	$(1.4 \pm 0.03) \times 10^2$	—
90	—	10	—	$(1.8 \pm 0.1) \times 10^2$	$(3.56 \pm 0.02) \times 10^3$
85.5	4.5	10	5.5	$(1.4 \pm 0.1) \times 10^2$	$(2.96 \pm 0.02) \times 10^3$
70	—	30	—	$(5.6 \pm 0.3) \times 10$	
66.5	3.5	30	15.9	$(5.3 \pm 0.2) \times 10$	
73	—	27			$(5.3 \pm 0.2) \times 10^2$
69.35	3.65	27			$(4.5 \pm 0.1) \times 10^2$

The results in Figs. 3–6 were described by simple polynomials

$$p/p_0 = A_0 + A_1 x + A_2 x^2 + \cdots , \tag{6}$$

where

$$A_1 = F \tag{7}$$

for $x \to 0$.

Table 1 shows the values of F for different compositions and an examination of them is useful.

Solutions of PEA and PEAc in water both display extreme deviations from the values for ideal solutions; PEAc more so than PEA. This is expected; the OH group has a significantly greater interaction with water than an acetate group as shown by the difference in solubility parameters of the two groups. The values for ethanol and ethylacetate illustrate the difference with the three Hansen parameters for dispersion (δ_d), polar (δ_p) and hydrogen bond (δ_h) interactions at 15.81. 8.8, 19.4 (EtOH) and 15.8, 5.3, 7.2 (EtAc) [42]. Both the polar and hydrogen bond parameters, especially the latter, are decisively greater for the alcohol.

The reduction in vapor pressure after addition of the polymer is due to association of the PEA and PEAc with the polymer. This "binding" of a fragrance to polymers is an essential phenomenon in food science [43] and is, of course, of importance for the interaction between odorants and odorant binding proteins [44]. It should be emphasized that the phenomenon is entirely different from the two-phase polymer–solvent systems of interest for controlled release [45].

In the present case, the influence on the fragrance vapor pressure emanates from several sources; interaction of the fragrance molecule with the hydrotrope monomer, with the eigenassociation structures of the hydrotrope and with the polymer/hydrotrope association structures. The numbers in Table 1 combined with those of the surface tension, Fig. 7, provide information about the relative importance of the different structures.

At first the values in Table 1 may be used to estimate the fraction of added fragrance bound to the polymer. The reduction in vapor pressure after adding polymer to the aqueous solution of the fragrance is assumed to reflect a reduction in the concentration of "free" fragrance; a reasonable assumption considering the curves in Figs. 3 and 4.

With such an assumption, one finds an approximate equilibrium

$$C_P^F = kC_{tot}^F , \tag{8}$$

where C_P^F is the amount adsorbed onto the polymer and C_{tot}^F is the total concentration. The value of k becomes 0.21 for PEA and 0.32 for PEAc; an expected difference considering the greater hydrophobocity of PEAc.

For the PEA, the F values for hydrotrope and polymer/hydrotrope solutions follow a straightforward pattern in agreement with earlier discussion about hydrotrope/hydrotrope and polymer/hydrotrope association (vide retro). In the solution containing 5% SXS and no polymer, there is no significant association between the hydrotrope molecules as shown by the surface tension data (Fig. 7) and by the phase diagram (Fig. 1). Hence, there is no hydrotrope monomer/PEA interaction and, consequently, the F constant is as high as that in pure water, 3.3×10^2 and 3.4×10^2, respectively (Table 1). The polymer/hydrotrope combination on the other hand, shows a significant reduction of the F factor; a 41% reduction from the value in the aqueous polymer solution demonstrates the influence of the hydrotrope/polymer association on the vapor pressure in agreement with the interpretation of the surface tension curves (Fig. 7). These show that 5% of SXS ($\log c = -1.30$) is a concentration far in excess of the primary saturation of the polymer by SXS. This saturation of polymer by hydrotrope electrostatically bound is complete at 5% of SXS and at 4.75% of polymer a considerable amount of SXS, of the order of 2–2.5%, forms an association structure with the polymer. This association structure solubilizes the fragrance compound and the reduction in vapor pressure is a rational and expected consequence. It is essential to realize that information about the exact structure of this association is not available.

At 7.5% SXS the PEA–hydrotrope association structures have become established as demonstrated by the phase diagram (Fig. 1). Hence, the F constants for the hydrotrope/PEA and hydrotrope/polymer associations are at a lower level; the fragrance is now solubilized also in the PEA/hydrotrope association structure. Ten percent SXS shows similar values, but at 30% SXS another factor becomes significant; the hydrotrope is now heavily eigen-associated and analyzing the separate contributions to the lowering of the vapor pressure by different associations is not possible. In addition, the solution is by now concentrated to a degree that its general structure is modified.

The values for the PEAc are different. The reduction of the F constant after addition of 10% hydrotrope is significantly less than the one for PEA. This result is expected from the phase diagram (Fig. 2), which does not show a distinct onset of solubilization; as a matter of fact, the solubilization at 10% is extremely low reflecting the poor interaction between PEAc and the hydrotrope monomers. This result is in agreement with those in surfactant systems [46] where the difference in solubilization between alcohols and esters is pronounced. The change in the PEAc vapor pressure from 10 to 27% SXS is greater than that of the PEA compositions. This difference is a direct consequence of the stepwise association of SXS; the larger aggregates at high SXS concentrations are more efficient to solubilize the more hydrophobic PEAc.

Acknowledgement The authors express their gratitude to Prof. P. Zuman for his valuable advice on electrochemical phenomena. The research was financed in part by the New York State Science and Technology Foundation at the Center for Advanced Materials Processing at Clarkson University, Potsdam, NY.

References

1. Yamaguchi Y, Musata T (December 1994) Jpn Patent 06, 336, 443
2. Chung SL, Tan CT, Tuhill IM, Sharpt LG (February 1994) US Patent 5 283 056
3. Tokuoka Y, Abe M (1996) Nihon Yukagakkgishi 45:635
4. Behan JM, Perring KD (1987) Int J Cosmetic Sci 9:261
5. Abe M, Tokuoka Y, Uchiyama H, Ogino K (1990) J Jpn Oil Chem Soc 39:565
6. Tokuoka Y, Uchiyama H, Abe M, Ogino K (1992) J Colloid Interface Sci 152:402
7. Abe M, Mizuguchi K, Kondo Y, Ogino K, Uchiyama H, Scamehorn JF, Tucker EE, Christian SD (1993) J Colloid Interface Sci 160:16
8. Kayali I, Heisig C, Friberg SE (1995) In: Tan CT, Tong CH (eds) Physical Chemistry of Flavors, Chap. 12, ACS Symp. Series, Vol 610. Marcel Dekker, New York, 154
9. Friberg SE, Huang T, Fei L, Vona Jr SA, Aikens PA (1996) Progr Colloid Polym Sci 101:18
10. Friberg SE, Int J Cosmet Sci (in press)
11. Roberts DB, St. Elmore J, Langley KR, Bakker J (1994) J Agric Food Chem 44:132
12. O'Neill TE (1996) In: McGorrin RJ, Leland JV (eds) Flavor-Food Interactions, ACS Symp. Series, Vol 633. Marcel Dekker, New York, p 59
13. Carr J, Balogg D, Guinard JX, Lawler L, Marty C, Squire C (1996) In: McGorrin RJ, Leland JV (eds) Flavor–Food Interactions, ACS Symp. Series, Vol 633. Marcel Dekker, New York, 98
14. Thalberg K, Lindman B, Karlström G (1990) J Phys Chem 94:10
15. Thalberg K, Lindman B (1991) Langmuir 7:2
16. Piculell L, Lindman B (1992) Adv Colloid Interface Sci 41:149
17. Thalberg K, Lindman B, Karlström G (1991) J Phys Chem 95:15
18. Zhang KW, Karlström G, Lindman B (1992) Colloids Surf 67:147
19. Lindman B, Khan A, Marques E, daGraca Miguel M, Piculell L, Thalberg K (1993) Pure Appl Chem 65:953
20. Balusubramanian D, Friberg SE (1993) In: Matijević E (ed) Surface & Colloid Science, Vol 15. Plenum Press, New York, p 197
21. Friberg SE, Fei L, Campbell S, Yang H, Lu Y (in press) Colloids Surf
22. Danielsson I, Stenius P (1971) J Colloid Interface Sci 37:264
23. Saleh AM, El-Khordaugi LK (1985) Int J Pharm 24:231
24. Goddard ED, Ananthapadanabhan KP (eds) (1993) Interactions of Surfactants with Polymers and Proteins. CRC Press, Boca Raton, FL
25. Siegel G, Walter A, Lindman B (1984) J Physique 45, C-595
26. Carlsson A, Karlström G, Lindman BL (1986) Langmuir 2:536
27. Piculell L, Guillemet F, Thuresson K, Shubin V, Ericsson O (1996) Adv Colloid Interface Sci 63:1
28. Anthony O, Zana R (1996) Langmuir 12:1967
29. Holbmerg Ch, Nilsson S, Sundelöf LO (1997) Langmuir 13:1392
30. Hansson P, Almgren M (1996) J Phys Chem 100:9038
31. Faes H, de Schryver FC, Sein A, Bijima K, Kevelam J, Engberts JBFN (1996) Macromolecules 29:3875
32. Antonietti M, Maskos M (1996) Macromolecules 29:4199
33. Ponomarenko EA, Waddon AJ, Bakeev KN, Tirrel DA, MacKnight WJ (1996) Macromolecules 29:4340
34. Ranganathan S, Kwak JTC (1996) Langmuir 12:1381
35. Vollmer D, Vollmer J, Stuehn B, Wehrli E, Eicke HF (1995) Phys Rev E:Stat Phys Plasmas, Fluids Relat Interdiscip Top 52:5146
36. Sakai M, Satoh N, Tsujii K, Zhang YQ, Tanaka T (1995) Langmuir 11:2493
37. Bergfelt K, Piculell L (1996) J Phys Chem 100:5935
38. Ruokalainen J, Tanner J, tenBrinke G, Ikkala O, Torkkeli M, Serimaa R (1995) Macromolecules 28:7780
39. Breuer MM, Robb ID (1972) Chem Ind 530:35

40. Conway BE (1952) In: Electrochemical Data. Elsevier, London, p 145
41. Gustavsson H, Lindman B (1975) J Am Chem Soc 97:3923
42. Barton ATM (1983) Handbook of Solubility Parameters and Other Cohesion Parameters. CRC Press, Boca Raton, FL
43. Hau MYM, Gray DA, Taylor AJ (1996) In: McGorrin RJ, Leland JV (eds) Flavor–Food Interactions ACS Symp Series, Vol 633, p 109
44. Hérent MF, Collin S, Pelosi P (1996) Chem Senses 21:601
45. Peppas NA, Brannon-Peppas L (1996) J Control Release 40:245
46. Ekwall P (1975) In: Brown GH (ed) Advances in Liquid Crystals, Vol 1. Academic Press, New York, p 1

Progr Colloid Polym Sci (1998) 109:101–117
© Steinkopff Verlag 1998

M.J. Rang
C.A. Miller

Spontaneous emulsification of oil drops containing surfactants and medium-chain alcohols

Received: 2 December 1997
Accepted: 3 December 1997

M.J. Rang · C.A. Miller (✉)
Department of Chemical Engineering
Rice University
6100 Main Street
Houston, Texas 77005-1892
USA

Abstract Spontaneous emulsification of mixtures of *n*-hexadecane, *n*-octanol, and the pure nonionic surfactant $C_{12}E_6$ brought into contact with water was observed using videomicroscopy. Relevant aspects of equilibrium phase behavior in this system were determined, and information on emulsion drop size and stability was obtained by turbidity measurements. It was found that small, uniform oil droplets and the most stable emulsions occurred when the entire original oil drop was converted to another phase – in this case a microemulsion – which subsequently became supersaturated in oil owing to diffusion of octanol into the aqueous phase. Spontaneous emulsification was also studied for mixtures of *n*-decane, *n*-heptanol, and tetradecyldimethylamine oxide contacting water. In this case, small droplets were formed when both the microemulsion and an intermediate lamellar liquid crystalline phase became supersaturated with oil.

Key words Emulsions emulsification – spontaneous self-emulsification

Introduction

Many emulsions are made by mechanical mixing of oil and water phases. Generally, vigorous mixing is required to obtain small drop sizes. However, in some applications such mixing is impossible or undesirable, and spontaneous emulsion formation is preferred. That is, compositions of the initial oil and water phases are not in equilibrium and are chosen such that small drops form spontaneously when the phases are brought into contact. The term spontaneous emulsification is usually limited to situations where no external energy of agitation is supplied. "Self-emulsification" generally refers to situations where a small amount of energy is supplied to achieve gentle mixing. Sometimes, e.g., when interfacial tension is very low, such mixing may be sufficient to produce drops and form an emulsion. In other cases the drops form spontaneously, and mixing simply serves to disperse them throughout a large volume and to bring together portions of the oil and water phases which were not near the initial surface of contact.

An extensive literature exists on spontaneous emulsification. Various mechanisms have been suggested, for instance spontaneous expansion and breakup of an oil-water interface produced by transient local negative values of interfacial tension. Davies and Rideal [1] reviewed the older literature on this mechanism, while Granek et al. [2] have recently presented an improved model of how such an instability could develop. In this paper we are interested in another mechanism of spontaneous emulsification, viz., formation of drops of one phase in another that has become locally supersaturated as a result of diffusion. Miller [3] summarized knowledge of this mechanism with examples from oil–water–alcohol and oil–water–surfactant systems. In the latter case the examples mainly involved spontaneous emulsification of water drops in an oil phase when the surfactant diffused from the aqueous phase into an oil in which it was highly soluble. For instance, such behavior occurs in a nonionic surfactant/hydrocarbon/water system above its phase inversion temperature (PIT). Clearly, understanding when local supersaturation can be expected requires knowledge both of equilibrium phase behavior and of the diffusion processes which take place.

Self-emulsification of an oil phase in water is of interest, for example, in use of "emulsifiable concentrates" of agricultural chemicals [4, 5] and in some methods for drug delivery [6, 7]. Basically, a suitable surfactant or surfactant mixture is added to an oil phase containing a pesticide or drug to enable it to disperse when added to water under conditions where gentle mixing occurs. The mechanism of such self-emulsification is not well understood. Lee and Tadros [8] found that development of low interfacial tensions could not fully explain the emulsification phenomena they observed. Sometimes self-emulsification has been attributed to the formation of liquid crystalline phases near the surface of contact between the two phases [7, 9] although the precise role of the liquid crystal was not made clear by these authors. Accordingly, selection of surfactants for self-emulsification and determination of the required minimum surfactant concentration remain somewhat empirical.

In this paper we consider spontaneous emulsification in situations where a drop of oil containing both a surfactant and a medium-chain alcohol is injected into water. Two systems with different surfactants are discussed which have differences in phase behavior leading to some differences in the emulsification mechanism. In one system the lamellar liquid crystalline phase is formed during the process, but in the other spontaneous emulsification usually occurs with no liquid crystal present. Whether or not a liquid crystalline phase forms, a necessary but not sufficient condition for obtaining uniformly small droplets is that the original oil phase be completely converted by diffusional processes into another phase which subsequently becomes supersaturated, causing oil drops to be nucleated. The phase which becomes supersaturated may be either a microemulsion or the lamellar phase.

Experimental

Tetradecyldimethylamine oxide (C_{14}DMAO) was obtained from Hoechst and recrystallized twice from acetone at the University of Bayreuth. Linear alcohol ethoxylates $C_{12}E_4$ and $C_{12}E_6$ with reported purities of about 99% were purchased from Nikkol Chemical Co., Japan, and used without further purification. Reagent grade *n*-decane and *n*-hexadecane were obtained from Humphrey Chemical. Reagent grade *n*-heptanol and *n*-octanol were obtained from Sigma. Deionized water was prepared by distilling and deionizing with a SYBRON Barnstead glass still and Nanopure II system.

Samples for phase behavior studies were prepared as described previously [10] and equilibrated for times up to one month as required in an environmental room maintained at 30 °C. Oil drop and vertical cell contacting experiments utilizing a thermal stage to maintain constant temperature and a videomicroscopy system to observe and record dynamic behavior were conducted as in previous work [11, 12].

Information on emulsions formed by "self-emulsification" was obtained using an apparatus which was operated in the environmental room at 30 °C. A hydrocarbon/alcohol/surfactant mixture of 250 μl was injected with a syringe over a period of about 2 s into 75 ml of water in a 100 ml Pyrex beaker while applying gentle mixing with a magnetic stirrer. The turbidity of the mixture was monitored as a function of time using a PC800 colorimeter with 600 nm filter (Brinkman Instruments, Inc.), and the results were stored in a computer. Both the needle for injecting the oil and the probe were located halfway between the water surface and the bottom of the beaker. A microswitch connected to the syringe plunger provided a signal to the computer indicating the time of injection. After 30 min mixing was stopped, but monitoring of turbidity continued for another 30 min. Thereafter, part of the sample was maintained at 30° without further mixing and checked at least daily for evidence of significant creaming, flocculation, or separation. Further details are given elsewhere [13].

Results

$C_{12}E_6$/*n*-octanol/*n*-hexadecane/water system. Oil drop and vertical cell contacting experiments with surfactant in the oil

Experiments were performed in which drops having initial diameters of 50–100 μm and containing *n*-hexadecane, *n*-octanol, and the pure linear alcohol ethoxylate $C_{12}E_6$ were injected into water at 30 °C. The results for various drop compositions are summarized in Table 1. In most cases interfacial tension was low enough that the original drop broke up to some extent when subjected to shear during the injection process. For hexadecane/octanol ratios of 100/0, 96/4, and 97/3 emulsification occurred so rapidly that its mechanism could not be directly observed in these experiments. However, as discussed below, the vertical cell contacting experiments demonstrated that at least some oil droplets were formed by local supersaturation, not drop breakup.

Table 1 Behavior of drops of hexadecane/octanol/$C_{12}E_6$ mixtures contacting pure water at 30 °C

Weight ratio of oil to $C_{12}E_6$	Weight ratio of oil (C_{16}/C_8OH)					
	100/0	96/4	93/7	90/10	85/15	75/25
95/5	No activity	—	—	A*	A*	A*
90/10	—	—	—	H(A*)	H(A*)	H(A*)
85/15	A	A	A*	H*(B)	H*(B)	H*(B)
80/20				H*(B)	H*(B*)	H*(B*)
75/25	A	A*	A*	B*	I	I
65/35	A*	—	—	—	—	—

Note. A, A*, B, B*: Immediate conversion into drops having size distributions similar to those shown in accompanying video frames. H: Drop initially decreases in size and then increases. Small droplets form on the surface of the increasing drop and coalesce into larger drops. The original drop becomes miscible with the aqueous phase. The final size distribution is as indicated in parentheses. H*: Same as H, but with much less coalescence. I: Drop initially decreases in size and then increases. Eventually it becomes miscible with the aqueous phase. Very few or no droplets are observed.

A

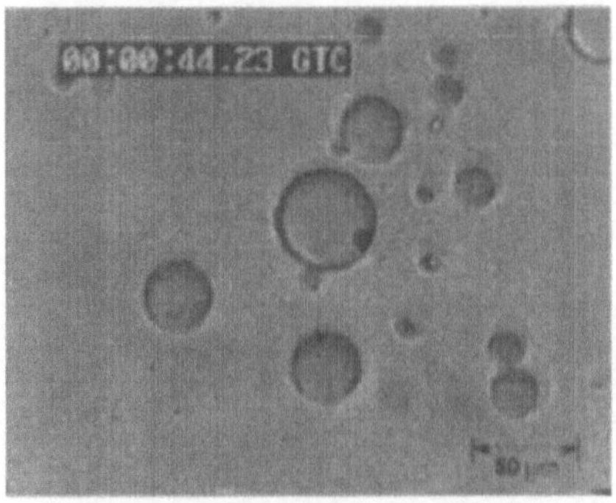

A*

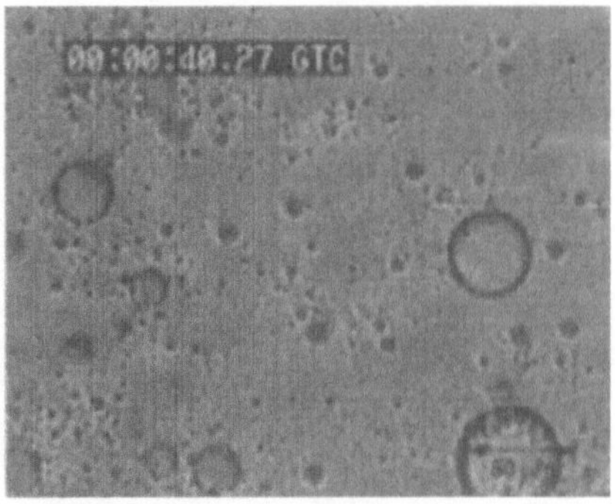

B

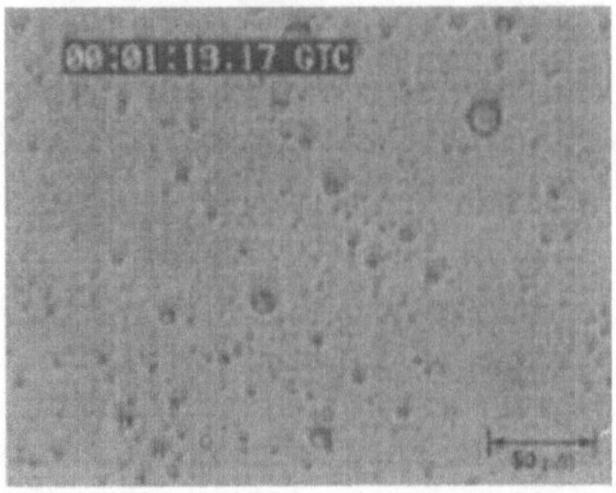

B*

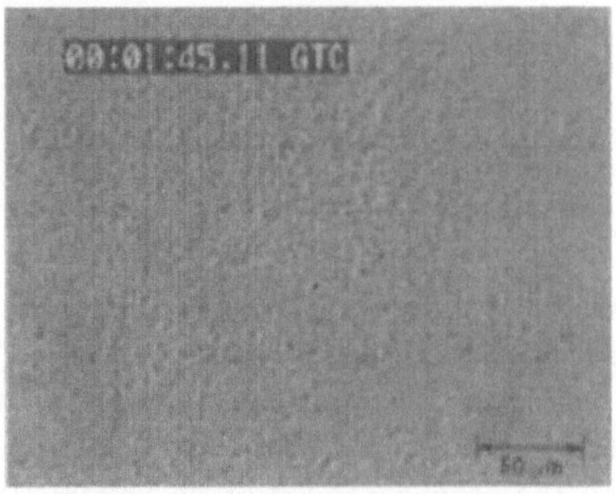

When alcohol and surfactant concentrations were sufficiently large (entries designated H and H* in Table 1), it was possible to observe the emulsification process. Figure 1 shows the behavior for a drop containing 10 wt% $C_{12}E_6$ and 90 wt% of an oil which itself is made up of 90 wt% *n*-hexadecane and 10 wt% *n*-octanol. For convenience, we designate this composition [90(90/10)10], the first number indicating the percentage by weight of the hydrocarbon/alcohol mixture in the drop, and the expression in parentheses giving the composition of the surfactant-free oil. This nomenclature will be used throughout the paper for drops of various compositions. About 2 min after contact many tiny droplets (about 1 μm diameter) are seen on the surface of the injected drop (Fig. 1b). Considerable coalescence of the droplets occurs, yielding a few large drops and some smaller drops (Fig. 1c). The contact angle these drops make with the original drop increases until some separate from it completely (Figs. 1d and 1e). Finally, the original drop becomes miscible with the aqueous phase, leaving in this case one large drop and many smaller

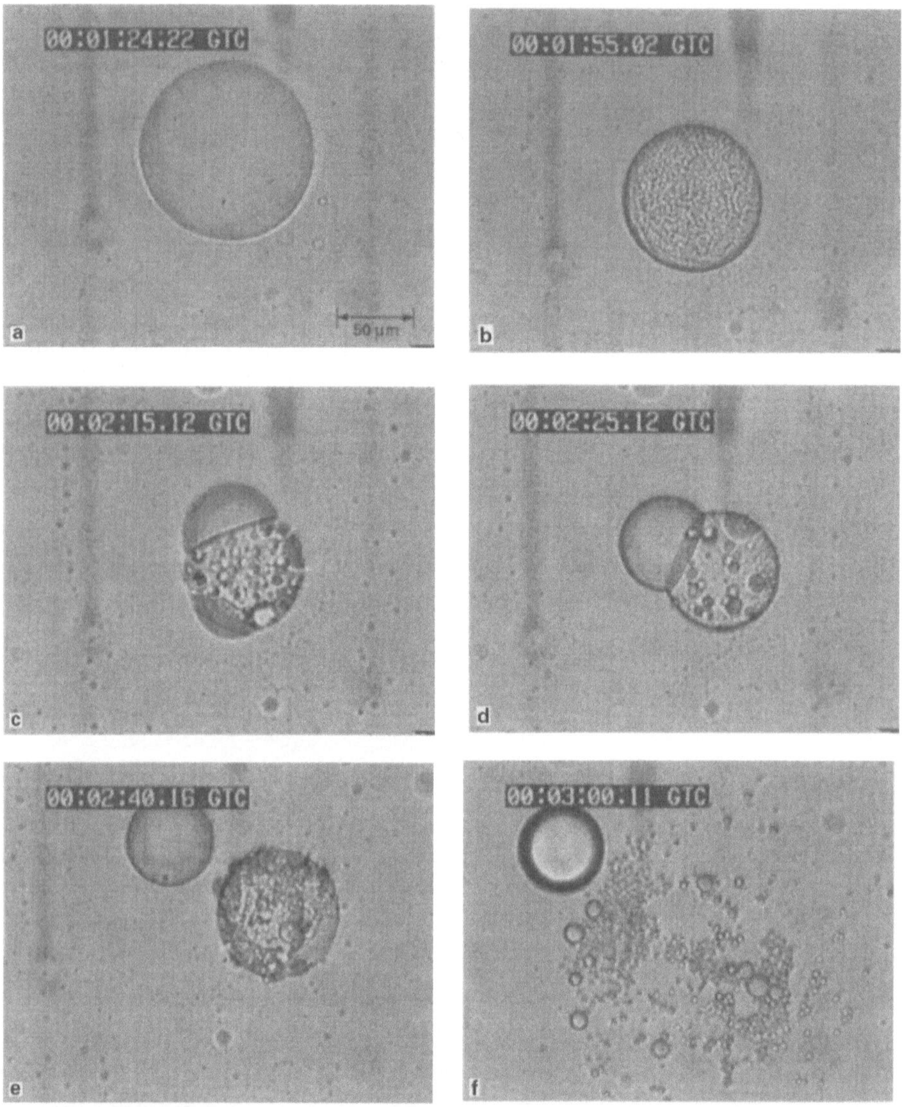

Fig. 1 Behavior of a drop having initial composition [90(90/10)10] in the hexadecane/octanol/$C_{12}E_6$ system after injection into water at 30 °C

drops of different sizes (Fig. 1f). Similar behavior was seen for the other entries denoted by H(A*) in Table 1.

Figure 2 shows behavior when surfactant concentration is doubled with the same hydrocarbon/alcohol ratio, so that the composition is [80(90/10)20]. Clearly, fewer droplets are formed, and much less coalescence occurs, so that only small droplets remain after the original drop has become miscible with the aqueous phase. Other items designated H*(B) and H*(B*) in Table 1 exhibited similar behavior. As surfactant concentration is increased further, the number of droplets formed continues to decrease. For the entries denoted by I in Table 1, e.g., [75(85/15)25], very few or no droplets were observed although the injected drop again became miscible with the aqueous phase.

Experiments were also conducted where mixtures of hexadecane, octanol, and $C_{12}E_6$ were contacted with water in the vertical-stage microscope at 30 °C. The initial behavior observed in these experiments corresponds to that which occurs immediately upon contact during the oil drop experiment while some portion of the drop remains at its initial composition. The results are summarized in Table 2. When the hexadecane/octanol ratio was 85/15 or smaller, no intermediate phase formed at the surface of contact and no spontaneous emulsification occurred during early stages of the contacting process for surfactant concentrations up to 25 wt% although convection near the interface was observed which presumably arose from interfacial tension gradients. For somewhat less lipophilic oils, i.e., those containing less octanol, both the lamellar liquid crystal and a microemulsion formed as intermediate phases. Figure 3 shows behavior some 9 min after contact for an oil with initial composition [75(90/10)25]. At this time the microemulsion and aqueous phases have become miscible. Spontaneous emulsification of oil in the micro-

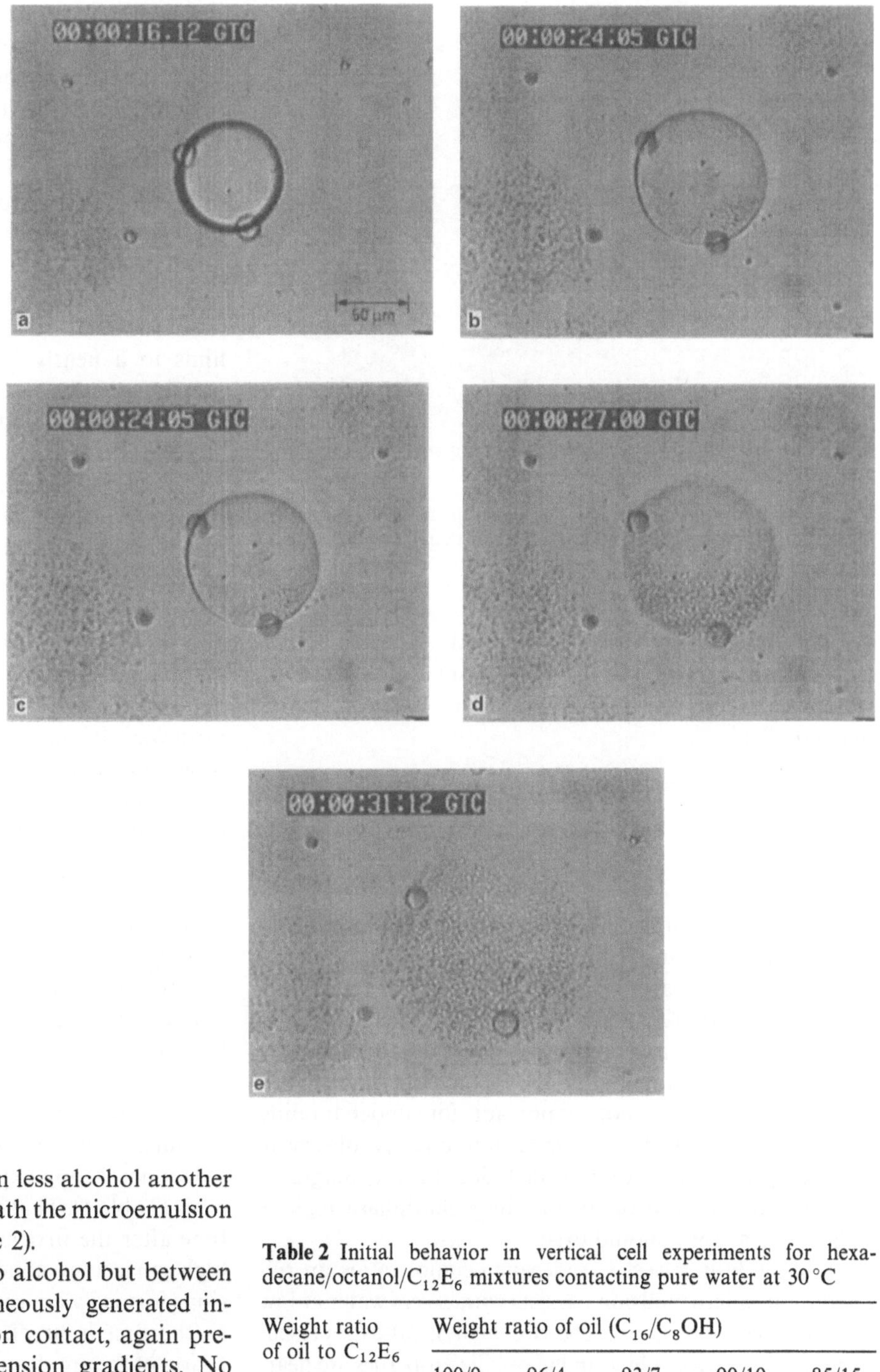

Fig. 2 Behavior of a drop having initial composition [80(90/10)20] in the hexadecane/octanol/$C_{12}E_6$ system after injection into water at 30 °C

emulsion can also be seen. With even less alcohol another birefringent layer L'_α developed beneath the microemulsion (compositions designated 2 in Table 2).

When the oil phase contained no alcohol but between 15 and 25 wt% surfactant, spontaneously generated interfacial convection was observed on contact, again presumably the result of interfacial tension gradients. No intermediate phase formed, but some spontaneous emulsification of oil in the aqueous phase was seen. Because the temperature was well below the PIT, surfactant was diffusing into a phase in which it was highly soluble, and local supersaturation produced the emulsification. This behavior is the counterpart of that described previously (3) where spontaneous emulsification of water in the oil phase occurred above the PIT when the surfactant was initially in the aqueous phase but preferentially soluble in oil. It is

Table 2 Initial behavior in vertical cell experiments for hexadecane/octanol/$C_{12}E_6$ mixtures contacting pure water at 30 °C

Weight ratio of oil to $C_{12}E_6$	Weight ratio of oil (C_{16}/C_8OH)				
	100/0	96/4	93/7	90/10	85/15
90/10	—	—	3	4	4
85/15	1	2	3*	4	4
80/20	—	—	—	3	4
75/25	1	2	2	3*	4

Note. 1: Convection, no intermediate phase formation, some spontaneous emulsification. 2: Intermediate L_α, ME and L'_α phases, spontaneous emulsification. 3: Intermediate L_α and ME phases; For 3* some spontaneous emulsification in ME. 4: Convection, no intermediate phase formation or spontaneous emulsification.

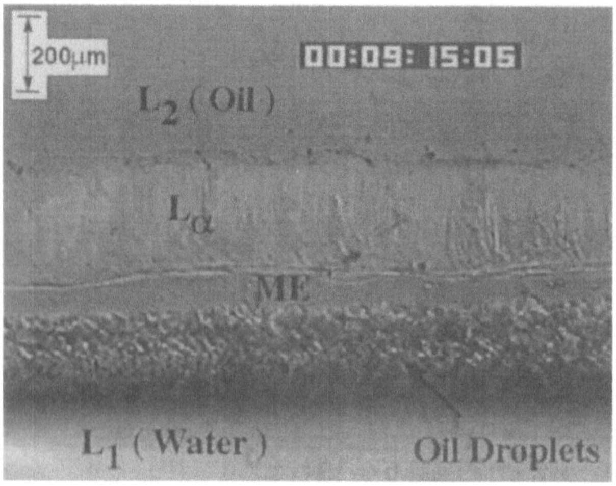

Fig. 3 Video frame from vertical cell experiment for an oil having initial composition [75(90/10)25] placed in contact with water at 30 °C

noteworthy that in the present situation, the oil drop experiments showed that while some small droplets were produced by this mechanism, numerous larger drops remained (Table 1).

$C_{12}E_6$/*n*-octanol/*n*-hexadecane/water system. Phase behavior and turbidity experiments

The phase behavior of this system was studied at 30 °C to help in interpretation of the contacting experiments. Figure 4a–f show partial phase diagrams at constant hexadecane/octanol ratios between 75/25 and 95/5 by weight. Each diagram is based on about 110 samples. Even so, a few multiphase regions present over only small composition ranges but not important for understanding spontaneous emulsification were not directly observed. Accordingly, they were not included in the diagrams although the pattern of surrounding multiphase regions indicates that they should exist.

For the four lowest hexadecane/octanol ratios an oil-continuous microemulsion coexists with excess water at low surfactant concentrations. The two highest ratios represent systems whose hydrophilic and lipophilic properties are nearly balanced at low surfactant concentrations, as indicated by the regions in Figs. 4e and 4f where a microemulsion phase coexists with both oil and water. The "fish" diagram for this system with equal amounts of water and oil present has the usual pattern of phases (Fig. 5) with the minimum surfactant concentration required to form a balanced microemulsion for these conditions being about 8 wt%.

Another region at relatively high water content (80 wt%) where three liquid phases coexist is seen in Fig. 4c, as is an adjacent four-phase region where the lamellar phase is also present. The phase volumes and the bluish appearance of the least dense phase in the former region suggest that it is not simply oil with some dissolved surfactant but instead a microemulsion containing solubilized water. Further investigation of phase behavior in this region is a possible topic for future work.

The single-phase region of the lamellar liquid crystalline phase L_α is found at higher surfactant concentrations at lower hexadecane/octanol ratios. The reason is that more surfactant is needed to bring the surfactant/alcohol films to a nearly balanced state when more alcohol is present initially.

The oil drop and vertical cell contacting experiments allowed observation of the emulsification process on a microscopic scale in the absence of mixing. Macroscopic self-emulsification experiments were also carried out to provide some information on emulsion formation and stability in the presence of gentle mixing. Figure 6 shows variation of turbidity with time when oils with a hexadecane/octanol weight ratio of 90/10 and with various amounts of added surfactant were injected into water at 30 °C in the apparatus described above. The water-to-oil ratio was 300 on a volume basis or about 365 on a weight basis. With only 5–15 wt% surfactant added to the hydrocarbon/alcohol mixture, turbidity increased rapidly until the amount of light passing through the system was only a few percent of that for pure water. In contrast, when the amount of added surfactant was 20 or 25 wt%, turbidity exhibited a peak immediately upon oil injection, then quickly dropped and reached a constant value only a few seconds after injection. The lower turbidity for these higher surfactant concentrations indicates that the drops formed were smaller than for the lower surfactant concentrations, and indeed these solutions were translucent with a bluish tinge. The decrease in drop size with increasing surfactant content is consistent with the results of the videomicroscopy experiments for the same oil compositions (Table 1). That turbidity remained constant for some time after the first few seconds of the experiment for the high surfactant concentrations indicates that the emulsions formed were rather stable. Although Fig. 6 is limited to the first hour after oil injection, no noticeable separation of the emulsions with 20 and 25 wt% surfactant was observed on standing for about six days. These were the most stable emulsions found for this hydrocarbon/alcohol/surfactant system at 30 °C (see Fig. 7a).

Fig. 4 Partial phase diagrams for water/hexadecane/octanol/$C_{12}E_6$ system at 30 °C. Ratios of hexadecane/octanol are fixed at (a) 75/25, (b) 80/20, (c) 85/15, (d) 90/10, (e) 93/7, (f) 95/5 by weight. L_1, L_2: aqueous and oil phases. L_α: lamellar phase. ME: microemulsion. E: o/w emulsion

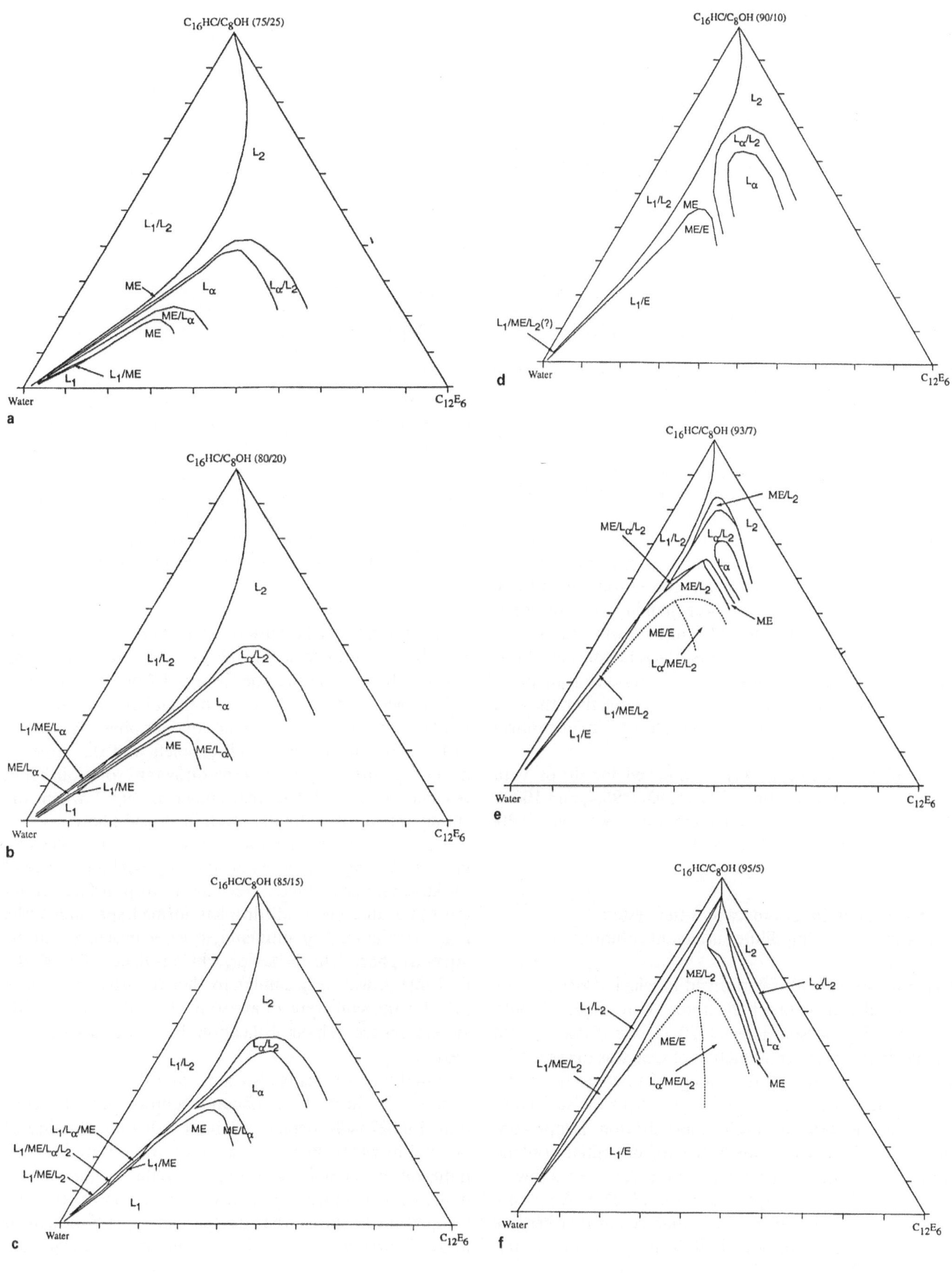
C16HC/C8OH (75/25)
L2
L1/L2
ME
Lα
Lα/L2
ME/Lα
ME
L1/ME
L1
Water
C12E6
a
C16HC/C8OH (90/10)
L2
Lα/L2
Lα
L1/L2
ME
ME/E
L1/E
L1/ME/L2(?)
Water
C12E6
d
C16HC/C8OH (80/20)
L2
L1/L2
Lα/L2
Lα
L1/ME/Lα
ME
ME/Lα
ME/Lα
L1/ME
L1
Water
C12E6
b
C16HC/C8OH (93/7)
ME/L2
ME/Lα/L2
L1/L2
Lα/L2
L2
Lα
ME/L2
ME
ME/E
Lα/ME/L2
L1/ME/L2
L1/E
Water
C12E6
e
C16HC/C8OH (85/15)
L2
L1/L2
Lα/L2
Lα
L1/Lα/ME
ME
ME/Lα
L1/ME/Lα/L2
L1/ME
L1/ME/L2
L1
Water
C12E6
c
C16HC/C8OH (95/5)
L2
ME/L2
Lα/L2
L1/L2
ME/E
L1/ME/L2
Lα/ME/L2
Lα
ME
L1/E
Water
C12E6
f

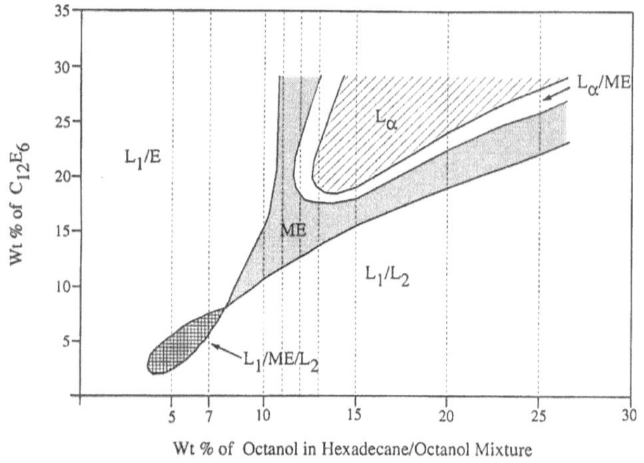

Fig. 5 "Fish" diagram for the system of Fig. 4 with equal weights of the hexadecane/octanol mixture and water

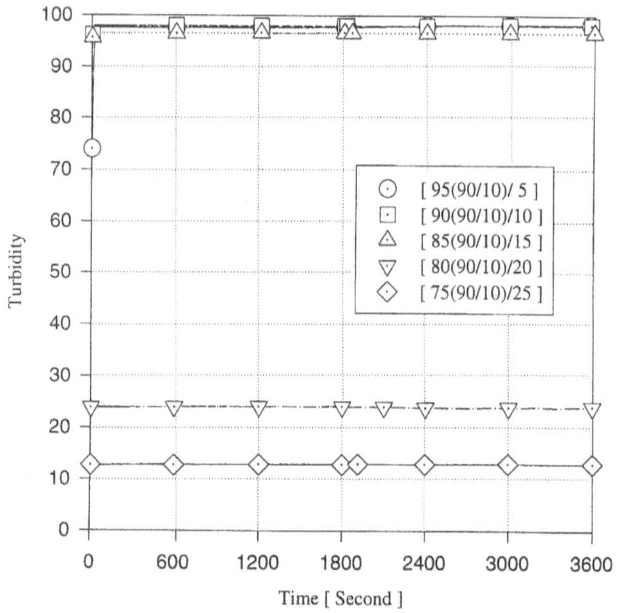

Fig. 6 Turbidity as a function of time for self-emulsification experiments with oils having a hexadecane/octanol ratio of 90/10 and various surfactant contents. Gentle mixing was applied during the first 1800 s with no mixing during the following 1800 s

Figure 8 shows the corresponding results for oils where the weight ratio of hexadecane/octanol was 85/15. Here too turbidity rose rapidly to high values for oils having low surfactant contents. And here too turbidity (after the initial peak corresponding to injection had disappeared) was much lower for higher surfactant concentrations. Indeed, the minimum turbidity was lower than in the corresponding experiments of Fig. 6 for 20 and 25 wt% surfactant. However, turbidity at the high surfactant concentrations in Fig. 8 increased gradually over a period of an hour, indicating that the emulsions were less stable than those of Fig. 6. The longer-term stability results of Fig. 7a confirm this conclusion.

Similar experiments were conducted for drops with hexadecane/octanol ratios of 75/25, 93/7, 96/4, and 100/0. As Figs. 7a and b show, these emulsions were generally rather turbid after one hour and not very stable.

$C_{12}E_6$/*n*-octanol/*n*-hexadecane/water system. Oil drops contacting dilute surfactant solutions

Experiments were also performed in which drops of various hexadecane/octanol mixtures were injected into 0.05 wt% $C_{12}E_6$ solutions at 30 °C (see Table 3). No change in drop size as a function of time was observed for drops of pure hexadecane although the oil was doubtless being solubilized at some very slow rate as the surfactant was above its critical micelle concentration. Drops containing 4–10 wt% octanol broke up on initial contact, mostly into drops larger than 5 μm in diameter, a few of them much larger. Drops containing 15–25 wt% octanol initially shrank, then increased in diameter as they became microemulsions. Many small oil droplets (1–3 μm in diameter) formed at the surface of the growing drop and rapidly coalesced to form one or more large oil drops, which subsequently became detached from the interface. Oil droplets continued to form on the microemulsion drop until it became miscible with the aqueous phase. At the end of the experiment some small droplets and the large oil drops remained dispersed in the surfactant solution. Thus, behavior was similar to that shown in Fig. 1 where the surfactant was initially dissolved in the oil phase.

At even higher alcohol concentrations (35–45 wt%) greater swelling was observed, and tiny myelinic figures of the lamellar phase could be seen at the drop surface. About the same time many oil droplets formed spontaneously, and the original drop ultimately became miscible with the aqueous phase. This behavior, which is denoted by $K(B^*)$ in Table 3, was very similar to that reported previously [10] for the same type of experiments in the amine oxide system whose self-emulsification behavior is discussed next.

Similar experiments also summarized in Table 3 were conducted where the surfactant solution initially contained 0.027 wt% octanol – about half the solubility of octanol in water. Because the concentration gradient for diffusion of octanol into water was reduced, the various dynamic phenomena generally occurred later and developed more slowly than in the experiments described above. Moreover, myelinic figures were seen for drop com-

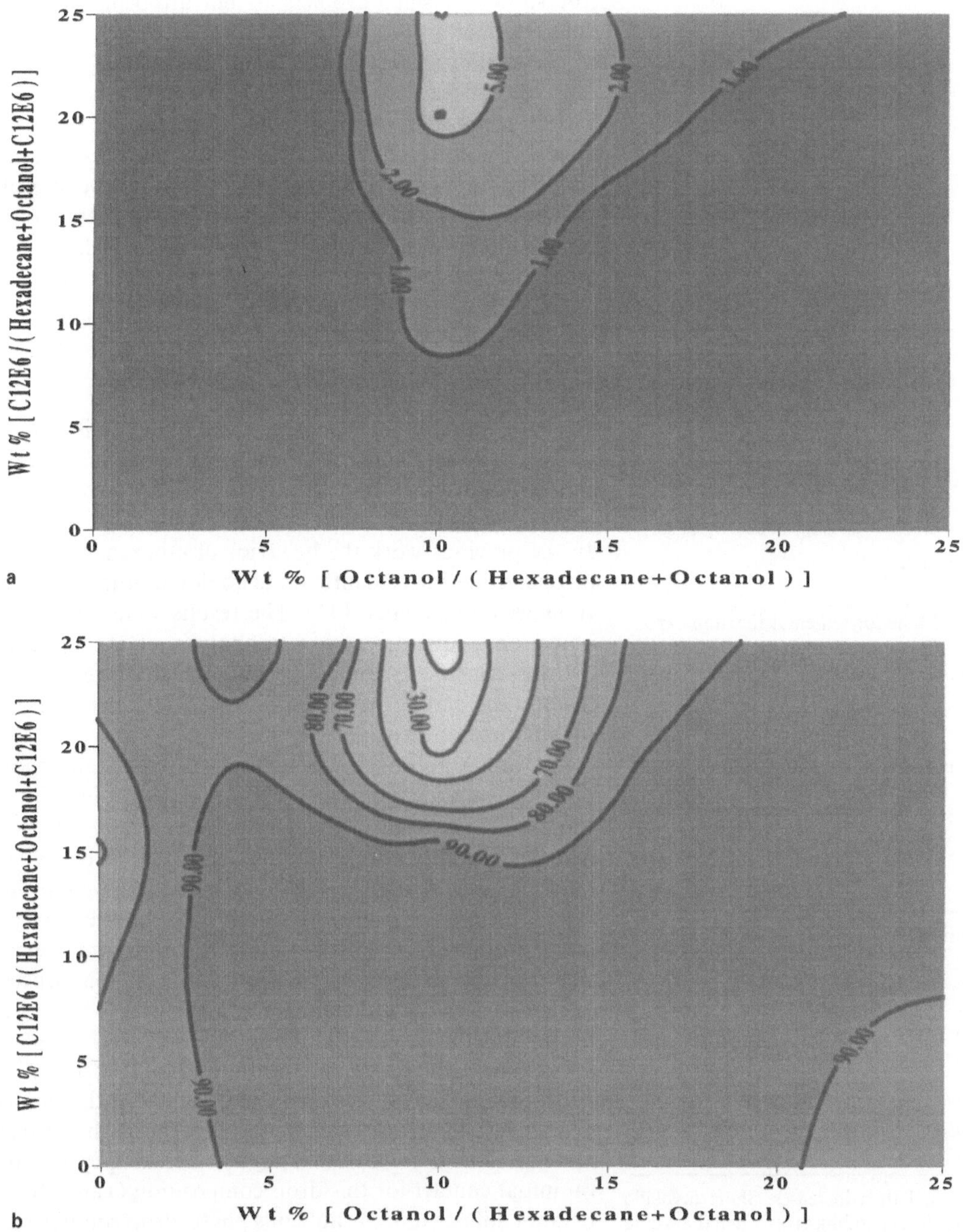

a

b

Fig. 7 (a) Time in days until emulsions of various compositions showed significant creaming or separation. (b) Turbidity after 1 h as a function of emulsion composition with mixing sequence as in Fig. 6

positions which exhibited no liquid crystal for the alcohol-free surfactant solutions. For example, Fig. 9 shows the behavior for a drop whose initial ratio of hexadecane/octanol was 70/30. The rough surface with small myelinic figures of Fig. 9c is clearly different from that of Fig. 1b where oil droplets are forming in the absence of a liquid crystal. Note also that only small droplets but no large oil drops develop (Fig. 9e).

Amine oxide/*n*-decane/alcohol/water system

Drops containing mixtures of *n*-decane, *n*-heptanol, and C_{14}DMAO were injected into deionized water at 30 °C. When the amount of heptanol was at least 10 wt% of the amount of decane and when surfactant concentration in the drop exceeded 2 wt%, the drops expanded rapidly, exhibited short myelinic figures, and broke up explosively

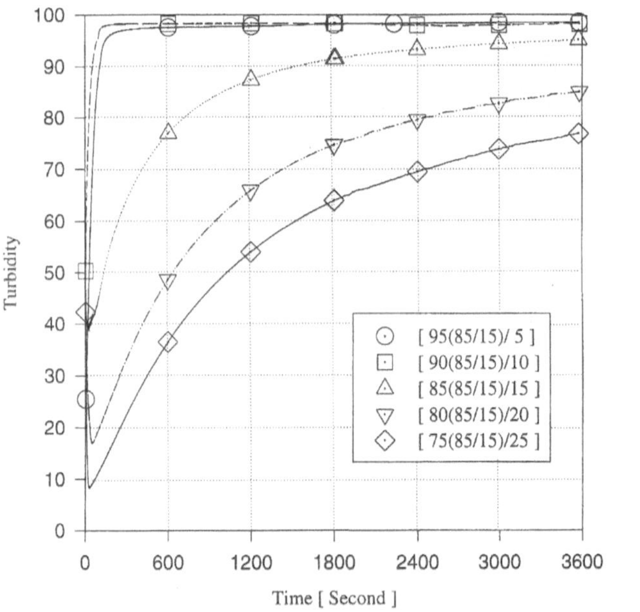

Fig. 8 Turbidity as a function of time for self-emulsification experiments with oils having a hexadecane/octanol ratio of 85/15 and various surfactant contents. Mixing sequence as in Fig. 6

Table 3 Behavior of drops of hexadecane/octanol mixtures contacting 0.05 wt% $C_{12}E_6$ aqueous solution at 30 °C

Weight ratio of C_{16} to C_8OH	Solution: 0.05 wt% $C_{12}E_6$	Solution: 0.05 wt% $C_{12}E_6$ +0.027 wt% C_8OH
100/0	No activity	—
96/4	A	—
93/7	A	—
90/10	A	A
85/15	H(A*)	H*(B)
80/20	H(A*)	H*(B)
75/25	H(A*)	K(B*)
70/30	H*(B)	K(B*)
65/35	K(B*)	—
55/45	K(B*)	—

Note. K: Drop size decreases and then increases. Short myelinic figures develop on the surface of the expanding drop. Then the whole drop is converted into many tiny droplets (B*). Other symbols are defined in Table 1

into many small droplets about 1 μm in diameter although in some cases a few large drops (up to about 30 μm) were also observed. The results of experiments with various initial drop compositions are summarized in Table 4. Only the small droplets were seen for those experiments designated K(B*) in the table. Figure 10 shows the behavior for one such drop with composition [95(90/10)5]. Figure 10a, taken only 13 s after injection, shows both the myelinic figures and the oil droplets within the original drop. Figure 10b shows the small droplets as they are being dispersed only 9 s later.

Several experiments were conducted where *n*-octanol was used instead of *n*-heptanol. The behavior was similar but occurred somewhat more slowly because of the lower rate of dissolution of octanol into water. Also less octanol than heptanol was required to achieve significant emulsification. Table 4 also summarizes these results. Here too the compositions designated K(B*) and K(C) yielded only small droplets having diameters of about 1 μm.

Discussion

$C_{12}E_6$/*n*-octanol/*n*-hexadecane/water system. Oil drop and vertical cell contacting experiments with surfactant in the oil

In the previous work the behavior of *n*-hexadecane/oleyl alcohol drops injected into dilute nonionic surfactant solutions was investigated [11]. The results were interpreted based on a quasi-steady-state analysis of the diffusion process, which was justified because the characteristic diffusion time within the drops was much shorter than the overall time of the experiments.

A similar approach can be used to understand spontaneous emulsification when the surfactant is initially in the oil drop instead of the surfactant solution. Suppose, e.g., that the initial drop composition is that of Fig. 1, i.e., [90(90/10)10]. Although this composition is well within the L_2 region, water quickly enters, hydrates the ethylene oxide groups of the surfactant, and becomes the internal phase of the water-in-oil microemulsion. If this process occurs so fast that little alcohol can diffuse into the water during this time, the path followed to the L_1/L_2 surface is approximately $m_0 - m_1$, as shown in Fig. 11. In actuality, a little octanol is lost so that the path intersects the surface at a slightly higher hexadecane/octanol ratio. The vertical cell experiments confirm that no intermediate phase forms on initial contact for this drop composition (Table 2). As octanol diffuses into the aqueous phase, drop composition moves along this surface to higher hydrocarbon/alcohol ratios, e.g., to point m_2 in Fig. 11. Little surfactant leaves the drop because for these somewhat lipophilic conditions, its solubility in water is less than the CMC. Accordingly, the ratio of hydrocarbon to surfactant in the drop remains nearly constant. Eventually, drop composition reaches a point such as m_3 at the boundary of the L_1/ME/L_2 three-phase region, the microemulsion becomes supersaturated, oil droplets form spontaneously and coalesce, and the microemulsion ultimately becomes miscible with the aqueous phase. This behavior is shown schematically in Fig. 12.

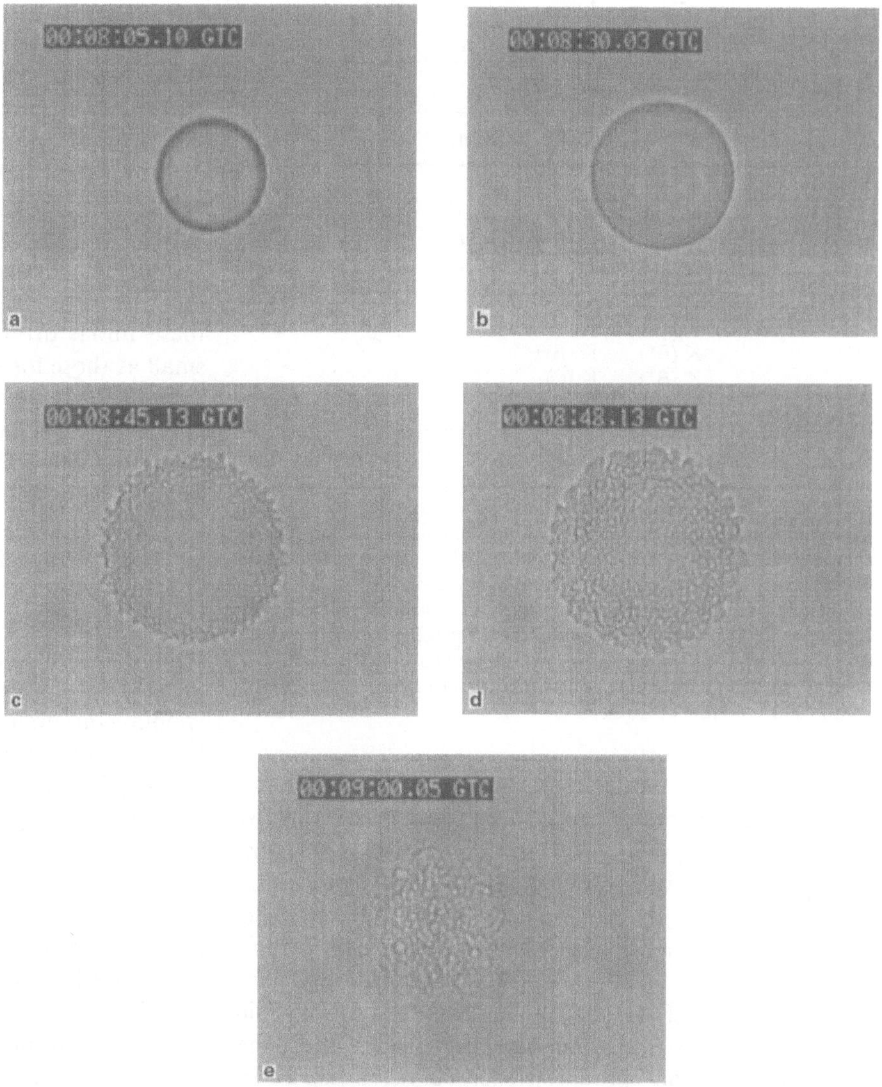

Fig. 9 Behavior of a drop initially containing 70 wt% hexadecane and 30 wt% octanol injected into an aqueous solution containing 0.05 wt% $C_{12}E_6$ and 0.027 wt% octanol at 30 °C

If the initial surfactant concentration in the drop increases so that the composition is [80(90/10)20], the line $q_0 - q_1$ connecting the initial composition and the water vertex passes through the L_α/L_2 region (see Fig. 11). That some L_α forms on contact is confirmed by the vertical cell experiment for this composition (see Table 2). However, the L_α phase apparently does not persist during the drop experiment as it was not observed during the experiment of Fig. 2. Since, as before, little oil or surfactant should leave the drop during the period immediately after injection, the overall composition of the drop should meet the L_1/L_2 surface at q_1 as shown. After this time behavior is similar to that described in the preceding paragraph except that, because the microemulsion has a larger water-to-oil ratio when it reaches the three-phase region at a point such as q_2, fewer droplets form in a larger volume, as shown in Fig. 2. As a result, less coalescence takes place.

When surfactant concentration was increased further to yield an initial composition of [75(90/10)25], we were unable to observe details of the emulsification process, only that small droplets were produced. The phase diagram of Fig. 11 indicates that when water enters a drop having this composition (s_0), intermediate L_α (at s_1) and microemulsion phases should form, behavior that is confirmed by the vertical cell experiment of Fig. 3. Moreover, with its higher surfactant content, the microemulsion phase is more hydrophilic, takes up substantial water initially (to s_2) and should not need to lose much alcohol before becoming supersaturated with oil, which may account for the rapidity of the spontaneous emulsification. Apparently, the L_α phase was again converted to a microemulsion quickly because there was no evidence of it in the videotape.

For drops with higher initial alcohol contents behavior is very similar. Indeed, with the assumption made above

Table 4 Behavior of drops of decane/alcohol/C_{14}DMAO mixtures contacting pure water at 30 °C

% C_{14}DMAO	Weight ratio of oil				
	95/5	90/10	85/15	80/20	75/25
	C_{10}/C_7OH				
5.0	A	K (B*)	A*	A*	A*
10.0	A	K (B*)	K (C)	K (C)	K (C)
15.0	!!	A*	K (C)	K (C)	K (C)
20.0	!!	!!	A*	L (C)	L (C)
	C_{10}/C_8OH				
5	K (B*)	K (B*)			
10	K (B*)	K (C)			
15	!!	M (A*)			

Note. C: Droplets are initialy small (B*) but are solubilized or broken up and dispersed rapidly. K: As defined in Table 3. Similar to Fig. 1 of ref. [10]. L: Drop shrinks initially and changes to a viscous nonspherical polygon surrounded by myelinic figures. The polygon then becomes a cluster of tiny droplets, which are dispersed and solubilized into aqueous phase. Similar to Fig. 2 of ref. [10]. M: Drop shrinks initially and changes to a viscous nonspherical polygon. Small droplets form on the surface of the drop and later are dispersed into the aqueous phase, leaving a medium-sized drop in the center. ! !: Not all surfactant dissolves in the decane/alcohol mixture. Other symbols are defined in Table 1.

that the hydrocarbon-to-surfactant ratio remains approximately constant, drops with the same initial surfactant content should have approximately the same composition when they reach the three-phase region. For example, drops with initial compositions [90(85/15)10] and [90(75/25)10] should reach the L_1/L_2 coexistence surface by taking up water and then move along the surface while losing octanol until, at some time, they reach almost the same point m_1 on the coexistence curve of the 90/10 triangle in Fig. 11 as that for the [90(90/10)10] drop discussed above. Beyond this point all three drops follow the same path along the surface to the three-phase region. Of course, the time until the three-phase region is reached is longer for drops with higher initial alcohol contents. The videomicroscopy experiments generally confirm this picture although at high surfactant contents drop size is slightly smaller when initial alcohol content is high and the hydrocarbon/surfactant ratio slightly lower for a given initial surfactant concentration.

For drops having initial compositions [85(93/7)15] and [75(93/7)25], the phase diagram again indicates and vertical cell experiments confirm development of intermediate lamellar and microemulsion phases on initial contact. However, the droplets formed spontaneously for these initial drop compositions were not as uniformly small as those for the 90/10 hexadecane/octanol ratio (see Table 1), and the emulsions were not as stable (Fig. 7a). A possible explanation is that for these systems, which are very near the PIT, alcohol diffusion into the aqueous phase rapidly shifted the system to a condition below the PIT. Thus, not all of the initial oil was converted to either L_α or microemulsion phases, and the larger oil droplets were remnants of this unconverted oil.

As a matter of interest drops of *n*-hexadecane containing various amounts of $C_{12}E_4$ and no alcohol were injected into water at 30 °C, very near the PIT. Some spontaneous emulsification was seen upon initial contact but only for surfactant concentrations exceeding about 15 wt%. The drops were somewhat smaller than in similar experiments with pure $C_{12}E_6$ listed in Table 1, probably because interfacial tension was lower. Nevertheless, they were not as small as obtained in Fig. 1 or other compositions with size designations B and B* in Table 1 for oils containing both $C_{12}E_6$ and octanol.

Figure 13 is the published phase diagram for $C_{12}E_4$/*n*-hexadecane/water at 30 °C [14] as modified slightly to incorporate results of experiments performed during the present work to better define the extent of the single-phase L_α region at relatively low water and surfactant contents. If we assume as before that water enters a $C_{12}E_4$/*n*-hexadecane drop much faster than either component can diffuse into the aqueous phase, overall drop composition follows a line connecting the initial oil

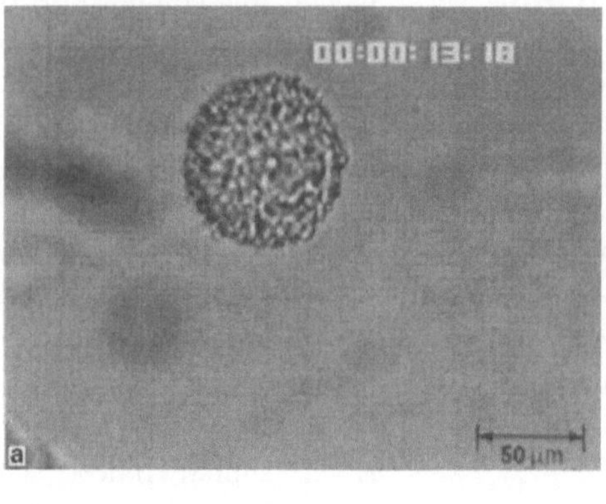

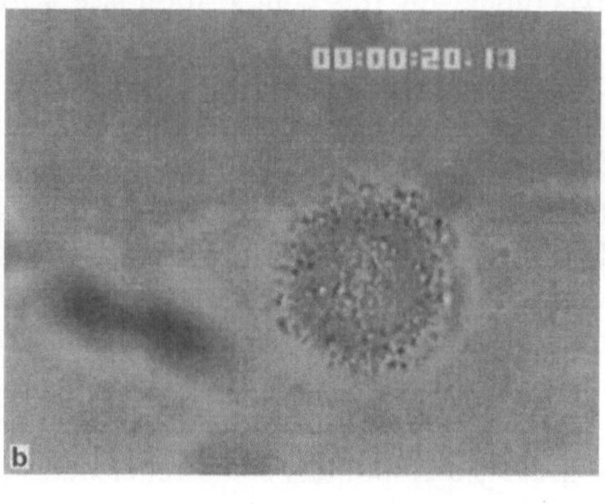

Fig. 10 Behavior of a drop having initial composition [95(90/10)5] in the decane/heptanol/C_{14}DMAO system after injection into water at 30 °C

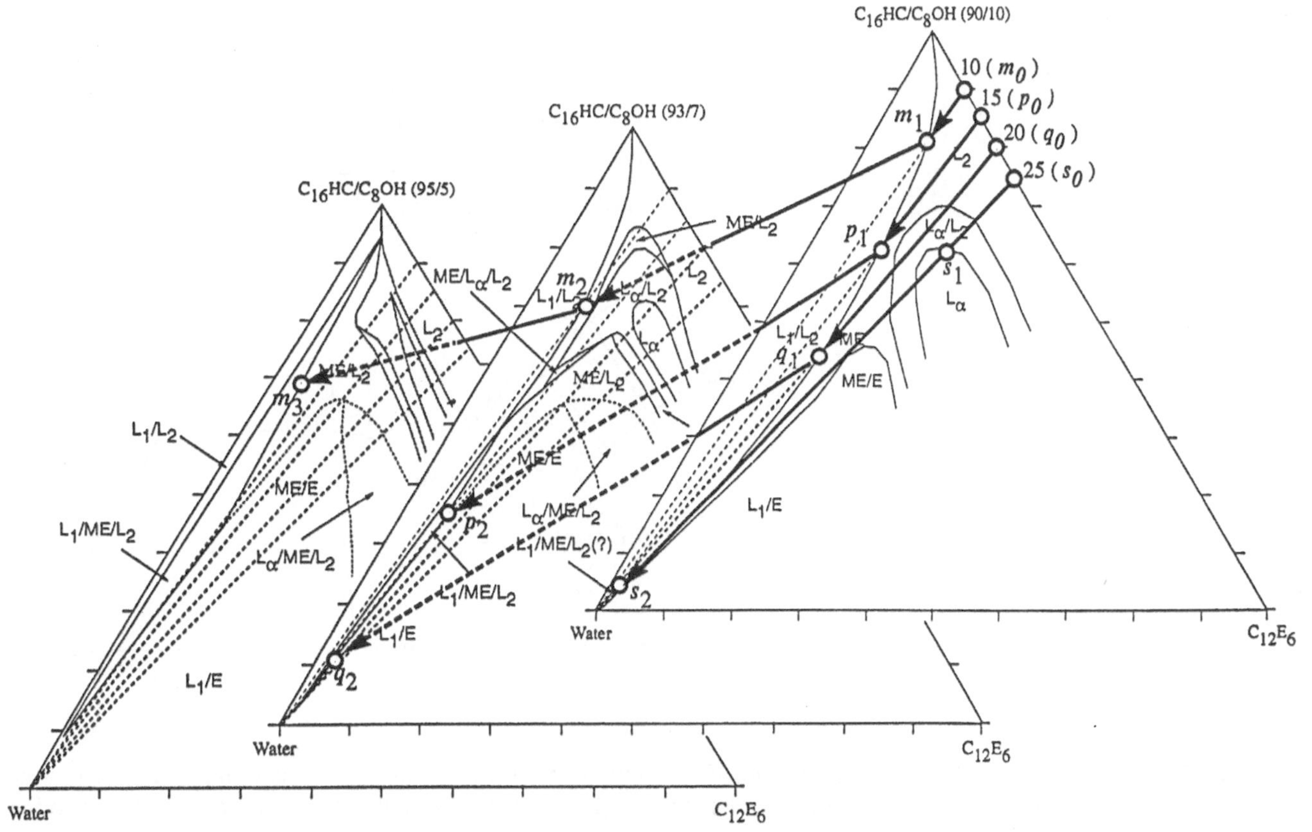

Fig. 11 Series of diagrams from Fig. 4 showing approximate composition paths of drops initially having a hexadecane/octanol ratio of 90/10 and various surfactant contents injected into water

Fig. 12 Schematic behaviour showing spontaneous emulsification process for drops of hydrocarbon/ alcohol/ surfactant contacting water

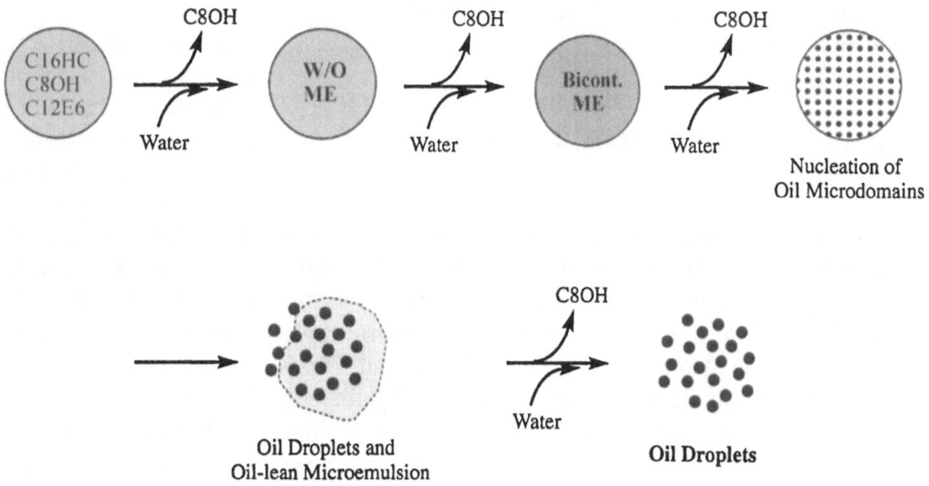

composition with the water vertex. As Fig. 13 indicates, when initial $C_{12}E_4$ content of the drop is less than about 25 wt%, all the multiphase regions on such a path involve the L_2 phase, i.e., the oil is never *completely* converted to other phases from which small droplets can be subsequently nucleated. Moreover, for higher $C_{12}E_4$ concentrations where the paths do pass through single-phase lamellar and/or microemulsion regions, the subsequent multiphase regions do not involve the L_2 phase, so that an oil-in-water emulsion is not expected in the final state. In

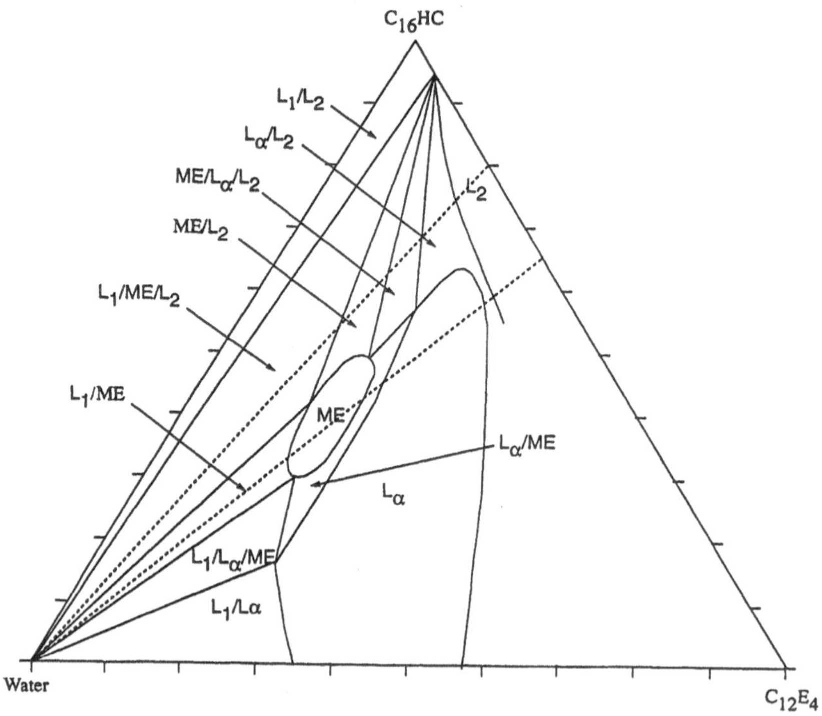

Fig. 13 Phase diagram for $C_{12}E_4$/hexadecane/water system at 30 °C. Adapted from ref. [14] with additional data from the present study

contrast, the phase behavior of the $C_{12}E_6$/octanol/hexadecane/water system is such that oil droplets are nucleated and an oil-in-water emulsion does form within a few minutes of initial contact (Figs. 1, 2, Table 1). Thus, even though both systems are near the PIT, a combination of hydrophilic and lipophilic species such as $C_{12}E_6$ and octanol provides phase behavior conducive to self-emulsification, while a single pure surfactant at the balanced condition does not.

$C_{12}E_6$/*n*-octanol/*n*-hexadecane/water system. Turbidity results

The turbidity results indicate that, for initial surfactant concentrations in the drop of at least 15 wt%, the emulsions formed from drops where the initial hexadecane/octanol ratio is 90/10 are considerably more stable than those from drops with ratios of 85/15 and 75/25 (Fig. 7a). The reason is that for the 90/10 situation, there is less alcohol in the surfactant/alcohol films in the final equilibrium state of the system, i.e., the films are more hydrophilic. As is well known, oil-in-water emulsions become more stable as their films become more hydrophilic. As mentioned previously, Fig. 8 shows that considerable coalescence occurred even during the first few minutes after mixing for the 85/15 oil.

The emulsions with hexadecane/octanol ratios exceeding 90/10 were apparently less stable because many of the oil drops formed by the emulsification process were rather large, so that less coalescence was required to produce macroscopic phase separation.

$C_{12}E_6$/*n*-octanol/*n*-hexadecane/water system. Oil drops contacting dilute surfactant solutions

The quasi-steady approach [11] is applicable here for *n*-hexadecane/*n*-octanol drops injected into dilute $C_{12}E_6$ solutions except that diffusion of alcohol into the surfactant solution must be accounted for. Let us consider drops with initial alcohol contents of at least 15 wt%. For these somewhat lipophilic conditions the analysis [11, 15] predicts that, after a brief initial transient, drop composition reaches the L_2 coexistence surface with L_1 and follows it as octanol is transferred to the aqueous phase and surfactant and water to the drop. The coexistence surface is made up of the various coexistence curves in diagrams such as Figs. 4a–f for different hexadecane/octanol ratios.

Figure 14 shows the bounding surface of the L_1/L_2 region, indicating where it is in contact with single-phase and multiphase regions. It also shows schematically paths followed by drops of various compositions along this surface until they reach a multiphase region. For example, path s_0 represents the behavior for a drop having an initial 75/25 hexadecane/octanol ratio. When the drop, which has become a microemulsion, reaches the boundary of the $L_1/ME/L_2$ three-phase region, its hexadecane-to-octanol

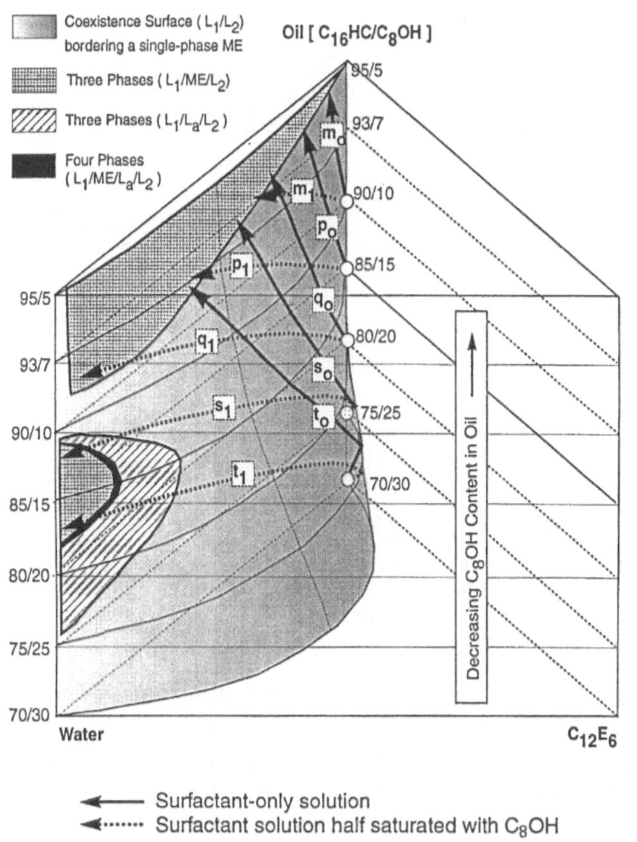

Fig. 14 Diagram showing schematic composition paths for drops of hexadecane/octanol contacting dilute $C_{12}E_6$ solutions with and without added octanol

ratio is considerably higher than the initial 75/25 value and indeed exceeds 90/10. At this time, the microemulsion has become so hydrophilic that it is supersaturated with oil, and oil droplets start to form. The low concentration sequence of Fig. 15 shows this and subsequent behavior of the drop schematically. The videomicroscopy experiment for this situation shows that the droplets do not spread at the interface between microemulsion and surfactant solution but instead form small lenses which coalesce to form larger lenses and eventually a large oil drop. As the micoremulsion continues to lose alcohol and gain surfactant and water, it becomes hydrophilic enough to be miscible with the aqueous phase, apparently passing through a critical end point. In terms of Fig. 14 the path exits the three-phase region at a relatively high overall water content and enters the L_1/L_2 region. During this time spontaneous emulsification of oil continues. Moreover, some octanol continues to diffuse into the aqueous phase from the oil drops. What remains at the end of the experiment is an oil-in-water emulsion, the oil being the large drop mentioned above and the small droplets which formed spontaneously, the aqueous phase being the initial surfactant solution with some solubilized octanol and hexadecane.

When the initial alcohol content of the surfactant solution is high or when it is partially saturated with octanol, the schematic path for the same initial drop composition is different. The behavior shown in Fig. 9, for example, is represented schematically by path t_1 of Fig. 14. The drop loses alcohol more slowly, experiences more swelling, and eventually reaches the L_1/L_α/ME three-phase region. Then the lamellar liquid crystal L_α begins to develop as an intermediate phase in the form of small myelinic figures. The high concentration sequence of Fig. 15 shows this behavior schematically. After the lamellar phase forms, the microemulsion becomes supersaturated in the L_2 phase, causing droplets to form spontaneously. This corresponds to the system entering the four-phase L_1/ME/L_α/L_2 region of Fig. 14. Perhaps the myelinic figures keep most of the droplets from contacting one another as there is much less coalescence than before. In any case the microemulsion eventually becomes miscible with the aqueous phase. The myelinic figures lose octanol and incorporate surfactant and are converted to oil droplets and the aqueous phase. At the end of the experiment an oil-in-water microemulsion and small oil droplets remain.

For drops having initial alcohol concentrations below about 10 wt%, octanol diffuses into the aqueous phase as before. The drops take up much less surfactant and hardly any water (path m_0 of Fig. 14). While the system may pass briefly through the three-phase region, thereby generating a few oil droplets by the local supersaturation scheme just described, the main result is that the surfactant solution becomes an oil-in-water microemulsion which solubilizes hydrocarbon from the drop. This process is slow, however, for the low surfactant concentrations of our experiments. Interfacial tension is apparently low enough that larger oil drops are broken up into several smaller drops during the injection process. The small drops are nevertheless not as small as the droplets formed spontaneously by the local supersaturation mechanism.

Amine oxide/*n*-decane/alcohol/water systems

The results of oil drop contacting experiments where drops of *n*-decane/*n*-heptanol were injected into 0.05 wt% solutions of C_{14}DMAO (slightly above the CMC) at 30 °C were presented previously along with partial phase diagrams for this system at selected ratios of hydrocarbon to alcohol [10]. For drops where this ratio was initially 85/15 by weight, behavior was similar to that denoted by $K(B^*)$ in Table 3, i.e., many tiny myelinic figures appeared at the surface of the expanding drop followed by spontaneous emulsification of oil droplets. Because the phase behavior

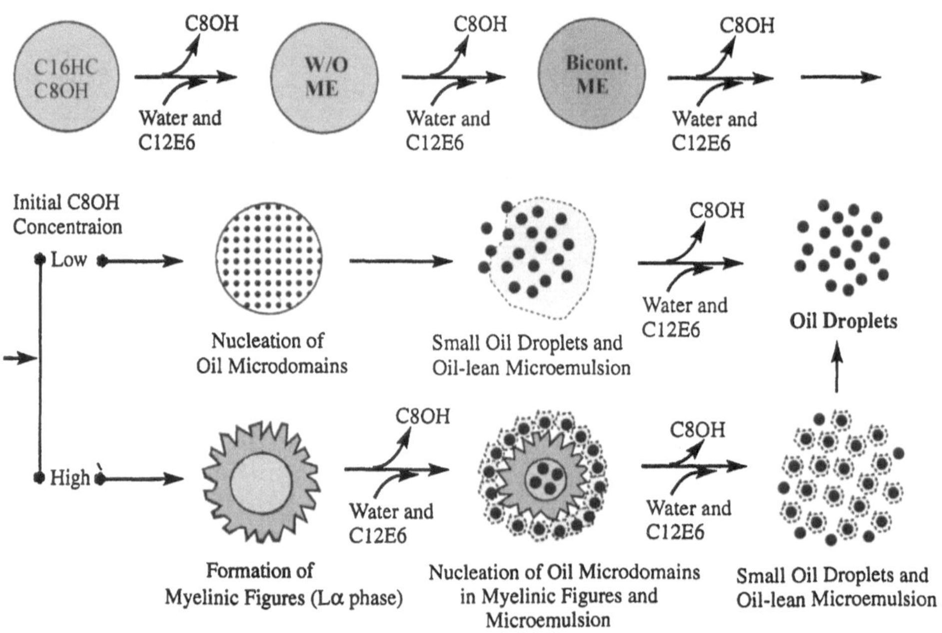

Fig. 15 Schematic behavior showing spontaneous emulsification process for drops of hydrocarbon/alcohol contacting dilute surfactant solutions. The behavior is different at high and low octanol contents in hexadecane drops contacting $C_{12}E_6$ solutions as indicated

for the two systems is qualitatively similar, the composition path followed by the 85/15 decane/C_{14}DMAO drop is similar to that of path s_1 of Fig. 14, and the behavior is as shown schematically in the high concentration sequence of Fig. 15. Existence of the four-phase region L_1/ME/L_α/L_2 has been experimentally confirmed in the C_{14}DMAO system [10]. As before, these four phases coexist simultaneously during the contacting experiment after both myelinic figures and oil droplets have formed but before the microemulsion has become miscible with water and the lamellar phase has been completely transformed into oil and microemulsion. As shown by a series of video frames in the original paper [10], all these phenomena took place within a few seconds, so that it appeared to an observer as if the original drop were exploding into many tiny oil droplets.

This behavior for a decane/heptanol drop injected into a surfactant solution is very similar to that shown in Fig. 10 for a decane/heptanol/surfactant drop injected into water. Indeed, the composition of the drop in Fig. 10, after taking up enough water to reach the L_2 coexistence surface with L_1, follows a path along the surface analogous to s_1 or t_1 of Fig. 14. The development of myelinic figures, spontaneous formation of oil drops, and miscibility between microemulsion and aqueous phases occur in the same manner as described above.

General remarks

The results of this work provide some insight as to when spontaneous emulsification can be expected when mixtures of hydrocarbon, alcohol, and surfactant are injected into water. In the systems investigated here the alcohols were more lipophilic than the surfactants but nevertheless diffused more rapidly into water because their concentrations were higher in the aqueous phase at the oil–water interface for systems initially above the PIT. Spontaneous emulsification occurred and produced the smallest oil droplets when composition of the injected drop was such that it took up a substantial amount of water shortly after contact, becoming a microemulsion. Continuing diffusion of alcohol from the drop to the aqueous phase assured that the microemulsion became more hydrophilic over time. First it became supersaturated in oil, leading to spontaneous formation of oil droplets. (In some cases a lamellar liquid crystalline phase was also involved, as discussed further below.) Later it became miscible with water, leaving the droplets dispersed in the aqueous phase. For this sequence of events to occur the initial drop had to contain enough alcohol to be slightly lipophilic and enough surfactant to assure that it would incorporate appreciable water and become a microemulsion. Also the overall system composition had to be sufficiently hydrophilic that the final equilibrium state was an aqueous micellar solution containing some solubilized oil in equilibrium with an excess oil phase. That is, the ratio of oil to surfactant obviously had to be large enough that not all of the oil was solubilized at equilibrium.

While interfacial tensions between the injected oil drops and water were doubtless low in many of the experiments reported here and indeed led to some breakup of the oil phase when it experienced shear forces during injection, this mechanical mechanism of emulsification invariably produced a mixture of large and small drops. Uniformly small oil droplets were observed only when the injected drop became a microemulsion that ultimately became supersaturated in oil as described above. However, nucleation of small droplets does not in itself assure that the emulsions formed are stable because rapid coalescence sometimes occurs with an accompanying increase in mean drop size.

This mechanism of self-emulsification is closely related to formation of an oil-in-water emulsion by cooling a microemulsion near its PIT condition or the lamellar liquid crystalline phase until it becomes supersaturated in oil [16, 17]. The difference is that supersaturation is produced spontaneously by diffusion in the self-emulsification case, so that is no need for removal of heat from the system.

As mentioned in the introduction, previous workers have observed formation of intermediate liquid crystalline phases during self-emulsification [7, 9], and some have conjectured that this phase is an essential part of the emulsification process. Our experiments with $C_{12}E_6$/octanol/hexadecane drops demonstrate that self-emulsification can occur in the absence of liquid crystal. However, for some drop compositions in this system and for the amine oxide system, spontaneous emulsification was observed when the phase behavior dictated that the lamellar liquid crystal formed as an intermediate phase. In some of these experiments it is probable that some of the oil droplets formed when the lamellar phase became supersaturated with oil although other droplets were likely the result of supersaturation of a microemulsion. In a future publication we will show that the liquid crystal plays a more important role when the medium-chain alcohols of the present work are replaced by long-chain alcohols with much lower solubilities in water.

Finally, we showed that phase behavior of pure nonionic surfactant/hydrocarbon/water systems is not favorable for self-emulsification yielding oil-in-water emulsions by the local supersaturation mechanism. A suitable mixture of hydrophilic and lipophilic species should be used instead.

References

1. Davies JT, Rideal EK (1963) Interfacial Phenomena, 2nd ed. Academic Press, New York
2. Granek R, Ball RC, Cates ME (1993) J Phys II (France) 3:829
3. Miller CA (1985) Colloids and Surfaces 29:89
4. Becher DZ (1985) In: Becher P (ed) Encyclopedia of Emulsion Technology. Vol 2. Marcel Dekker, New York, pp 239ff
5. Tadros TF (1995) In: Tadros TF (ed) Surfactants in Agrochemicals. Marcel Dekker, New York, pp 63ff
6. Constantinides PP (1995) Pharm Res 12:1561
7. Wakerly MG, Pouton CW, Meakin BJ, Morton FS (1986) In: Scamehorn JF (ed) Phenomena in Mixed Surfactant Systems ACS Symp Ser #311, pp 242ff
8. Lee GWJ, Tadros TF (1982) Colloids and Surfaces 5:105,117,129
9. Groves MJ (1978) Chemistry and Industry 17 June issue, p 417
10. Rang MJ, Miller CA, Hoffmann HH, Thunig C (1996) Ind Eng Chem Res 35:3233
11. Lim JC, Miller CA (1991) Langmuir 7:2021
12. Miller CA, Raney KH (1993) Colloids and Surfaces A 74:169
13. Rang MJ (1997) PhD Thesis, Rice University
14. Moucharafieh N, Friberg SE, Larsen DW (1979) Mol Cryst Liq Cryst 53:189
15. Rang MJ, Lim JC, Miller CA, Hoffmann HH, Thunig C (1995) J Colloid Interface Sci 175:440
16. Sagitani H (1992) In: Friberg S, Lindman B (eds) Organized Solutions. Marcel Dekker, New York, pp 259ff
17. Förster T, von Rybinski W, Wadle A (1995) Adv Colloid Interface Sci 58:119

Progr Colloid Polym Sci (1998) 109:118–125

H.-D. Dörfler

The influence of ethylene glycol on the phase behavior of aqueous microemulsions in quaternary systems

Received: 4 September 1997
Accepted: 15 September 1997

Prof. Dr. habil. H.-D. Dörfler
Technische Universität Dresden
Institut für Physikalische Chemie
u. Elektrochemie
01062 Dresden

Abstract The phase behavior of ternary and quaternary mixtures of water, ethylene glycol, *n*-dodecane and nonionic surfactant $C_{14}E_6$ was studied. Similarities and differences of the phase behavior between aqueous and nonaqueous systems were found. Replacing water by ethylene glycol in the quaternary system water + ethylene glycol/*n*-dodecane/$C_{14}E_6$ leads to a shift in the one-, two- and three-phase regions in the phase diagrams. The polar protic solvent ethylene glycol reduces the hydrophobic effect in the multicomponent systems. The result is a higher mutual solubility of ethylene glycol, a weaker adsorption at the internal interface and, hence, a lower solubilization capacity of the amphiphilics. The reduced amphiphilic strength of $C_{14}E_6$ in ethylene glycol or water + ethylene glycol mixtures corresponds to $C_{14}E_6$ in aqueous systems.

In the ternary and quaternary systems we found one-, two- and three-phase regions. However, in the waterless ethylene glycol/*n*-dodecane/$C_{14}E_6$ system only one- and two-phase regions appeared, but we have not observed the three-phase region.

Key words Nonaqueous microemulsions – phase diagrams – interfacial tension – *n*-alkyl polyglycol ether $C_{14}E_6$ – ethylene glycol

Introduction

The concept of microemulsions, micellar solutions and lyotropic mesophases (liquid crystals) has been advanced recently into the area of nonaqueous systems [1–70]. In the waterless microemulsion systems, water is replaced by polar organic liquids such as formamide, hydrazione, glycerol and ethylene glycol, sulfoxides, and so on. Some aspects about waterless microemulsions we have summarized in ref. [73].

The following studies were initiated to complete the experimental data and elucidate the properties of nonaqueous microemulsions from more systematic and phenomenological point of view. Several experiments help to classify the properties of selected systems and to assign each of them either to microemulsions or to "molecular solutions". It must be mentioned from the experimental point of view that it will be, in principal, not too easy to distinguish in the phase diagrams between the homogeneous regions of the so-called "microemulsion" and/or the "molecular solutions" in the multicomponent systems using protic and aprotic liquids.

In our previous studies of waterless microemulsions [60–62, 73] the phase diagrams and the appertaining homogeneous "microemulsion regions" were already discussed. But it seems to us that the use of the terminus "microemulsion" for the homogeneous regions are not without doubt and questions.

In order to make possible an assessment of the influences of media on structure formation of microemulsions, alterations in microemulsion phase regions caused by water addition were studied in the systems [73]. In addition, we tested the formation of the so-called three-phase regions that are typical of aqueous microemulsions [66–70] and the mixing properties of the ternary systems without surfactants [73].

Today one knows [70, 71] that the homogeneous mixture of water, oils, and nonionics may show various microstructure, ranging from "weakly structured solutions" to sprong-like structures with water and oil domains with diameter of the order of 10 nm, apparently separated by saturated monolayers of the surfactant and cosurfactant. This makes drawing a border line between "weakly structured mixtures" and microemulsions difficult.

This paper deals with liquid mixtures of water, ethylene glycol, *n*-dodecane and hexaethylene glycol tetradecylether ($C_{14}E_6$). The effect of replacing water by ethylene glycol in the quaternary system water + ethylene glycol/*n*-dodecane/$C_{14}E_6$ was studied. The properties of aqueous solutions of amphiphiles were interpreted by Tanford [74] to represent the consequence of competing forces, mainly, the hydrocarbon tails of amphiphiles and water, and the attractive hydrophilic interaction between their headgroups and water. This raises a question about interactions between amphiphiles and solvents other than water or alkanes. The modification of the polar protic solvent in microemulsion systems is the goal of our investigations.

Experimental

Materials

n-Dodecane (>99.5) and the *n*-alkyl polyglycol ether $C_{14}E_6$ (>99%) were purchased from Fluka (Switzerland). Ethylene glycol was purchased from PCK Schwedt (Germany). The substance were used as received. Water was triply distilled from dilute alkaline permanganate solution. The concentration of mixtures is given in wt%. The ethylene glycol content is expressed by the mixture ratio c_{EG} = ethylene glycol/(water + ethylene glycol) in wt%.

Phase diagram determination

Microemulsions were prepared by mixing weighed amounts of individual components. The model system chosen was water + ethylene glycol/*n*-dodecane/$C_{14}E_6$. Predetermined amounts of $C_{14}E_6$ were mixed with *n*-dodecane at fixed weight ratios and then titrated with water or water + ethylene glycol mixtures in order to detect phase transition characteristics of the system. Sections through the phase prism are measured at constant ratio water + ethylene glycol and oil varying, and recording the observed phases as function of temperature as described by Kahlweit et al. [66]. The mixtures are weighed into test tubes and sealed with Teflon stoppers. The mixtures are immersed in a so-called "Metallblockthermostat" (Techal DRI Block DB 2A; Fa. ThermoDux, Germany) thermostated to ± 0.2 K. A convenient procedure for obtaining precise phase separation temperatures has turned out. Phase equilibrium was determined by visual observation. Macroscopic phase separation was often formed to be very slow in the heterogeneous regions of the phase diagram. Details are discussed in ref. [75].

Interfacial tension measurement

It makes necessary to measure interfacial tension values of microemulsions in the region of the three-phase body down to a range of $\gamma \approx 10^{-5}$ mN m^{-1}. Such low values can be determined only by means of the spinning-drop method (SDM).

Therefore, equilibrium interfacial tension γ was measured under atmospheric pressure by SDM. Temperature was kept constant within ± 0.1 K. We have already described the details of the SDM measurements in ref. [76].

Results and discussion

Clouding temperatures of the pseudo-binary system water + ethylene glycol/$C_{14}E_6$

The aim of our investigations was to study the influence of ethylene glycol on the the miscibility in the ternary system water + ethylene glycol/$C_{14}E_6$. The ratio water + ethylene glycol was changed systematically. In Fig. 1 is shown the result of the clouding temperatures (T_c) at constant surfactant concentration ($c_T = 10$ wt%) in dependence of the water + ethylene glycol ratio (c_{EG}). In the case of the aqueous system without addition of ethylene glycol we have found for the clouding temperature $T_c = 317$ K. This value is in a good agreement with $T_c = 315$ K published by Mitchel et al. [77]. As shown in the T_c/c_{EG} plot (see Fig. 1), the clouding temperatures are increasing, if water is replaced by ethylene glycol. The water + ethylene glycol/$C_{14}E_6$ mixtures separate into two phases. This means that the upper miscibility gap shrinks in the pseudo-binary system.

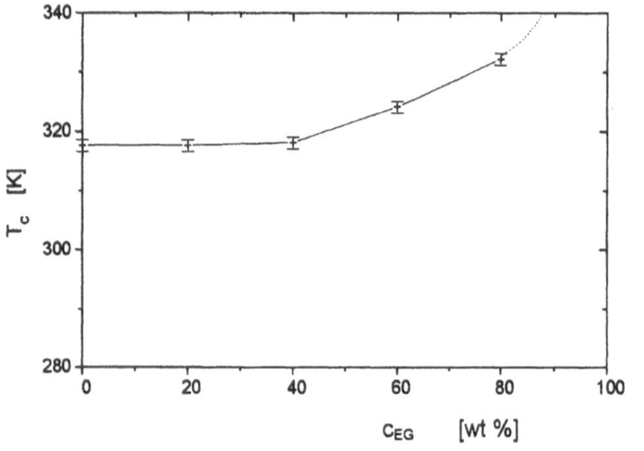

Fig. 1 Clouding temperatures (T_c) of the pseudo-binary system water + ethylene glycol/$C_{14}E_6$ in dependence on the ethylene glycol concentration c_{EG} at constant surfactant concentration (c_T = 10 wt%)

The effect of surfactant and ethylene glycol content on the phase behavior of quaternary systems water + ethylene glycol/*n*-dodecane/$C_{14}E_6$

For this type of quaternary system we have varied the content of amphiphile (c_T) and the ethylene glycol concentrations (c_{EG}) in dependence of the temperature. In Fig. 2a–f is shown sections through phase prisms at equal concentrations of polar (c_{EG}, c_W) and nonpolar (c_O) solvent ($c_{W+EG}/c_O = 1$). The comparison of the six phase diagrams leads to following results:

- According to Fig. 2a–e, up to concentrations $c_{EG} < 80$ wt% the three-phase body (3Φ), the so-called "fish", appears in the phase diagrams. Noteworthy is the fact that the "fish" always remains isometric and lies horizontally. Consequently, water and ethylene glycol truly form a pseudocomponent. Increasing ethylene glycol concentration up to $c_{EG} = 80$ wt% leads to a shift in the mean temperature of the "fish". The "fish" at $c_{EG} = 60$ wt% is bigger than the "fish" of the pure water system ($c_{EG} = 0$ wt%). In the waterless system (see Fig. 2f) no three-phase body was observed.
- After the high concentration of the three-phase body in the phase diagrams follows the one-phase region 1Φ, the so-called "fish-tail". The temperatures of the transition from the region of the three-phase body 3Φ to the isotropic microemulsions 1Φ is to identify with the HLB temperature (T_{HLB}) of the quaternary system [66]. In the c_T/T-diagrams at higher surfactant concentrations the lamellar L_α-phase is located. We have observed two types of textures by polarizing microscopy: oily streaks and mosaic textures. In Fig. 3 the pictures of these textures are shown.
- Below the "fishes" in the phase diagrams a second two-phase region (2ϕ_W) appeared: w/o microemulsions and the water containing excess phase.
- The comparison of the phase diagram in Fig. 2a–e leads to the conclusion that the replacement of water by ethylene glycol in the concentration region between $c_{EG} = 0$ wt% upt to $c_{EG} = 80$ wt% causes a shift in the one-, two- and three-phase regions to lower temperatures. The reason for the shift is the fact that the HLB temperature decreases. In Fig. 4 is plotted the T_{HLB} values against the mixture ratio water + ethylene glycol (c_{EG}).
- In the phase diagram of Fig. 2f no three-phase body appeared. We have found only the two-phase region 2ϕ_W and the crystalline phase *C*. In the surfactant concentration range $c_T < 20$ wt% and above melting temperatures T_m of the mixtures 270 K < T_m < 283 K only w/o microemulsions and the water containing excess phase are formed. This fact is in agreement with our results using other nonaqueous polar liquids [60–62].

Effect of dodecane content on the phase behavior of quaternary systems water + ethylene glycol/*n*-dodecane/$C_{14}E_6$

By our investigations we have varied as well the oil content c_O at constant surfactant concentration ($c_T = 20$ wt%) as the water + ethylene glycol ratio from pure water ($c_{EG} = 0$ wt%) to pure ethylene glycol ($c_{EG} = 100$ wt%). Our experimental results are summarized in the pseudo-binary phase diagrams in Fig. 5a–f. The comparison of the c_O/T-diagrams leads to following conclusions:

- In these systems we again observe two-phase regions 2ϕ_W (upper miscibility gap) at the top of the phase diagrams. The microemulsions are of o/w type. The other phase is the water containing excess phase.
- At higher oil concentrations downwards in the c_O/T-diagrams the region of the isotropic microemulsion (1Φ) is located.
- Below the two-phase regions 2Φ_W in the c_O/T-diagrams appears the lamellar L_α-phase. Again we observed the oily streaks and the mosaic textures (see Fig. 3).
- Some experimental difficulties arised from the phase separation in the phase regions at the bottom of the phase diagrams. From the literature [77] it is well known that in this region of the phase diagrams the low miscibility gap is formed. Because of the high viscosity in the mixtures the phase separation could not take place. During cooling of the mixtures only macroemulsions were formed.
- The comparison of the phase diagrams in Fig. 5a–e leads to the conclusion that the replacement of water by

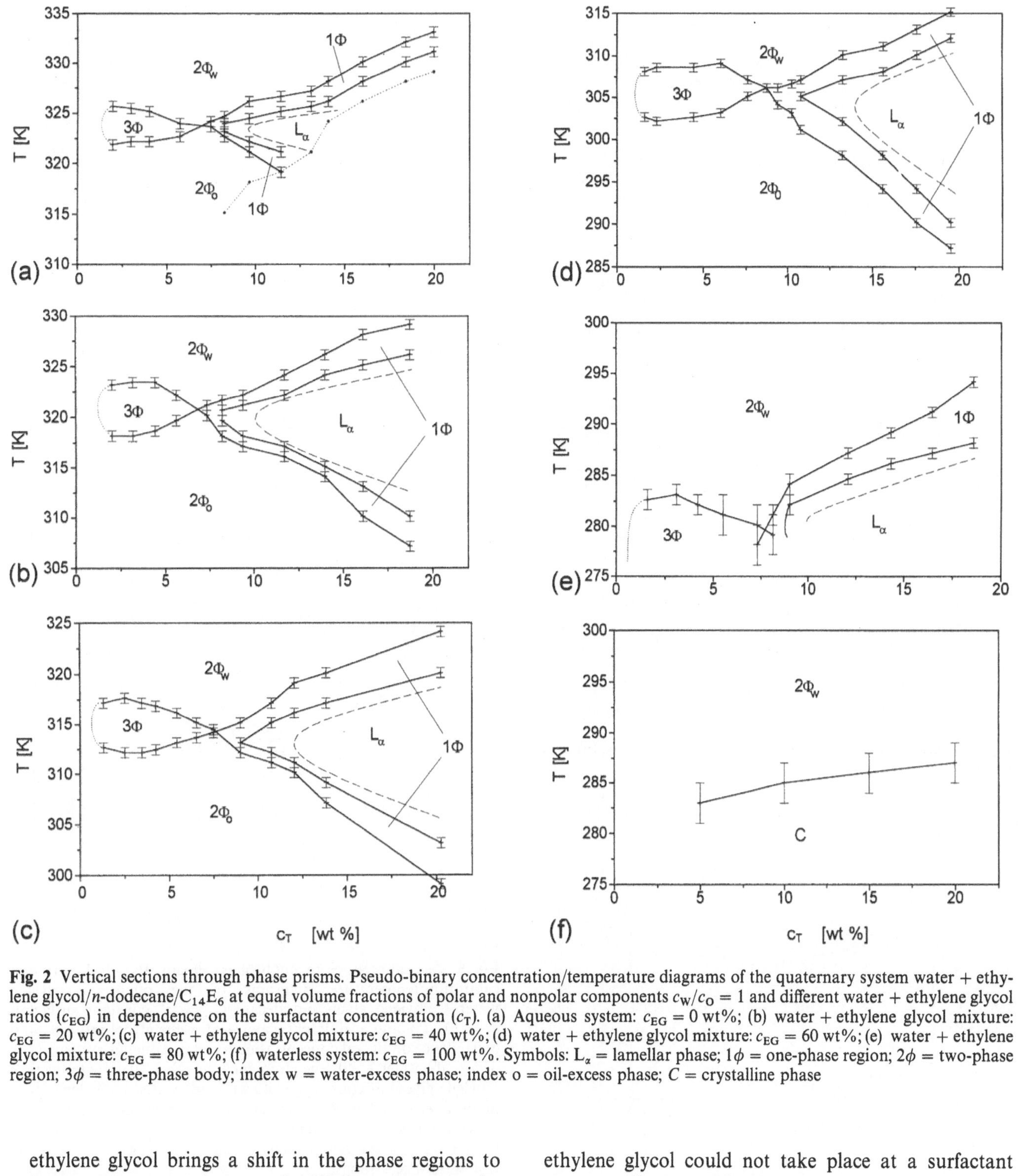

Fig. 2 Vertical sections through phase prisms. Pseudo-binary concentration/temperature diagrams of the quaternary system water + ethylene glycol/*n*-dodecane/$C_{14}E_6$ at equal volume fractions of polar and nonpolar components $c_W/c_O = 1$ and different water + ethylene glycol ratios (c_{EG}) in dependence on the surfactant concentration (c_T). (a) Aqueous system: $c_{EG} = 0$ wt%; (b) water + ethylene glycol mixture: $c_{EG} = 20$ wt%; (c) water + ethylene glycol mixture: $c_{EG} = 40$ wt%; (d) water + ethylene glycol mixture: $c_{EG} = 60$ wt%; (e) water + ethylene glycol mixture: $c_{EG} = 80$ wt%; (f) waterless system: $c_{EG} = 100$ wt%. Symbols: L_α = lamellar phase; 1ϕ = one-phase region; 2ϕ = two-phase region; 3ϕ = three-phase body; index w = water-excess phase; index o = oil-excess phase; C = crystalline phase

ethylene glycol brings a shift in the phase regions to lower temperatures because the HLB temperature decreases.

- In the waterless system (see Fig. 5e) we have found only the isotropic microemulsion 1Φ, the two-phase region $2\Phi_W$ and the crystalline phase C. Probably in this waterless system the solubilization of *n*-dodecane in ethylene glycol could not take place at a surfactant concentration $c_T = 20$ wt%.

Temperature dependence of interfacial tension

The interfacial tension γ between pure water and oil is about 50 mN m^{-1}. It is well known that the addition of

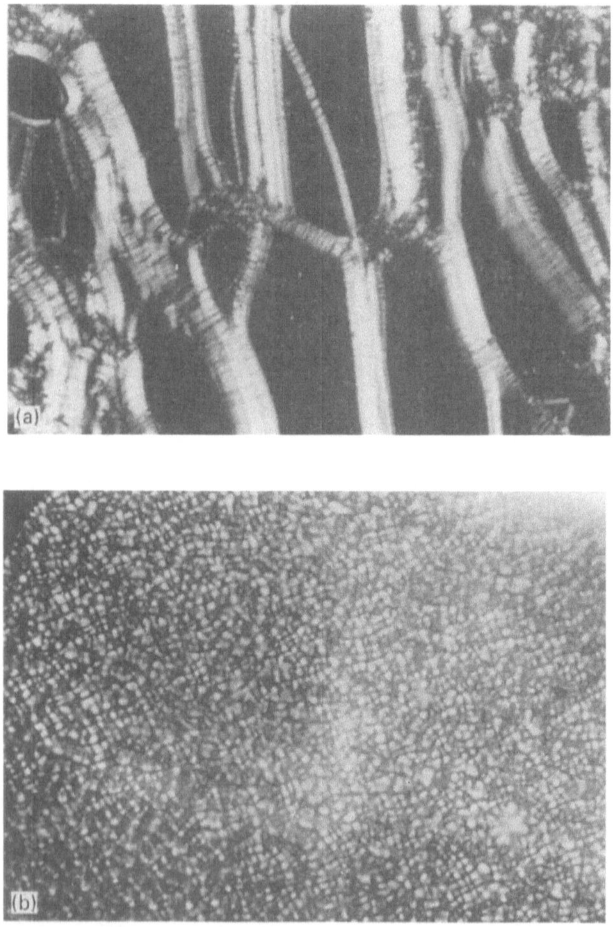

Fig. 3 Textures of the lamellar phase L_α: (a) oily streaks; (b) mosaic texture

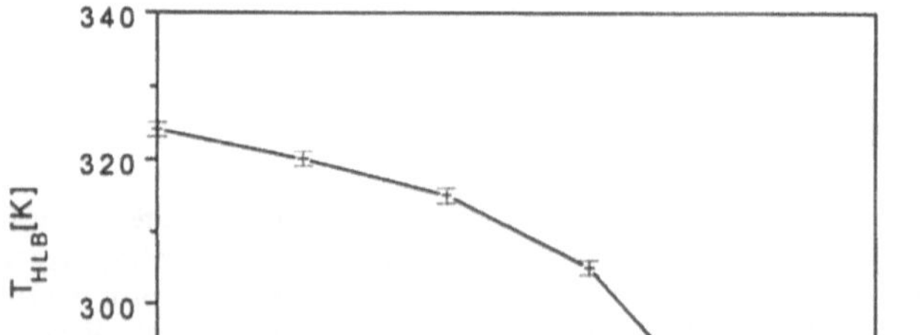

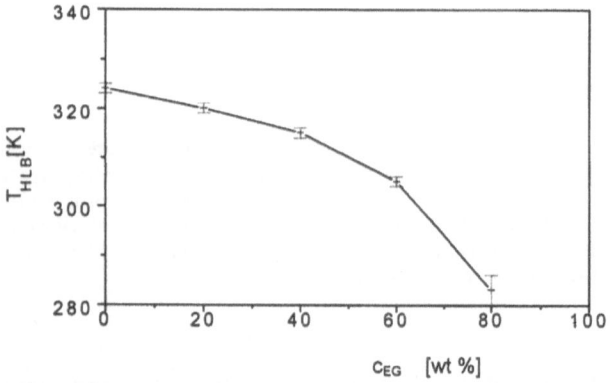

Fig. 4 HLB temperature (T_{HLB}) of the quaternary system water + ethylene glycol/*n*-dodecane/$C_{14}E_6$ in the dependence on the water + ethylene glycol ratio

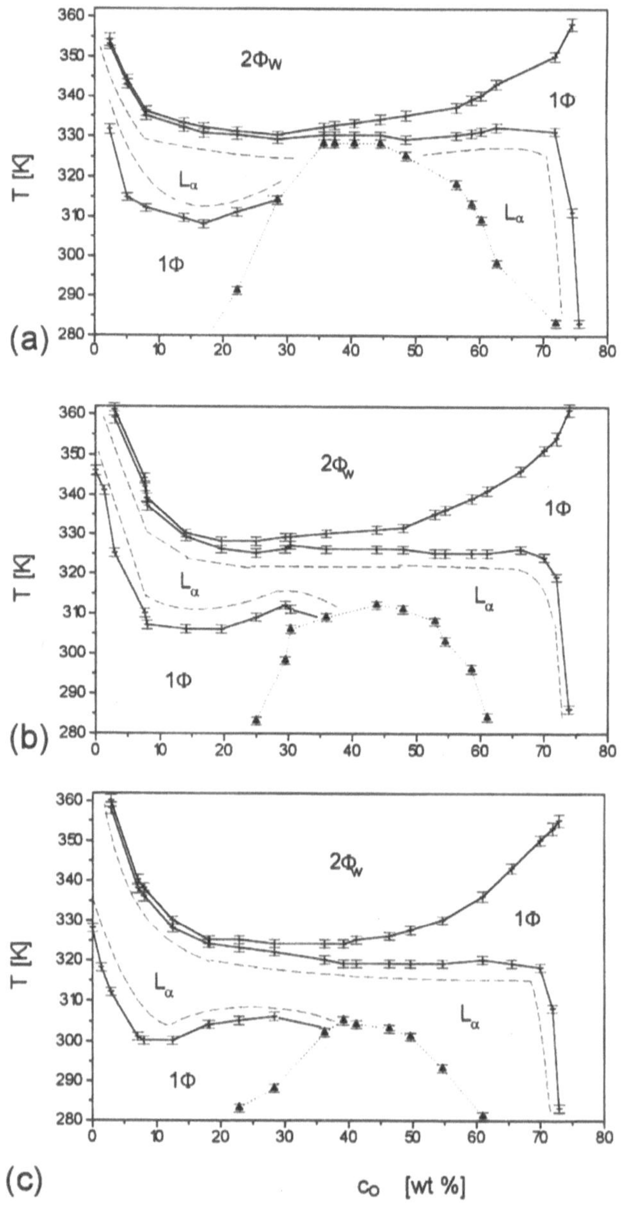

Fig. 5 Vertical sections through phase prisms. Pseudo-binary concentration/temperature diagrams of the quaternary system water + ethylene glycol/*n*-dodecane/$C_{14}E_6$ at constant surfactant $c_T = 20$ wt% and at different water + ethylene glycol ratios (c_{EG}) in dependence on the oil concentration (c_O). (a) Aqueous system: $c_{EG} = 0$ wt%; (b) water + ethylene glycol mixture: $c_{EG} = 20$ wt%; (c) water + ethylene glycol mixture: $c_{EG} = 40$ wt%; (d) water + ethylene glycol mixture: $c_{EG} = 60$ wt%; (e) water + ethylene glycol mixture: $c_{EG} = 80$ wt%; (f) waterless system: $c_{EG} = 100$ wt%. Symbols: L_α = lamellar phase; C = crystalline phase; $2\phi_W$ = two-phase region; index w = water-excess phase

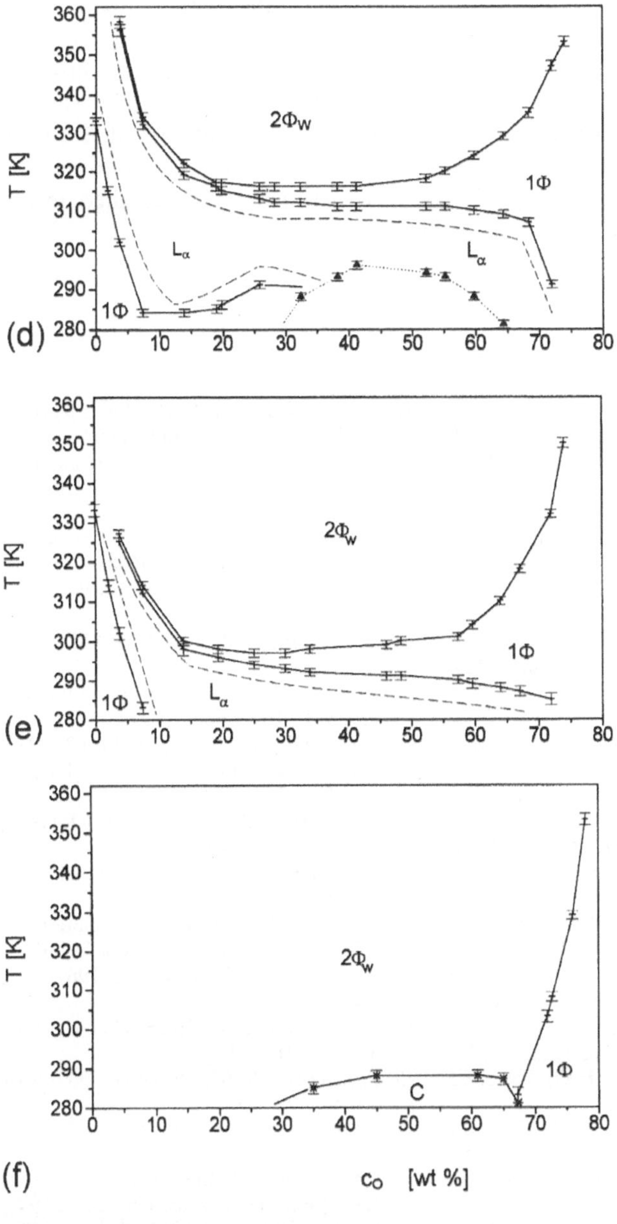

Fig. 5 (Continued)

surfactants leads to a strong decrease in interfacial tension until it becomes nearly zero at the plait points of the ternary miscibility gap [66]. If the ternary system separates into three phases, one must consider the interfacial tensions among the three phases.

The goal of our investigation is to prove the influence of ethylene glycol on the interfacial properties of the surfactant, because one of the criteria for the existence of microemulsion is the existence of low interfacial tension values in the range of three-phase body. We have measured the interfacial tension of the system water + ethylene glycol/*n*-dodecane/$C_{14}E_6$ between the water-excess phase and the

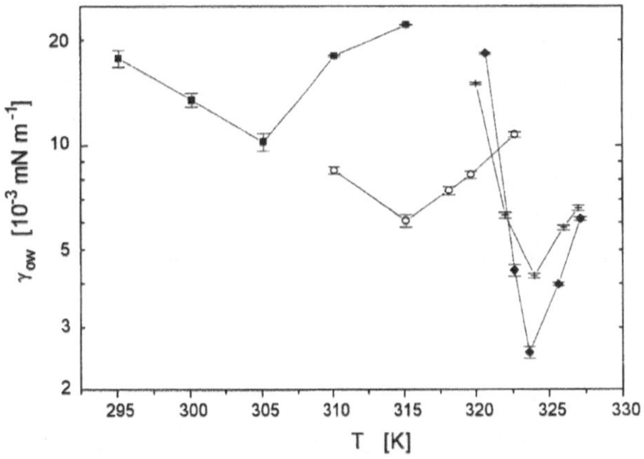

Fig. 6 Interfacial tension (γ_{OW}) in the region of the three-phase body of the quaternary system water + ethylene glycol/*n*-dodecane/$C_{14}E_6$ in dependence of the temperature at different water + ethylene glycol ratios (c_{EG}). Symbols: ◆ $c_{EG} = 0$ m-%; + $c_{EG} = 20$ m-%; ○ $c_{EG} = 40$ m-%; ■ $c_{EG} = 60$ m-%

oil-excess phase (see Fig. 2). The composition of the mixtures was: $c_O = 49$ wt% *n*-dodecane; $c_T = 2$ m-% $C_{14}E_6$. In the mixtures the ratio were varied water + ethylene glycol: $c_{EG} = 0$ wt%; $c_{EG} = 20$ wt%; $c_{EG} = 40$ wt%; $c_{EG} = 60$ wt%. The results of the interfacial tension measurements are shown in Fig. 6. We have studied the interfacial tension γ_{WO} in dependence on the temperature T.

The comparison of the curvature of the γ_{OW}/T plots in Fig. 6 leads to the conclusion that by replacing water by ethylene glycol effects an increase of the γ_{WO}-values. All of the curves run through a minimum. The γ_{min}-values of the aqueous system lies about $\gamma_{min} = 2.5 \times 10^{-3}$ mN m^{-1}. The stepwise substituion of water by ethylene glycol effects a shift in the γ_{OW}/T-curves. As already mentioned in the case of waterless system ethylene glycol/*n*-dodecane/$C_{14}E_6$ (see Fig. 4), we have not found the three-phase body in the temperature range 278 K < T < 358 K. In the two-phase region we have measured $\gamma_{WO} = 0.836$ mN m^{-1} at $T = 323$ K. The result is a higher mutual solubility of ethylene glycol and $C_{14}E_6$, a weaker adsorption at the internal interface with increasing c_{EG} concentrations.

Conclusions

There are similarities and differences of aqueous and nonaqueous microemulsions in ternary and quaternary systems. The same qualitative feature in phase behavior of water + ethylene glycol systems as for aqueous systems are observed. Since ethylene glycol reduces the hydrophobic effects the surfactants are less efficient in ethylene glycol or ethylene glycol/water mixtures. At ambient temperatures, the nonionic amphiphiles are more soluble in the polar phase than in the oil-rich phase. The opposite is observed at elevated temperatures. In between, the mixtures may separate into three coexisting phases. The experiment shows that the repulsive interaction between the hydrocarbon group of surfactants and ethylene glycol is considerably weaker than with water as polar solvents, which result in reduced surface activity of the amphiphiles. As a consequence, the mean temperature of the three-phase bodies increases, and the efficiency of the surfactant decreases upon replacing water by ethylene glycol. Probably, the microstructure in water/*n*-dodecane/$C_{14}E_6$ mixtures is to be comparable to the water + ethylene glycol/*n*-dodecane/$C_{14}E_6$ mixtures. This is confirmed by SAXS experiments. The detailed results are published in ref. [78].

References

1. Rand RP (1976) In: Friberg SE (ed) Food Emulsions. Marcel Dekker, New York
2. Friberg SE, Venable RL (1983) Encyclopedia of Emulsion Technology 1:287
3. Lindmann B, Danielsson J (1981) Colloid and Surfaces 3:391
4. Friberg SE (1982) Colloid and Surfaces 4:201
5. Friberg SE (1971) J Am Oil Chem Soc 48:578–581
6. Friberg SE, Shinoda K (1975) Adv Colloid Interface Sci 4:281–300
7. Moucharafieh N, Friberg SE (1979) Mol Cryst Liq Cryst 49:231
8. Larsen DW, Friberg SE, Christenson H (1980) J Am Chem Soc 102:6565
9. El Nokaly MA, Ford LD, Friberg SE, Larsen DW (1981) J Colloid Interface Sci 84:228
10. Larsen DW, Rananavare SB, El Nokaly MA, Friberg SE (1982) Finn Chem Lett: 96
11. Ganzuo L, El Nokaly MA, Friberg SE (1982) Mol Cryst Liq Cryst 72:183
12. El Nokaly MA, Friberg SE, Larsen DW (1984) Liq Cryst Ordered Fluids 4:441
13. El Nokaly MA, Friberg SE, Larsen DW (1984) J Colloid Interface Sci 98:274
14. Friberg SE, Podzimek M (1984) Colloid Polym Sci 262:252
15. Friberg SE (1985) Colloid Polym Sci 263:156
16. Friberg SE (1986) Nonaqueous Microemulsions. CRC Press, Boca Raton, FL, p 1
17. Friberg SE, Rananavare SB, Ward AJJ, Osoborne DW, Kaiser M (1988) J Phys Chem 92:18, 5181
18. Friberg SE (1988) In: Chane (ed) Flood Enhanced Oil Recovery. Smith, Washington DC, p 108
19. Friberg SE, Rong G (1988) Langmuir 4:796
20. Friberg SE, Liang YC (1989) In: Martelucci S, Chester AN (eds) Progress in Microemulsions, Plen Pub Cor
21. Friberg SE, Liang YC (1990) In: Microemulsions: Structure and Dynamics; Chap 3 "Nonaqueous Microemulsions", p 79
22. Friberg SE, Quamheyek (1990) The Structure, Dynamics and Equilibrium Properties of Colloidal Systems. Kluwer Academic Publishers, Netherlands, p 221
23. Friberg SE, Sun WM (1990) Colloid Polym Sci 8:155
24. Rico I, Lattes A (1984) Nouv J Chim 8:424
25. Escoulaa B, Hajjaji N, Rico I, Lattes A (1984) J Chem Soc Chem Comm: 1233
26. Rico I, Lattes A (1984) J Colloid Interface Sci 124:285
27. Rico I, Lattes A (1984) Nouv J Chim 8:429
28. Rico I, Lattes A (1984) J Colloid Interface Sci 102:285
29. Gautier M, Rico I, Ahmed-Zadeh Samii A, De Savignac A, Lattes A (1986) J Colloid Interface Sci 112:484
30. Rico I, Lattes A (1986) J Phys Chem 90:5870
31. Rico I, Condrec F, Perez E, Lawal JP, Lattes A (1987) J Chem Soc Chem Comm: 1205
32. Lattes A, Rico I, De Savignac A, Ahmad-Zaddeh Samii A (1987) Tetrahedron 43:1725
33. Auvray X, Anthore R, Petipas C, Rico I, Lattes A, Ahmed-Zadeh Samii A, De Savignac A (1987) Colloid Polym Sci 265:925
34. Belmajdoub A, Marchal JP, Canet D, Rico I, Lattes A (1987) New York J Chem 11:415
35. Rico I, Lattes A, Belmajdoub A, Marchal JP, Canet D (1987) Nouv J Chem 11:415
36. Auvray X, Anthore R, Petipas C, Rico I, Lattes A (1988) CRAS Paris 308:695
37. Marti MJ, De Savignac A, Rico I, Lattes A (1988) J Com Esp Det 19:531
38. Lattes A, Rico I (1989) Colloid Surfaces 35:221
39. Rico I, Lattes A (1989) J Am Chem Soc 111:7266
40. Auvray X, Petipas C, Anthore R, Rico I, Lattes A (1989) J Phys Chem 93:7458
41. Gautier M, Rico I, Lattes A (1990) J Org Chem 55:1500
42. Auvray X, Petipas C, Percha T, Anthore R (1990) J Phys Chem 94:8604
43. Evans DF, Yamauchi A, Roman R, Casassa EZ (1982) J Colloid Interface Sci 88:89
44. Ramadau MW, Evans DF, Lunweg RJ (1983) J Phys Chem 87:4538
45. Evans DF, Kater EV, Benton WJ (1983) J Phys Chem 87:5335
46. Evans DF, Ninham BW (1983) J Phys Chem 87:5025
47. Beesley AM, Evans DF, Laughlin BG (1988) J Phys Chem 92:791
48. Ray A (1971) Nature (London) 231:313
49. Ionescue LG, Fung DS (1981) J Chem Soc Faraday Trans I 77:2907
50. Singh HN, Saleem SN, Singh RP (1980) J Phys Chem 84:2191
51. Winsor PA (1974) Liq Cryst Plast Cryst 1:199
52. McDaniel RV, McIntosh TJ, Simon SA (1983) Biochim Biophys Acta 731:97
53. Reinsborough VC, Bloom H (1967) Aust J Chem 20:2583
54. Reinsborough VC, Bloom H (1968) Aust J Chem 21:1525
55. Reinsborough VC, Bloom H (1969) Aust J Chem 22:519
56. Reinsborough VC (1970) Aust J Chem 23:1471
57. Couper A, Gladden GP, Ingram B (1975) Faraday Discuss Chem Soc 59:63
58. Fletcher PDE, Galal MF, Robinson BH (1984) J Chem Soc Faraday Trans I 80:3307
59. Dörfler HD (1978) Tenside Surf Det 15:232
60. Dörfler HD, Nestler E (1990) Tenside Surf Det 27:168
61. Dörfler HD, Borrmeister E (1990) Tenside Surf Det 28:167
62. Dörfler HD, Borrmeister E (1992) Tenside Surf Det 29:154
63. Dörfler HD, Knape M (1993) Tenside Surf Det 30:196
64. Dörfler HD, Knape M (1993) Tenside Surf Det 30:359
65. Dörfler HD, Senst A (1993) Colloid Polym Sci 271:173
66. Kahlweit M, Strey R (1985) Angew Chem 97:655

67. Kahlweit M, Strey R, Busse G (1990) J Phys Chem 94:3881
68. Kahlweit M, Strey R, Firman P (1986) J Phys Chem 90:671
69. Kahlweit M, Strey R, Firman P, Haase D, Jen J, Schomäcker R (1988) Langmuir 4:499
70. Schubert KV, Strey R (1991) J Phys Chem 95:8532
71. Schubert KV, Strey R (1992) Progr Colloid Polym Sci 89:263
72. Smith GD, Barden RE (1982) Solution Behavior of Surfactants, Vol 2. Plenum, New York
73. Dörfler HD, Swaboda C (1993) Colloid Polym Sci 271:586
74. Tanford C (1980) In: The Hydrophobic Effect, 2. Wiley, New York
75. Dörfler HD, Große A, Tenside Surf Det (in press)
76. Dörfler HD, Große A, Tenside Surf Det (in press)
77. Mitchell JD, Tiddy GJT, Waring L, Bostock T, McDonald MP (1983) J Chem Soc Faraday Trans I 79:975
78. Dörfler HD, Große A, Tenside Surf Det (in press)

Progr Colloid Polym Sci (1998) 109:126–135

W. von Rybinski
Th. Förster

Fundamentals of the development and applications of new emulsion types

Received: 4 November 1997
Accepted: 14 November 1997

Dr. W. von Rybinski (✉) · Dr. Th. Förster
Henkel KGaA
Henkelstraße 67
D-40191 Düsseldorf
Germany

Abstract New types of emulsions have been developed in recent years which are based on specific characteristics of the phase behavior of water/oil/surfactant systems. The temperature-induced phase inversion of oil-in-water emulsions containing ethoxylated nonionic surfactants allows the preparation of fine-disperse, long-term stable systems. In contrast to this mixtures of oil, water and alkyl polyglycosides undergo a phase inversion when the ratio of the components of the emulsifier mixture is varied. The resulting microemulsion phase has a high temperature stability and is therefore particularly interesting for various applications. Among the numerous fields of applications cosmetic emulsions and washing and cleaning processes are of special importance. Specific examples of the exceptional properties of the emulsions are given.

Key words Emulsions – microemulsions – alkyl polyglycosides – phase inversion – phase behavior

Introduction

Emulsions are so attractive because they consist of at least one oil and one water phase, so that they are suitable solvents for water-soluble and hydrophobic substances. The good solubilizing properties of emulsions with regard to substances of different polarity is exploited mainly in the pharmaceutical and agricultural sectors, but also in detergency and in the field of soil remediation. In other applications the interactions of emulsions with solid surfaces play a central role. Cooling lubricants, rolling oil emulsions, fiber and textile auxiliaries and other lubricant emulsions have the primary task of preventing undesirable frictional effects during machining processes, or at least keeping them within acceptable limits. The water phase, with its thermal capacity, dissipates the frictional heat generated, while oil, emulsifiers and other auxiliary substances are adsorbed on the treated materials, making the surface hydrophobic and therefore causing a lubricating effect.

The scope of this article is to give an insight into the fundamentals of recent developments of emulsions and a detailed explanation of the connection between the structure and use of an emulsion. Special emphasis is laid on the current state in the various fields of application and on trends and future developments.

Fundamentals

In recent years new types of emulsions have been developed which have aroused considerable interest. Among these PIT emulsions (PIT = phase inversion temperature) and microemulsions are of special importance.

Emulsions consisting of oil, water and an ethoxylated nonionic surfactant undergo a temperature-induced phase inversion, during which a microemulsion is formed [1, 2].

Fig. 1

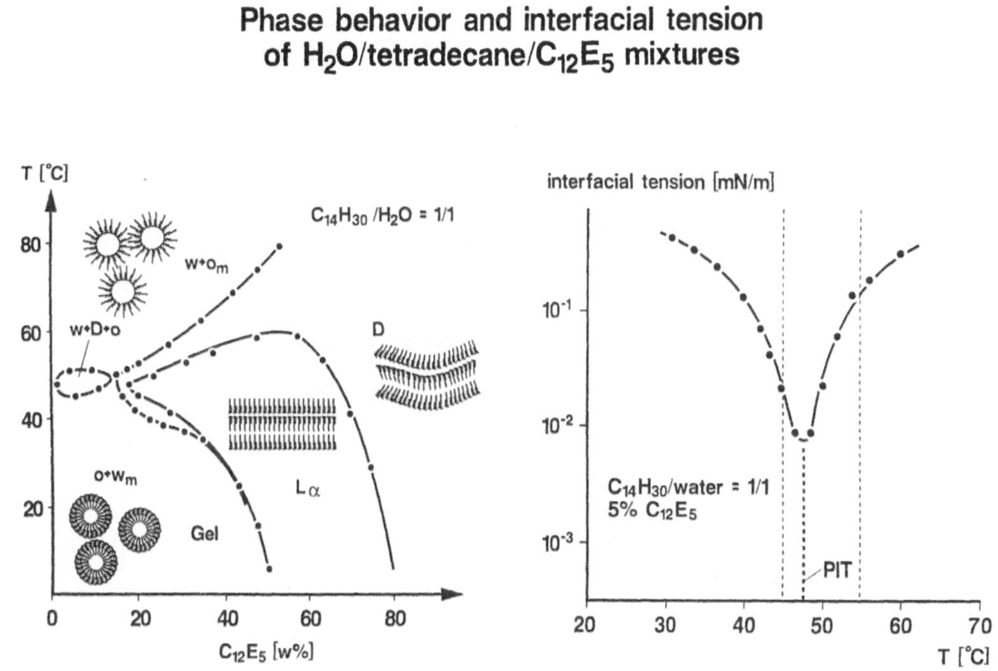

A system consisting of tetradecane, water and dodecyl polyethyleneglycol-(5)-ether ($C_{12}E_5$) inverts from an oil-in-water to a water-in-oil emulsion in the temperature range between 45 and 55 °C (Fig. 1). A microemulsion is formed in the phase inversion zone, resulting in the so-called Kahlweit fish in the phase diagram. At low emulsifier concentrations (below 15%) the microemulsion phase is in equilibrium with an oil phase and a water phase, and is therefore called three-phase microemulsion (w + D + o). Emulsifier concentrations of more than 15% are sufficient to solubilize the whole volume of water and oil in the form of a single-phase microemulsion (D) or a lamellar phase L_α.

In the PIT zone the hydrophilic and lipophilic properties of the fatty alcohol ethoxylates are in equilibrium, and this is manifested in a clear minimum on the interfacial tension curve [3].

The phase-inversion temperature process has a number of advantages. Fine-disperse emulsions can be created with a simple hot-cold process [4], during which a microemulsion is passed through at higher temperature range (Fig. 2). The fine-disperse nature of the microemulsion is partially retained after cooling [5]. Emulsions with particle sizes of about 100 nm exhibit long-term stability simply as a consequence of the Brownian molecular motion of the oil droplets [6], so low-viscosity, sprayable emulsions can also be produced.

The PIT phenomenon is predictable by parameters. With the help of characteristic variables for oil and emulsifiers, new formulations with desired components can be

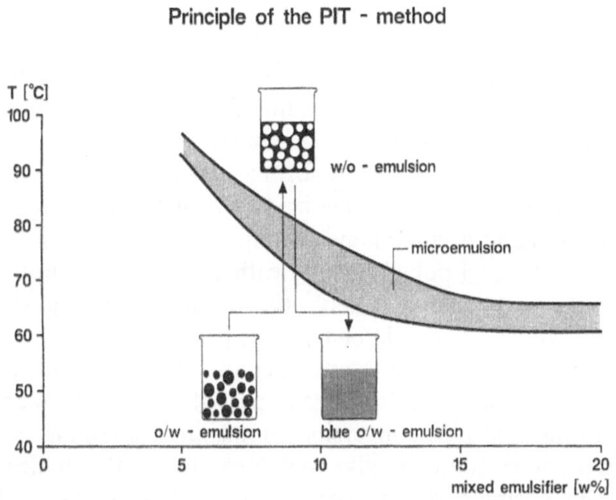

Fig. 2

calculated by means of the CAPICO procedure (calculation of phase inversion in concentrates), so that development times can be dramatically reduced [7, 8].

Microemulsions have been the subject of intensive research efforts for many years. This research has been concentrated above all on microemulsions with ethoxylated nonionic surfactants [9–13]. The main disadvantage is probably the very high concentration of surfactants – often in combination with large amounts of short

Fig. 3

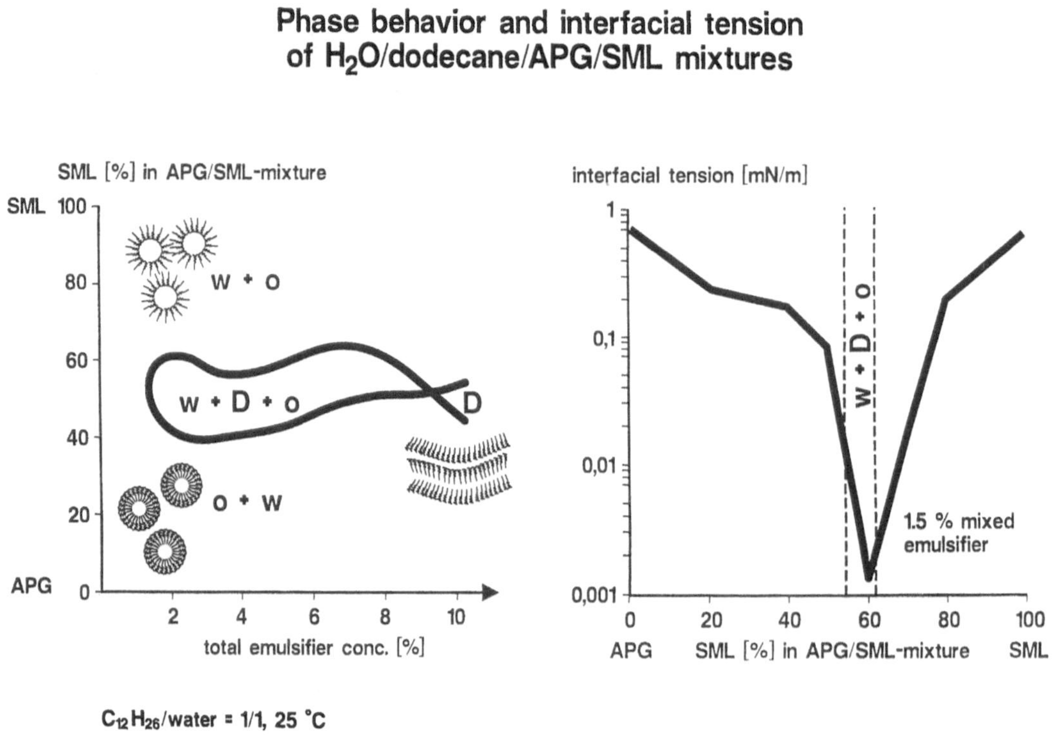

or long-chain alcohols [11–13] – that are needed to obtain stable microemulsions. In addition to this temperature stability is often within narrow limits.

The alkyl polyglycosides form a new class of nonionic surfactants that are characterized by their excellent environmental and dermatological compatibility [14]. Combinations of alkyl polyglycosides with certain hydrophobic cosurfactants are extremely effective in expediting the formation of stable microemulsions [15–17].

Whereas emulsions consisting of oil, water and an ethoxylated nonionic surfactant pass through a temperature-induced phase inversion, emulsions containing oil, water and a mixture of alkyl polyglycosides with a hydrophobic coemulsifier undergo inversion when the ratio of the components in the emulsifying mixture is varied. A system consisting of dodecane, water and lauryl glycoside, with sorbitan laurate (SML) as a hydrophobic coemulsifier, forms microemulsions when the ratio of alkyl polyglycoside to SML is in the mixing range between 4:6 and 6:4 [16, 18] (Fig. 3). W/o emulsions are formed when a higher proportion of SML is present, and o/w emulsions are formed if the proportion of alkyl polyglycosides is higher. If the total emulsifier concentration is varied the Kahlweit fish appears in the phase diagram again, with three-phase microemulsions in its body and one-phase microemulsions in its tail. The greater emulsifying power of the alkyl polyglycoside/SML mixture in comparison with the fatty alcohol ethoxylate system (Fig. 1) is demonstrated by the fact that only 10% of the mixed emulsifier suffices to form a one-phase microemulsion.

The similarity of both types of surfactant in the phase inversion zone is not limited to their phase behavior but is also apparent in the interfacial tension of the mixed emulsifier. At an alkyl polyglycoside/SML ratio of 4:6 the hydrophilic–lipophilic properties of the mixed emulsifier are in equilibrium and the interfacial tension is at a minimum. It is worth remarking that the alkyl polyglycosides/SML mixture has a very low minimum interfacial tension (about 10^{-3} mN/m), which is an order of magnitude lower than that of the comparable fatty alcohol ethoxylate system (Fig. 1).

In the case of alkyl polyglycosides-containing microemulsion, the high level of interfacial activity is attributable to the fact that the hydrophilic alkyl polyglycoside with its large polyglycoside head group is present in exactly the correct ratio with a hydrophobic coemulsifier with a smaller head group at the oil–water interface. Unlike the ethoxylated nonionic surfactants, the hydration – and therefore the effective size of the head group – is almost independent of temperature [15], so that a smaller degree of temperature dependence is to be expected for the phase behavior of the emulsion, just as for the interfacial tension [14]. This is indeed the case, as Fig. 4 illustrates for a system consisting of equal parts dioctyl cyclohexane and

Fig. 4

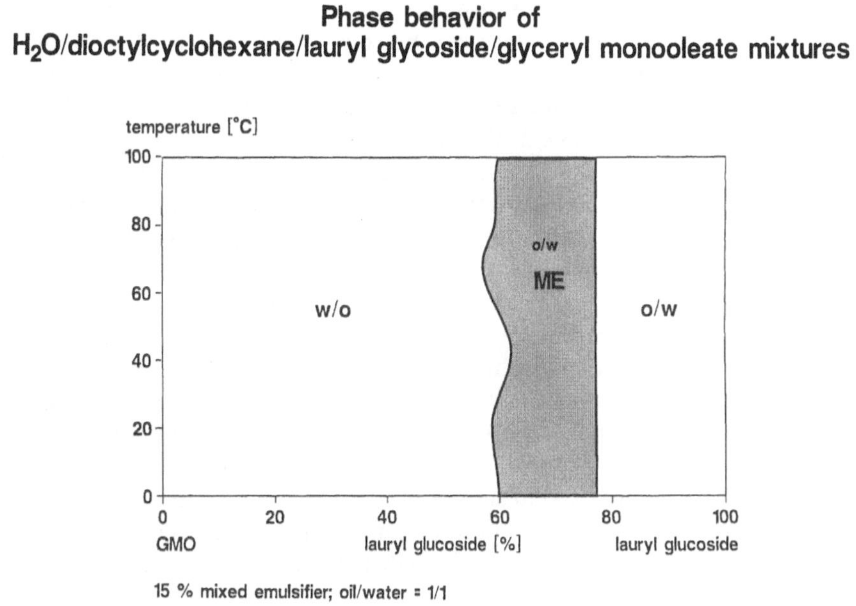

water as well as 15% of an emulsifying mixture of lauryl glycoside and glyceryl monooleate (GMO) [18]. Irrespective of temperature, at alkyl polyglycoside/GMO ratios between 60:40 and 75:25 the system forms transparent microemulsions or very fine-disperse blue emulsions (with particle sizes of about 100 nm and smaller) of the o/w type, which was determined by electrical conductivity measurements.

Applications

Cosmetics

All skin creams and lotions and many other toiletry products are emulsions. The applications of an aqueous solution on its own has no protective or caring effect on the outermost hydrophobic layer of the skin, the stratum corneum, whereas the application of an oil film makes a definite contribution to skin care but is experienced as too greasy in most cases. Only the skillful combination of water and oil in the right proportions, together with emulsifiers and lipids, imparts a pleasant sensation to skin care. The aqueous phase contains water-soluble active components such as moisturizers, for example, glycerol, urea, amino acids and sugar, or cell growth stimulators such as α-hydroxy acids or special oligopeptides. The oil phase usually consists of a number of oil components and active substances such as oil-soluble vitamins, etc. [19].

A thoroughly studied potential application for alkyl polyglycoside microemulsions in the toiletries sector is cleansing facial care products, which combine good cleansing performance with a refatting effect. Table 1 shows a model formulation for an alkyl polyglycoside microemulsion, which was studied as a 2-in-1 facial cleanser in comparison with an alkyl polyglycoside surfactant solution, which served as a standard [18].

Table 1 Model facial cleanser formulations (in wt.% active substance)

	Microemulsion	Surfactant solution
Lauryl glycoside	1.7	1.7
Coco glycoside	1.1	1.1
Glyceryl oleate	0.7	—
Dicaprylyl ether	4.0	—
Octyldodecanol	1.0	—
Perfume oil	1.0	—
1,2-Propylene glycol	5.0	5.0
Methylparaben	0.2	0.2
Water, dist.	85.3	92.0

The cleansing performance of both products in the standardized in vitro cleansing test on pig's skin with black model soiling was excellent (Table 2). Even in the presence of oil, the cleansing performance of the microemulsion remained good.

The refatting was measured in vitro and in vivo (in vitro on the pig's skin model and in vivo on the forearms of volunteers) by quantitative analysis of the marker substance dicaprylyl ether, which does not occur naturally in the skin [18]. In both models the microemulsion facial cleanser was associated with a clearly detectable release of oil. These tests therefore confirmed the 2-in-1 concept for the alkyl polyglycoside microemulsion, i.e. that very good cleansing performance can be combined with care effects in one product.

Modern skin care products contain active substances that have not only a physical effect on the skin but also a biological effect inside it. These products need to penetrate the uppermost layers of the skin to a sufficient degree before they can realize their full potential (Fig. 5).

Figure 6 illustrates this, using the example of the oil-soluble, antioxidative vitamin E to show how different galenic vehicles – in this case an oil solution, a w/o cream and two o/w creams with different particle sizes – influence penetration into the uppermost layer of the skin. The penetration studies were carried out on the most recently developed in vitro skin model, i.e. perfused bovine udder skin [20, 21]. In comparison to simple oil solutions, emulsions generally enhance penetration, because at the same total concentration the effective active substance concentration in the solvent phase – in this case in oil – is higher [22]. Moreover it can be seen that smaller amounts penetrate the skin from the two o/w emulsions than from the w/o emulsion. The cause is the lamellar gel network, which is responsible for the viscosity of the external water phase in the two o/w emulsions but simultaneously hinders the free diffusion of the vitamin E from the oil droplets into the skin [23].

As well as the direct influence of the emulsion type and structure on skin penetration, attention has recently been focused on effects, which are associated with changes in the emulsion structure after topical application. When emulsions are applied to the skin, they are spread out over it in the form of a thin film. The emulsions heat up to skin temperature, and components with a high vapour pressure begin to evaporate [22, 24–26]. Drying experiments on thin emulsion films show that – depending on atmospheric humidity and layer thickness – almost all of the water escapes from the emulsion within 5–10 min. In this window of time there is a change in the viscosity behavior of the emulsion, and in some cases also in its structure. Such phase changes may have dramatic effects on the thermodynamic activity of active substances, which can be exploited to expedite their liberation. This can be of considerable interest for not only the cosmetics but also the pharmaceutical sector [27].

So-called temperature/water-content maps give an impression of the changes in the galenic vehicles during the drying process. Emulsions with a defined water content are prepared for this purpose, and are characterized with

Table 2 Cleansing performance and refatting effect of the facial cleanser

	Microemulsion	Surfactant solution
Cleansing performance [%] in vitro	95 ± 3	93 ± 1
Refatting in vitro Marker: Dicaprylyl ether	300 μg/100 cm²	0
Refatting in vivo (12 volunteers) Marker: Dicaprylyl ether	270 μg/100 cm²	0

Fig. 5

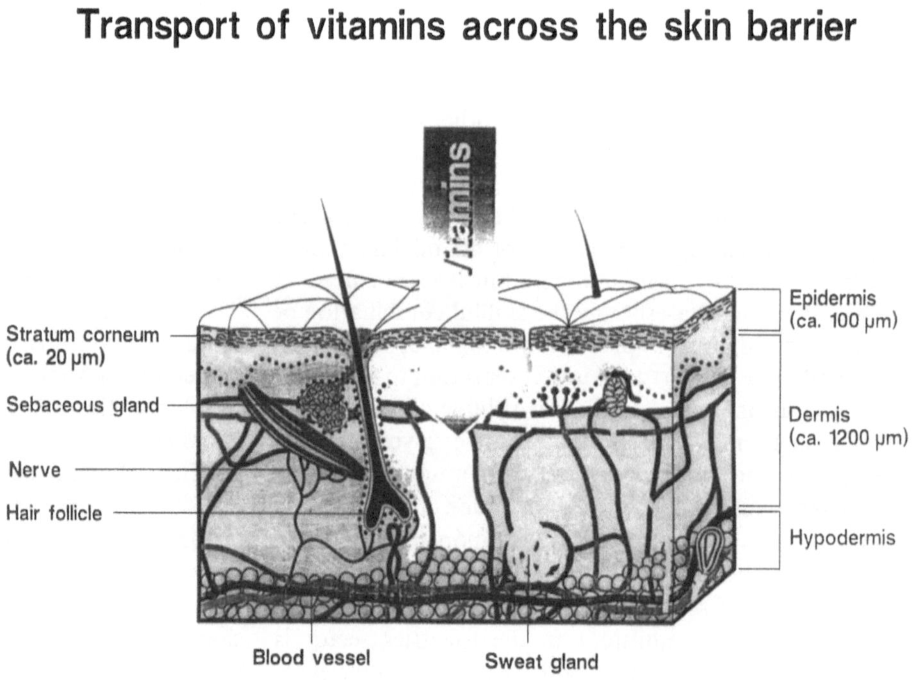

Fig. 6

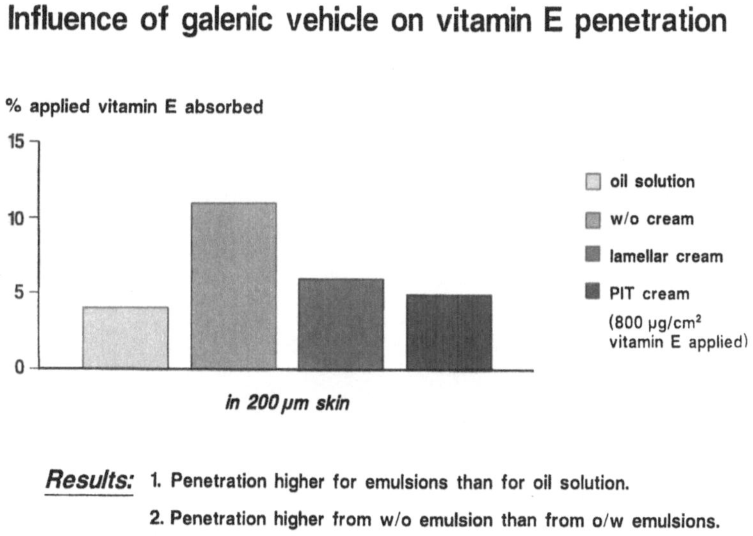

Fig. 7

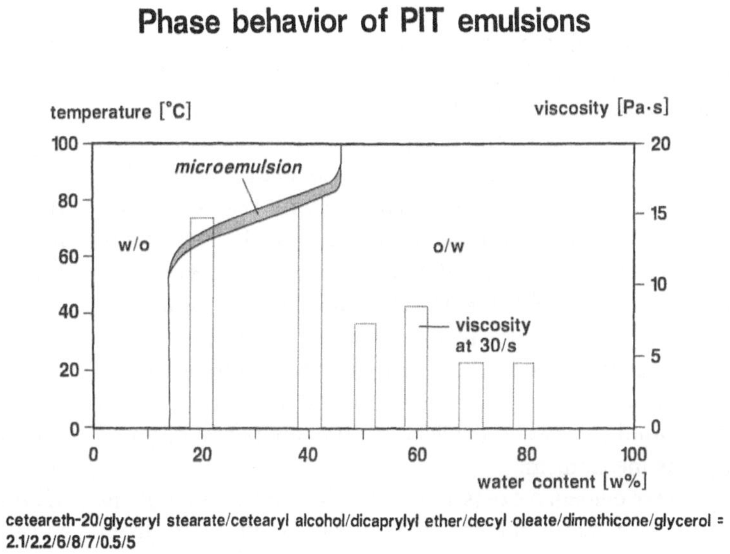

regard to their phase behavior and viscosity [22]. In Fig. 7 the PIT cream is used as an example to illustrate the viscosity and phase changes that can be expected in o/w emulsions.

The PIT cream passes through a phase inversion from o/w to w/o when the water content is in the range from 20 to just below 50% and the temperature is between 60 and 80 °C. A microemulsion forms in the phase inversion zone, as described above. After topical application, however, the cream only reaches the skin temperature of about 30 °C, so that the microemulsion zone is not reached. Just like the lamellar emulsion, the PIT emulsion is thickened by a lamellar fatty alcohol gel. During drying the viscosity increases as the lamellar fatty alcohol gel becomes more densely packed. When the water content is low (about 15% and lower) the o/w cream undergoes a phase inversion to a w/o emulsion, caused not by the temperature but solely by the geometrical packing relationships when the water content is so low.

The situation is different in the case of the w/o cream. Here the external phase is oil and the viscosity is adjusted by increasing the proportion of the internal phase. In contrast to the o/w systems, the viscosity therefore decreases during drying.

Washing and cleaning processes

The cleaning action of aqueous detergent solutions is based above all on the action of surfactants. By rolling up

Fig. 8

Phase behavior and interfacial tension of the ternary system water-$C_{12}E_4$-decane

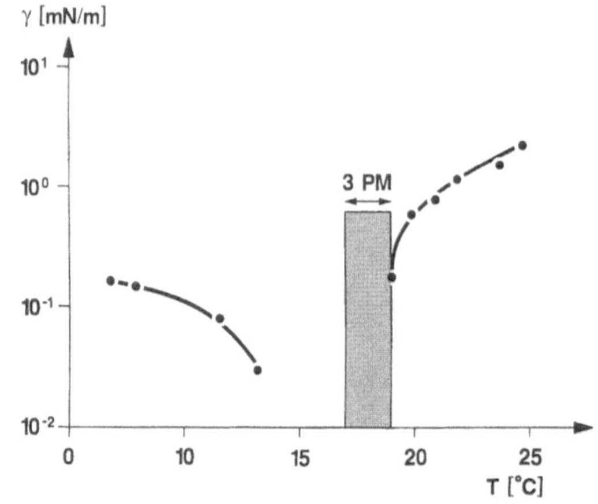

and lifting off the soil particles, then emulsifying and solubilizing them in the cleaning solution, they make the main contribution to removing soil from fabrics. A decisive factor in this context is the lowest possible interfacial tension between cleaning solution and soil or fabric [28].

Several factors have been studied with regard to their effect on the emulsification mechanism for the removal of mixtures of mineral oil and polar organic alcohols or acids from polyester [29]. The rate of emulsification of mineral oil/oleic acid mixtures from polyester films was found to change as the oleic acid content was varied. Other factors such as electrolyte concentration and temperature was also found to have large effects on the rate of soil removal by this mechanism.

Clear evidence exists that solubilization and emulsification are major factors in removal of oily soils from hydrophobic, synthetic fabrics [30, 31]. Unlike roll-up, in which the interaction of the fabric with the oily soil and water is the most critical factor, the solubilization–emulsification mechanism occurs primarily at the oil/detergent solution interface and is therefore directly influenced by the phase behavior of the corresponding oil–water–surfactant system.

An important feature of the phase behavior of systems containing water, surfactants, and hydrocarbon soils is the existence of microemulsions, which are thermodynamically stable liquid phases containing substantial amounts of both water and oil.

The condition for which the hydrophilic and lipophilic properties are exactly balanced and the surfactant films have no spontaneous tendency to curve in either direction has been called the phase inversion temperature (PIT) or hydrophile–lipophile balance (HLB) temperature by Shinoda and Friberg [32] for the case of nonionic surfactants for which temperature is usually the variable of greatest interest. For ionic surfactants it is more common to speak of "optimal" conditions, e.g. optimal salinity [33].

Schambil et al. compared the phase behavior of water/oil/surfactant systems and detergency [34]. When three-phase microemulsions are formed, extremely low interfacial tensions between the two phases are observed. Figure 8 shows the change of the interfacial tension at equilibrium in the proximity of the three-phase range for the system water/decane/$C_{12}E_4$ [35]. Because the interfacial tension is generally the restraining force with respect to the removal of liquid soil in the washing and cleaning process, it should be as low as possible for optimal soil removal. For instance, in mixtures of alkylbenzene sulfonate and alkyldiglycol ether sulfate, the minimum interfacial tension coincides with the optimum oil removal. Other quantities such as the wetting energy and the contact angle on polyester, as well as the emulsifiability of olive oil, also show optima at the same mixture ratio at which the minimum interfacial tension is observed [36].

Figure 9 (right) represents the three-phase temperature intervals for $C_{12}E_4$ and $E_{12}E_5$ vs. the number (n) of carbon atoms of n-alkanes [37]. The left side of Fig. 9 shows the results of detergency performance tests [38]. Comparison of both diagrams indicates that the maximum oil removal is in the three-phase interval of the soil used (n-hexadecane). This means that not only the solubilization capacity of the concentrated surfactant phase, but also the minimum interfacial tension existing in the zone of the three-phase body are responsible for the maximum oil removal. Further details about the influence of the polarity of the oil, the type of surfactant and the addition of salt are summarized in the review of Miller [39].

These results can be utilized to optimize surfactant systems in detergents, and in particular to improve the removal of oily soils. The formation of microemulsions is also described in the context of the pretreatment of oil-stained textiles with a mixture of water, surfactants and cosurfactants [40–42].

Aqueous cleaning liquors represent by far the best solution for the majority of practical problems encountered in the field of cleaning technology. However, some stains are very difficult to remove from textiles with aqueous cleaning liquors, e.g. bad oil stains or high-molecular lipophilic compounds (resins, waxes, paints). Such soil components can only be extracted by the organic solvents used in dry-cleaning [43]. In practice, pure organic solvents are only rarely used to clean textiles. Usually small amounts of surfactants, so-called cleaning boosters, are also present. These enable the water in the fabric to be dissolved in inverse micelles. Often small amounts of water are added, to promote the removal of hydrophilic soil and impart a higher soil-carrying capacity to the cleaning solution. The cleaning media used in dry cleaning are therefore, strictly speaking, oil-rich w/o microemulsions.

In contrast to the formation of microemulsions from aqueous surfactant systems and oily soils during the cleaning process, very little basic research has been carried out on microemulsions as a cleaning medium. Initial studies of textile cleaning with microemulsions were published by Solans et al. [44, 45]. At washing temperatures between 296 and 307 K, homogeneous microemulsions obtained from the system water/$C_{12}E_4$/n-hexadecane and systems with technical nonionic surfactant mixtures removed 1.5–2 times more soil from wool, cotton and cotton-polyester blended fabrics stained with oily and particulate soils than a highly concentrated commercial liquid detergent (Fig. 10). Soil removal by the microemulsions was increased by 20–25% by adding 0.05 mol/l of electrolytes sodium triphosphate and sodium citrate, which act as builders. The microemulsions also proved superior to the liquid detergent, in that they could be used seven times without losing any of their cleaning effectiveness.

Dörfler et al. systematically studied the phase behavior of quaternary systems, consisting of water, nonionic surfactants, a cosurfactant and a hydrocarbon, with regard to

Fig. 9

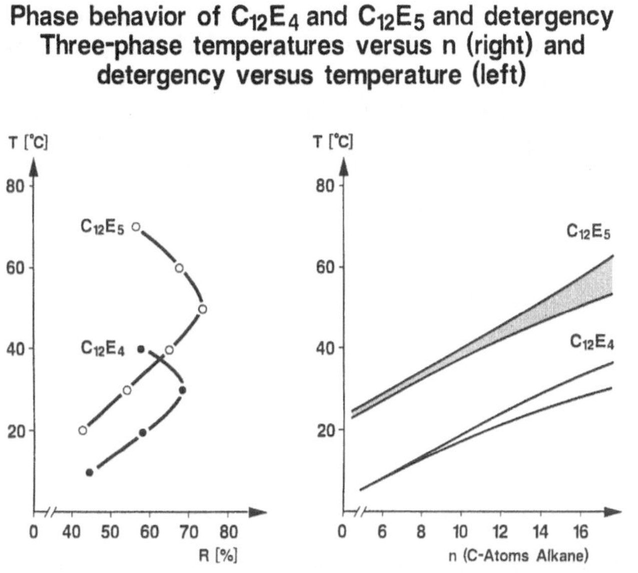

Fig. 10

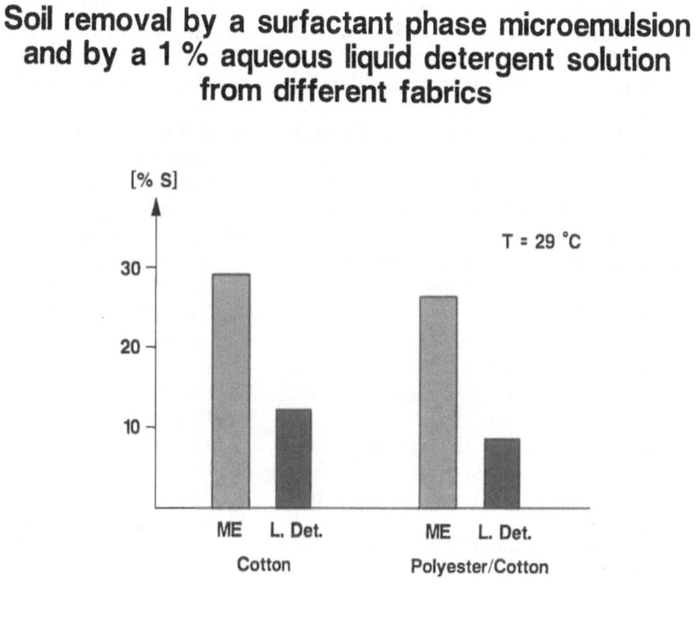

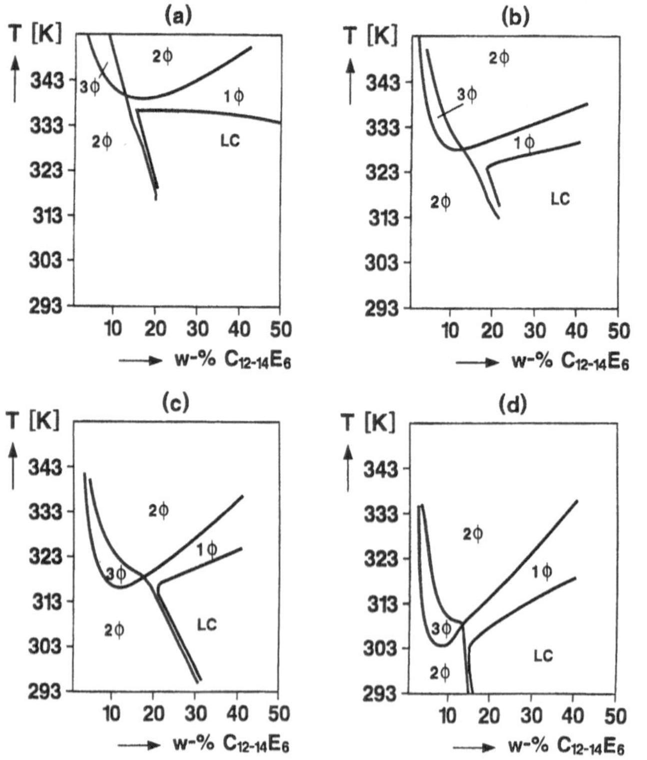

Fig. 11

possible applications in the textile cleaning sector [46]. As an example, Fig. 11 shows the influence of the cosurfactant on the phase behavior of the water-oil-surfactant system. In this case the phase inversion range decreases by an average of about 5 K per added mol-% cosurfactant ((a) without cosurfactant, (b) +2% n-pentanol, (c) +4% n-pentanol, (d) +6% n-pentanol). The extent of the three-phase zone is scarcely affected.

The influence of the degree of ethoxylation j of the nonionic surfactants ($C_{12-14}E_j; j = 5, 4, 6, 7, 8$) and added cosurfactant (n-pentanol) on the phase inversion temperature was quantified in the form of a phase diagram [46]. When microemulsions were subjected to model washing experiments with oil-stained test fabrics, the best cleaning results for oily soil were obtained in the low-water w/o microemulsion range. Besides the mentioned applications for cosmetics as well as washing and cleaning the new emulsion types have found use or are currently being discussed for many other fields of applications, e.g. metal processing [47], textile finishing [48], drilling fluids [49] and soil remediation [50]. This shows the high potential for the new fine-disperse emulsions.

References

1. Kahlweit M, Strey R (1985) Angew Chem 97:665
2. Förster Th, von Rybinski W, Wadle A (1995) Adv Colloid Interface Sci 58:119
3. Aveyard R, Binks BP, Fletcher PDI (1989) Langmuir 5:1210
4. Förster Th, Tesmann H (1991) Cosmetics Toiletries 106:49
5. Förster Th, Schambil F, von Rybinski W (1992) J Disp Sci Technol 13:183
6. Förster Th, von Rybinski W, Tesmann H, Wadle A (1994) Int J Cosmetic Sci 16:84
7. Hofmann R, Förster Th, von Rybinski W, Wadle A (1995) Progr Colloid Polym Sci 98:106
8. Wadle, Tesmann H, Leonard M, Förster Th (1997) In: Rieger MM, Rhein LD (eds) Surfactants in Cosmetics: Marcel Dekker, New York, 9:207
9. Shinoda, Kunieda H (1983) In: Becher P (ed) Encyclopedia of Emulsion Technology, Vol 1, Marcel Dekker, New York, p 337
10. Nürnberg E, Pohler W (1984) Progr Colloid Polym Sci 69:48
11. Sagitani H (1988) J Disp Sci Technol 9:115
12. Comelles F, Megias V, Sanchez J, Parra JL, Coll J, Balaguer F, Pelejero C (1989) Int J Cosmet Sci 11:5
13. Suzuki T, Nakamura M, Sumida H, Shigeta A (1992) J Soc Cosmet Chem 43:21
14. Hill K, von Rybinski W, Stoll G (1997) Alkyl Polyglycosides. VCH, Weinheim
15. Fukuda K, Söderman O, Lindman B, Shinoda K (1993) Langmuir 9:2921
16. Förster Th, Guckenbiehl B, Hensen H, von Rybinski W (1996) Progr Colloid Polym Sci 101:105
17. Guckenbiehl B, von Rybinski W, Tesmann H (1997) Proc 20th IFSCC Congr, Budapest
18. Föster Th, Guckenbiehl B, Ansmann A, Hensen H (1996) Seife Öle Fette Wachse 122:746
19. Umbach W (1995) Kosmetik. Georg Thieme Verlag, Stuttgart
20. Kietzmann M, Löscher W, Arens D, Maaß P, Lubach D (1993) J Pharmacol Toxicol Meth 30:75
21. Pittermann, Kietzmann M, Jackwerth B (1995) ALTEX 12:196
22. Förster Th, Jackwerth B, Pittermann W, von Rybinski W, Schmitt M (1997) Cosmetics Toiletries 112:73
23. Bodde HE, De Vringer T, Junginger HE (1986) Progr Colloid Polymer Sci 72:37
24. Friberg SE, Langlois B (1992) J Disp Sci Tech 13:223
25. Langlois B, Friberg SE (1993) J Soc Cosmet Chem 44:23
26. Friberg SE (1994) J Disp Sci Tech 15:359
27. Friberg SE, Brin AJ (1995) J Soc Cosmet Chem 46:255
28. Dörfler H-D (1994) Grenzflächen- und Kolloidchemie. VCH, Weinheim
29. Dillan KW, Goddard ED, McKenzie DA (1980) J Am Oil Chem Soc 57:230
30. Raney KH, Benton WJ, Miller CA (1987) J Colloid Interface Sci 117:282
31. Solans C, Azemar N (1992) In: Friberg S, Lindman B (ed) Organized Solutions. Marcel Dekker, New York, p 19
32. Shinoda K, Friberg S (1986) Emulsions and Solubilization. Wiley, New York

33. Reed RL, Healy RN (1997) In: Shah DO, Schechter RS (eds) Improved Oil Recovery by Surfactant and Polymer Flooding. Academic Press, New York, p 383
34. Schambil F, Schwuger MJ (1987) Colloid Polymer Sci 265:1009
35. Findenegg GH, Föllner B (1986) unpublished data
36. Schwuger MJ (1984) In: Rosen MJ (ed) Structure/Performance Relationships in Surfactants, AC Symp Ser Vol 253, Washington DC, p 3
37. Kahlweit M, Strey R (1985) In: Rosano HL (ed) Proc 5th Int Conf on Surface and Colloid Sci, Potsdam, New York
38. Benson HL, Cox KR, Zweig JE (1985) Happi, p 50
39. Miller CA, Raney KH (1993) Colloids Surf A: Physico-Chemical Eng Aspects 74:169
40. Kurzendörfer C-P, Wichelhaus W (1988) Henkel KGaA, EP 0288 858
41. Krüßmann H, Bercovici R (1991) J Chem Tech Biotechnol 50:399
42. Krüßmann H, Bercovici R (1993) Tenside Surf Det 30:99
43. Vaeck SV (1975) Tenside Det 12:107
44. Solans C, Garcia Dominguez J, Frieberg SE (1985) J Dispersion Sci Technol 6:523
45. Comelles F, Solans C, Azemar N, Sánchez Leal J, Porra JL (1985) Tenside Det 22:323
46. Dörfler H-D, Grosse A, Krüßmann H (1995) Tenside Surf Det 32:484
47. Paesold D, van Drunen A, Höglinger M, Niedermayr H, Cornely R, Köhler M (1994) Stahl und Eisen 114:75
48. Lautenschläger HJ, Bindl J, Huhn KG (1992) Textil Praxis Int 47:460
49. Bailey L, Denis JH, Maitland GC (1991) In: Ogden PH (ed) Chemicals in the Oil Industry: Developments and Applications. R Soc Chem Cambridge, p 53
50. Clemens WD, Haegel FH, Stickdorn K, Schwuger MJ, Webb L (1993) In: Arendt F, Annokée GJ, Bosman R, van den Brink WJ (eds) Contaminated Soil '93. Kluwer Academic Publishers, Dordrecht, p 1315

Progr Colloid Polym Sci (1998) 109:136–141
© Steinkopff Verlag 1998

V. Raicu
A. Băran
D.F. Anghel
S. Saito
A. Iovescu
C. Rădoi

Electrical conductivity of aqueous polymer solutions

2. Effects of octaethylene glycol mono(*n*-dodecyl) ether upon the poly(acrylic acid) coil

Received: 14 July 1997
Accepted: 16 July 1997

V. Raicu
Biotechnology Group
Research and Development Institute
for Radioactive and Rare Metals
Str. Atomistilor 51
76920 Bucharest
Romania

A. Băran · Dr. D.F. Anghel (✉) · A. Iovescu
C. Rădoi
Department of Colloids
Institute of Physical Chemistry
"I.G. Murgulescu", Spl. Independentei 202
77208 Bucharest
Romania

S. Saito
Nigawa-Takamaru 1-12-15
Takarazuka 665
Japan

Abstract The model previously proposed to account for electrical conductivity of charged and uncharged aqueous polymer solutions [Raicu et al. (1997) Colloid Polym Sci 275:372] is now applied to polymeric acid–nonionic surfactant systems. The polymer tested was the poly(acrylic acid) (PAA) and octaethylene glycol mono(*n*-dodecyl) ether ($C_{12}E_8$) was the surfactant. The work provides information about the shape and volume of polymer coil, and correlates their changes with the interaction critical points (T_n). It was established that the shape of the polymer coil changes during the interaction from one almost spherical (at T_1) to another corresponding to an oblate spheroid (after T'_2). The study also gives a picture of the polymer–surfactant complex.

Key words Electrical conductivity – poly(acrylic acid) – octaethylene glycol (*n*-dodecyl) ether ($C_{12}E_8$) – interaction critical points – complex model

Introduction

There are many studies about the interaction between nonionic surfactants and polymeric acids, especially between polyethoxylated nonionic surfactants and poly (acrylic acid) (PAA) and poly(methacrylic acid) (PMA) [1–10]. These studies revealed that the interaction has some critical points (T_1, T'_2, T_2) and the binding occurs cooperatively by hydrophobic attraction and by hydrogen bonds.

Based on combined measurements of surface tension, pH, viscosity, etc., it has been previously shown [8] that the interaction of PAA with the $C_{12}E_8$ nonionic surfactant may cause changes in the polymer coil. The conclusions were that, owing to the hydrogen bonding between the poly(ethylene oxide) (PEO) moieties of the surfactant and PAA, the polymer coils shrunk until the surfactant concentration reaches the so-called "T_2 critical point", (2.8 mM, for 10.2 mM PAA). As the surfactant concentration was further increased, a swelling of the polymer coil was necessary to be hypothesized in order to explain the viscosity data.

In a more recent paper [9] we proposed a model (based on the theory of Boned and Peyrelasse [11]) to account for the electrical conductivity of aqueous polymer solutions. This model was successfully applied to simple PAA/water and PEO/water solutions. If this model could also be applied to PAA/$C_{12}E_8$/water systems, it will predict a decrease in conductivity as the volume of polymer coil increase. However, the present conductivity data for 10.2 mM PAA exhibits a plateau for surfactant concentrations higher than 2.8 mM (i.e., molar fraction higher than 0.215), and consequently, the hypothesis of swelling the polymer coil above the T_2 critical point (suggested by viscosity data [8]) does no longer hold after analyzing the data. It is the aim of the present paper to reconcile the two studies. The paper also aims to get detailed (and quantitative) information regarding the coil shape and volume, and to correlate their changes (induced by the nonionic surfactant) with the interaction critical points.

Experimental

Octaethylene glycol mono(*n*-dodecyl) ether ($C_{12}E_8$) was a product of Nikko Chemicals Co., Tokyo, Japan. The poly(acrylic acid) (PAA) used was supplied by Wako Pure Chem. Co., Osaka, Japan and had $M_w \sim 150\,000$. Both substances, the surfactant and the polyacid, were used without further purification. The doubly distilled water used to prepare the solutions had an electrical conductivity lower than 2 μS/cm. The surfactant concentration was expressed as $x = C_s/(C_s + C_m)$, where C_s is the molar surfactant concentration, and C_m the unit mol polymer concentration.

The viscosity measurements were made with an Oswald viscometer having the same characteristics as previously described [8]. The conductivity measurements were carried out with an open-ended coaxial probe connected to an impedance analyzer (Hewlett-Packard Model 4194A). The measuring probe has been calibrated with saline solutions of known conductivity and permittivity, according to the method previously described [12]. The measurements were made at 2.8 MHz. The accuracy of conductivity data was within 1% for all samples. All the measurements were performed at 25 °C.

Theoretical assessment

In order to obtain detailed quantitative information on the influence of $C_{12}E_8$ surfactant upon the polymer coil, we tried to apply the theory published in the previous paper [9]. This theory provided good results when applied to pure PAA/water systems.

The electrical conductivity of PAA/water systems may be properly described [9] by the equation of Boned and Peyrelase [11]:

$$\sigma = \sigma_m(1 - \phi)^{1/L} , \quad (1)$$

where σ_m is the conductivity of continuum phase, Φ is the volume fraction of dispersed phase, $L = 6A(1 - A)/(1 + 3A)$ and A is the shape dependent depolarization factor.

For a polymer in water, having the degree of polymerization, g, and the monomer molar concentration, C_m, the volume fraction is given by

$$\Phi = C_m N_A V g , \quad (2)$$

where N_A is the Avogadro number, and V the hydrated volume of the polymer coil.

For spheroidal particles, having the semiaxes $a \neq b = c$, the depolarization factor, A, depends on the semiaxis's ratio (a/b) in a rather complicated way [9]. It is of interest here only to give their range of variation, so as to allow one to find the particle shape corresponding to the value of A obtained from the analysis of data. For prolate spheroids $(a > b = c)$ A lies in between 0 and 1/3, and for oblate spheroids $(a < b = c)$ A ranges from 1/3 to 1. The 1/3 value is the limiting case of both types of spheroids for $a = b = c$ (i.e., spherical particles).

For the PAA/water systems, the conductivity of continuum phase, σ_m, is accounted for by the electrical migration of protons and charged PAA coils [9]. For simplicity reasons, and based on the assumption of low electrical mobility of polymer coils (as compared to the proton mobility), σ_m can be well approximated by the following equation:

$$\sigma_m = C_H^+ F \mu_H^+ , \quad (3)$$

where C_H^+ is the proton concentration (in mole/m^3), μ_H^+ their mobility, and F the Faraday's number. To further simplify the analysis of conductivity data, instead of measuring the pH of each sample and subsequently calculating the σ_m (as in the first paper of this series [9]), we chose here to express C_H^+ by one of the mass action law formulations, namely:

$$C_H^+ = K_a/2[(4C_m/K_a + 1)^{1/2} - 1] . \quad (4)$$

The acidity constant, K_a, is a parameter which can be derived from the analysis of conductivity data.

In order to apply the theory derived for pure polymer solutions to the data on PAA/$C_{12}E_8$/water systems, the continuum phase was assumed to consist of water, free surfactant monomers, and protons, while the dispersed phase is made up of polymer/surfactant aggregates. The contribution of aggregates to the system overall conductivity may be quantified, via Eq. (1), by the volume fraction, Φ (see Eq. (2)), in which V is now called "the aggregate volume". At the same time, the free monomer contribution to the conductivity of continuum phase is considered as vanishingly small, and σ_m may be completely described by Eqs. (3) and (4). This assumption was verified by making conductivity measurements in the absence of polymer. The results were that, within ± 1% experimental errors, the surfactant has no influence upon the conductivity, when dispersed both, in doubly distilled water and solutions with relatively high ionic strength (20 mM KCl in water). This happens at least for surfactant concentrations lower than 6 mM, which represents the highest concentration used in this paper. The effect is more evident in the presence of PAA, since a great part of surfactant molecules is aggregated on the polymer coils.

The next step in our study was to ascertain whether the method used in the first paper for determining the geometrical characteristics of the PAA polymer coils (based on the C_m variation [9]) may be applied to the polymer/surfactant aggregates as well. This is especially necessary if one considers that the PAA–$C_{12}E_8$ interaction (and also the aggregate geometry [8]), depends on both surfactant and polymer concentration. Fortunately, the answer is affirmative and it will be argued next.

Results and discussion

Figure 1 shows the effect of $C_{12}E_8$ upon the electrical conductivity of PAA/water system at various polymer concentrations. As one may see, irrespective of polymer level, the conductivity significantly decreases by increasing the surfactant concentration, and has plateaus at low and respectively high concentration level.

When plotted against the molar fraction of $C_{12}E_8$, the relative viscosity curves of PAA–$C_{12}E_8$ systems having different PAA concentrations, are very similar in shape. This is clearly illustrated by the data in Fig. 2. Our preliminary investigations (data not shown) revealed that the same is true for surface tension and pH data. In terms of the "surfactant critical concentrations" (customarily used in relation to the PAA-nonionic surfactant interaction [8]) this means that the T_1, T_2', and T_2 critical points do not depend on the polymer concentration, when expressed as molar fraction of surfactant. Thus, it becomes conceivable to assume that, for a given value of the surfactant mole fraction, the polymer coils preserve their shape and dimensions, irrespective of the C_m value.

To obtain information about the polymer coil dimension and shape, we fitted the conductivity data on the basis of the theory mentioned above. The results (the coil volume and the shape-dependent depolarization factor, A) are presented in Figs. 3 and 4, respectively. After the T_2 critical point, according to the viscosity data, the polymer coil

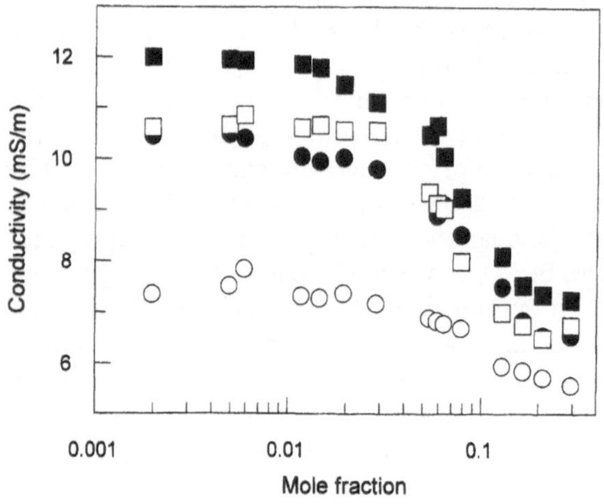

Fig. 1 Conductivity at 25 °C of PAA/$C_{12}E_8$/water systems versus the surfactant molar fraction, and various PAA concentrations: 6.0 mM (○), 10.2 mM (●), 12.0 mM (□) and 13.6 mM (■). The molecular weight of polymer was 150 000

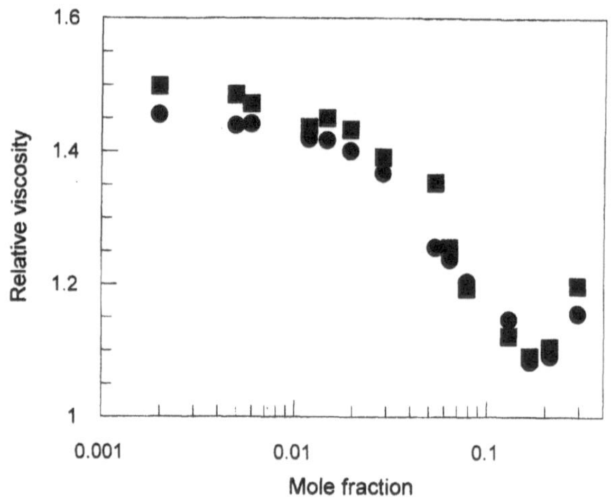

Fig. 2 Relative viscosity of PAA/$C_{12}E_8$/water system versus the surfactant molar fraction. PAA concentrations were: 10.2 mM (●), and 13.6 mM (■) at 25 °C

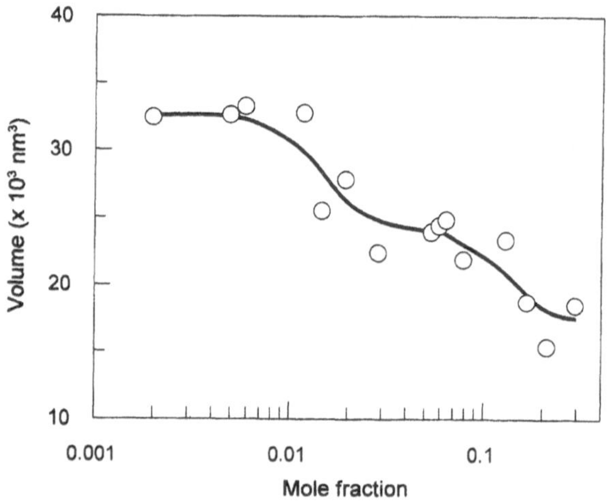

Fig. 3 The effect of $C_{12}E_8$ upon the volume of polymer coil

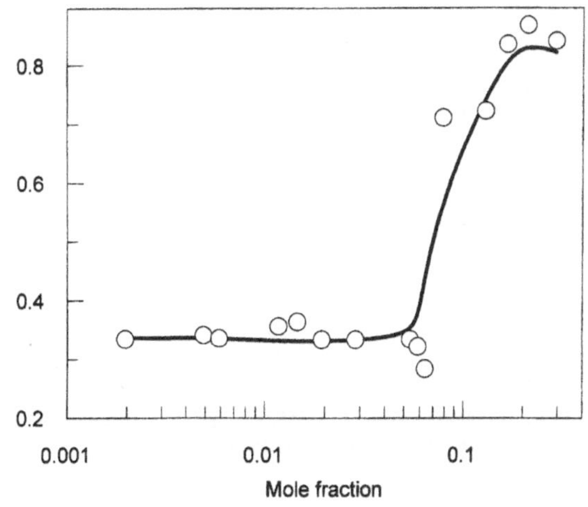

Fig. 4 The effect of $C_{12}E_8$ on the depolarization factor of polymer coil

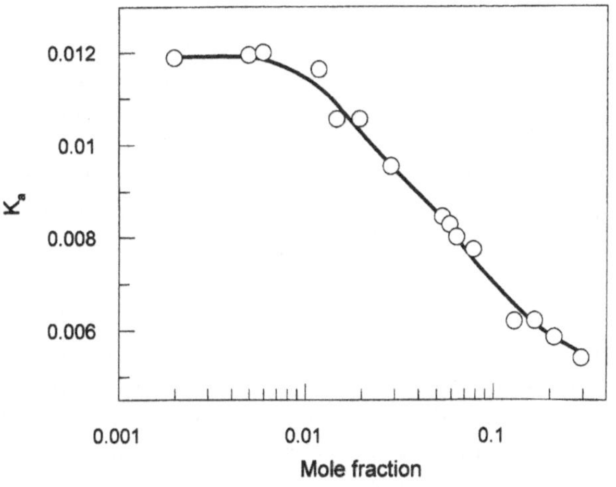

Fig. 5 The changes induced by the $C_{12}E_8$ upon the acidity constant of PAA

volume would increase (the viscosity increases), but as one can see in Fig. 3, it slowly decreases. Thus, it is necessary to make another hypothesis regarding the increase of viscosity: the surfactant alters or modifies not only the volume but also the shape of the polymer coil. The hypothesis is supported by the data concerning the surfactant effect upon the depolarization factor. For $A = 1/3$ the coil is spherical in shape, while for $A > 1/3$ it is an oblate spheroid [9]. The data presented in Fig. 4 show a steep transition in the polymer shape at a surfactant concentration of about 0.65 mM, that is rather similar to the value of T_2' critical point previously found by surface tension measurements [8].

Another parameter obtained from fitting the theory to the data was the acidity constant, K_a. The results are shown in Fig. 5. From these K_a values and on the basis of Eq. (4) one may calculate the pH of the system. The calculated pH values (for the polymer concentration of 10.2 mM) versus surfactant concentration are shown in Fig. 6. The values agree well with those from the previous paper [8], except for a downward shift of about 0.2 pH units in the present study. This is probably due to systematic errors occurring in the former study, although the ignoring of polymer coil mobility, in the present paper, could partially explain the observed shift.

Looking to the Figs. 2–4, and 6, one may easily identify the critical points T_1, T_2', and T_2. The values are presented in Table 1 and agree very well with those from the previous study [8]. Below T_1, the surfactant does not influence the polymer coil. Viscosity, volume coil, A, and pH remain constant. The polymer coil has an almost spherical shape (A is a little higher than 1/3), and its radius is 19.7 nm, that agrees well with the value of 22.2 nm from previous study [9]. At T_1, the interaction between the surfactant and

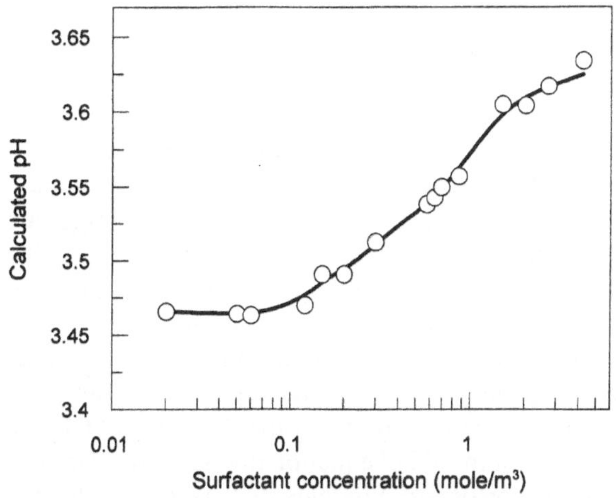

Fig. 6 The calculated pH of PAA/$C_{12}E_8$/water system versus the surfactant concentration. Calculations were based on the Eq. (4) and the data in Fig. 5. PAA concentration was of 10.2 mM

Table 1 Critical points of interaction between $C_{12}E_8$ and PAA obtained by various methods

Method	$T_1 \times 10^5$ [M]	$T_2' \times 10^4$ [M]	$T_2 \times 10^3$ [M]
Surface tension*)	6.3	8.4	2.8
Viscometry	6:6	—	2.4
A	—	6.5	2.0
pH	6.5	—	2.6

*) Data from Ref. [8].

polymer begins cooperatively by hydrophobic attraction and by hydrogen bondings. Beyond the T_1 critical point, the polymer coil begins to shrink (the viscosity and volume coil decrease; pH starts to increase) without changing its shape (A remains constant). This happens until the T_2' critical point is reached. From that point on, the polymer coil continues to shrink (viscosity and volume coil still decrease), but concomitantly it starts to change the shape (A increases) from an almost spherical one to another corresponding to an oblate spheroid. The change is not visible from viscosity data because the coil volume still decreases as a result of the interaction with the nonionic surfactant, and the number of hydrogen bondings becomes greater (see Fig. 6). Between T_2' and T_2, the decrease of volume coil covers the depolarization factor increasing (viscosity decreases). At the T_2 critical point, the interaction through hydrogen bondings ends, and the further viscosity rising may be connected to an important change occurring in the shape of the polymer coil.

It was previously shown that poly(acrylic acid) begins to interact with $C_{12}E_8$ at T_1 and free surfactant micelles appear in the system at T_2 [8]. Thus, the system will contain polymer-bound micelle-like aggregates or clusters within the $T_1 - T_2$ surfactant concentration range. The time resolved fluorescence data also unveiled that PAA wraps around the surfactant clusters and the number of $C_{12}E_8$ clusters per chain of PAA equals two above T_1, and respectively three around T_2 [10]. Now, we have to imagine how those clusters and polymer strings accommodate to each other? The following approach will resume to the complex containing three surfactant clusters per polymer chain which is by far the most complicate case encountered here. To calculate the complex dimensions, one may use the shape-dependent depolarization factor and the volume of polymer coil. By using for A a value of 0.8 (see Fig. 4), and with the aid of equation that gives the depolarization factor for oblate spheroids [9], one obtains for the ratio of major to minor axis, $a/b = 6.67$. This figure together with a value for the spheroid volume of 17 000 nm^3 (see Fig. 3) gives $a = 30$ nm and $b = 4.5$ nm. For ethoxylated nonionic surfactants, the minor axis (b_m) of the micelle lays in between 1.8 and 3.0 nm, whereas the major axis (a_m) is of 3.2–5.1 nm [13, 14]. It means that the micelle "thickness", b_m, is a little smaller than the complex minor axis (b). This is a reasonable finding because the PAA chain wraps around the micelle-like aggregate and increases the complex "thickness". As the micelle major axis (a_m) is 6–10 times shorter than the polymer coil major axis (a), there is enough room in the polymer coil to accommodate three micelle-like aggregates of the complex. Figure 7 outlines a possible arrangement of the micellar clusters within the PAA coil. For simplicity reasons, the hydrogen bondings between the PAA carboxyls and the oxygen atoms of POE chains are not shown, but the coupling is similar to that in the model recently proposed by one of the present authors [6].

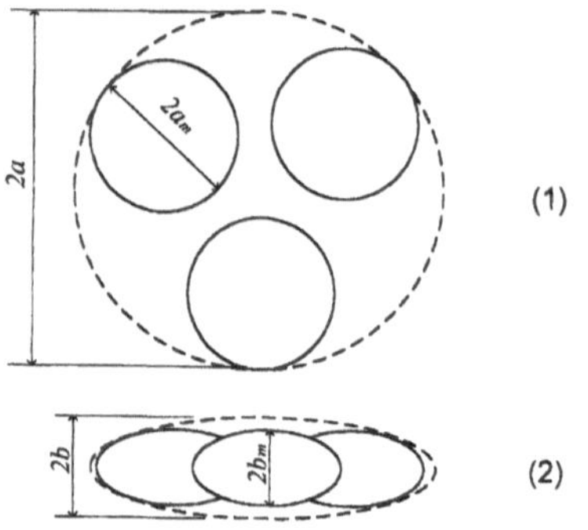

Fig. 7 Schematic representation of PAA/$C_{12}E_8$ complex in the proximity of T_2: (1) Frontal view, (2) lateral view

Conclusions

Conductivity measurements were carried out on PAA/$C_{12}E_8$/water systems. By using a suitable electrical model, the effect of $C_{12}E_8$ surfactant upon the geometrical dimensions of PAA coil has been assessed for a wide range of surfactant concentrations. It has been shown that the behavior of the system around the critical surfactant concentrations (known from previous reported studies as T_1, T_2' and T_2 critical points [8]) can be explained by changes in the polymer coil volume and/or shape. Finally, based on the present results and on the geometrical data on micelles of nonionic surfactants reported in the literature, the paper proposes the most probable model describing the polymer/surfactant complex in the neighborhood of the T_2 critical point. The complexes were depicted as oblate spheroids containing three micelles of the same spheroidal shape.

References

1. Saito S, Taniguchi T (1973) J Colloid Interface Sci 44:114–120
2. Saito S (1979) Colloid Polym Sci 257: 266–272
3. Saito S (1987) In: MJ Schick (Ed) Nonionic Surfactants Physical Chemistry. Marcel Dekker, New York, pp 881–926
4. Saito S (1989) J Am Oil Chem Soc 66: 987–993
5. Saito S (1990) Rev Roum Chim 35: 821–830

6. Saito S (1993) J Colloid Interface Sci 158:77–84
7. Maloney C, Huber K (1994) J Colloid Interface Sci 164:463–476
8. Anghel DF, Saito S, Iovescu A, Baran A (1994) Colloid Surf A 90:89–94
9. Raicu V, Baran A, Iovescu A, Anghel DF, Saito S (1997) Colloid Polym Sci 275:372–377
10. Vasilescu M, Anghel DF, Almgren M, Hansson P, Saito S (1997) Langmuir 13:6951–6955
11. Peyrelasse J, Boned C (1985) J Phys Chem 89:370–379
12. Raicu V (1995) Meas Sci Technol 6: 410–414
13. Mandal AB, Ray S, Biswas AM, Moulik SP (1980) J Phys Chem 84:856–859
14. Mandal AB, Gupta S, Moulik SP (1985) Indian J Chem 24A:670–673

Progr Colloid Polym Sci (1998) 109:142–152

M. Borkovec
G.J.M. Koper

Towards a quantitative description of ionization properties of linear and branched polyelectrolytes

Received: 24 July 1997
Accepted: 5 September 1997

Dr. Michal Borkovec (✉)
Department of Chemistry
Clarkson University
Potsdam
New York 13699-5814

Dr. Ger J.M. Koper
Leiden Institute of Chemistry
Leiden University
Gorlaeus Laboratories
P.O. Box 9502
2300 RA Leiden
The Netherlands

Abstract Ionization properties of various polymeric amines are shown to be explained in terms of a simple Ising model with short range interactions. This model is commonly used in statistical mechanics to describe magnetic systems. The basic parameters of the model are microscopic dissociation constants of each ionizable site and a small number of interaction parameters. For a system with a finite number of ionizable sites, this model reduces to the classical description of polyprotic acids and bases in terms successive equilibria. However, the model is equally well applicable to systems with a large number of ionizable sites and thus the protonation behavior of polyelectrolytes can be studied within the same framework. We shall demonstrate that one can model the macroscopic dissociation constants of a number of low molecular weight polyprotic amines, as well as the titration behavior of linear and branched poly(ethylene imine) and dendritic poly(propylene imine) with a single set of model parameters. The Ising model further provides a rationalization of the differences in ionization properties of linear and branched polyelectrolytes.

Key words Potentiometric titration – dissociation constants – aliphatic amines – Ising model

Introduction

Ionization properties of various natural polyelectrolytes represent a central topic in various applied disciplines. Such systems include proteins, polycarboxylates, cellulose or humic acids [1–6]. Proper assessment of the proton binding characteristics of such systems is a prerequisite for the understanding of the action of such macromolecules in various processes, as for example, binding of metal ions, protein folding, antigen–antibody interactions, adsorption to surfaces, colloid stability and many more. The progress in the understanding of ionization properties of such systems is commonly hampered by the complex and often unknown chemical structures of such materials; an exception being several proteins where the geometrical structure is known from X-ray diffraction. Even in such favorable cases, prediction of the ionization properties of all amino acid residues in a protein is a difficult undertaking. Ionization of proteins is usually treated in the spirit of the approach proposed by Tanford and Kirkwood [7]. Thereby, one considers the most stable conformation and performs the thermal averaging over all possible protonation states by taking electrostatic site-site interactions into account. These interactions are evaluated on the Poisson–Boltzmann or the primitive model level [1, 4]. In spite of the encouraging results obtained from this kind of approach, substantial uncertainties still persist on how to treat the dielectric discontinuity at the water–macromolecule boundary and how to include averages over possible conformational degrees of freedom [4, 8]. For rather compact macromolecules such as proteins, these methods are expected to mature in the near future, but an analogous treatment of less compact and flexible macromolecules, such as polycarboxylates [5] or humic acids [6], poses serious problems at present.

As an alternative approach to the problem, one may consider the classical framework for the description of ionization of polyelectrolytes discussed by Marcus [9] and Katchalsky et al. [10]. These authors have shown that the problem of the ionization of a linear polyelectrolyte is closely related to the thermal statistics of a linear arrangement of magnetic dipoles – a well-known problem in statistical mechanics introduced by Ising [11]. However, the model discussed within the Marcus–Katchalsky framework applies in a very specialized situation. The model assumes that all sites are arranged along a chain and only nearest neighbor pairs of sites interact. Even though this model captures the essential features of the ionization process of linear polyelectrolytes qualitatively, the inherent assumptions are too restrictive to allow a quantitative description of the titration curves of most linear polyelectrolytes. For example, for poly(maleic acid) or poly(vinyl amine) the description fails since the tacticity of the polyelectrolyte introduces a variation of the interaction parameter along the chain [13]. For poly(ethylene imine), additional types interactions are necessary [14]. Nevertheless, the Marcus–Katachalsky approach has a great advantage; the conformational degrees of freedom are averaged out and do no longer enter the calculation explicitly.

Linear poly(ethylene imine) is an example of a polyamine [15, 16]. Other common polyamines are shown in Fig. 1, and include a variety of branched structures as well. From the fundamental point of view, polyamines represent very interesting model systems since their ionization properties are known for a number of different structures [14, 17–22]. However, the number of various types of ionizable groups is limited, a fact which simplifies the problem substantially. There are three different kinds of groups – primary, secondary and tertiary amine groups. As will become apparent further below, a consistent and quantitative description of the ionization properties of all these systems is possible, provided the branching is taken into account explicitly and additional types interactions beyond the nearest neighbor pair interactions are introduced. For a linear chain, the present model description reduces the original Marcus–Katchalsky model if all ionizable groups are taken to be equal and only nearest neighbor pair interactions are taken into account. As we shall see, an extension of the Marcus–Katchalsky model to branched structures provides a framework to understand the basic features of the ionization behavior for this class of molecules as well.

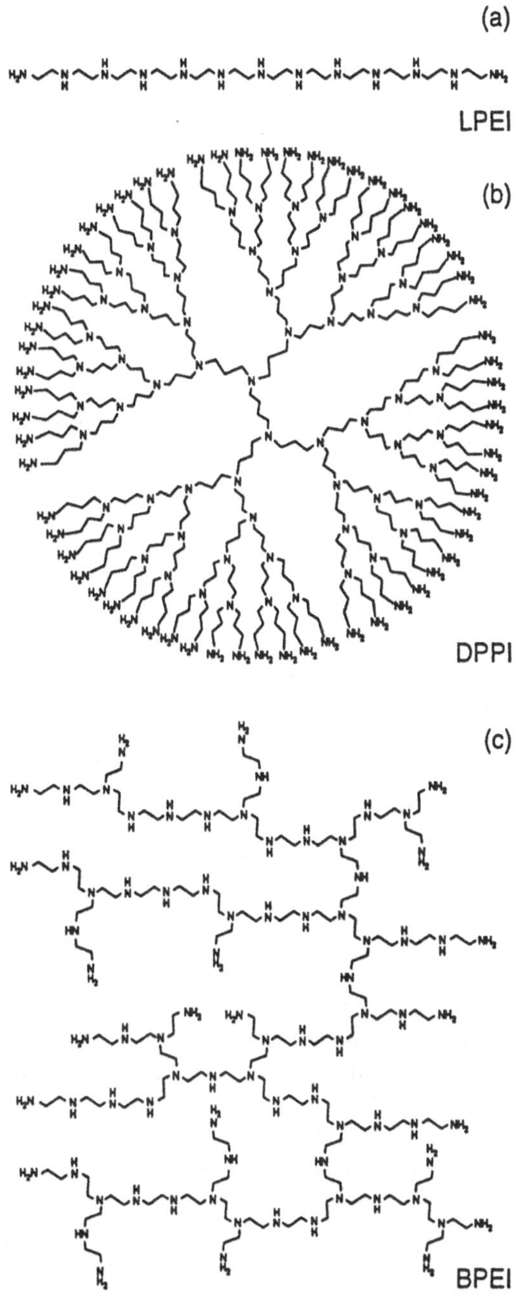

Fig. 1 Various high molecular weight polyamines. (a) Linear poly(ethylene imine) (LPEI), (b) dendritic poly(propylene imine) (DPPI), and (c) branched poly(ethylene imine) (BPEI)

Ising model of proton binding

Ionization properties of polyelectrolytes and their oligomeric analogs can be described with an Ising model

[9, 10, 23–25]. One introduces a set of state variables $s_1, s_2, \ldots, s_N$ where $s_i = 0$ if site i is deprotonated and $s_i = 1$ if this site is protonated and N is the total number of ionizable sites. The entire set specifies a given microscopic configuration with a corresponding free energy. This free energy can be written as a sum of contributions of increasing order

$$\frac{\beta\mathscr{F}(s_1, \ldots, s_N)}{\ln 10} = \sum_i (\mathrm{p}H - \mathrm{p}\hat{K}_i)s_i + \sum_{i>j} \varepsilon_{ij}s_i s_j + \sum_{i>j>k} \lambda_{ijk}s_i s_j s_k + \cdots , \tag{1}$$

where $1/\beta = kT$ is the thermal energy. The first term in Eq. (1) describes the ionization of the independent sites whereby pH and p$\hat{K}_i$ are the negative common logarithms of the proton activity and the microscopic dissociation constant given all other sites are deprotonated. The second term introduces interactions between the pairs of sites where ε_{ij} parametrizes the interaction strength between site i and j. Similarly, the third term incorporates interactions between triplets of sites where λ_{ijk} parametrizes the interaction strength between sites i, j and k. The pair interaction terms reflect the well documented group additivity relationships for small molecules [29].

Most quantities of interest can be obtained by evaluating the partition function

$$\Xi = \sum_{s_1, \ldots, s_N} \mathrm{e}^{-\beta\mathscr{F}(s_1, \ldots, s_N)} , \tag{2}$$

where the sum extends over all possible microscopic ionization states. The titration curve can be obtained by observing that this curve is equivalent to the average of the thermal expectation values of all state variables. This expectation value can be obtained by taking the derivative of Eq. (2) with respect to the chemical potential of the protons. The average degree of protonation thus becomes

$$\theta = \frac{z}{N}\frac{\partial \ln \Xi}{\partial z} , \tag{3}$$

where z is the activity of the protons (p$H = -\log_{10} z$).

If the number of sites N is finite, the Ising model is precisely equivalent to the classical description of polyprotic acids and bases in terms of successive protonation equilibria [23]. This equivalence can be made specific by considering the fugacity expansion

$$\Xi = \sum_{n=0}^{N} \bar{K}_n z^n , \tag{4}$$

where $\bar{K}_n$ are the expansion coefficients which follow from Eq. (2). Inserting Eq. (4) into e.g. Eq. (3), the familiar expression for the titration curve of a polyprotic acid or base is obtained [23]. One observes that the coefficient $\bar{K}_n$ is nothing but the macroscopic formation constant for the protonation step n. From these constants one finds the familiar macroscopic pK values, which are the negative common logarithms of the macroscopic dissociation constants of the corresponding protonation steps.

If the number of sites N is very large, however, the classical description in terms of polyprotic acids and bases is no longer practical. The necessary number of macroscopic dissociation constants and thus the number of adjustable parameters becomes very large. However, the Ising model description is equally well applicable and the titration curves can be obtained within this framework. The basic advantage of the Ising description is that the problem now involves a limited number of parameters. These parameters are the microscopic pK values of all ionizable sites and the various interaction parameters. Even a system with a large number of sites can be modeled with a small number of adjustable parameters. However, analytical solutions of the Ising model with a large number of sites are known in special cases only. In the general case, one can always obtain solutions by Monte Carlo simulation techniques. These techniques are standard in statistical mechanics and have been described in the context of ionization problems elsewhere [3, 14, 23–25, 27]. Here we shall not dwell on the computational details, but rather show how one can understand the ionization behavior of various kinds of polyprotic systems within a unified framework.

Illustrative model examples

Before we discuss real polyamines, consider simple model systems first. Let us mimic the ionization behavior of polyamines by assuming that all ionizable sites have the same microscopic value p$\hat{K} = 10$ and nearest neighbor pair interactions only. These interactions will be characterized with the parameter $\varepsilon = 2$. This model mimics poly(ethylene imines) in a semi-quantitative fashion. Poly(propylene imines) behave similarly with an interaction parameter of $\varepsilon \simeq 1$.

For a symmetric molecule with two sites, the Ising model predicts the following relation for the macroscopic pK values

$$\mathrm{p}K_1 = \mathrm{p}\hat{K} + \log_{10} 2 ,$$

$$\mathrm{p}K_2 = \mathrm{p}\hat{K} - \log_{10} 2 - \varepsilon , \tag{5}$$

where p$\hat{K}$ is the microscopic pK value given the other site is deprotonated and ε is the neighbor pair interaction parameter. These expressions were already known to Bjerrum [28]. For p$\hat{K} = 10$ and $\varepsilon = 2$ the corresponding pK values are given in Fig. 2a. Beside the statistical factor, the

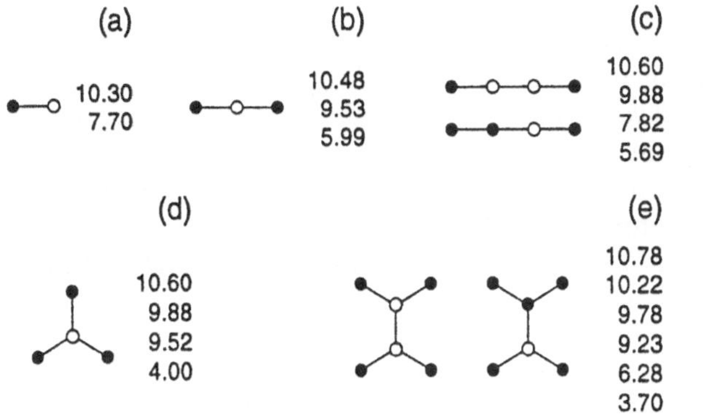

Fig. 2 Macroscopic pK values of small molecules with equivalent sites and nearest neighbor pair interactions only. The microscopic ionization constant is given by $\mathrm{p}\hat{K} = 10$ and nearest neighbor pair interactions with $\varepsilon = 2$

splitting of the macroscopic pK values is given by the pair interaction parameter ε. One site protonates around $\mathrm{p}H \simeq \mathrm{p}\hat{K}$ which results in the intermediate protonation state displayed in Fig. 2a. For the protonation of the second site, one pair interaction must be overcome, and the second site protonated around $\mathrm{p}H \simeq \mathrm{p}\hat{K} - \varepsilon$.

The same model can be also applied to more complicated molecules. A few additional results are given in Fig. 2. Before these results will be discussed in detail, let us emphasize a very characteristic sequence of pK values. For the linear molecules shown, two pK values are high and the remaining ones lower. For the branched structures, the number of high pK values is either three or four and the remaining pK values are generally lower than the ones for the linear molecules.

The macroscopic pK values of a linear arrangement of three sites are given in Fig. 2b. The two singly coordinated sites protonate around $\mathrm{p}H \simeq \mathrm{p}\hat{K}$ and result into the protonation state shown. For the protonation of the doubly coordinated site two pair interactions must be overcome and this site is then protonated around $\mathrm{p}H \simeq \mathrm{p}\hat{K} - 2\varepsilon$. The first two macroscopic pK values are mainly split due to statistical effects. A linear arrangement of four sites is shown in Fig. 2c. Two sites protonate around $\mathrm{p}H \simeq \mathrm{p}\hat{K}$ and may result into the protonation state shown in Fig. 2c (top). Next site can be protonated by overcoming one pair interaction at $\mathrm{p}H \simeq \mathrm{p}\hat{K} - \varepsilon$ and results in the state shown in Fig. 2c (below). Protonation of the last site involves two pair interactions happens around $\mathrm{p}H \simeq \mathrm{p}\hat{K} - 2\varepsilon$.

The macroscopic pK values of the simplest arrangement with one triply coordinated site are given in Fig. 2d. All three singly coordinated sites protonate around $\mathrm{p}H \simeq \mathrm{p}\hat{K}$ and the given protonation state results. For the protonation of the triply coordinated site three pair interactions must be overcome and the last site is then protonated around $\mathrm{p}H \simeq \mathrm{p}\hat{K} - 3\varepsilon$. Statistical effects are again responsible for the slitting of the three higher pK values. In the final example shown in Fig. 2e, four singly coordinated sites are protonated around $\mathrm{p}H \simeq \mathrm{p}\hat{K}$ first and one obtains the protonation state displayed in Fig. 2e (left). The first triply coordinated site protonates at $\mathrm{p}H \simeq \mathrm{p}\hat{K} - 2\varepsilon$ since only two pair interactions must be overcome. This process results in a protonation state shown in Fig. 2e (right), where only one triply coordinated site remains unprotonated. This last triply coordinated site protonates around $\mathrm{p}H \simeq \mathrm{p}\hat{K} - 3\varepsilon$.

This site binding model was originally solved for an infinite linear chain by Ising [11] and applied to polyelectrolytes by Marcus [9] and Katchalsky et al. [10]. The analytical solution will not be reproduced here. We shall rather discuss the overall appearance of the resulting titration curve shown in Fig. 3. The parameters are the same as before, namely for $\mathrm{p}\hat{K} = 10$ and $\varepsilon = 2$. The titration curves shows two steps and an intermediate plateau at $\theta = 1/2$. This characteristic behavior can be understood in the same way as the macroscopic pK values discussed further above. During the first step around $\mathrm{p}H \simeq \mathrm{p}\hat{K}$, basically every second site along the chain is protonated. As shown in Fig. 3 (top), the system minimizes the interaction energy around $\theta = 1/2$ by protonating every second site. Further protonation happens only around $\mathrm{p}H \simeq \mathrm{p}\hat{K} - 2\varepsilon$ since

Fig. 3 Titration curve of a linear chain with equivalent sites and nearest neighbor pair interactions. The microscopic ionization constant is given by $\mathrm{p}\hat{K} = 10$ and nearest neighbor pair interactions with $\varepsilon = 2$. The intermediate plateau reflects the stability of the protonation state, where every second site is protonated. Thin line refers to the case of no interactions

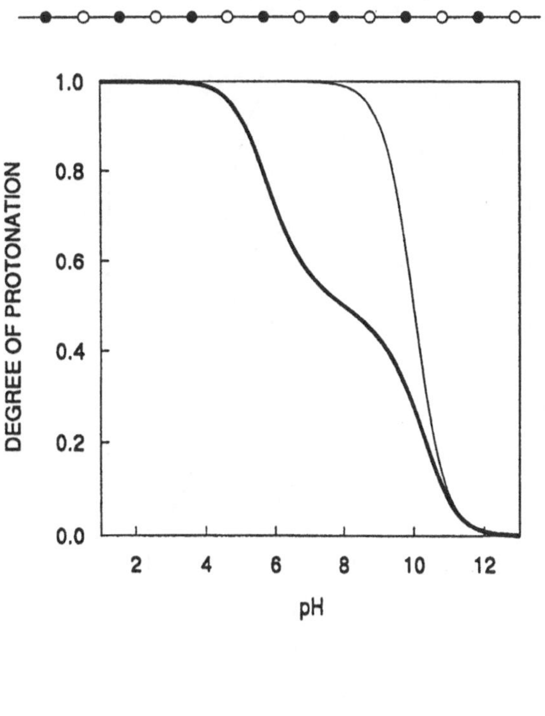

each unprotonated site has two occupied sites as neighbors and thus two pair interactions must be overcome.

For a dendritic structure, an entirely analogous model can be considered. Unfortunately, no simple analytical solution for Ising model on this so-called Cayley tree has been obtained so far [31]. However, the corresponding titration curve can be calculated numerically and the result for a dendrimer with 126 sites is shown in Fig. 4. The parameters $p\hat{K} = 10$ and $\varepsilon = 2$ are the same as before. Addition of further shells to the dendrimer does not change the resulting titration curve substantially. The titration curve displays the same qualitative features as for the linear chain, but the result is quantitatively different. The intermediate plateau arises at higher degree of protonation and the second protonation step occurs in rather acidic conditions. During the first step around $pH \simeq p\hat{K}$, all singly coordinated sites and all additional odd shells of triply coordinated sites protonate which results in the protonation state shown in Fig. 4. In this structure, there are always two protonated sites for each deprotonated site and thus the intermediate plateau arises at $\theta \simeq 2/3$. Further protonation occurs around $pH \simeq p\hat{K} - 3\varepsilon$ since each deprotonated site has three protonated neighbors [25].

As a final example consider a randomly branched structure. Let us assume a structure without closed loops (tree) and take the fraction of singly to doubly to triply coordinated groups as 1:2:1. (Since we neglect the presence of closed loops, the number of singly and doubly coordinated groups is the same). The resulting titration curves are neither very sensitive on the ratio between primary and secondary groups, nor do they depend much on other structural properties of the branched molecule. The titration curve shown in Fig. 5 can therefore be

Fig. 4 Titration curve of a dendrimer with equivalent sites and nearest neighbor pair interactions. The microscopic ionization constant is given by $p\hat{K} = 10$ and nearest neighbor pair interactions with $\varepsilon = 2$. The intermediate plateau reflects the stability of the onion-line protonation state, where every second shell is protonated. Thin line refers to the case of no interactions

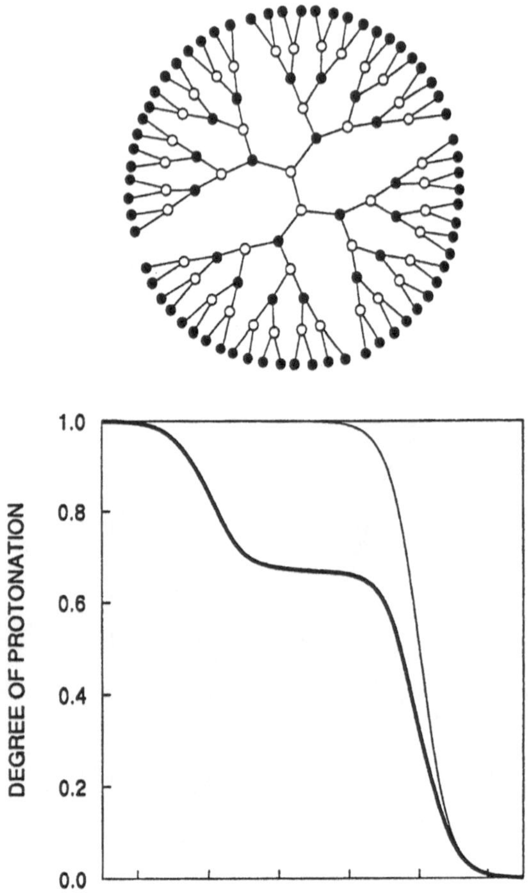

Fig. 5 Titration curve of a randomly branched polyelectrolyte with equivalent sites and nearest neighbor pair interactions. The microscopic ionization constant is given by $p\hat{K} = 10$ and nearest neighbor pair interactions with $\varepsilon = 2$. The break in the curve results from the possibility that one half of all sites can be protonated without invoking any interactions. Thin line refers to the case of no interactions

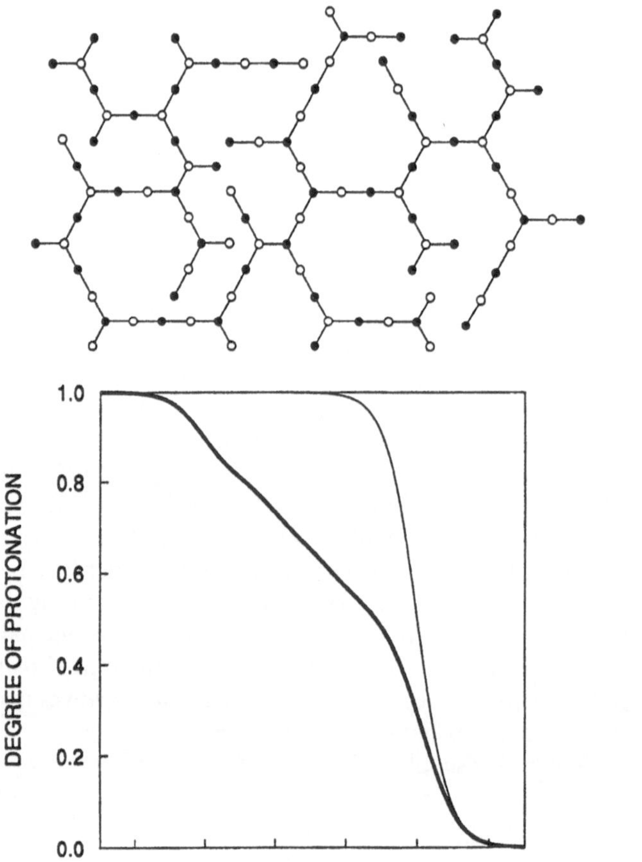

considered to be generic for a randomly branched polyelectrolyte. We again use the parameters $p\hat{K} = 10$ and $\varepsilon = 2$. The titration curve has a rather different appearance than in the two previous cases. First the curve rises quickly, but later very gradually and no intermediate plateau is now apparent. The titration curve rises quickly around $pH \simeq p\hat{K}$ but levels off around $\theta = 1/2$. As shown in Fig. 5, every second site can be protonated without invoking any pair interactions in such a structure. Further protonation proceeds with sites where one pair interaction must be overcome around $pH \simeq p\hat{K} - \varepsilon$, then with sites with two pair interactions around $pH \simeq p\hat{K} - 2\varepsilon$ and finally with three interactions around $pH \simeq p\hat{K} - 3\varepsilon$. Due to the random structure, these steps are rather broad and overlap substantially which results in a gradually rising titration curve [25].

Ionization properties of polyamines

Focus now on the quantitative description of the ionization behavior of various types of polyamines. We shall see that the titration behavior of such systems can be modeled within a single framework. To model the polyamines quantitatively, one must introduce different pK values for various types of groups and interactions which act beyond nearest neighbors.

The macroscopic pK values of various low molecular weight amines are well documented [17, 18]. From this information, one can derive using the Ising model the corresponding microscopic pK values and the various interaction parameters [23, 25]. Here we shall not describe the details of such a procedure, but rather go into the opposite direction; we show how to predict the ionization properties of various polyamines from a common parameter set. Because most experimental data are available for the ionic strength of 0.5 M, we shall only focus on this ionic strength in the following.

The pK value of ethylamine is 10.66 at an ionic strength of 0.5 M [17]. Due to the relatively high electronegativity of nitrogen, the microscopic pK values of amine groups lower with increasing number of neighboring amine groups and their proximity. The corresponding values are given in Table 1. The microscopic pK value of the primary amine groups in ethylenediamine is 9.64, of a secondary group coordinated with ethyleneamines 8.59 and of a tertiary group with the same coordination around 7.50. These systematic changes in pK values are consistent with group additivity relationships [29]; there is a decrease of about one pK unit for each additional ethyleneamines coordinating an amine group. This trend is similar for propylamines and can be used to estimate microscopic pK values for most entries given in Table 1.

Table 1 Microscopic pK_i values of primary, secondary and tertiary amine groups in aliphatic polyamines at an ionic strength of 0.5 M

Molecule	pK_i
$H_2N\text{–}(CH_2)_2N$	9.64
$H_2N\text{–}(CH_2)_3N$	10.24
$H_2N\text{–}(CH_2)_4N$	10.57
$NH\ [\text{–}(CH_2)_2N]_2$	8.59
$NH\ [\text{–}(CH_2)_3N]_2$	9.97
$N(CH_2)_2\text{–}NH\text{–}(CH_2)_3N$	9.23
$N\text{–}\ [(CH_2)_2N]_3$	7.50
$N(CH_2)_3\text{–}N\ [\text{–}(CH_2)_2N]_2$	7.70
$[N(CH_2)_3\text{–}]_2N\text{–}(CH_2)_3N$	7.90
$N\ [\text{–}(CH_2)_3N]_3$	8.90
$N(CH_2)_4\text{–}N\ [\text{–}(CH_2)_3N]_2$	9.40

Table 2 Interaction parameters of various carbon chains in aliphatic polyamines at an ionic strength of 0.5 M

Molecules	ε	λ
$N\text{–}(CH_2)_2\text{–}N$	1.85[a]	—
$N\text{–}(CH_2)_3\text{–}N$	1.00[a]	—
$N\text{–}(CH_2)_4\text{–}N$	0.58[a]	—
$N\text{–}(CH_2)_2\text{–}N\ [\text{–}(CH_2)\text{–}N]_2$	0.27[b]	0.42[c]

[a] Nearest neighbor interactions.
[b] Next nearest neighbor interactions between amine groups coordinating a tertiary amine group.
[c] Nearest neighbor triplet interaction.

The nearest neighbor interaction parameter correlates with the number of carbon atoms in between the nitrogen [23]; each additional atom reduces this parameter by a factor of about 2. Higher order interactions are negligible in propylamines, butylamines and higher homologues. In ethylamines, higher order interactions must be introduced because of the short distances between the ionizable groups. For tertiary amines, all neighboring amine groups give rise to a next nearest neighbor interaction of 0.27. The same kind of interaction can be shown to be negligible for a secondary amine group [23]. Furthermore, the neighboring amine groups give rise to a triplet interaction around 0.4. All other interactions can be neglected to a first approximation.

This model is obviously simplified. Nevertheless, we shall now demonstrate that this Ising model framework with the common parameter set given in Tables 1 and 2 has substantial predictive capabilities. These predictive capabilities will be illustrated in the following.

Let us first demonstrate that this model reproduces the macroscopic pK values of various low molecular weight amines to reasonable accuracy. Tables 3 and 4 compare experimental and calculated macroscopic pK values for

Table 3 Comparison of experimental macroscopic p*K* values of low molecular weight primary and secondary amines at an ionic strength of 0.5 M [17, 18] with calculated values obtained from the Ising model. Model parameters are summarized in Tables 1 and 2. Structural analogs are shown in Fig. 1a–c

Molecule	Structural analog	pK_i Exp.	pK_i Calc.
Ethylenediamine (en) $H_2N(CH_2)_2NH_2$	a	10.03 7.31	9.94 7.49
Trimethylenediamine $H_2N(CH_2)_3NH_2$	a	10.65 8.95	10.54 8.94
Tetramethylenediamine $H_2N(CH_2)_4NH_2$	a	10.87 9.69	10.87 9.69
1,4,7-Triazaheptane (dien) $H_2N(CH_2)_2NH(CH_2)_2NH_2$	b	9.88 9.09 4.47	9.95 9.36 4.47
1,5,8-Triazaoctane $H_2N(CH_2)_2NH(CH_2)_3NH_2$	b	10.44 9.36 6.37	10.36 9.52 6.36
1,5,9-Triazanonane[a)] $H_2N(CH_2)_3NH(CH_2)_3NH_2$	b	10.71 9.72 7.96	10.64 9.88 7.92
1,4,7,10-Tetraazadecane (trien) $H_2N(CH_2)_2NH(CH_2)_2NH(CH_2)_2NH_2$	c	9.87 9.21 6.87 3.71	9.97 9.37 6.96 3.74
1,4,8,11-Tetraazaundecane $H_2N(CH_2)_2NH(CH_2)_3NH(CH_2)_2NH_2$	c	10.25 9.50 7.28 6.02	10.08 9.45 7.44 6.06
1,5,8,12-Tetraazadodecane $H_2N(CH_2)_3NH(CH_2)_2NH(CH_2)_3NH_2$	c	10.66 9.96 8.53 5.84	10.58 9.98 8.44 6.08
1,5,9,13-Tetraazatridecane[a)] $H_2N(CH_2)_3NH(CH_2)_3NH(CH_2)_3NH_2$	c	10.56 10.02 8.74 7.51	10.73 10.09 8.95 7.65

[a)] Estimated from experimental data at other ionic strengths.

various low molecular weight aliphatic amines. The structural analogs of these molecules were shown in Fig. 2. One quickly observes that the macroscopic p*K* values given for the simplified model in Fig. 2 follow the same trends as the experimental data. The linear amines have two high p*K* values, while for branched structures the number of high p*K* values is either three or four. The remaining p*K* values are generally lower for the branched structures than for the linear ones.

The agreement between the experimental data and model predictions is rather satisfactory. Focusing on the data for linear structures given in Table 3, the model is even able to predict the right trends for the p*K* values for the various structural analogs for the tetraprotic amines. For the branched structures, the model is slightly less successful but still satisfactory. While the model is quite reasonable for butylamines, additional interactions might be present in ethylamines. These interactions are difficult to identify since the available experimental data on small branched molecules are limited. Thus this part of the model must be considered as tentative and might need some revision in the future. Poor predictions of the p*K* values for the last two protonation steps for both dinitrilotetrakis (2-ethylamines) reflect this deficiency of the model.

Nevertheless, we have observed rather good agreement between model and experiment at this stage, but this agreement is not necessarily a demonstration of the predictive powers of this kind of description. The ionization properties of some of these molecules were in fact used to construct the parameter set used above, and the agreement merely demonstrates the overall consistency of the description. Actual predictive power of the model should become apparent with the consideration of high molecular weight polyamines. The titration curves of polyelectrolytes were not used for model calibration.

Linear poly(ethylene imine) represents the first example where the predictive powers of this framework can be demonstrated. For the linear chain we need to know the p*K* of the secondary amine group, and the nearest neighbor pair and triplet interactions parameters only. The next nearest neighbor pair interactions appear to be absent. The titration curve in the long chain limit can be calculated with the transfer matrix techniques [24] or with combinatorial methods [14]. The result of the model calculation is compared with experimental data [14] in Fig. 6. In light of the fact that no parameter adjustment was made, the agreement between model prediction and experiment is certainly satisfactory.

The titration curve displays the typical two step behavior with the intermediate plateaus at $\theta = 1/2$ which was already apparent in the simple Marcus–Katchalsky model (cf. Fig. 3). In contrast to this case, however, the titration curve of LPEI is not quite symmetrical. This asymmetry originates from the triplet interaction contribution.

Dendritic poly (propylene imine) will serve as the second example to demonstrate the predictive power of the Ising model. This highly branched structure contains primary and tertiary amine groups only [30]. Since we deal with a propylamine, only nearest neighbor pair interactions enter the description. The prediction of the Ising model for the fifth generation dendrimer with 126 ionizable sites is compared with experimental titration data [20] in Fig. 7. Again, no parameter adjustment was made, and

Table 4 Comparison of experimental macroscopic pK values of low molecular weight tertiary amines at an ionic strength of 0.5 M [17, 18] with calculated values obtained from the Ising model. Model parameters are summarized in Tables 1 and 2. Structural analogs are shown in Fig. 1d and e

Molecule	Structural analog	pK_i	
		Exp.	Calc.
Nitrilotris(2-ethylamine) (tren) $N[(CH_2)_2NH_2]_3$	d	10.14	10.12
		9.68	9.36
		8.64	8.62
		< 3	0.69
Nitrilotris(3-propylamine)[a] $N[(CH_2)_3NH_2]_3$	d	10.61	10.72
		9.95	10.24
		9.30	9.76
		5.92	5.90
Ethylenedinitrilotetrakis(2-ethylamine) (penten) $[NH_2(CH_2)_2]_2N(CH_2)_2N[(CH_2)_2NH_2]_2$	e	10.23	10.24
		9.83	9.74
		9.54	9.26
		8.85	8.76
		1–2	3.14
		< 1	− 0.99
Trimethylenedinitrilotetrakis(2-ethylamine)[a] $[NH_2(CH_2)_2]_2N(CH_2)_3N[(CH_2)_2NH_2]_2$	e	10.26	10.24
		9.66	9.74
		9.50	9.27
		8.74	8.77
		2.90	3.88
		< 1	2.28
Ethylenedinitrilotetrakis(3-propylamine)[b] $[NH_2(CH_2)_3]_2N(CH_2)_2N[(CH_2)_3NH_2]_2$	e	10.67	10.84
		10.27	10.41
		9.91	10.06
		9.15	9.63
		6.72	6.20
		3.42	3.75
Tetramethylenedinitrilotetrakis(4-butylamine)[c] $[NH_2(CH_2)_3]_2N(CH_2)_4N[(CH_2)_3NH_2]_2$	e	11.02	10.87
		10.40	10.43
		10.00	10.06
		9.41	9.62
		7.65	7.68
		6.51	6.52

[a] Estimated from experimental data for other ionic strengths.
[b] Experimental data from Ref. [19].
[c] From Ref. [20].

the model predicts the ionization behavior in a satisfactory fashion.

The dendrimer titration curve shows the typical two step behavior with the intermediate plateau at $\theta = 2/3$ in agreement with the simpler Marcus–Katchalsky model discussed above (see Fig. 4). In spite of the fact that we deal with propylamine as the repeating unit, the second protonation step occurs at rather low pH. This observation can be explained by the fact that each group has three nearest neighbors and three pair interactions must be overcome to achieve full protonation. The actual titration curve of the dendrimer is broadened due to the differences in the pK values of the primary amine groups and two kinds of tertiary amine groups.

Branched poly(ethylene imine) is the final example to demonstrate the predictive capabilities of this model. The ratio of primary to secondary to tertiary groups is assumed to be 1:2:1 [21]. The prediction of the Ising model is compared with experimental data [22] in Fig. 8. Again no parameter adjustment was made, and the agreement is quite satisfactory. The model underestimates the degree of protonation somewhat, possibly because some additional interactions have been neglected.

The overall shape of the titration curve was already captured by the simpler Ising model with pair interactions only (see Fig. 5). The next nearest neighbor pair interactions are essential to lower the degree of protonation into the regime of experimental values. However, the

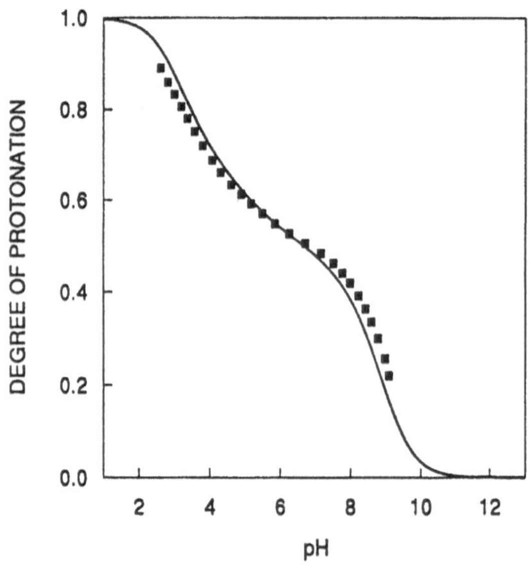

Fig. 6 Titration curve of linear poly(ethylene imine) (LPEI) at an ionic strength of 0.5 M. Experimental data points are from Ref. [14]. The solid line is the prediction based on the Ising model with parameters given in Tables 1 and 2

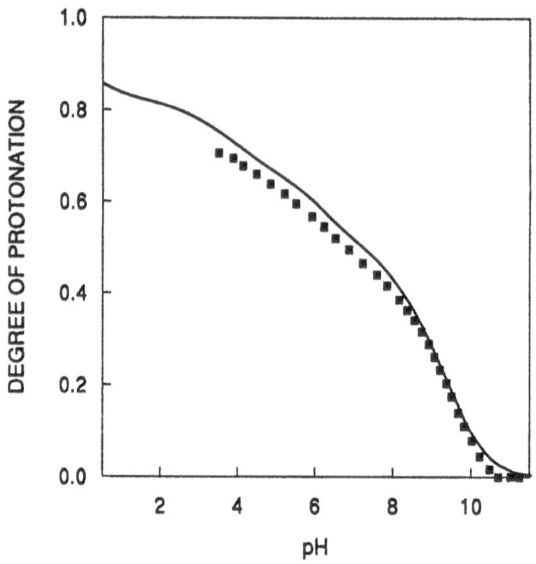

Fig. 8 Titration curve of randomly branched poly(ethylene imine) (BPEI) at an ionic strength of 0.5 M. Experimental data points are from Ref. [22]. The solid line is the prediction based on the Ising model with parameters given in Tables 1 and 2

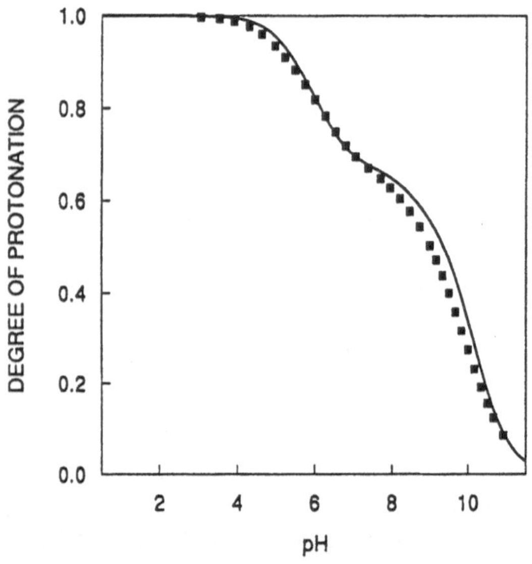

Fig. 7 Titration curve of dendritic poly(propylene imine) (DPPI) at an ionic strength of 0.5 M. Experimental data points are from Ref. [20]. The solid line is the prediction based on the Ising model with parameters given in Tables 1 and 2

protonation mechanism is slightly different for BPEI than for the model system shown in Fig. 5. In contrast to the model system, in BPEI the p*K* values of the primary amine groups are somewhat higher than the p*K* values of the tertiary groups.

In BPEI, all primary groups will protonate first at high p*H* and then a fraction of the secondary ones. Only at lower p*H* values the protonation of the tertiary groups will set in. In the simpler Marcus–Katchalsky model, on the other hand, a small fraction of tertiary amine groups would also protonate within the first protonation step. For the same reason, the initial rise in the titration curve of BPEI is less steep than for the curve shown in Fig. 5.

Discussion

The aim of the present communication was to show that a simple Ising model with short range interactions provides a comprehensive framework for the quantitative understanding of the ionization properties for various polymeric amines. The model parameters are the microscopic p*K* values of various kinds of primary, secondary and tertiary amine groups and a limited number of interaction parameters. These parameters are derived from the macroscopic p*K* values of various low molecular weight amines. The model is then able to reproduce the macroscopic p*K* values for a number of low molecular weight polyprotic amines to a reasonable degree of accuracy. The basic advantage of the Ising model framework is that the model can be equally well applied to polyelectrolytes with high molecular weight. The model is reasonably successful to predict the titration behavior of linear and branched poly(ethylene imine) and of dendritic poly(propylene imine). Based on this observation we expect that the model will be equally reliable in predicting the ionization behavior of other polyamines.

The basic protonation pattern can be understood in a semi-quantitative fashion by assuming that all sites have the same microscopic pK values and the interactions are of nearest neighbor type characterized with an interaction parameter ε. For propylene imines $\varepsilon \simeq 1$ while for ethylene imines $\varepsilon \simeq 2$. Linear polyelectrolytes protonate in two steps with an intermediate plateau at $\theta \simeq 1/2$ since the nearest neighbor pair interactions stabilize a microscopic protonation state of alternating protonated and deprotonated sites (see Fig. 3). The splitting between both protonation steps is roughly 2ε since two pair interactions must be overcome for further protonation. The dendritic structure protonates in two steps as well, but the intermediate plateau lies at $\theta \simeq 2/3$ since the pair interactions stabilize a protonation state of alternating shells of protonated and deprotonated sites (see Fig. 4). The splitting between these two steps is larger than for the linear structure, and it is roughly 3ε since now three pair interactions must be overcome. For a randomly branched structure, the first half of the site protonate in a rather narrow range, but further protonation is gradual and extends over the entire range of 3ε (see Fig. 5). In a semi-quantitative way, the actual titration curves for poly(propylene imines) and poly(ethylene imines) are governed by the same patterns. However, their quantitative description calls for a more detailed parametrization of the problem.

This analysis demonstrates unequivocally that the protonation states of individual sites are strongly correlated in highly charged polyelectrolytes, and necessarily invoke short range site-site interactions. This observation is in marked contrast to the behavior of some polyelectrolytes of low charge density, interfaces, and possibly some proteins where sites protonate independently and a mean-field description is sufficient [32]. Proper understanding of the crossover between these two extremes is lacking at present, and can only be addressed by investigations of site-site interaction potentials in such systems.

The success of this model shows that description of other kinds of polyelectrolytes should be possible within a similar framework. Polycarboxylates can be approached in an analogous fashion, but fewer experimental data are available. However, the tacticity of these molecules incorporates another unknown into the problem [13]. The description of various kinds of natural polyelectrolytes, such as proteins, polycarboxylates, cellulose or humic acids, which contain various types of ionizable groups, represent the next step in the analysis. It remains to be seen how the present type of approach bridges into the established description of proteins where the pair interactions are mostly weak but long-ranged, and usually calculated with electrostatic Poisson–Boltzmann type models. Possibly, a hybrid approach might become useful, where long range interactions are addressed with an electrostatic model and short range interactions incorporated within the currently discussed framework. In spite of some problems of this kind, further research along these lines will lead to a comprehensive description of proton binding for many types of natural polyelectrolytes. Quantitative understanding of these processes is one of the essential prerequisites for the assessment of the many important topics in contemporary applied colloid chemistry mentioned in the introduction where ionization of macromolecules play a key role.

Acknowledgements We thank U. Steuerle, A. Hettche and D. Horn for interesting discussions and for pointing out Ref. [19] to us. This communication is dedicated to Milan Schwuger on the occasion of his 60th birthday.

References

1. Bashford D, Karplus M (1990) Biochemistry 29:10219
2. Honig B, Nicholls A (1995) Science 268:1144
3. Beroza P, Fredkin DR, Okamura MY, Feher G (1991) Proc Natl Acad Sci 88:5804
4. Kesvatera T, Jönsson B, Thulin E, Linse S (1996) J Mol Biol 259:828
5. Cleland RL (1982) Macromolecules 15:386
6. Milne CJ, Kinniburgh DG, De Wit JCM, Van Riemsdijk WH, Koopal LK (1995) Geochim Cosmochim Acta 59:1101
7. Tanford C, Kirkwood JC (1957) Am J Chem Soc 79:5333
8. Schaefer M, Sommer M, Karplus M (1997) J Phys Chem 101:1663
9. Marcus RA (1954) J Phys Chem 58:621
10. Katchalsky A, Mazur J, Spitnik P (1957) J Polym Sci 23:513
11. Ising E (1925) Zeitsch f Physik 31:253
12. Ullner M, Woodward CE, Jönsson B (1996) J Chem Phys 105:2056
13. Kitano T, Kawaguchi S, Ito K, Minakata A (1987) Macromolecules 20:1598
14. Smits RG, Koper GJM, Mandel M (1993) J Phys Chem 97:5745
15. Horn D (1980) In: Goethals EJ (ed) Polymeric Amines and Ammonium Salts. Pergamon Press, Oxford
16. von Zelewsky A, Barbosa L, Schläpfer CW (1993) Coord Chem Rev 123:229
17. Smith RM, Martell AE (1975) Critical Stability Constants, Vol 2. Plenum Press, New York
18. Smith RM, Martell AE (1989) Critical Stability Constants, Vol 6. Plenum Press, New York
19. Garcia-España E, Micheloni M, Paoletti P, Bianchi A (1986) Inorg Chem 25:1435
20. van Duijvenbode R, Koper GJM, Borkovec M (1997) submitted

21. Bloys van Trenslong CJ, Staverman AJ (1974) Recl Trav Chim Pays-Bas 93:171
22. Balif JB, Lerf C, Schläpfer CW (1994) Chimia 48:336
23. Borkovec M, Koper GJM (1994) J Phys Chem 98:6038
24. Borkovec M, Koper GJM (1994) Langmuir 10:2863
25. Borkovec M, Koper GJM (1997) Macromolecules 30:2151
26. King EJ (1965) Acid–Base Equilibria. Pergamon Press, Oxford
27. Chandler D (1987) Introduction to Modern Statistical Mechanics. Oxford University Press, New York
28. Bjerrum N (1923) Z Physik Chemie 106:219
29. Perrin DD, Dempsey B, Serjeant EP (1981) pK_a Prediction for Organic Acids and Bases. Chapman & Hall, London
30. Tomalia DA, Baker H, Dewald J, Hall M, Kallos G, Martin S, Roeck J, Ryder J, Smith P (1985) Polymer J 17:117
31. Runnels LK (1967) J Math Phys 8:2081
32. Borkovec M, Daicic J, Koper GJM (1997) Proc Natl Acad Sci 94:3499

Progr Colloid Polym Sci (1998) 109:153–160

L.H. Torn
A. de Keizer
L.K. Koopal
W. Blokzijl
J. Lyklema

Polymer adsorption on a patchwise heterogeneous surface

Received: 1 October 1997
Accepted: 13 October 1997

L.H. Torn · A. de Keizer
L.K. Koopal · Prof. Dr. J. Lyklema (✉)
Laboratory for Physical Chemistry
and Colloid Science
Wageningen Agricultural University
Dreijenplein 6
NL-6703 HB Wageningen
The Netherlands

W. Blokzijl
Unilever Research Laboratory
Olivier van Noortlaan 120
NL-3133 AC Vlaardingen
The Netherlands

Abstract The adsorption of the uncharged polymer poly(vinylpyrrolidone) (PVP) on a homoionic Na-kaolinite has been studied. Potentiometric acid–base titrations of kaolinite were performed on samples at different concentrations of sodium chloride. An interpretation in terms of the contributions of the individual surface types has been given. Protons are strongly favored over sodium ions at the basal planes. Some striking similarities were observed between the results of the acid–base titrations and the PVP adsorption experiments. PVP adsorbs readily on at least part of the kaolinite surface showing a high affinity character and an adsorbed amount at the plateau of about 1 mg m^{-2} total area. The influence of the pH, electrolyte concentration and multivalent ions on the amount adsorbed at the plateau has been investigated. Increasing the pH or the electrolyte concentration leads to a decrease in adsorption. A model is proposed in which PVP adsorbs on edges and basal planes by different mechanisms. The adsorption of PVP on the edges is strongly pH dependent, that of the plates only weakly. Specifically adsorbed protons at the plates act as anchor sites for PVP segments. Multivalent ions do not influence the proposed adsorption mechanism directly but primarily change the surface area accessible for PVP.

Key words Kaolinite – poly(vinylpyrrolidone) (PVP) – polymer adsorption – heterogeneous surface – surface charge

Introduction

Clay minerals are used in a number of industrial processes, e.g., as paper fillers and coating pigments, to improve the properties of the material. For these purposes, stable dispersions are needed and therefore the adsorption of polymers on clays becomes of interest.

The adsorption of uncharged polymers on clay minerals is very complex due to the heterogeneous character of the clay surface. The overall interaction is the result of a subtle balance of forces determined by polymer–surface, polymer–solvent, and surface–solvent interactions. Most of the published work of uncharged polymer adsorption on clay minerals is restricted to polyvinylalcohol (PVA) and polyacrylamide (PAM) on kaolinite and montmorillonite [1–6]. Two French research groups [5, 6] were the first to interpret results of PAM adsorption on kaolinite by accounting for the different types of surfaces. Their main discussion point was the extent of adsorption of PAM on the basal planes of the gibbsite surface.

Less attention has been paid to the adsorption of polyvinylpyrrolidone (PVP), although it is an often used dispersant. Due to the presence of both hydrophilic and hydrophobic functional groups, PVP is soluble in water and a wide variety of organic solvents [7]. In solution it probably occurs as a random coil. The adsorption of PVP on hydrophilic [8–12] (mostly mineral oxides) and hydrophobic surfaces [13–16] has been reported. PVP hardly adsorbs on metal oxides, except for silica.

Clay minerals consist of sheets of silicon–oxygen tetrahedra and aluminium- or magnesium oxygen–hydroxyl octahedra [17]. They can be classified according to the arrangement of these layers. The kaolinite group represents clay minerals with a 1 : 1 unit layer structure consisting of a Si-tetrahedron sheet and an Al-octahedron sheet. These minerals are non-swelling and form flat, hexagonal particles. Three types of surfaces can be distinguished: siloxane plates, gibbsite plates, and edges. Much research has been carried out to reveal the surface chemical and charge characteristics of kaolinite [17–24]. It is generally accepted that the (surface) charge of the clay particles can be separated in a permanent and a variable part. A permanent negative charge is present on the basal planes due to isomorphous substitution of Al^{3+} and Si^{4+} ions inside the solid by ions of lower valency, while the edges possess a pH dependent charge caused by (de)protonation of surface hydroxyl groups. The two basal planes show a much lower affinity for the H^+ and OH^- ions [25, 26].

For homo-ionic kaolinite samples many values determined by different techniques have been reported for the point of zero charge of the edges (epzc) and the overall isoelectric point (iep). The epzc is difficult to establish by titration because only sums of H^+ and OH^- consumptions on the three types of surfaces are measurable. Measured values for the epzc fall in the range 5–9; most of them are around 7. The papers of Rand and Melton [19], and Herrington et al. [22] can be consulted for a discussion of the observed discrepancies. The overall iep is in principle measurable but difficult to interpret in terms of edge and plate properties. Values around five [22, 27] and below two have been reported [28, 29]. In general it can be stated that important causes of these discrepancies involve problems in obtaining reproducible pure samples and finding suitable experimental techniques. Most data are obtained by electrophoretic measurements [22, 27, 28, 30]. The difficulty of converting mobilities into zeta-potentials for kaolinite samples is well known; it is caused by nonuniformity of charge and shape, and the occurrence of large surface conductance around the particles [30–32]. To our knowledge, the literature contains only two examples of the adsorption of PVP on kaolinite. In the early seventies Francis [33] studied it on reference clay minerals by gravimetry. Hardly any adsorption of PVP on kaolinite was detectable and no adsorption isotherms were given. Since then, researchers seemed to have lost their interest in this system for a long time. However, very recently Hild et al. [34] studied it again, emphasizing the solution side of the system. Their findings will be discussed below.

The aim of the present paper is to advance our understanding of the interaction between PVP and the three types of surfaces of kaolinite by comparing it with the uptake of protons, which also differs between these faces. To that end, the PVP adsorption measurements are extended by potentiometric titrations at different electrolyte concentrations. The influence of multivalent ions on the adsorption of PVP on kaolinite is also studied.

Materials

Kaolinite was obtained from Sigma Company. Extensive characterisation of this sample by Mehrian [28, 35] showed it to be very pure. According to Mehrian, the particle size range of the sample is 0.1–4 μm and its BET (N_2) surface area amounts to 17.7 $m^2 g^{-1}$. For the cation exchange capacities (CEC) she found: 30 $\mu mol\,g^{-1}$ (determined by the silver thiourea method) or 57 $\mu mol\,g^{-1}$ (measured by the ammonium acetate method). It was assumed that the latter values also include the surface sites of the edges. An edge/plate area ratio of 0.25 was found by argon adsorption.

Poly(vinylpyrrolidone) with a number average molar mass of $17.4 \times 10^3\,g\,mol^{-1}$ ($M_w/M_n = 1.9$) was obtained from BASF and used as received.

HCl, NaOH, and NaCl were all of analytical grade. Water is purified by passing it over a mixed bed ion exchanger, a carbon column, and a microfilter.

Methods

Potentiometric titrations were performed on Na-kaolinite samples prepared according to the procedure described by Mehrian [28]. The titration vessel is filled with 0.5 g of clay dispersed in 30 ml electrolyte solution. First, the pH is lowered to pH 4 and five titration curves were measured between pH 4 and 10 (three upwards, two downwards). After completion the pH is lowered to around 7, the electrolyte concentration is raised, the resulting change in pH recorded, and the same procedure is followed at the higher concentration. Blank titrations were performed under similar conditions.

The surface charge was calculated from the difference between the amount of H^+ and OH^- adsorbed, taken up by the samples, and the blank experiments. The charge obtained in this way is the relative surface charge which is the change in the total surface charge. The absolute surface charge cannot be determined unambiguously because the absolute amount of permanent negative charge due to isomorphous substitution is not well established. It can be estimated to be 30 μmol g^{-1}. Curves at different electrolyte concentrations are mutually positioned by accounting for the effect of increasing electrolyte concentration on the surface charge.

Adsorption isotherms of PVP were determined by depletion measurements at 25 °C. The PVP-concentration was determined by UV-adsorption at 204 nm. Centrifuge tubes are filled with 0.4 g kaolinite, 30 ml demineralised water, and 5 ml polymer solution of the desired concentration. The tubes were shaken end-over-end for 16 h at 30 rpm. Preliminary kinetic experiments showed that an almost constant adsorbed amount is reached within one hour; thereafter the adsorption increases slightly. After six hours, there is no detectable change in the adsorbed amount anymore. The pH of the solution is repeatedly adjusted. The solids were separated by centrifugation for 30 min at 20 000 rpm.

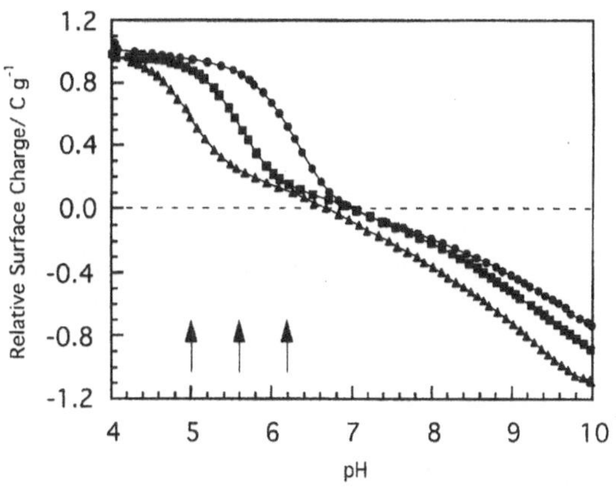

Fig. 1 Relative surface charge of Na-kaolinite at different NaCl concentrations: ●: 10^{-3} M, ■: 10^{-2} M, ▲: 10^{-1} M; the arrows are explained in the text

Results and discussion

Potentiometric titrations

Characteristic acid–base titration curves of homo-ionic Na-kaolinite samples at three electrolyte concentrations are shown in Fig. 1. This set of relative surface charge curves is arbitrarily referred to the charge being zero for the curve with the lowest electrolyte concentration at the pH where the effect of indifferent electrolyte is smallest: $\Delta\sigma \equiv 0$ at pH 7 for the 10^{-3} M curve.

Successive titrations at one electrolyte concentration showed a small hysteresis effect. In accordance with Mehrian [28], this phenomenon can be attributed to a retardation effect in the formation and destruction of the card-house structure. This open structure is formed by coagulation of negatively charged plates and positively charged edges. The fact that the hysteresis increases slightly with decreasing electrolyte concentration supports this explanation. The hysteresis is too small to influence the essential characteristics of the surface charge curves.

To explain the observations in Fig. 1, it is useful to divide the total relative surface charge into additive contributions of edges and plates, ignoring overspill at the edge–plate border:

$$\Delta\sigma_{H^+/OH^-,\mathrm{total}} = \Delta\sigma_{H^+/OH^-,\mathrm{plates}} + \Delta\sigma_{H^+/OH^-,\mathrm{edges}}\ . \qquad (1)$$

The negative charge of the plates is compensated by an excess of counterions and a deficit of co-ions in the electrical double layer. The contribution $\Delta\sigma_{H^+/OH^-,\mathrm{plates}}$ is the result of counterion exchange against specifically adsorbed protons [18, 28] and depends on the electrolyte concentration and pH. At high pH the proton concentration is too low (compared to the counterion concentration) to contribute to the charge on the plates. Therefore, any further proton desorption stems from the edges. The charge on the edges is caused by either adsorption or desorption of protons or hydroxyl ions, resulting in a variable-charge surface. For isolated variable charge surfaces, the absolute value of the charge increases with increasing concentration of indifferent electrolyte. Curves at different concentrations of indifferent electrolyte cross each other at a common intersection point (cip) which corresponds to the pzc [36].

In Fig. 1, no common intersection point can be observed. Recently, this was also reported by Braggs et al. [23] but subsequently ignored in their discussion. On the other hand, Herrington et al. [22] did find common intersection points for different kaolinite samples with potassium chloride as the indifferent electrolyte. A reason for this discrepancy may be found in their pretreatment, using acid and hydrogen peroxide.

In our experiments, it is observed that with decreasing pH the curves approach each other, although the cip is masked by the superimposed exchange on the plates. Extrapolation of the edge part of the charges suggest a cip of about 7, which may be identified with the epzc which agrees well with literature values [22, 27, 28].

Below pH 7, compared to the behaviour of variable-charge surfaces, an additional contribution to the proton uptake can be observed, which sets in at higher pH at lower electrolyte concentrations. Considering the Na^+/H^+ exchange, the preference of the surface for either protons or sodium ions manifests itself in the non-diffuse part of the double layer. The resulting change in the Gibbs free energy, ΔG^*, due to this exchange can be expressed as

$$\Delta G^* = RT \ln K \qquad (2)$$

with

$$K = \frac{\phi_{Na^+} c_{H^+}}{\phi_{H^+} c_{Na^+}}, \tag{3}$$

where ϕ_{H^+} and ϕ_{Na^+} are the volume fractions of the ions adsorbed at the surface, and c_{H^+} and c_{Na^+} are their bulk concentrations. A more detailed interpretation of K is deferred to a future publication. Values of K can be determined from the steepest part of the curves of Fig. 1 (the arrows in Fig. 1 mark these points) where $\phi_{H^+} = \phi_{Na^+} = 0.5$, assuming $c_{Na^+} \approx$ the initial electrolyte concentration. Table 1 shows the corresponding pH ($pH_{1/2}\Delta\sigma$) and K for each electrolyte concentration.

It follows that $\Delta G^* = -7$ to -9 RT, which expresses the specific preference of the plate surface for protons over sodium ions. Apparently, protons start to displace sodium ions at the plates if $[H^+] > 10^{-4}$ $[Na^+]$.

It can be concluded that no cip or intersecting curves are found when (1) the contribution of the H^+/Na^+ exchange is significant and (2) this exchange occurs close enough to the epzc. The titration curves can be interpreted very well as reflecting the consecutive titration of edges and plates. The question to which extent the two types of plates contribute remains to be addressed.

Adsorption of PVP

Adsorption isotherms of PVP on kaolinite at three different pH values in 10^{-2} M NaCl are shown in Fig. 2. PVP shows in all cases a fairly strong affinity for at least part of the kaolinite surface. The graphs show a steep initial rise followed by a pseudo plateau. The polydispersity of the samples is reflected in the bending of the curves and the reluctance to attain the plateau [37, 38]. The amount adsorbed at the plateau is somewhat low for uncharged polymers (the adsorbed amount can be 1.5–2.5 mg m^{-2}). However, the amount adsorbed at pH 5.5 is comparable to values reported for PVP on silica [8, 10] and PAM on kaolinite [1, 3, 39].

A significant pH-dependence is observed in the present system, while in earlier studies, there was hardly a pH-dependence [3, 8] or the influence of the pH was not investigated [1, 10, 39]. In order to gain more insight into the adsorption mechanism, adsorption at the plateau was determined as a function of pH at different electrolyte concentrations. To assure plateau adsorption, the initial PVP concentration is chosen sufficiently high (0.5 g l^{-1}). The results are presented in Fig. 3.

Table 1 $pH_{1/2}\Delta\sigma$ and K at different Na^+ concentrations

$[Na^+]$ (M)	$pH_{1/2}\Delta\sigma$	K
10^{-3}	6.2	$10^{-3.2}$
10^{-2}	5.6	$10^{-3.6}$
10^{-1}	4.9	$10^{-3.9}$

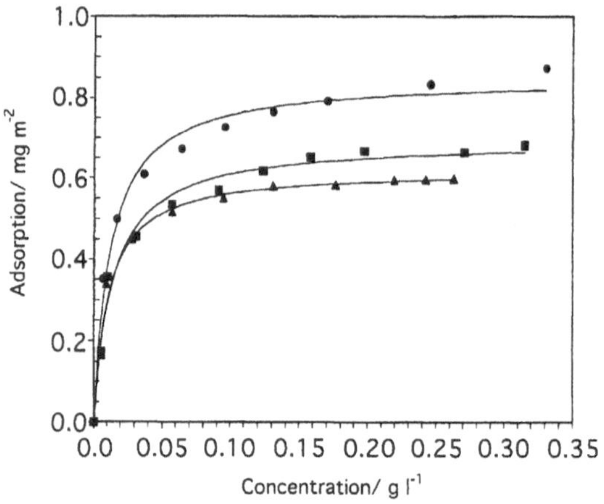

Fig. 2 Adsorption of PVP on kaolinite: ●: pH 5.5, ■: pH 7, ▲: pH 8; $I = 10^{-2}$ M NaCl

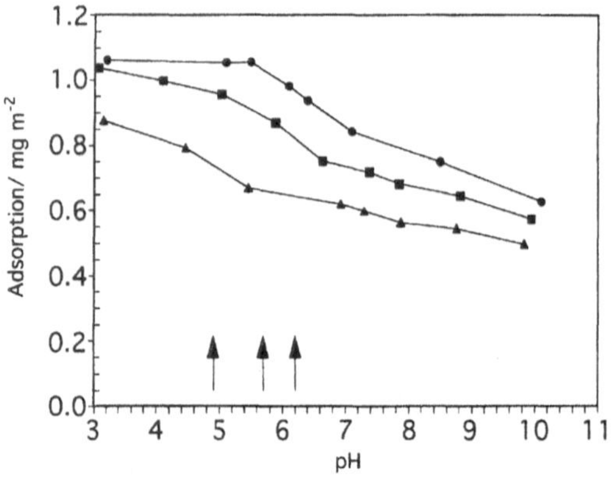

Fig. 3 Plateau adsorption of PVP on kaolinite as a function of pH at different NaCl-concentrations: ●: 2×10^{-3} M, ■: 10^{-2} M, ▲: 10^{-1} M; $[PVP]_0 = 0.50$ g l^{-1}, $[PVP]_{eq} = 0.29$–0.40 g l^{-1}; the arrows are explained in the text

Three observations can be made. First, there is a monotonous but discontinuous decrease of the adsorption with pH. At 10^{-3} M NaCl, the adsorbed amount is constant between pH 6 and pH 3. Perhaps here complete coverage of the adsorbing surface type(s) has been attained. Second, even at high pH values (pH > epzc) a significant adsorption persists. Third, the addition of electrolytes decreases the adsorbed amount. In order to explain these observations, two questions must be answered:

1) What is/are the driving force(s) for adsorption?
2) On which surface type(s) does adsorption occur?

Since PVP is not charged in the pH-range 4–10, coulombic interactions cannot be responsible for the adsorption and other physical or chemical interactions must be considered. In studies of PAM adsorption on clay minerals, hydrogen bonding was considered to be the major driving force for adsorption [2, 3, 39]. It could in principle take place in the actual system between surface hydroxyl groups of the clay and the carbonyl group of the pyrrolidone ring. Furthermore, PVP is known to be strongly attracted by hydrophobic surfaces [13–15, 40], so hydrophobic interactions are also possible. With these possible mechanisms in mind, it is useful to look separately at the different surface types of kaolinite (a siloxane and a gibbsite as plate surfaces, and metal oxide type edges) for possible adsorption sites.

First, the siloxane surface is considered. PVP adsorbs readily on silica [8, 10]. In both studies, hydrogen bonding to silanol groups as well as hydrophobic interactions contributed to the adsorbed amount. The siloxane surface of the kaolinite particles carries no silanol groups, its composition is expected not to vary with pH and it has a permanent negative charge. Adsorption of PVP on this surface type is expected to occur by hydrophobic bonding. As a consequence, this contribution will be independent of pH.

Let us now look at the gibbsite surface part. PVP hardly adsorbs on gibbsite [41, 42] although pure gibbsite possesses many hydroxyl groups. This general observation for metal oxide surfaces other than silica is probably caused by their strongly hydrophilic character: hydrogen bonding with water molecules is preferred over that with the carbonyl groups of the pyrrolidone ring. Otherwise stated, the critical exchange Gibbs energy for polymer adsorption [43] is not surpassed on these surfaces. It is therefore expected that substantial adsorption onto the gibbsite surface of the clay is absent or very small.

Finally, the edges are considered. They consist of exposed aluminol and silanol groups. The silanols have an acidic character (the pzc of silica is around 2–3) [36] with a variable charge. The aluminols are basic groups (the pzc of gibbsite is about 9) [36] also behaving like a variable-charge surface. Their behaviour is different from that of the aluminols on the basal faces, due to a different coordination of these surface groups [26]. As a consequence, a pH dependent (de)protonation takes place at both the silanol and the aluminol sites. Figure 4 shows schematically some surface characteristics of the edges at different pH values [6, 23]. Adsorption on the edges can occur by hydrogen bonding to (di)protonated surface oxygens. The number of these groups increases with decreasing pH. It is therefore likely that adsorption on the edges will gradually increase with decreasing pH.

Summing up the expected effects of edges and plates on the adsorption of PVP, some insight into the curves of Fig. 3 can be obtained. At high pH when all surface types of kaolinite are negatively charged and the edge hydroxyl groups are ionized, adsorption can only take place by hydrophobic bonding on the siloxane plates. This amount will be independent of pH and the electrolyte concentration. When the curves in Fig. 3 are extrapolated, they merge around pH 13 at an adsorbed amount of about 0.45 $mg\,m^{-2}$ of the total surface. Converting this value to $mg\,m^{-2}$ siloxane surface leads to an adsorbed amount of about 1.2 $mg\,m^{-2}$, in good agreement with PVP adsorption values on a silica surface [8, 10]. Upon decreasing the pH, in addition to the adsorption on the siloxane plates, hydrogen bonding to (di)protonated surface oxygens will lead to further adsorption. This amount increases gradually from zero at high pH till about 0.45 $mg\,m^{-2}$ total area at pH 3 which is about 1.8 $mg\,m^{-2}$ edge area. These contributions are estimated from the 10^{-2} M curve in Fig. 3. In Fig. 5 the two contributions are schematically shown by the curves a and b, and their sum is represented by curve c. Comparing curve c of Fig. 5 with the 10^{-2} M curve in Fig. 3, it follows that the PVP adsorption is not yet fully explained. In each curve of Fig. 3, a stepwise contribution to the adsorption can be seen around pH 5 to 7. This contribution is more pronounced at lower electrolyte concentrations and it shifts to lower pH values at increased electrolyte concentrations. The pH halfway these additional contributions is marked by arrows in Fig. 3 and given in Table 2.

Although we are aware that the $pH_{1/2}\Delta\Gamma_{PVP}$ values are not highly accurate, comparison with the $pH_{1/2}\Delta\sigma$ values of Table 1, reveals a close correspondence: the exchange of sodium ions for protons at the plates (Fig. 1 and Table 1) parallels the additional increase in PVP adsorption (Fig. 3 and Table 2). Apparently, the specific adsorption of protons on the plates creates adsorption sites for PVP. It is most likely that this additional adsorption occurs on the

Fig. 4 Surface charge characteristics of the edges of kaolinite at different pH values [6, 23]

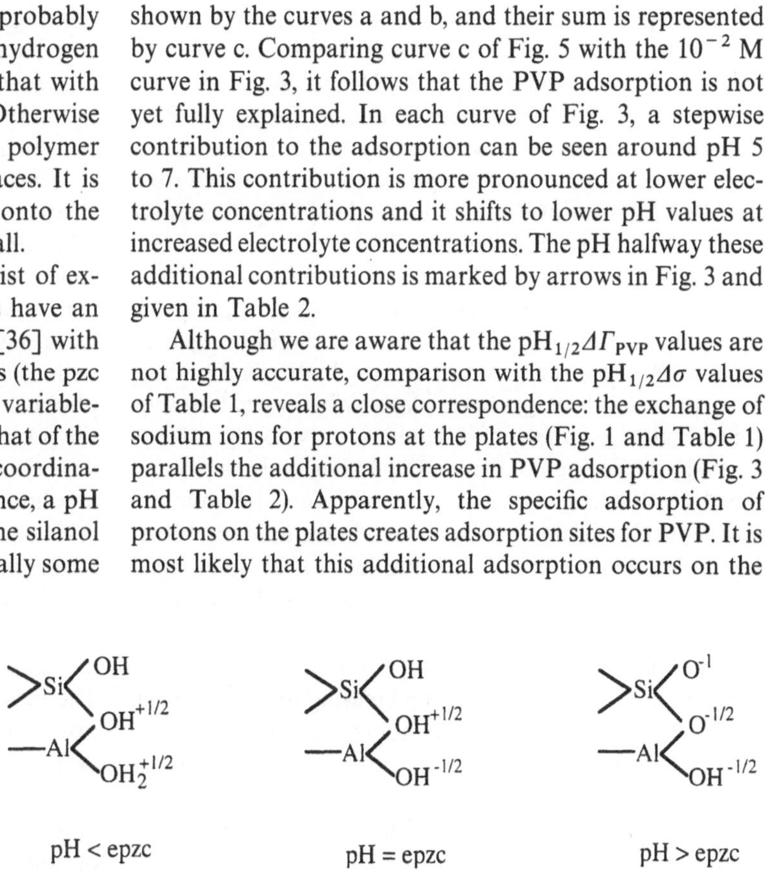

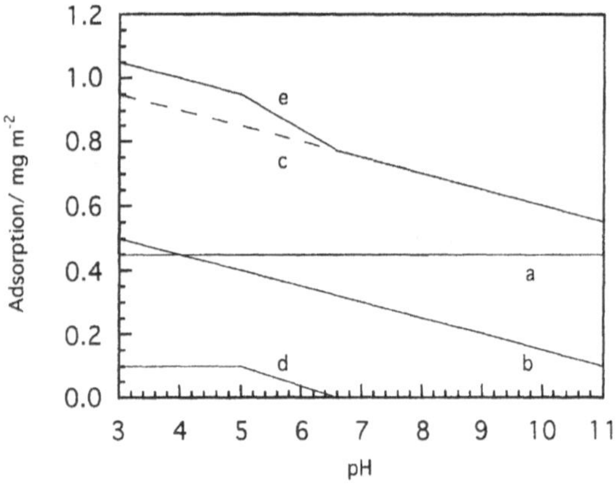

Fig. 5 Schematic contributions of the different surface types of kaolinite to the plateau adsorption of PVP as a function of pH; a: siloxane contribution; b: edge contribution; c: a + b; d: additional siloxane contribution; e: overall contribution = a + b + d

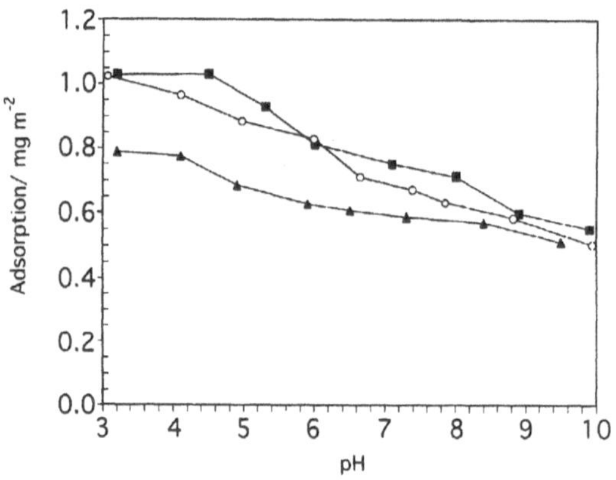

Fig. 6 Plateau adsorption of PVP on kaolinite in the presence of multivalent ions: ○: 10^{-2} M NaCl, ■: 10^{-2} M NaCl, 10^{-2} M Na_2HPO_4, ▲: 10^{-2} M NaCl, 2×10^{-2} M $CaCl_2$; $[PVP]_0 = 0.5$ gl^{-1}

Table 2 $pH_{1/2}\Delta\Gamma_{PVP}$ at different Na^+ concentrations

$[Na^+]$ (M)	$pH_{1/2}\Delta\Gamma_{PVP}$
2×10^{-3}	6.2
10^{-2}	5.6
10^{-1}	4.9

siloxane plates, for example to newly formed silanol groups. According to the titration curves, the Na^+/H^+ exchange occurs over about one pH-unit. When the process is completed, the number of anchor points for PVP adsorption will also not increase any more. The contribution of these additional adsorption sites is estimated from the 10^{-2} M curve of Fig. 3 and included in Fig. 5 as curve d; the resulting curve is curve e.

The shape of the PVP adsorption curves is now well explained. Adsorption occurs on the siloxane plates by hydrophobic bonding and to a lesser extent (if bridged by specifically adsorbed protons) also by hydrogen bonding; adsorption on the edges takes place by hydrogen bonding. As a result, the surface of kaolinite is only partly covered by PVP, the degree of coverage depending on both pH and the electrolyte concentration. An indication of the incomplete coverage could already be seen in the amount adsorbed at the plateau which is quite low for an uncharged polymer. This finding will have consequences for the stability of PVP-coated kaolinite.

The influence of the electrolyte concentration on the adsorption is twofold. First, an increasing number of edge surface hydroxyl groups will be ionized at higher electrolyte concentration. As a result, fewer sites are available for hydrogen bonding, so adsorption on the edges will decrease and also the total adsorbed amount. Second, at a higher electrolyte concentration, less protons will be adsorbed at the plates leading to fewer additional adsorption sites for PVP. Overall, increasing the electrolyte concentration lowers the adsorption both at the edges and the plates.

The experimental results of Hild et al. [34] concern mainly the adsorbed amount and the layer thickness measured by microelectrophoresis. With respect to the adsorbed amount their results are in good agreement with the present ones. The plateau values are comparable and by looking carefully the additional contributions we found in Fig. 3 can also be seen in their results. However in their interpretation they consider a complete, averaged coverage of the kaolinite surface by PVP which is not supported by our observations.

Plateau adsorption experiments for PVP were also carried out in the presence of phosphate and calcium ions, additional to 10^{-2} M NaCl. The results are shown in Fig. 6. At high pH, the differences between the curves are very small, indicating that the adsorption mechanism in this range is hardly influenced by multivalent ions. This is in line with our proposed adsorption mechanism of hydrophobic bonding at the siloxane plates. At lower pH values, the presence of calcium ions leads to a decrease in adsorption whereas phosphate ions slightly increase the adsorbed amount. In the latter case, the adsorbed amount below pH 4.5 is constant at about 1 mgm^{-2}, indicating saturation of adsorption sites.

Calcium ions are known to adsorb specifically on kaolinite [4, 18] but no systematic study of their adsorption on the different surface types of kaolinite could be found. The results of Atesok et al. [4] suggest a major contribution of the edges to the total adsorbed amount, and ion exchange between calcium and sodium ions. Calcium ions are expected to adsorb on negatively charged edge sites and on the basal planes. Edge adsorption of calcium ions on ionized hydroxyl groups is not expected to hinder the

PVP adsorption on the edge hydroxyl groups by hydrogen bonding substantially. Adsorbed calcium ions at the plates may compete more strongly with protons than Na^+ ions leading to a reduction in the amount of polymer adsorbed at the plates. However, an additional property of calcium ions is their valency effect on coagulation. Two particles can be coagulated by calcium ions, thereby reducing the total available surface area for PVP. We observed that the presence of calcium ions greatly destabilizes the kaolinite dispersions. This reduced surface accessibility effect may be another cause of the decreased adsorption for PVP.

Phosphate ions adsorb specifically on the edges of kaolinite [44, 45]. It is therefore likely that these ions will influence only the adsorption of PVP on the edges. Specifically adsorbed phosphate ions compete with hydrogen bonded PVP segments. Figure 6 shows overall a slight increase in the adsorbed amount as a function of pH. Obviously, the presence of phosphate ions does not substantially hinder adsorption of PVP, and even seems to create slightly more adsorption sites. This can be attributed to an increased electrostatic repulsion between the kaolinite particles due to the presence of (adsorbed) phosphate ions. This results in an increased surface area accessible for PVP and therefore in an enhanced adsorption.

Conclusions

The combination of potentiometric titrations and uncharged polymer adsorption measurements yields information that helps to unravel the heterogeneity of the kaolinite surface. The key element is that specifically adsorbed protons on the basal planes act as anchor sites fo·r PVP-segments, so that the pH and electrolyte concentration dependence of the PVP adsorption runs parallel to that of protons.

Potentiometric titrations show that surface oxygens of the edges are titrated in the entire pH range. At $pH < 7$, sodium ions adsorbed at the basal planes are exchanged for protons leading to an additional proton charge.

Poly(vinylpyrrolidone) adsorbs on the individual faces of kaolinite by different mechanisms. Adsorption occurs on the edges by hydrogen bonding. This amount increases with decreasing pH. Adsorption takes place on the siloxane plates by hydrophobic bonding and below pH 7 also by hydrogen bonding to specifically adsorbed protons. The first contribution is independent of pH, whereas the latter decreases with increasing electrolyte concentration. The influence of multivalent ions is rather indirect and mainly caused by changes in the surface accessibility for PVP.

Acknowledgements Mr. P.L.M. Theunissen is greatly acknowledged for performing the potentiometric titration experiments.

References

1. Theng BKG (1979) Formation and Properties of Clay-polymer Complexes. Elsevier, Amsterdam
2. Pefferkorn E, Nabzar L, Varoqui R (1987) Colloid Polym Sci 265:889
3. Hollander AF, Somasundaran P, Gryte CC (1981) J Appl Polym Sci 26:2123
4. Atesok G, Somasundaran P, Morgan LJ (1988) Colloids Surf 32:127
5. Lee LT, Rahbari R, Lecourtier J, Chauveteau G (1991) J Colloid Interface Sci 147:351
6. Pefferkorn E, Nabzar L, Carroy A (1985) J Colloid Interface Sci 106:94
7. Meza R, Gargallo L (1977) Eur Polym J 13:235
8. Cohen Stuart MA, Fleer GJ, Bijsterbosch BH (1982) J Colloid Interface Sci 90:310
9. Esumi K, Oyama M (1993) Langmuir 9:2020
10. Parnas RS, Claimberg M, Taepaisitphongse V, Cohen Y (1989) J Colloid Interface Sci 129:441
11. Sato T, Kohnosu S (1994) Colloids Surf 88:197
12. Ishimaru Y, Lindström T (1984) J Appl Polym Sci 29:1675
13. Kawaguchi M, Hayashi I, Takahashi A (1981) Polym J 13:783
14. Smith JN, Meadows J, Williams PA (1996) Langmuir 12:3773
15. Kellaway IW, Najib NM (1980) Int J Pharm 6:285
16. Sugiura M (1970) Bull Chem Soc Japan 43:2604
17. Van Olphen H (1977) An Introduction to Clay Colloid Chemistry. 2nd ed. Wiley Interscience, New York
18. Ferris AP, Jepson WB (1975) J Colloid Interface Sci 51:245
19. Rand B, Melton IE (1977) J Colloid Interface Sci 60:308
20. Bolland MDA, Posner AM, Quirk JP (1980) Clays Clay Minerals 28:412
21. Zhou ZH, Gunter WD (1992) Clays Clay Minerals 40:365
22. Herrington TM, Clarke AQ, Watts JC (1992) Colloids Surf 68:161
23. Braggs B, Fornasiero D, Ralston J, Smart RS (1994) Clays Clay Minerals 42:123
24. Brady PV, Cygan RT, Nagy KL (1996) J Colloid Interface Sci 183:356
25. Wierer KA, Dobias B (1988) J Colloid Interface Sci 122:171
26. Hiemstra T, Van Riemsdijk WH, Bolt GH (1989) J Colloid Interface Sci 133:105
27. Alince B, Van de Ven TGM (1993) J Colloid Interface Sci 155:465
28. Mehrian Isfahany T (1992) Thermodynamics of the adsorption of organic cations on kaolinite. PhD thesis, Wageningen Agricultural University, Wageningen
29. Jarnstrom L, Stenius P (1990) Colloids Surf 50:47
30. Williams DJA, Williams KP (1978) J Colloid Interface Sci 65:79
31. Fair MC, Anderson JL (1989) J Colloid Interface Sci 127:388
32. O'Brien RW, Rowlands WN (1993) J Colloid Interface Sci 159:471
33. Francis CW (1973) Soil Sci 115:40
34. Hild A, Sequaris JM, Narres HD, Schwuger M (1997) Colloids Surf A 123–124:515
35. Mehrian T, de Keizer A, Lyklema J (1991) Langmuir 7:3094
36. Lyklema J (1995) Fundamentals of Interface and Colloid Science. Vol 11, Academic Press, London
37. Cohen Stuart MA, Scheutjens JMHM, Fleer GJ (1980) J Polym Sci Polym Phys Ed 18:559
38. Koopal LK (1981) J Colloid Interface Sci 83:116
39. Dodson PJ, Somasundaran P (1984) J Colloid Interface Sci 97:481

40. Esumi K, Takamine K, Ono M, Osada T, Ichikawa S (1993) J Colloid Interface Sci 161:321
41. Ishiduki K, Esumi K (1997) J Colloid Interface Sci 185:274
42. Torn LH (1997) Unpublished results
43. Fleer GJ, Cohen Stuart MA, Scheutjens JMHM, Cosgrove T, Vincent B (1993) Polymers at Interfaces. Chapman & Hall, London
44. De Haan FAM (1965) The interaction between certain inorganic anions, clays and soils. PhD thesis, Wageningen Agricultural University, Wageningen
45. Lyons JW (1964) J Colloid Interface Sci 19:399

Progr Colloid Polym Sci (1998) 109:161–169
© Steinkopff Verlag 1998

D. Bauer
E. Killmann
W. Jaeger

Adsorption of poly (diallyl-dimethyl-ammoniumchloride) (PDADMAC) and of copolymers of DADMAC with N-methyl-N-vinyl-acetamide (NMVA) on colloidal silica

Received: 15 July 1997
Accepted: 16 July 1997

D. Bauer · E. Killmann (✉)
Lehrstuhl für Makromolekulare Stoffe
Institut für Technische Chemie
TU München
Lichtenbergstraße 4
D-85747 Garching
Germany

W. Jaeger
Fraunhofer-Institut für Angewandte Polymerforschung
Kantstraße 55
D-14513 Teltow
Germany

Abstract The adsorption of cationic polyelectrolytes on colloidal silica-particles is investigated. The adsorbed amounts and hydrodynamic layer thicknesses of the cationic polyelectrolytes poly(diallyl-dimethyl-ammoniumchloride) PDADMAC of different molar masses and statistic copolymers of DADMAC and N-methyl-N-vinyl-acetamide (NMVA) of various contents have been measured. Furthermore, the surface charge densities of the bare and polyelectrolyte covered silica particles have been determined by potentiometric titration. The adsorbed amounts as well as the hydrodynamic layer thicknesses of the adsorbed polyelectrolyte layer increase with the ionic strength and the pH of the suspension and at high salt concentration with the molar mass of the macroions. At low salt concentrations the adsorbed amounts rise with decreasing charge density of the polyelectrolyte. During polyelectrolyte adsorption the surface charge density of colloidal silica is increased because of the dissociation of silanol groups.

Key words Polyelectrolyte adsorption – colloidal silica – poly(diallyl-dimethyl-ammoniumchloride) – copolymers of diallyl-dimethyl-ammoniumchloride and N-methyl-N-vinyl-acetamide – polyelectrolyte titration – photon correlation spectrometry – potentiometric surface charge density titration

Introduction

Polyelectrolytes find widespread application in many industrial processes and in numerous products of our daily life [1]. An important property of polyelectrolytes is their tendency to adsorb on solid surfaces. This is used to modify solid surfaces and interfaces by polyelectrolyte-based coatings and adhesives [2–4].

The conformation of the adsorbed polyelectrolytes and the charges of the polyelectrolyte covered particles have a large influence on the stability and the flocculation behaviour of suspensions. Because of this polyelectrolytes are used either as stabilizers e.g. of dye dispersions or as flocculants e.g. in waste water treatment or in paper producing industries [5, 6].

The adsorption of polyelectrolytes onto charged surfaces depends not only on the chemical and molecular structure of the macroions but also on their electrochemical character and the surface charge density of the substrate [7]. Electrostatic interactions between the polymer segments and the adsorbent and between the polymer segments themselves influence the conformation of the adsorbed macromolecules and the thickness of the adsorbed layer [8]. They are strongly dependent on the ionic strength and on the pH of the solution. Thus the structure and the adsorbed amounts depend on the charge density and molar mass of the polyelectrolytes, the charge density

Scheme 1

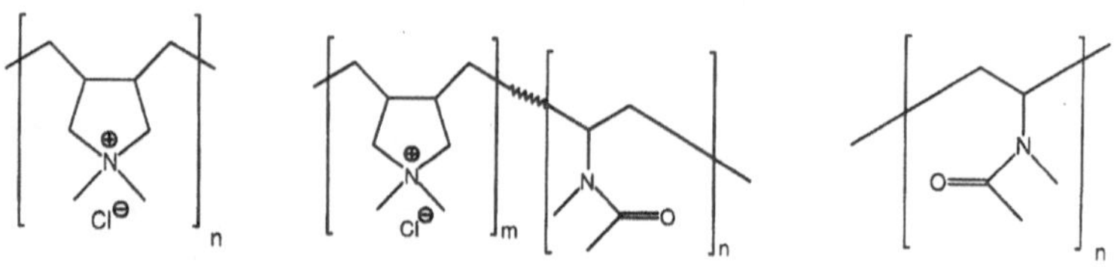

PDADMAC **1** P(DADMAC-co-NMVA) (m:n) **2** PNMVA **3**

of the substrate and the ionic strength and the pH of the solution.

In this study the influences of the molar mass of highly charged cationic poly(diallyl-dimethyl-ammonium-chloride) (PDADMAC), of the molar ratio in the copolymers of DADMAC with N-methyl-N-vinyl-acetamide (NMVA), of the ionic strength and of the pH of the suspension on the adsorbed amounts and on the hydrodynamic layer thicknesses of the adsorbed macroions are investigated. As substrate we use precipitated colloidal silica which is suspended by ultrasonification. By comparing the surface charge density of the bare and the polyelectrolyte covered silica particles we get insight into the adsorption mechanism of the investigated system. Parallel to the investigations in this study measurements on the colloidal stability of the same systems are accomplished [9].

Experimental part

Materials

Precipitated silica particles were prepared by the method of Stöber et al. [10]. The characterization was carried out by photon correlation spectroscopy (PCS) [11] (see Table 1), where D_{90} and D_0 are the diffusion coefficients at scattering angles of 90° and 0°, d_{90} and d_0 are the diameters of the particles calculated from D_{90} and D_0 by the Einstein–Stokes equation [Eq. (1)], ρ is the density of the particles.

Usually a OH-group density of 4–5 OH-groups/nm^2 on a dehydrated but completely hydroxylated silica surface is found [12, 13].

To prepare a 1 g/l silica suspension, 50 mg of precipitated silica were weighed into a 100 ml breaker and 30 ml of doubly distilled water were added. After ultrasonification for 15 min the suspension was filled into a 50 ml glass measuring flask and shaken by hand. After standing for about 16 h in order to settle down metal particles, adsorption measurements were accomplished.

Poly(diallyl-dimethyl-ammonium-chloride) (PDADMAC) **1**, copolymers of DADMAC and N-methyl-N-vinyl-acetamide (NMVA) **2** and poly(N-methyl-N-vinylacetamide) (PNMVA) **3** were produced by radical polymerisation [14].

Their characteristics are given in Tables 2 and 3, where $[\eta]$ is the intrinsic viscosity, $\bar{M}_n$ and $\bar{M}_w$ are the number and the weight average of the molar mass of the polymers, respectively.

The projected distance of the charges of a linear PDADMAC in aqueous solution is estimated to be 0.5–0.55 nm [15].

All chemicals were of the highest purity grade commercially available and were used without further purification. Solutions were prepared with doubly distilled water.

NaCl was employed as the supporting electrolyte. The bulk pH was adjusted using either chloride acid or sodium hydroxide. The pH was measured using a Metrohm pH Meter, Model 691, equipped with a Metrohm pH Electrode, No. 6.0239.100.

Procedures

Adsorption isotherms were measured at room temperature (23 ± 2 °C) using the depletion method. One ml of the silica-suspension was added to 9 ml of a polymer solution of certain concentration at different pH and ionic strength and shaken by hand. After 12 h silica was separated from the solution at 2500 rpm in a Heraeus Labofuge A. In the case of PDADMAC and of the P(DADMAC-co-NMVA) (1:1) and (1:3) the concentration of the supernatant polyelectrolyte was determined by polyelectrolyte titration with the potassium salt of polyvinylsulfate using toluidine blue as indicator [16, 17]. The titration could be accomplished at a salt concentration smaller than 0.05 mol/l. At higher ionic strength the supernatant solution had to be dialyzed before measuring in order to reduce the salt concentration. At a NMVA content in the polymer higher than (3:1) the concentration of the polymer in the supernatant was determined by UV-absorption.

Table 1 Characterization of precipitated silica

D_{90} [10^{12} m^2 s^{-1}]	D_0 [10^{12} m^2 s^{-1}]	d_{90} [nm]	d_0 [nm]	Second moment [%]	ρ [g cm^{-3}]
3.37	2.8	140	175	6	2.05

Table 2 Molecular characterization of PDADMAC **1**

Polymer	$[\eta]$[a)] [ml g^{-1}]	$\bar{M}_n$[b)] [10 g mol^{-1}]	$\bar{M}_n$[c)] [10^3 g mol^{-1}]	$\bar{M}_w$[c)] [10^3 g mol^{-1}]	$\bar{M}_w$[d)] [10^3 g mol^{-1}]
PDADMAC 428.000	320	373	428	643	709
PDADMAC 372.000	324		372	580	574
PDADMAC 117.000	118	101	117	162	143
PDADMAC 30.000	33	31	30	44	45
PDADMAC 5.000	12	5[e)]			

[a)] Determined at 30 °C in 1 M NaCl.
[b)] Determined by membrane osmometry.
[c)] Determined by size exclusion chromatography; solvent: 0.5 M $NaNO_3$ in water.
[d)] Determined by light scattering spectroscopy.
[e)] Determined by viscosimetry.

Table 3 Molecular characterization of P(DADMAC-co-NMVA) (*m*:*n*) **2** and of PNMVA **3**

Polymer	Content [mol %]	$[\eta]$[a)] [ml g^{-1}]	$\bar{M}_a$[b)] [10^3 g mol^{-1}]	$\bar{M}_a$[c)] [10^3 g mol^{-1}]	$\bar{M}_w$[c)] [10^3 g mol^{-1}]
			DADMAC:NMVA		
P(DADMAC-co-NMVA) (3:1)	75:25	73	71	74	107
P(DADMAC-co-NMVA) (1:1)	53:47	63	58	66	98
P(DADMAC-co-NMVA) (1:3)	24:76	64	51	65	92
P(DADMAC-co-NMVA) (17:83)	17:83	67		76	100
P(DADMAC-co-NMVA) (14:86)	14:86	74	79	83	121
P(DADMAC-co-NMVA) (6:94)	6:94	73		87	116
PNMVA 120.000	0:100	124	111	124	201
PNMVA 260.000	0:100	187	269	260	349

[a)] Determined at 30 °C in 1 M NaCl.
[b)] Determined by membrane osmometry.
[c)] Determined by size exclusion chromatography; solvent: 0.5 M $NaNO_3$ in water.

Surface charge density–pH curves of the bare and the polymer covered silica were determined at room temperature (23 ± 2 °C) using an automatic titration setup of Metrohm. For the measurements with the bare silica particles 40 ml of a 10 g/l silica suspension, prepared as described above, were mixed with 10 ml of a NaCl-solution of the desired concentration and purged with nitrogen for at least 30 min. The silica suspension was titrated between pH 3 and pH 10 using 0.01 M NaOH or 0.01 M HCl. Blank titrations in the absence of silica were performed additionally. The reference values were subtracted from titration data in the presence of silica to calculate the surface charge density [18].

To determine the surface charge density of the polymer covered silica particles 30 ml of the 10 g/l silica suspension were mixed with 10 ml of a polymer solution of a certain concentration and with 10 ml of a salt solution. At every pH-value the polymer concentration was higher than the concentration needed for saturation of the surface. The polymer concentrations were 1 g/l in the case of PDADMAC 5.000 and 2 g/l in the case of P(DADMAC-co-NMVA) (1:3), (6:94) and PNMVA 120.000.

The hydrodynamic layer thicknesses of the adsorbed polymers were determined by measuring the diffusion coefficients of bare and covered particles by photon correlation spectroscopy (PCS) using an argon ion laser of

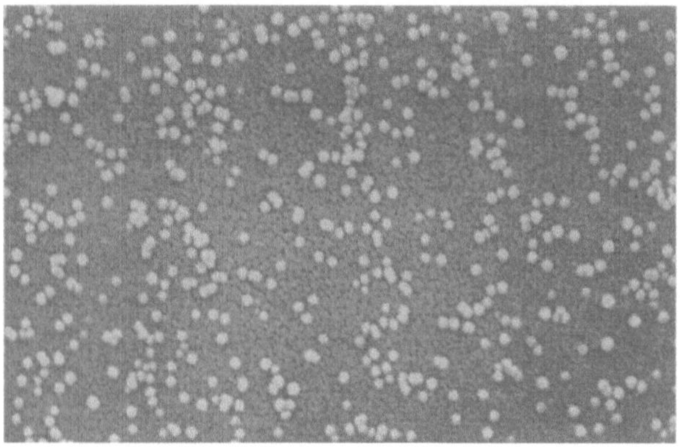

Fig. 1 Electron-micrograph of bare silica in water without any added salt

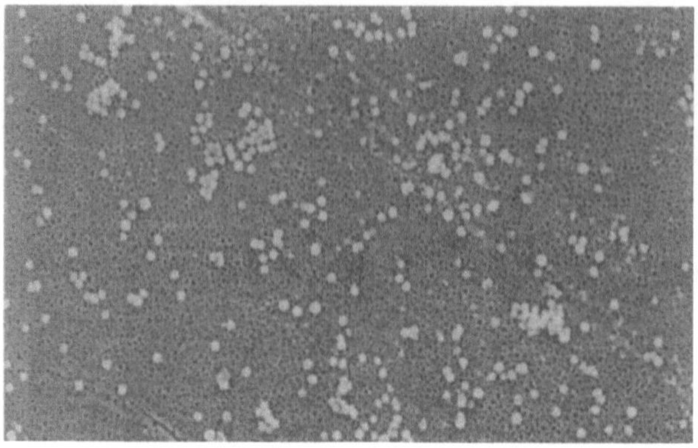

Fig. 3 Electron-micrograph of silica fully covered with PDADMAC 428.000 in 0.01 M NaCl at pH 5.8

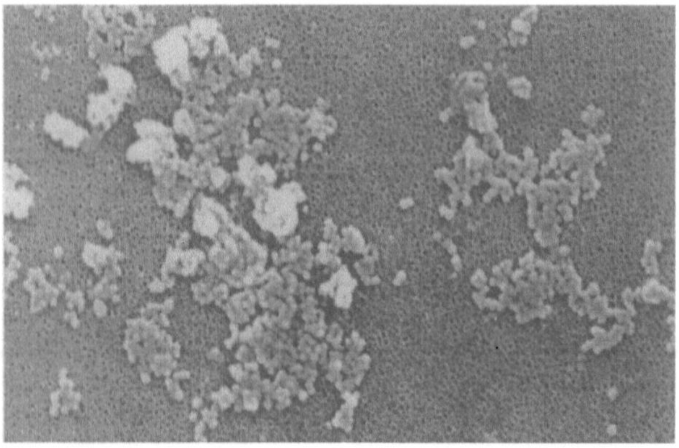

Fig. 2 Electron-micrograph of bare silica in 1 M NaCl at pH 5.8

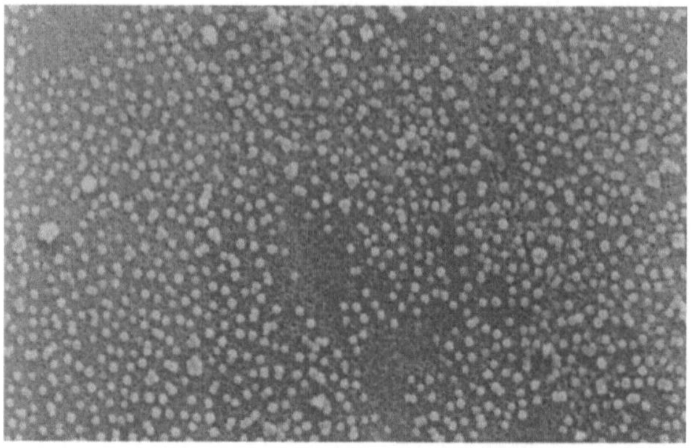

Fig. 4 Electron-micrograph of silica fully covered with PDADMAC 428.000 in 1 M NaCl at pH 5.8

Spectraphysics and an autocorrelator (system 4700) of Malvern Instruments [19]. The scattering angle for the PCS measurements was usually 90°. The layer thicknesses were measured at 25 °C, 1 h after the addition of 1 ml of a silica dispersion to 1 ml of a mixture of polyelectrolyte and salt. The resulting concentration of silica was 0.05 g/l. The hydrodynamic layer thicknesses were obtained by applying the Einstein–Stokes equation [Eq. (1)] to the diffusion coefficients of the pure and modified particles and subtracting the radii from each other.

$$a = \frac{kT}{6\pi\eta/D}, \qquad (1)$$

where a is the radius of the particles, k the Boltzmann constant, T the temperature in K, η the dynamic viscosity of the suspension and D the diffusion coefficient of the particles.

To be sure that a decrease of the diffusion coefficient was due to an increase of the hydrodynamic layer thickness of the adsorbed polymers we measured the diffusion coefficients as functions of time and scattering angle. Further we made some scanning electron micrographs of the bare and fully polymer covered silica particles. In the absence of polymer the suspension is only stable at low ionic strength (see Fig. 1), while at high salt concentration aggregation is observed (see Fig. 2). If the particles are fully covered with polymer, we obtain stable suspensions (see Figs. 3 and 4). So the hydrodynamic layer thicknesses can be measured in the plateau of the adsorption isotherms.

Results and discussion

All adsorption isotherms of PDADMAC and of the copolymers of DADMAC and NMVA on precipitated silica are of the high affinity character [20].

At low ionic strength the adsorption of the polyelectrolyte onto the oppositely charged surface is dominated by the electrostatic attraction. In solution the macroions

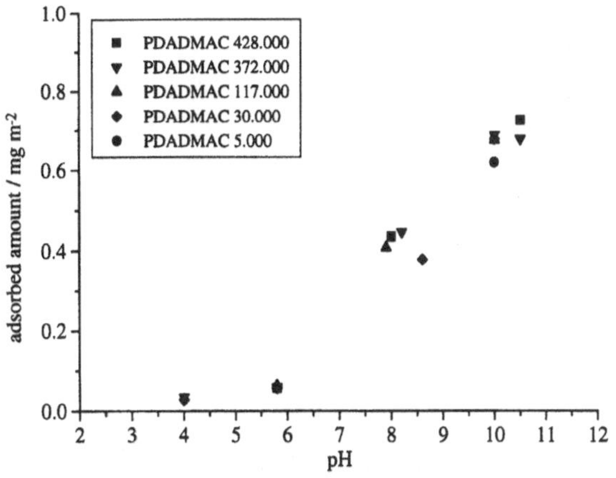

Fig. 5 Adsorbed amounts of PDADMAC of various molar masses as a function of the pH without any added NaCl

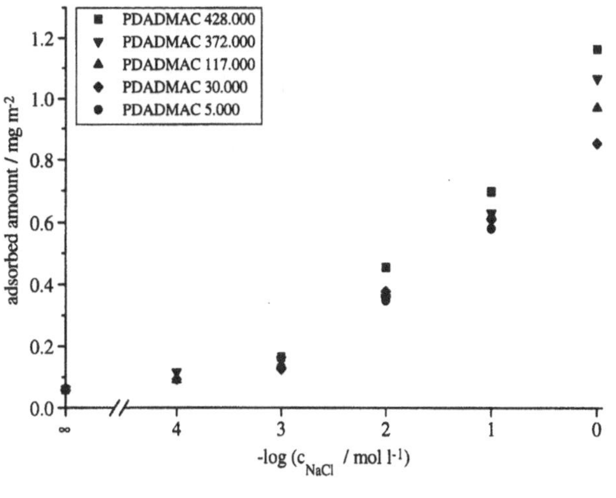

Fig. 6 Adsorbed amounts of PDADMAC of various molar masses as a function of the salt concentration at pH 5.8

assume a voluminous elongated conformation because of the mutual electrostatic repulsion of the charged chain segments. This also leads to a flat conformation of the polymer at the solid interface. The polyvalent ions have to assume the role of the counterions and are kept in the double layer near the surface. So at low ionic strength the adsorbed amount of saturation on the substrate is dominantly determined by charge compensation [21].

With increasing pH-value of the suspension the adsorbed amounts of PDADMACs of various molar masses rise (Fig. 5). This is due to the increasing surface charge density induced by the dissociation of the silanol groups with rising pH. For charge compensation at high pH a large amount of macroions is adsorbed. The adsorption of PDADMAC also takes place at pH 2–4, where the bare silica particles are uncharged as proved by potentiometric titration. Because of this result we assume that the driving force for polyelectrolyte adsorption is not only the surface charge compensation but also additional chemical interactions between the silica substrate and the polyelectrolyte.

The adsorption is independent of the molar mass of the PDADMACs. This indicates a flat conformation of the adsorbed polyelectrolytes because of the mutual electrostatic repulsion of the charged polyelectrolyte segments.

With increasing ionic strength of the suspension at pH 5.8 an increase of the adsorbed amounts is observed (Fig. 6). This is due to the formation of loops and tails because of the screening of the polyelectrolyte charges. The adsorption becomes influenced by the chemical interaction between the surface groups and the adsorbing polymer segments in contrast to pure electrosorption, where the adsorbed amount should decrease with increasing ionic strength [22]. At very high ionic strength beginning with 0.1 M the adsorbed amounts increase with molar mass due to the formation of longer loops and tails by the high molar mass macromolecules.

An experimental evidence for this explanation is given by Bauer et al. [20] where the hydrodynamic layer thicknesses in the plateau of the isotherms are drawn as a function of salt concentration. With increasing ionic strength the hydrodynamic layer thicknesses for all investigated polymers increase because of the screening of the polyelectrolyte segment charges by the NaCl ions. At high salt concentration the hydrodynamic thicknesses of the adsorbed layer rise with the molar mass of the macroion because the longer polymer molecules form more extended tails than shorter ones.

In Fig. 7 the hydrodynamic layer thickness of PDADMACs of the molar masses 428.000, 372.000 and 30.000 g/mol are plotted as a function of the adsorbed amounts at various salt concentrations in the plateau-region of the adsorption isotherms. At low adsorbed amounts the increase of hydrodynamics layer thicknesses with the adsorbed amounts is independent of the molar mass of the macroions. There are only short tails and the hydrodynamic layer thicknesses are small. With rising adsorbed amounts longer tails are formed, the hydrodynamic layer thicknesses increase. This is predicted theoretically for uncharged polymers in a θ-solvent [23]. At high salt concentration longer tails are formed by polymers with higher molar masses leading to an increase of the hydrodynamic layer thicknesses with the chain length of the adsorbed PDADMACs. At other pH-values we find the same adsorption behaviour of PDADMAC by varying the salt concentration [20].

In Fig. 8 the surface charge density of bare silica as a function of the pH is compared with the surface charge

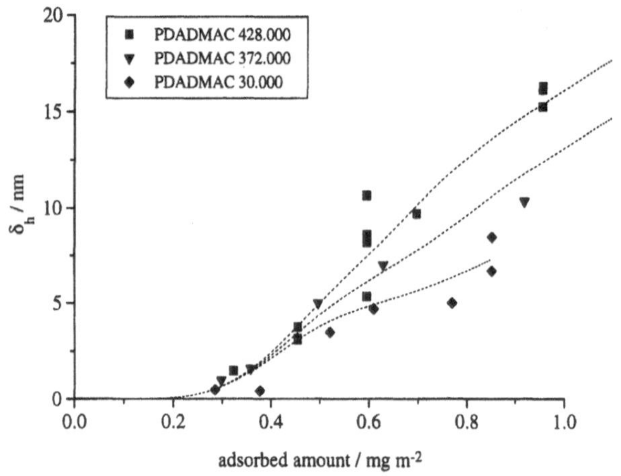

Fig. 7 Hydrodynamic layer thicknesses as a function of the adsorbed amounts for PDADMACs of various molar masses at pH 5.8

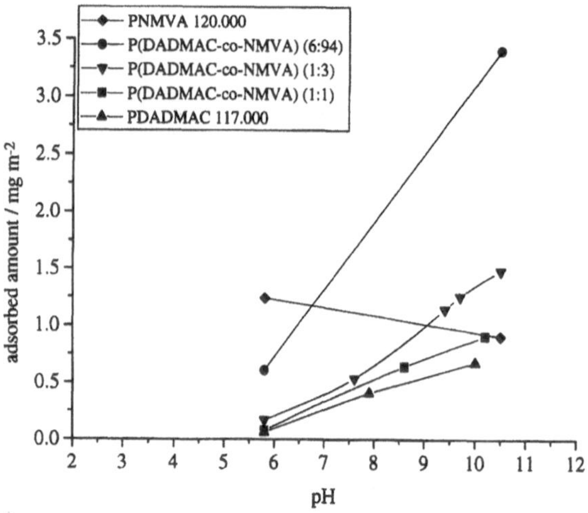

Fig. 9 Adsorbed amounts of various copolymers as a function of the pH without any added salt

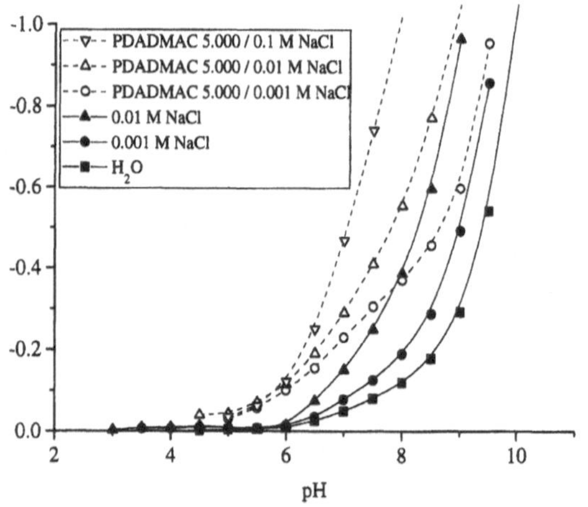

Fig. 8 Surface charge densities of bare silica and silica covered with PDADMAC 5.000 as a function of the pH for various salt concentrations

density of silica fully covered with PDADMAC of the molar mass 5.000 g/mol at various salt concentrations.

At every investigated ionic strength the surface charge density both of the bare and the polyelectrolyte covered silica particles increase with the pH because of the dissociation of the silanol groups. The surface charge densities of the polyelectrolyte covered silica are higher than those of the bare silica until a pH of about 9 at 0.001 M NaCl and 8 at 0.01 M NaCl.

The charges on the bare silica surface have larger distances in comparison to the charges of the polycations. Therefore we postulate that during the adsorption process of the polyelectrolytes further dissociation of the SiOH-groups occurs to compensate the charges of the adsorbed polyelectrolytes in order to minimize the free energy of the system. This idea is supported by results of Goloub et al. [18], who obtained an increase of the surface charge density of Aerosil by surfactant adsorption at low salt concentration. They suppose that the charge adjustment is strongly influenced by the presence of the hydrocarbon tails of the surfactant molecules. At pH values higher than 9 the charge density measured by potentiometric titration does not reveal the real surface situation because of the dissociation of SiOH-groups or adsorption of OH^--ions inside the porous surface [24] and because of counter ion condensation [25].

The surface charge density of the bare and polyelectrolyte covered silica particles also increase with the salt concentration of the suspension especially at high pH because of the screening of the SiO^--groups by the salt ions.

In Fig. 9 the adsorbed amounts of various copolymers of DADMAC and NMVA are shown as a function of the pH. At every pH the adsorbed amounts increase with decreasing charge density of the copolymers. As shown above there should be a further dissociation of silanol groups during the adsorption of polyelectrolytes with smaller charge distances. Because of the neutral segments the charge distance in the copolymers is larger than in the pure PDADMACs. On the other hand, the copolymeric polycations can occupy a shorter distance at the surface because of the lower charge density. So the adsorption is determined by an equilibrium between a low electrostatic repulsion between the macroions and a large electrostatic attraction between the polyelectrolytes and the surface leading to the same number of adsorbed charges of the polyelectrolytes and therefore to a corresponding increase of the adsorbed amounts with decreasing polyelectrolyte charge density at pH 5.8.

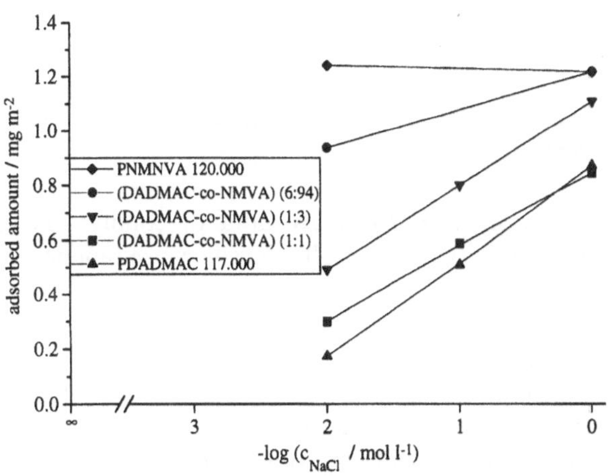

Fig. 10 Adsorbed amounts of various copolymers as a function of the salt concentration at pH 2

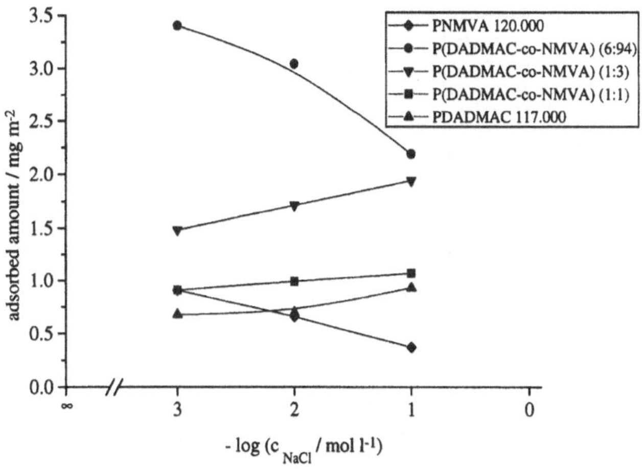

Fig. 12 Adsorbed amounts of various copolymers as a function of the salt concentration at pH 10.5

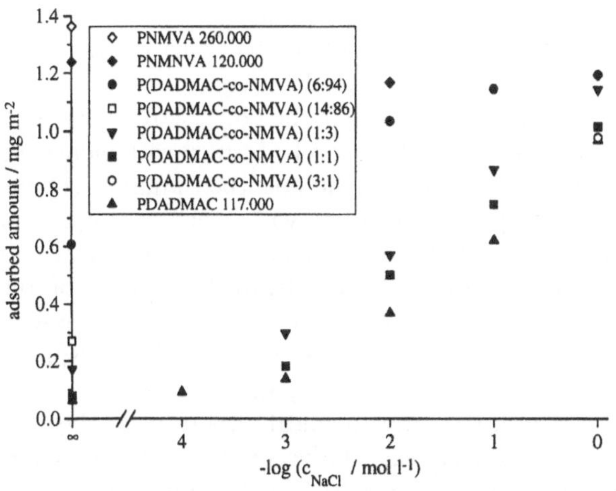

Fig. 11 Adsorbed amounts of various copolymers as a function of the salt concentration at pH 5.8

The adsorbed amounts of PDADMAC and of the copolymers increase with the pH of the suspension. Also at high pH the adsorbed amounts increase with decreasing charge density of the copolymers but the number of the adsorbed charges decrease. Because of this we postulate that the adsorption energy between the NMVA segments and the highly charged silica surface is low and the polymer molecules are displaced by the salt salt ions.

The adsorbed amounts of the neutral PNMVA decreases with increasing pH. This is due to the displacement of the polymer segments by the salt ions. We assume that the adsorption energy of the NMVA segments decreases with rising pH. This may be due to less hydrogen bonding between the carboxy groups of NMVA and the silanol groups because of dissociation of the silanol groups or due to less hydrophobic interaction between NMVA and the hydrophobic siloxan groups because of adsorption of OH^--ions.

The adsorbed amounts of every copolymer rise with the ionic strength at pH 2 and 5.8 (see Figs. 10 and 11) because of screening of the polyelectrolyte charges and formation of loops and tails. At high ionic strength the adsorption is dominated by the chemical affinity between the polyelectrolytes and the silica surface. This seems to be almost the same for NMVA and DADMAC segments, because the adsorbed amounts for the various copolymers are similar at very high ionic strength.

At pH 10.5 a different behaviour is observed (see Fig. 12). Only the adsorbed amounts of PDADMAC and of the copolymers P(DADMAC-co-NMVA) (1:1, 1:3) increase with the salt concentration. The adsorbed amounts of copolymers with a larger content of NMVA and of PNMVA decrease with rising salt concentration. The chemical affinity between NMVA and silica seems to be very low at high pH. Therefore the salt ions are able to displace the polymer segments.

Measuring the surface charge density of silica covered with P(DADMAC-co-NMVA) (1:3) (see Fig. 13) similar curves as for silica covered with PDADMAC 5.000 are obtained (see Fig. 8). At pH 5–6 in 0.001 M NaCl the surface charge densities are the same for particles covered with both polyelectrolytes. This is due to the same number of adsorbed polymer charges. At high pH there are less numbers of adsorbed charges of the copolymer (1:3) than of pure PDADMAC. Less dissociation of the silanol groups and a lower surface charge density of the polyelectrolyte covered silica is observed.

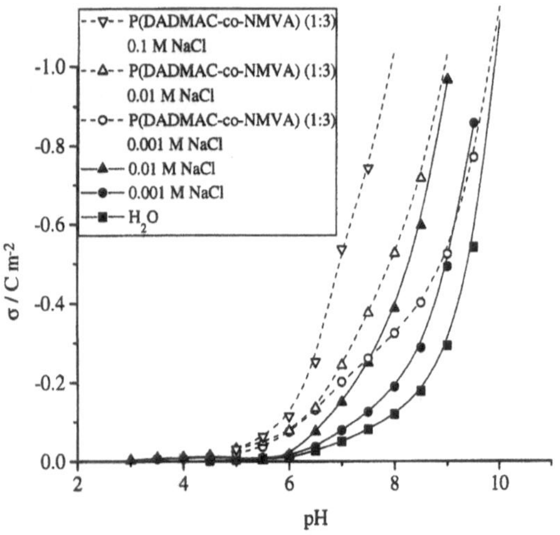

Fig. 13 Surface charge densities of bare silica and silica covered with P(DADMAC-co-NMVA) (1:3) as a function of the pH for various salt concentrations

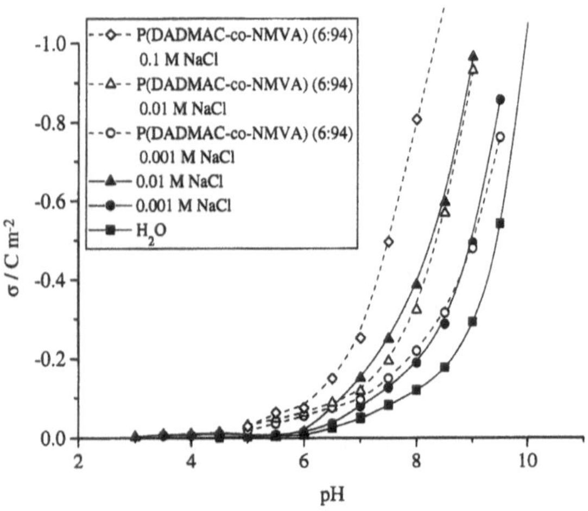

Fig. 14 Surface charge densities of bare silica covered with P(DADMAC-co-NMVA) (6:94) as a function of the pH for various salt concentrations

This behaviour is even more pronounced for P(DADMAC-co-NMVA) (6:94) (see Fig. 14). The adsorption of PNMVA (not shown) does not change the surface charge-density of silica because no further silanol dissociation has to take place in order to compensate adsorbed polymer charges.

Conclusions

The adsorption of PDADMAC and of copolymers of DADMAC and NMVA on colloidal silica as a function of ionic strength, pH of the suspension and of charge density and molar mass of the cationic polyelectrolytes was investigated.

At low ionic strength the adsorption is dominated by the electrostatic interactions. Because of the electrostatic repulsion of the segments the polyelectrolytes adsorb in a flat conformation. The adsorbed amount increases with decreasing charge density of the polyelectrolytes at every pH. The distance of the polyelectrolyte charge is smaller than the distance of the charges at the bare silica surface. So further dissociation of silanol groups during the adsorption and a charge compensation between the surface and the adsorbed macroions occurs. This is proved by measuring the surface charge densities.

Because of the screening of the polyelectrolyte charges by salt ions loops and tails can be formed with increasing ionic strength . Therefore the adsorbed amounts of the investigated polyelectrolytes with the exception of P(DADMAC-co-NMVA)(6:94) at high pH increases with the salt concentration. The adsorbed amounts of the copolymer (6:94) and of PNMVA 120.000 decrease with rising ionic strength at high pH because of the displacement of the NMVA segments by salt ions.

At pH 2 and 5.8 the chemical affinity of DADMAC and NMVA to silica seems to be equal because the adsorbed amounts at high ionic strength are almost the same.

At high salt concentration the polyelectrolytes behave like uncharged polymers. The adsorbed amounts and the hydrodynamic layer thicknesses of the adsorbed polyelectrolytes increase with the molar masses of the macroions.

Acknowledgements The authors are indebted to the Arbeitsgemeinschaft Industrieller Fördervereinigungen e.V. (AIF-project No. 10016). D. Bauer thanks the Stiftung Stipendien-Fonds des Verbandes der chemischen Industrie e.V. for financial support.

References

1. Dautzenberg H, Jaeger W, Kötz J, Philipp B, Seidel Ch, Stscherbina D (1994) In: Polyelectrolytes: Formation, Characterization and Application. Carl Hanser Verlag, Munich
2. Bárány S, Baran AA, Solomentseva I, Velichanskaya L (1994) In: Schmitz KS (ed) Macroion Characterization. American Chemical Society, Washington DC, pp 406–420
3. Bárány S, Baran AA, Gregory D (1996) Colloid 58:9–14
4. Napper DH (1977) J Colloid Interface Sci 58:390–406

5. Sato T, Ruch R (1980) In: Stabilization of Colloidal Dispersions by Polymer adsorption. Marcel Dekker, New York
6. Laskowski J (1982) In: Matijevic E (ed) Surface Colloid Science, Vol 12. Plenum Press, New York, pp 315–320
7. Cohen Stuart MA (1988) J Phy France 49:1001
8. Onabe F (1978) J Appl Polym Sci 22:3495–3510
9. Bauer D, Killmann E, Jaeger W (1998) Colloid Polym Sci, accepted
10. Stöber W, Fink A, Bohn E (1968) J Colloid Interface Sci 26:62
11. Killmann E, Sapuntzjis P, Maier H (1992) Macromol Chem, Macromol Symp 61:42–58
12. Hair ML (1994) In: Bergna HE (ed) Infrared Spectroscopy in Surface Chemistry. American Chemical Society, Washington
13. Iler RK (1979) In: The Chemistry of Silica: Solubility, Polymerization, Colloid and Surface Properties and Biochemistry. Wiley, New York
14. Buchhammer HM, Lunkwitz K, Schwarz S, Bauer D, Fuchs A, Rehmet R, Killmann E, Jaeger W. In preparation
15. Dautzenberg H (verbal information)
16. Terayama H (1952) J Polymer Sci 8:243
17. Horn D (1978) Progr Colloid Polym Sci 65:251
18. Goloub TP, Loopal LK, Bijsterbosch BH (1996) Langmuir 12:3188–3194
19. Killmann E, Sapuntzjis P (1994) Colloids and Surfaces A 86:229–238
20. Bauer D, Killmann E (1997) In: Macromol Chem, Macromol Symp, to be accepted
21. Cohen Stuart MA, Fleer GJ, Lyklema J, Norde W, Scheutjens JMHM (1991) Adv Colloid Interface Sci 34:477–533
22. Wang TK, Audebert R (1987) J Colloid Interface Sci 121:32–41
23. Cohen Stuart MA, Waajen FHWH, Cosgrove T, Vincent B, Crowley TL (1984) Macromolecules 17:1825
24. Lyklema J (1968) J Electroanal Chem 18:341–348
25. Manning GS (1978) Quart Rev Biochem 11:179–246

Progr Colloid Polym Sci (1998) 109:170–184
© Steinkopff Verlag 1998 SURFACE ENERGETICS

D.Y. Kwok
A.W. Neumann

Contact angles and surface energetics

Received: 10 August 1997
Accepted: 5 September 1997

Prof. A. Wilhelm Neumann (✉)
D.Y. Kwok
University of Toronto
Department of Mechanical and
Industrial Engineering
5 King's College Road
Toronto, Ontario
Canada M5S 3G8

Abstract Recent progress in the correlation of contact angles with solid surface tensions are summarized. The measurements of meaningful contact angles in terms of surface energetics are also discussed. It is shown that the apparent controversy with respect to measurement and interpretation of contact angles are due to the fact that some (or all) of the assumptions made in all energetic approaches [7–14] are violated when contact angles are measured and processed. For a large number of polar and non-polar liquids on different solid surfaces, the values of $\gamma_{lv} \cos\theta$ are shown to depend only on γ_{lv} and γ_{sv} when the appropriate experimental techniques and procedures are used. An equation which follows these experimental patterns and which allows the determination of solid surface tensions from contact angles is discussed.

Key words Surface tension of solid – surface tension from contact angle – contact angle complexity – contact angle pattern

Introduction

It is generally believed that contact angle measurement is the best approach to determine solid surface tensions. The contact angle of a liquid drop on a solid surface is defined by the mechanical equilibrium of the drop under the action of three interfacial tensions: solid–vapour, γ_{sv}, solid–liquid, γ_{sl}, and liquid–vapour, γ_{lv}. This equilibrium relation is known as Young's equation:

$$\gamma_{lv} \cos\theta_Y = \gamma_{sv} - \gamma_{sl} , \qquad (1)$$

where θ_Y is the Young contact angle, i.e. a contact angle which can be used in conjunction with Young's equation. While there are a number of thermodynamic equilibrium contact angles θ_e, they are not necessarily equal to θ_Y in Young's equation, see below.

Equation (1) implies a single, unique contact angle; in practice, however, contact angle phenomena are complicated [1–3]. For example, the contact angle made by an advancing liquid (θ_a) and that made by a receding liquid (θ_r) are not identical; nearly all solid surfaces exhibit contact angle hysteresis, H (the difference between θ_a and θ_r):

$$H = \theta_a - \theta_r , \qquad (2)$$

Contact angle hysteresis can be due to roughness and hetereogeneity of a solid surface. If roughness is the primary cause, then the measured contact angles are meaningless in terms of Young's equation. On rough surfaces, contact angles are larger than on chemically identical smooth surfaces [4]. Obviously, interpreting such angles in terms of Eq. (1) would lead to erroneous results.

In general, the experimentally observed apparent contact angle, θ, may or may not be equal to the Young contact angle, θ_Y [3]:

(1) On ideal solid surface, there is no contact angle hysteresis and the experimentally observed contact angle is equal to θ_Y.

(2) On smooth, but chemically heterogeneous solid surfaces, θ is not necessarily equal to the thermodynamic equilibrium angle. Nevertheless, the experimental advancing contact angle, θ_a, can be expected to be a good approximation of θ_Y. Therefore, care must be exercised to ensure that the experimental apparent contact angle, θ, is the advancing contact angle in order to be inserted into the Young equation.

(3) On rough solid surfaces, no such equality between θ_a and θ_Y exists: all contact angles on rough surfaces are meaningless in terms of Young's equation.

Moreover, there are as yet no general criteria to answer the question of how smooth a solid surface has to be. This and similar problems are linked to line tension, which has its own complexities [5]. It is, therefore, of utmost importance to prepare solid surfaces as smooth as possible so that the experimental advancing angles can be a good approximation of θ_Y. In addition to these complexities, penetration of the liquid into the solid, swelling of the solid by the liquid, and chemical reactions can all play a role. For example, swelling of solid by the liquid [6] can change the chemistry of the solid in an unknown manner and hence affect the values of γ_{sv}, and/or γ_{sl}. Therefore, it is also important to ensure the solid surfaces to be as inert as possible in order to minimize such effects.

Several approaches [7–14], of current interest, were largely inspired by the idea of using Young's equation for surface energetics. While these approaches are, logically and conceptually, mutually exclusive, they share, nevertheless, some of the basic assumptions:

(1) All approaches rely on the validity of Young's equation for surface energetics from experimental contact angles.

(2) Pure liquids are always used; surfactant solutions or mixtures of liquids should not be used, since they would introduce complications due to preferential adsorption.

(3) The values of γ_{lv}, γ_{sv}, (and γ_{sl}) are assumed to be constant during the experiment, i.e. there should be no chemical reaction between the solid and the liquid.

(4) The values of γ_{sv} going from liquid to liquid are also assumed to be constant, i.e. independent of the liquids used.

With respect to the first assumption, one requires the solid surfaces to be rigid and smooth so that Young's equation is applicable; the experimentally observed contact angles should also be advancing contact angles. However, many attempts have been made in the literature to interpret surface energetics of solids, which are not rigid (e.g. gels [13]) and not smooth (e.g. biological surfaces), in conjunction with Young's equation. Clearly, these results are open to question, since Young's equation may not be valid in these situations. With respect to the other assumptions, the solid surfaces should be as inert as possible so that effects such as swelling and chemical reactions are minimized.

In order to assure that the experimentally measured contact angles do not violate any of the above assumptions, one requires careful experimentation and suitable methodology. However, contact angles are typically measured simply by depositing a drop of liquid on a given solid surface, and manually placing a tangent to the drop at its base using a so-called goniometer-sessile drop technique. Apart from the subjectivity of the technique, it normally yields contact angle accuracy of no better than $\pm 2°$. More important, the technique cannot be expected to reflect the complexities of solid–liquid interactions. We believe that much of the controversy with respect to the interpretation of contact angles in terms of surface energetics lies in the fact that not enough attention is given to the above assumptions.

In this review, we summarize recent progress in the correlation of contact angles with solid surface tensions. We will first discuss the experimental contact angle patterns using an automated axisymmetric drop shape analysis, and contrast them with the patterns obtained using a conventional goniometer-sessile drop technique. It will be shown that the discrepancy indeed comes from the fact that some (or all) of the readily accepted assumptions are violated when contact angles are measured. An equation which follows these experimental patterns and which allows the determination of solid surface tensions from contact angles will be sought.

Experimental contact angle patterns

On carefully prepared solid surfaces, Li et al. [15, 16] have performed static (advancing) contact angle experiments using an automated axisymmetric drop shape analysis-profile (ADSA-P). ADSA-P [17, 18] is a technique to determine liquid–fluid interfacial tensions and contact angles from the shape of axisymmetric menisci, i.e., from sessile as well as pendant drops. Assuming that the experimental drop profile is axisymmetric and Laplacian, i.e. is determined only by surface tension and gravity, ADSA-P finds the theoretical profile that best matches the drop profile extracted from the image of a real drop, from which the surface tension, contact angle, drop volume, surface area and the three-phase contact radius can be computed. It has been found [15, 16] that a contact angle accuracy of better than $\pm 0.3°$ can be obtained on well-prepared solids.

In the experiments of Li et al., static (advancing) contact angles were measured by supplying test liquids from below the surface into the sessile drop, using a motor-driven syringe device. A hole of about 1 mm in the centre of each solid surface was required to facilitate such procedures. Liquid was pumped slowly into the drop from below until the three-phase contact radius was about 0.4 cm. After the motor was stopped, the sessile drop was allowed to relax for approximately 30 s to reach equilibrium. Then 3 pictures of this sessile drop were taken successively at intervals of 30 s. More liquid was then pumped into the drop until it reached another desired size, and the above procedure was repeated [15]. These procedures ensure that the measured static contact angles are indeed the advancing contact angles. In the literature, it is customary to first deposit a drop of liquid on a given solid surface using a syringe or a teflon needle; the drop is then made to advance by supplying more liquid from above using a syringe or a needle immersed into the drop. Such experimental procedures cannot be used for ADSA-P since ADSA determines the contact angles and surface tensions based on the complete and undisturbed drop profile.

Three carefully prepared solids were used: they are FC-721-coated mica, Teflon (FEP) heat-pressed against quartz glass slides and poly(ethylene terephthalate) (PET). The FC-721 surface was prepared by a dip-coating technique. FC-721 is a 3M "Fluorad" brand antimigration coating designed to prevent the creep of lubricating oils out of bearings. Teflon FEP (fluorinated ethylene propylene) surfaces were prepared by a heat-pressing method. The material was cut to 2 × 4 cm, sandwiched between two glass slides, and heat-pressed by a jig in an oven. Poly (ethylene terephthalate) (PET) is the condensation product of ethylene glycol and terephthalic acid. The surfaces of PET films were exceedingly smooth as recieved and cleaned before measurement. Details of the solid surface preparation can be found elsewhere [15].

Figure 1 shows these contact angle results, by plotting the values of $\gamma_{lv}\cos\theta$ vs. γ_{lv} for a large number of pure liquids with different molecular properties. The curves in Fig. 1 are smooth so that one has to conclude that the values of $\gamma_{lv}\cos\theta$ depend only on γ_{lv} and γ_{sv} [15, 16, 19]:

$$\gamma_{lv}\cos\theta = f(\gamma_{lv}, \gamma_{sv}) \; . \tag{3}$$

Because of Young's equation, these experimental contact angles imply that γ_{sl} can be expressed as a function of only γ_{lv} and γ_{sv}:

$$\gamma_{sl} = F(\gamma_{lv}, \gamma_{sv}) \; . \tag{4}$$

This is in agreement with thermodynamics [20] and the thermodynamic phase rule for capillary systems [21–24] which states that there are only two degrees of freedom for such solid–liquid systems. Thus, one can change the contact angle by changing either γ_{lv} or γ_{sv}. While intermolecular forces determined the surface tensions, they do not have any additional, independent effects on the contact angles. This point becomes apparent by focusing on the

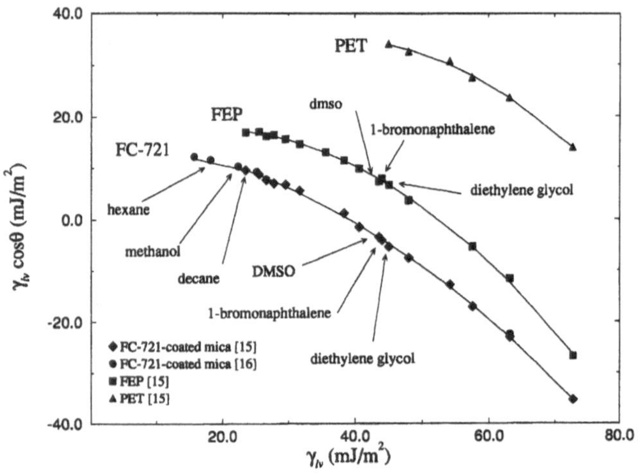

Fig. 1 $\gamma_{lv}\cos\theta$ vs. γ_{lv} for FC-721-coated mica, heat pressed Teflon (FEP), and poly(ethylene terephthalate) (PET). The contact angles are static angles by axisymmetric drop shape analysis-profile (ADSA-P). The smoothness of the curves suggests that $\gamma_{lv}\cos\theta$ depends only on γ_{lv} and γ_{sv}. Data are from Refs. [15, 16]

experimental contact angle data of hexane, methanol and decane on the FC-721 surface in Fig. 1. The surface tension of methanol is intermediate between the surface tensions of hexane and decane; methanol is polar and the alkanes are non-polar. Because of the difference in the polarity or non-dispersive property, the surface tension component approaches [7, 11–14] stipulate a contact angle for methanol which is significantly different from those of hexane and decane. Within the framework of these surface tension component models [7, 11–14], it stands to reason that for purely dispersive liquids such as alkanes, the contact angle changes smoothly as the liquid surface tension changes. It is therefore interesting to establish, by interpolation between hexane and decane, what the contact angle for a purely dispersive liquid of the same surface tension as methanol, i.e. $\gamma_{lv} = 22.3$ mJ/m^2, would be. This interpolated contact angle is found to be $\theta = 62.67°$, in excellent agreement with the experimental results for methanol, i.e. $\theta = 62.39 \pm 0.25°$. The fact that methanol has the same contact angle as a non-polar liquid of the same surface tension indicates that the surface tension component approaches [7, 11–14] are not tenable. Since FC-721 is a totally non-polar solid, it should, within the framework of these approaches, interact only with the putative dispersion component of the surface tensions of methanol, which would have to be much smaller than its physical surface tension, leading to a quite different (larger) contact angle. Similar conclusions can be drawn for DMSO, 1-bromonaphthalene and diethylene glycol on the FC-721 and FEP surfaces [19].

It should be noted that the experimental contact angle patterns shown in Fig. 1 are not always observed in the literature: curves far less smooth or no unique curves at all are sometimes reported. Such patterns can have a variety of causes. Accurate contact angle measurements require extreme experimental care. Even very minor vibrations can cause advancing contact angles to decrease, resulting in errors of several degrees. Surface roughness can affect the contact angles and make Young's equation inapplicable. Swelling of the solid by the liquid can change the chemistry of the solid and hence the values of γ_{sv} and θ in an unpredictable manner. Non-constancy of γ_{lv}, γ_{sv} and γ_{sl} during the experiment, and non-constancy of γ_{sv} from liquid to liquid can produce scatter in Fig. 1 easily. Therefore, scatter due to such causes prohibits the application of all contact angle approaches [7–14] to the determination of γ_{sv}.

The present controversy with respect to the experimental contact angle patterns is believed to arise from the fact that these patterns are often complex, and cannot be unraveled by the simple goniometer-sessile drop technique. We shown in Fig. 2 the contact angle results for two copolymers measured by a goniometer technique, by plotting $\gamma_{lv} \cos\theta$ vs. γ_{lv}. The copolymers are poly(propene-*alt*-*N*-(*n*-hexyl)maleimide), and poly(propene-*alt*-*N*-(*n*-propyl) maleimide). The copolymer-coated surfaces were prepared by a solvent-casting technique [25]: a few drops of the 2% copolymer/tetrahydrofuran solution were deposited on dried silicon wafers inside glass dishes overnight; the solution spread and a thin layer of the copolymer formed on the wafer surface after tetrahydrofuran evaporated. This preparation produced good quality coated surfaces, as manifested by light fringes, due to refraction at these surfaces, suggesting that surface roughness is in the order of nanometers or less.

The procedures to measure the contact angles using a goniometer are as follows: a sessile drop of about 0.4–0.5 cm radius was formed from above. The three-phase contact line of the drop was then slowly advanced by supplying more liquid from above through a capillary which was always kept in contact with the drop. The maximum (advancing) contact angles were measured carefully from the left and right side of the drop and subsequently averaged. The above procedures were repeated for 5 drops on 5 new surfaces. All readings were then averaged to give an averaged contact angle.

From Fig. 2, one might argue that $\gamma_{lv} \cos\theta$ cannot be a function of only γ_{lv} and γ_{sv}. Therefore, because of Young's equation, this scatter seems to favour the stipulation of the surface tension component approaches [7, 11–14] that γ_{sl} depends not only on γ_{lv} and γ_{sv}, but also on the specific intermolecular forces. While such contact angle patterns can be easily found in the literature, they do not necessarily support the stipulation of the surface tension component approaches [7, 11–14]. In the next section, it will be shown that this scatter indeed comes from the fact that many of the experimental contact angles in Fig. 2 have violated some (or all) assumptions made in all contact angle approaches [7–14]. Thus, the apparent additional degrees of freedom (inferred from the scatter) do not come from the putative independent effect of intermolecular forces on the contact angles: $\gamma_{lv} \cos\theta$ can be shown to change systematically with γ_{lv} when appropriate experimental procedures and contact angle techniques are used, see below.

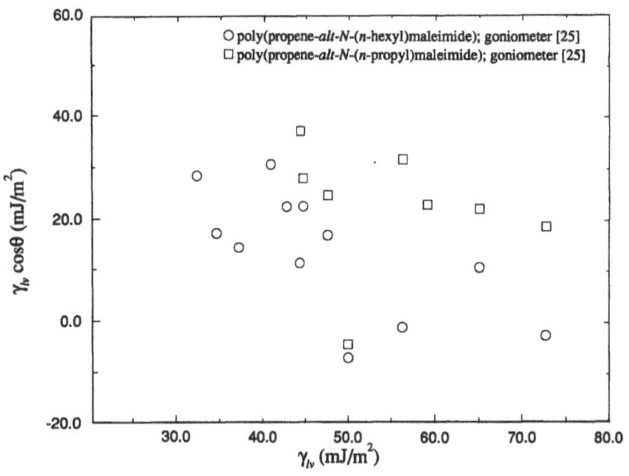

Fig. 2 $\gamma_{lv} \cos\theta$ vs. γ_{lv} for poly(propene-*alt*-*N*-(*n*-hexyl)maleimide) and poly(propene-*alt*-*N*-(*n*-propyl)maleimide) copolymers by a conventional goniometer technique. Because of the scatter, one might argue that $\gamma_{lv} \cos\theta$ depends not only on γ_{lv} and γ_{sv}, but also on the specific intermolecular forces. Data are from Ref. [25]. It will be shown that the scatter indeed comes from the fact that many of the experimental contact angles have violated some (or all) assumptions made in all contact angle approaches

Contact angle measurements

Low-rate dynamic (advancing) contact angles by ADSA-P

Inert (non-polar) surfaces: FC-722-coated mica surface

A recent study by Kwok et al. [26] to measure dynamic (advancing) contact angles at very slow motion of the three-phase contact line has reconfirmed the findings of Li et al. [15, 16] shown in Fig. 1. Seventeen polar and non-polar liquids were selected for the contact angle measurements on FC-722-coated mica surfaces. FC-722 is a fluorochemical coating available from 3M and is chemically very similar to the FC-721 used in other studies. They were prepared by the same dip-coating procedures described elsewhere [15].

In these low-rate dynamic contact angle experiments, images of an advancing drop (and hence information such

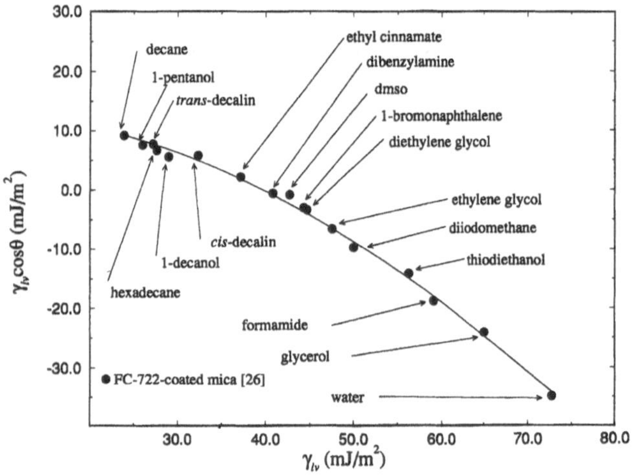

Fig. 3 $\gamma_{lv} \cos \theta$ vs. γ_{lv} for FC-72-coated mica. The contact angles are low-rate dynamic angles by axisymmetric drop shape analysis-profile (ADSA-P). The smoothness of the curves suggests that $\gamma_{lv} \cos \theta$ depends only on γ_{lv} and γ_{sv}. This result reconfirms the finding of Li et al. [15, 16] who worked with statistic drop, see Fig. 1. Data are from Ref. [26]

as surface tension and contact angle) are recorded continuously as drop volume is steadily increased from below the surface. The procedures used here are different from those by Li et al. [15] in that the contact angles measured by Li et al. were static angles, i.e. contact angles at zero velocity of the three-phase contact line. Since ADSA-P determines the contact angle and the three-phase contact radius simultaneously for each picture, the advancing dynamic contact angles as a function of the three-phase contact radius (i.e. location on the surface) can be obtained. In addition, the change in the contact angle, drop volume, drop surface area, and the three-phase contact radius can also be studied as a function of time. Such dynamic procedures allow direct observation of the contact angle behaviour as a function of surface location (i.e. the three-phase contact radius). It has been shown [25] that low-rate dynamic contact angles in the range from 0.1 to 1.0 mm/min are essentially identical to the static ones [15, 16] in Fig. 1, i.e. at zero velocity of the three-phase contact line. The results of this study are given in Fig. 3: it can be seen that the values of $\gamma_{lv} \cos \theta$ change smoothly with γ_{lv}, independent of the liquid properties.

Non-inert (polar) surfaces: poly(propene-alt-N(n-hexyl) maleimide) and poly(propene-alt-N-(n-propyl)maleimide)

Experience has shown that non-polar surfaces, such as teflon and fluorocarbons, often are quite inert with respect to many liquids; however, polar surfaces often are less inert and hence may show different contact angle patterns, due to such causes as chemical reaction and/or swelling and dissolution of the solid by the liquid. Since low-rate dynamic contact angle experiments when interpreted by ADSA-P have many advantages over the conventional way of manually putting tangents to the sessile drops, ADSA-P was employed recently [25] to measure low-rate dynamic contact angles on the copolymers, poly(propene-*alt*-*N*-(*n*-propyl)maleimide) and poly(propene-*alt*-*N*(*n*-hexyl)maleimide), in order to elucidate the discrepancies between the results in Figs. 1 and 3 on the one hand, and those in Fig. 2 by a goniometer on the other. It will become apparent that a goniometer technique is liable to produce a mixture of meaningful and meaningless angles (on non-inert surfaces), with no criteria to distinguish between the two. If one disregards the meaningless angles to be identified by dynamic ADSA measurements, the results are in harmony with those patterns shown in Figs. 1 and 3, as we shall show below.

Eight liquids were selected for the contact angle measurements on poly(propene-*alt*-*N*-(*n*-propyl)maleimide) and 13 liquids on poly(propene-*alt*-*N*-(*n*-hexyl)maleimide). These copolymer-coated surfaces were prepared by a solvent-casting technique [25], as described above. We reproduce in Fig. 4a a typical example of water on the poly(propene-*alt*-*N*-(*n*-propyl)maleimide) copolymer. In order to avoid that the drop hinges at the edge of the hole, a small drop is deposited from above, covering the hole completely. As can be seen in this figure, increasing the drop volume, V, linearly from 0.10 cm^3 to about 0.12 cm^3 increases the apparent contact angle, θ, from about 72°–77° at essentially constant three-phase contact radius, R. This is due to the fact that even carefully putting an initial water drop from above on a solid surface can result in a contact angle somewhere between advancing and receding. This effect can be pronounced for liquids, such as water, which evaporate fast. Thus, it takes time for the initial drop front to start advancing. Further increase in the drop volume causes the three-phase contact line to advance, with θ essentially constant as R increases. Increasing the drop volume in this manner ensures the measured θ to be an advancing contact angle. In this specific example, the measured contact angles are essentially constant as R increases. This indicates good surface quality of the surfaces used. A mean contact angle of $77.33 \pm 0.06°$ was obtained for this experiment. Additional experiments at different rates of advancing were performed (each on a newly prepared surface) [25]. The pattern observed in Fig. 4a is similar in all measurements summarized in Fig. 3.

Unfortunately, not all liquids yield essentially constant advancing contact angles. Figure 4b shows the results of formamide on the same copolymer. It can be seen that as drop volume increases initially, contact angle increases

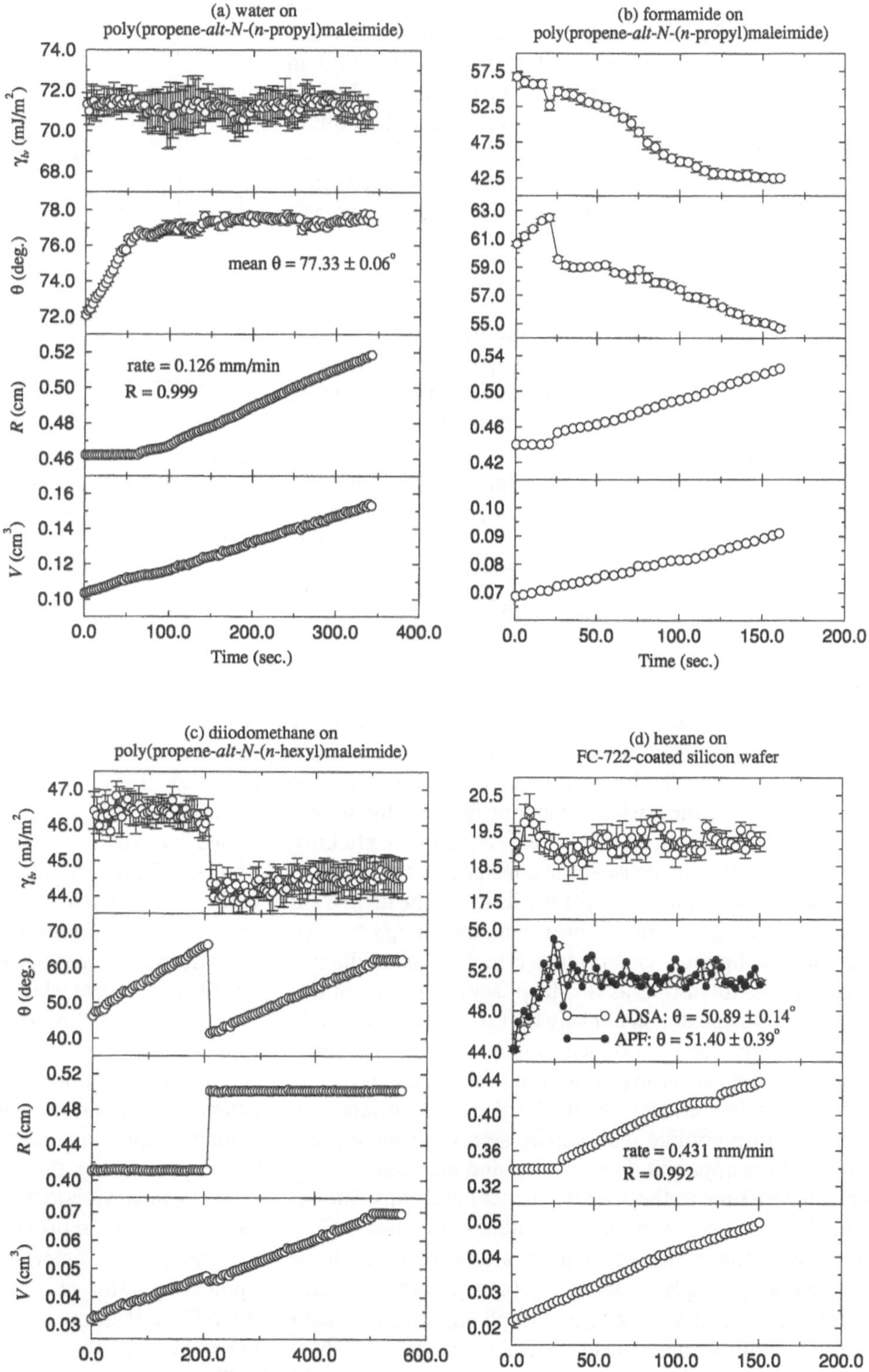

Fig. 4 Low-rate dynamic contact angles of (a) water, (b) formamide on a poly(propene-*alt*N-(*n*-propyl)maleimide) copolymer [25], and (c) diiodomethane on poly(propene-*alt*-*N*-(*n*-hexyl)maleimide) copolymer [25] by axisymmetric drop shape analysis-profile (ADSA-P); The decrease in the surface tension and contact angle indicates dissolution of the copolymer by formamide. The slip/stick behaviour of diiodomethane suggests that Young's equation may not be applicable (see text). These contact angles should not be used for the interpretation of surface energetics. (d) Low-rate dynamic contact angles of hexane on a FC-722-coated silicon wafer interpreted by ADSA-P and APF [27]. It can be seen that the APF contact angles are essentially the same as those from ADSA-P, but with more scatter

from 60° to 63° at essentially constant three-phase contact radius. As the drop volume continues to increase, θ suddenly decreases to 60° and the three-phase contact line starts to move. As R increases further, the contact angle decreases slowly from 60° to 54°. The surface tension-time plot indicates that the surface tension of formamide decreases with time. This suggests that dissolution of the copolymer occurs, causing γ_{lv} to change from that of the pure liquid. Similar behavior can also be observed in other experiments [25]. It is an important question to ask which contact angles one should use for the interpretation in terms of surface energetics. Since chemical or physical reactions such as polymer dissolution change the liquid–vapour, solid–vapour and solid–liquid interface (interfacial tensions) in an unknown manner, such a question is very difficult to answer. Because we are unsure whether or not γ_{sv} will remain constant and whether Young's equation is applicable, these contact angle data should be disregarded in the interpretation in terms of surface energetics. Obviously, it is virtually impossible for a conventional goniometer technique to detect complexities as those in Fig. 4b. Thus, the contact angle obtained from a goniometer for this and similar solid–liquid systems cannot be meaningful.

A different contact angle experiment is illustrated in Fig. 4c for diiodomethane on the poly(propene-*alt*-*N*-(*n*-hexyl)maleimide) copolymer. It can be seen that initially the apparent drop volume, as perceived by ADSA-P, increases linearly, and the contact angle increases from 88° to 96° at essentially constant three-phase contact radius. Suddenly, the drop front jumps to a new location as more liquid is supplied into the sessile drop. The resulting contact angle decreases sharply from 96° to 88°. As more liquid is supplied into the sessile drop, the contact angle increases again. Such slip/stick behavior could be due to non-inertness of the surface. Phenomenologically, an energy barrier for the drop front exists, resulting in sticking, which causes θ to increase at the constant R. However, as more liquid is supplied into the sessile drop, the drop front possesses enough energy to overcome the energy barrier, resulting in slipping, which causes θ to decrease suddenly. It should be noted that as the drop front jumps from one location to the next, it is unlikely that the drop will remain axisymmetric. Such a non-axisymmetric drop will obviously not meet the basic assumptions underlying ADSA-P, causing possible errors, e.g., in the apparent surface tension and drop volume. This can be seen from the discontinuity of the apparent drop volume and apparent surface tension with time as the drop front sticks and slips. Similar behavior can also be observed in other experiments [25]. Obviously, the observed contact angles cannot all be Young contact angles, since γ_{lv}, γ_{sv} (and γ_{sl}) are constants, so that because of Young's equation, θ ought to be a constant. In addition, it is difficult to decide unambiguously at this moment whether or not Young's equation is applicable at all because the lack of understanding of the slip/stick mechanism. Therefore, these contact angles should not be used for the interpretation in terms of surface energetics.

While pronounced cases of slip/stick behavior can indeed be observed by the goniometer, it is virtually impossible to record the entire slip/stick behaviour manually. In this case, the goniometer contact angle can be very subjective, depending on the skill of the expermentalist. It is expected that a contact angle thus recorded by the goniometer should agree with the maximum angles obtained by ADSA-P. Indeed, a contact angle of 98° was observed, in reasonable agreement with the maxima in the entire slip/stick pattern of the ADSA-P results ($\theta \approx 96°$). In cases where the liquid–vapour surface tension of the sessile drop decreases due to, e.g., dissolution of the surface, only ADSA can detect changes in the liquid–vapour surface tension. The distinctions and differentiations made by ADSA-P are not possible in a goniometer study. Thus, circumspection is necessary in the decision whether or not experimental contact angles can be used in conjunction with Young's equation; contact angles from a conventional goniometer-sessile drop technique may produce contact angles which violate the basic assumptions made in all surface energies approaches [7–14], e.g., constancy of γ_{sv}.

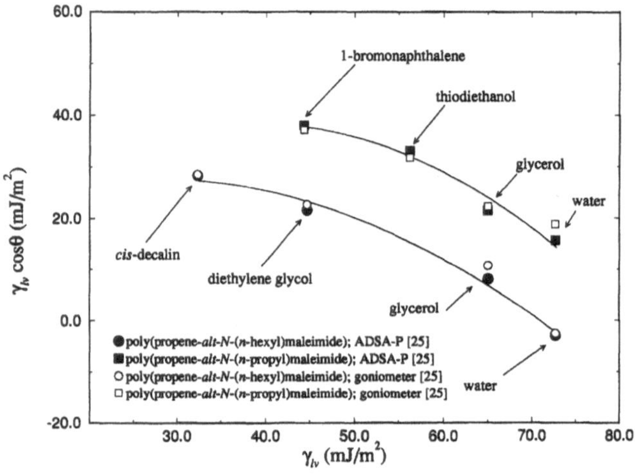

Fig. 5 $\gamma_{lv} \cos\theta$ vs. γ_{lv} for poly(propene-*alt*-*N*-(*n*-hexyl)maleimide) and poly(propene-*alt*-*N*-(*n*-propyl)maleimide) copolymers by ADSA-P and a goniometer technique. Neglecting the inconclusive contact angles by ADSA-P dynamic angles, $\gamma_{lv} \cos\theta$ changes regularly with γ_{lv} and γ_{sv} for poly(propene-*alt*-*N*-(*n*-hexyl)maleimide) and poly(propene-*alt*-*N*-(*n*-propyl)maleimide), contrary to those reported in Fig. 2. The data are from Ref. [25]

The picture emerging in Fig. 2 changes drastically upon elimination of the angles shown to be meaningless in the ADSA-P study, see Fig. 5. The curves in Fig. 5 are in harmony with the results obtained for more inert polar and non-polar surfaces. Thus, the experimental procedures and techniques used are crucial in the collection of contact angle data for surface energetics.

One might argue that the above comparison is misleading, in that the real difference between the two types of experiments is that of a fairly sophisticated and automated low-rate dynamic contact angle measurement and a very simple if not crude static measurement. To explore this thought, del Rio et al. [27] reported a novel, automated method to put tangents to the same drop images on which ADSA-P operates. Thus, low-rate dynamic contact angle experiments were performed and interpreted separately by ADSA-P and an automated polynominal fit program (APF). Two different types of solid surfaces were used: an inert (non-polar) FC-722-coated silicon wafer surface and a non-inert (polar) poly(propene-*alt*-*N*-methylmaleimide) copolymer surface.

An example is shown in Fig. 4d for hexane on the FC-722-coated wafer surface: increasing the drop volume, V, linearly from 0.22 cm^3 to about 0.03 cm^3 increases the apparent contact angle, θ, from about 44° to 54° at essentially constant three-phase contact radius, R. This increase in the contact angle has been explained above and is due to the fact that even carefully putting an initial hexane drop from above on a solid surface can result in a contact angle somewhere between advancing and receding. Further increase in the drop volume causes the three-phase contact line to advance, with θ essentially constant as R increases. A mean ADSA-P constant angle of $50.89 \pm 0.14°$ was obtained. The APF contact angles for this experiment are also shown in the same figure. It can be seen that the initial increase in the contact angles was also observed by the APF scheme. The mean APF contact angle is found to be $51.40 \pm 0.39°$, in good agreement with that from ADSA-P. Similar results have been obtained also for other liquids: glycerol, hexadecane, and 2-octanol. Plotting these contact angles as $\gamma_{lv} \cos\theta$ vs. γ_{lv} onto Fig. 3 suggests that $\gamma_{lv} \cos\theta$ also changes systematically with γ_{lv} (see Fig. 6), independent of which of the two techniques is used. It may be noted that there is essentially no difference between the results obtained for the FC-722 coated onto mica or onto silicon wafer surfaces.

On the non-inert poly(propene-*alt*-*N*-methylmaleimide) copolymer, a total of seven liquids with different molecular properties were chosen for the study. It was found [27] that ethanolamine and tetrabromoethane dissolved the copolymer on contact; formamide and thiodiethanol did not result in constant contact angles due to dissolution of the copolymer by the liquids. A general trend of such contact angle patterns was also observed by APF, but with larger scatter. If the inconclusive contact angles are omitted, essentially smooth curves emerge, when plotting the values of $\gamma_{lv} \cos\theta$ vs. γ_{lv}. Figure 7 shows this plot using the mean contact angles from both ADSA-P and APF schemes as well as the contact angles for other maleimide copolymers [25, 28]. It can be seen in this figure that the values of $\gamma_{lv} \cos\theta$ all change systematically with γ_{lv}, regardless of intermolecular forces. A consistent change in the contact angle patterns is observed as the length of the side chains decreases from hexyl to methyl groups.

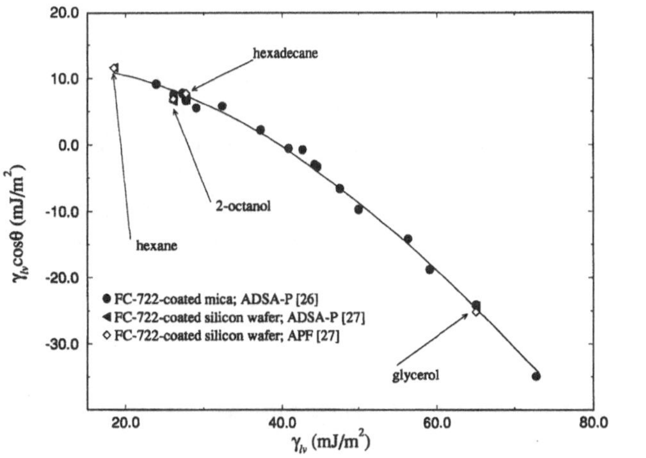

Fig. 6 $\gamma_{lv} \cos\theta$ vs γ_{lv} for FC-722-coated mica [26] and FC-722-coated silicon wafer [27]. The contact angles are low-rate dynamic angles by axisymmetric drop share analysis-profile (ADSA-P). The smoothness of the curves suggests that $\gamma_{lv} \cos\theta$ depends only on γ_{lv} and γ_{sv}. It may be noted that there is essentially no difference in the contact angles between the FC-722-coated mica and that coated onto silicon wafer

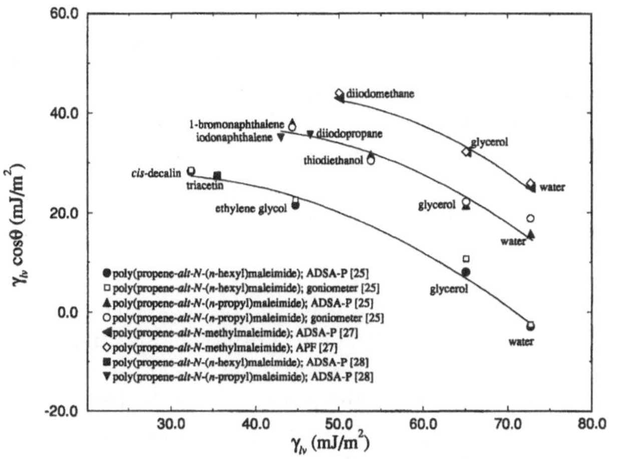

Fig. 7 $\gamma_{lv} \cos\theta$ vs γ_{lv} for poly(propene-*alt*-*N*-(*n*-hexyl)maleimide), poly(propene-*alt*-*N*-(*n*-propyl)maleimide), and poly(propene-*alt*-*N*-methylmaleimide) copolymers by ADSA-P, APF, and a goniometer technique. It can be seen that, neglecting the inconclusive angles, $\gamma_{lv} \cos\theta$ changes systematically with γ_{lv} and γ_{sv}. Data are from Refs. [25, 27, 28]

These patterns are in excellent agreement with the results in Figs. 1, 3, and 6.

It should be noted that the two schemes to determine contact angles are very different. APF puts tangents to the drop profile, whereas ADSA-P fits the best Laplacian shape to the entire profile of the drop. Given the fact that real solid surfaces show typically minor irregularities leading to minor deviations of the three-phase line from the circular shape, ADSA will average to some extent over such imperfections near the meridian section being imaged. In other words, ADSA, unlike APF, does not give a strictly two dimensional account of the meridian section but reflects, to some degree, properties of the solid surface away from this section. Thus, on real solid surfaces, we can expect less scatter in the ADSA data than the APF data. While the three-phase contact radius, drop volume and drop surface area can in principle be estimated, the liquid–vapour surface tension γ_{lv} cannot be obtained easily by APF; ADSA-P determines not only the contact angles, but also the liquid surface tension.

Universality of the contact angle patterns

In the vast majority of contact angle studies, the method used is direct measurement of sessile drops, e.g. by ADSA. Recent developments in image analysis and processing have increased the accuracy and reduced the subjectivity considerably. Nevertheless, there are still rooms for other techniques. For example, surface heterogeneity and/or roughness could well cause variations of the contact angle along the three phase line, as mentioned above.

The contact angle determination from the capillary rise at a vertical plate avoids such complications and offers advantages over the conventional goniometer-sessile drop technique. If a vertical plate is infinitely wide, the Laplace equation integrates to [29, 30]

$$\sin\theta = 1 - \frac{\Delta\rho g h^2}{2\gamma_{lv}} . \tag{5}$$

Knowing the density difference, $\Delta\rho$, between liquid and vapour, the acceleration due to gravity, g, and the liquid surface tension, γ_{lv}, the contact angle, θ, can be obtained from a measurement of the capillary rise, h. The task of measuring a contact angle has thus been reduced to a measurement of a length, which can be performed optically with a very high degree of accuracy by means of a cathetometer.

The technique of capillary rise at the vertical plate can also be used to determine γ_{lv} and θ separately. For example, in situations where the contact angles and surface tensions change with time, due to change in temperature or adsorption at the interfaces, both of these measurements can be performed by combining capillary rise with Wilhelmy plate technique. Thus, measurements of contact angles and surface tensions are reduced to the measurements of the capillary rise, h, and the change in weight of the vertical plate as a functional of time [31, 32].

The capillary rise at a vertical plate technique has been automated to perform various dynamic advancing and receding contact angle measurements at immersion speeds ranging from 0.008 to 0.9 mm/min. The advantages of this technique over the conventional goniometer-sessile drop technique are that the procedures are automated and any local variation of the contact angles along the three-phase contact line due to, e.g., roughness and/or heterogeneity can be observed directly. Recent study employing the automated capillary rise technique indicates that prolonged contact between the solid and liquid can affect the contact angles, e.g., due to swelling of the polymer by the liquid [6]. More important, measurements can be and have been performed under dynamic conditions closely resembling those used in ADSA. Only contact angle results meeting the stipulations established in connection with ADSA, above, will be considered below. In several instances, the capillary rise technique has been employed on relatively inert and well-prepared surfaces: FC-721-coated mica surface [33], heat pressed Teflon (FEP) [33], hexatriacontane [1, 34], and cholesteryl acetate [1, 35]. Hexatriacontane and cholesteryl acetate were produced by vapour deposition in a vacuum; the surface quality was found to be so good that there was no contact angle hysteresis when water was used. It has been found [30] that the contact angles determined from ADSA-P and the capillary rise at a vertical plate technique are virtually identical for the same solid–liquid systems. The capillary rise results are low-rate dynamic contact angles measured at different velocities of the three-phase contact line [33].

We plot in Fig. 8 the contact angle data from ADSA-P and those from the capillary rise technique for various polymers: FC-721-coated mica [15, 16, 33], FC-722-coated mica [26] and silicon wafer [27], heat-pressed

Fig. 8 $\gamma_{lv}\cos\theta$ vs. γ_{lv} for FC-721-coated mica [15, 16, 33], FC-722-coated mica [26] and coated silicon wafer [27], Teflon (FEP) [15, 33], hexatriacontane [1, 34], cholesteryl acetate [1, 35], poly(propene-*alt*-*N*-(*n*-hexyl)maleimide) [25, 28], poly(ethylene terephthalate) (PET) [15], poly(propene-*alt*-*N*-(*n*-propyl)maleimide) [25, 28], and poly(propene-*alt*-*N*-methylmaleimide) [27]. It can be seen that $\gamma_{lv}\cos\theta$ changes regularly with γ_{lv} and γ_{sv}: $\gamma_{lv}\cos\theta = F(\gamma_{lv}, \gamma_{sv})$

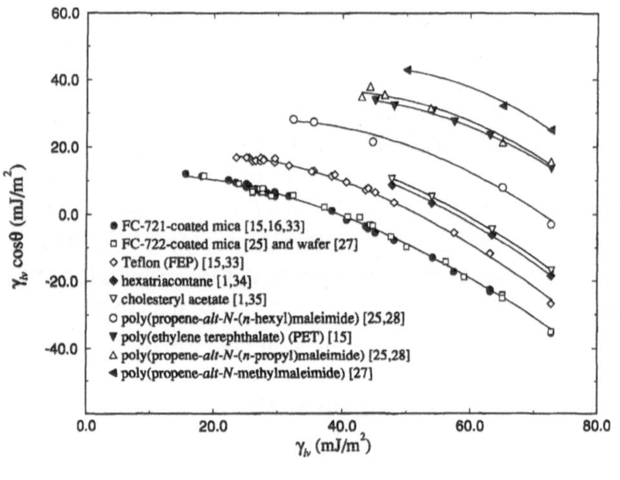

teflon (FEP) [15, 33], hexatriacontane [1, 34], cholesteryl acetate [1, 35], poly(propene-*alt*-*N*-(*n*-hexyl)maleimide) [25, 28], poly(ethylene terephthalate) (PET) [15], poly(prepene-*alt*-*N*-(*n*-propyl)maleimide) [25, 28], and poly(propene-*alt*-*N*-methylmaleimide) [27]. It can be seen that the values of $\gamma_{lv} \cos\theta$ change systematically with γ_{lv} in a very regular fashion, from the hydrophobic FC-721 surface to the hydrophilic poly(propene-*alt*-*N*-methylmaleimide) surface, and that the patterns are independent of experimental technique. Thus, one can change the contact angle, and because of Young's equation, the solid–liquid interfacial tension, by the simple mechanism of changing either the liquid or the solid. Obviously, these experimental results suggest that $\gamma_{lv} \cos\theta$ depends only on γ_{lv} and γ_{sv}. Because of Young's equation, γ_{sl} can be expressed as a function of only γ_{lv} and γ_{sv}, allowing us to search for an equation in the form of Eq. (4). This is called an equation-of-state relation.

Contact angle interpretation: determination of solid surface tensions from contact angles

The calculation of solid surface tension γ_{sv} from the contact angle of a liquid of surface tension γ_{lv} starts with Young's equation, Eq. (1). Of the four quantities in Young's equation, only γ_{lv} and θ are readily measurable. Thus, in order to determine γ_{sv}, further information is necessary. Conceptually, an obvious approach is to seek one more relation among the parameters of Eq. (1), such as an equation-of-state relation, of the form of Eq. (4) which we now know to exist from experimental facts. The simultaneous solution of Eqs. (1) and (4) would solve the problem.

Berthelot's (geometric mean) combining rule

An old equation-of-state relation can be obtained from the Berthelot combining rule [36]. It can be derived as follows. A thermodynamic relation of the free energy of adhesion per unit area of a solid–liquid pair is equal to the work required to separate an unit area of the solid–liquid interface [37]:

$$W_{sl} = \gamma_{lv} + \gamma_{sv} - \gamma_{sl} . \tag{6}$$

From the geometric mean combining rule (i.e., the Berthelot rule), the free energy of adhesion W_{sl} can be approximated in terms of the free energy of cohesion of the solid, W_{ss}, and the free energy of cohesion of the liquid, W_{ll}; [36, 38] i.e.,

$$W_{sl} = \sqrt{W_{ll} W_{ss}} . \tag{7}$$

Table 1 Calculated γ_{sv} values (mJ/m^2) of FC-722-coated mica from Berthelot's rule. The contact angles are low-rate dynamic angles measured by ADSA-P [26]

Liquid	γ_{lv} [mJ/m^2]	θ deg	γ_{sv} Berthelot's rule, Eq. (10)
Decane	23.88	67.36	11.45
1-Pentanol	26.01	72.95	10.87
trans-Decalin	27.19	73.38	11.24
Hexadecane	27.62	75.94	10.67
1-Decanol	28.99	78.84	10.32
cis-Decalin	32.32	79.56	11.27
Ethyl cinnamate	37.17	86.54	10.45
Dibenzylamine	40.80	90.70	9.95
DMSO	42.68	90.95	10.31
1-Bromonaphthalene	44.31	93.81	9.65
Diethylene glycol	44.68	94.22	9.59
Ethylene glycol	47.55	97.87	8.85
Diiodomethane	49.98	101.18	8.12
Thiodiethanol	56.26	104.56	7.88
Formamide	59.08	108.49	6.89
Glycerol	65.02	111.73	6.45
Water	72.70	118.69	4.91

By the definition of $W_{ll} = 2\gamma_{lv}$ and $W_{ss} = 2\gamma_{sv}$, Eq. (7) becomes

$$W_{sl} = 2\sqrt{\gamma_{lv}\gamma_{sv}} . \tag{8}$$

By combining Eq. (6) with Eq. (8), the solid–liquid interfacial tension γ_{sl} can be written as

$$\gamma_{sl} = \gamma_{lv} + \gamma_{sv} - 2\sqrt{\gamma_{lv}\gamma_{sv}} = (\sqrt{\gamma_{lv}} - \sqrt{\gamma_{sv}})^2 . \tag{9}$$

Combining this equation with Young's equation yields

$$\cos\theta_Y = -1 + 2\sqrt{\frac{\gamma_{sv}}{\gamma_{lv}}} . \tag{10}$$

Thus, the solid–vapour surface tension can be determined when experimental (Young) contact angle and liquid–vapour surface tension are known.

The question then arises as to what to expect when such an equation-of-state relation, in conjunction with Young's equation, is used to calculate solid surface tensions. As discussed before, there is one immediate criterion that the results obtained with this equation (or, for that matter, any other equation) must satisfy: when measuring contact angles with a number of liquids on a low-energy solid, the solid–vapour surface tension γ_{sv} is expected to be constant, independent of the liquid surface tension γ_{lv}. In other words, the different pairs of θ and γ_{lv} for one and the same solid should yield sensibly constant values of γ_{sv}. The results obtained from Eq. (10) are given in Table 1 for the FC-722-coated mica surface. It can be seen that that γ_{sv} values obtained from Berthelot's rule are not constant;

rather, they tend to decrease as the difference between γ_{lv} and γ_{sv} increases.

It should be noted that in the theory of intermolecular interactions and the theory of mixtures, combining rules are used to evaluate the parameters of unlike-pair interactions in terms of those of the like interactions. However, as for many other combining rules, the Berthelot rule [36], i.e., the geometric mean combining rule,

$$\varepsilon_{ij} = \sqrt{\varepsilon_{ii}\varepsilon_{jj}} \tag{11}$$

is only a useful approximation and does not provide a secure basis for the understanding of the unlike-pair interactions; ε_{ij} is the energy parameter of unlike-pair interactions; ε_{ii} and ε_{jj} are the energy parameters of like-pair interactions. For the interactions between two very dissimilar types of molecules or materials, where there is an apparent difference between ε_{ii} and ε_{jj}, it has been demonstrated [39, 40] that the geometric mean combining rule generally overestimates the strength of the unlike-pair interactions, i.e. the geometric mean value is too large an estimate. Clearly, this is why the geometric mean combining rule does not work for situations of large differences $|W_{ll} - W_{ss}|$ or $|\gamma_{lv} - \gamma_{sv}|$.

In the next section, we shall discuss the modification of Berthelot's rule, leading to an equation-of-state approach for solid–liquid interfacial tensions.

Equation-of-state approach

In the study of mixtures, it has become common practice to introduce a factor $(1 - K_{ij})$ to the geometric mean combining rule,

$$\varepsilon_{ij} = (1 - K_{ij})\sqrt{\varepsilon_{ii}\varepsilon_{jj}}\,, \tag{12}$$

where K_{ij} is an empirical parameter quantifying deviations from the geometric mean combining rule. Since the geometric mean combining rule overestimates the strength of the unlike-pair interactions, the modifying factor $(1 - K_{ij})$ should be a decreasing function of the difference $\varepsilon_{ii} - \varepsilon_{jj}$ and the equal to unity when the difference $\varepsilon_{ii} - \varepsilon_{jj}$ is zero. On the basis of this thought, we may consider a modified combining rule of the form

$$\varepsilon_{ij} = \sqrt{\varepsilon_{ii}\varepsilon_{jj}}\,\mathrm{e}^{-\alpha(\varepsilon_{ii}-\varepsilon_{jj})^2}\,, \tag{13}$$

where α is an empirical constant; the square of the difference $(\varepsilon_{ii} - \varepsilon_{jj})$, rather than the difference itself, reflects the symmetry of this combining rule, and hence the anticipated symmetry of the equation-of-state. Correspondingly, for the cases of large differences $|W_{ll} - W_{ss}|$ or $|\gamma_{lv} - \gamma_{sv}|$, the combining rule for the free energy of adhesion of a solid–liquid pair can be written as

$$W_{sl} = \sqrt{W_{ll}W_{ss}}\,\mathrm{e}^{-\alpha(W_{ll}-W_{ss})^2} \tag{14}$$

or, more explicitly, by using $W_{ll} = 2\gamma_{lv}$ and $W_{ss} = 2\gamma_{sv}$,

$$W_{sl} = 2\sqrt{\gamma_{lv}\gamma_{sv}}\,\mathrm{e}^{-\beta(\gamma_{lv}-\gamma_{sv})^2}\,. \tag{15}$$

In the above, α and β are as yet unknown constants. Clearly, when the values of γ_{lv} and γ_{sv} are close to each other, Eq. (15) reverts to Eq. (8), the geometric mean combining rule. By combining Eq. (15) with Eq. (6), an equation-of-state for interfacial tensions can be written as

$$\gamma_{sl} = \gamma_{lv} + \gamma_{sv} - 2\sqrt{\gamma_{lv}\gamma_{sv}}\,\mathrm{e}^{-\beta(\gamma_{lv}-\gamma_{sv})^2}\,. \tag{16}$$

Combining Eq. (16) with Young's equation yields

$$\cos\theta_Y = -1 + 2\sqrt{\frac{\gamma_{sv}}{\gamma_{lv}}}\,\mathrm{e}^{-\beta(\gamma_{lv}-\gamma_{sv})^2}\,. \tag{17}$$

Thus, the solid surface tensions can be determined from experimental (Young) contact angles and liquid surface tensions when β is known. Obviously, for a given set of γ_{lv} and θ data measured on one and the same type of solid surface, the constant β and γ_{sv} values can be determined by a least-squares analysis technique. Starting out with arbitrary values for γ_{sv} and β, iterative procedures can be used to identify that pair of γ_{sv} and β values which provides the best fit of the experimental data to the set of experimental γ_{lv} and θ_Y pairs belonging to one and the same solid surface. From the experimental contact angles on the FC-721-coated mica, heat-pressed teflon (FEP), and poly(ethylene terephthalate) (PET) surfaces, β was found to be 0.0001207, 0.0001339, and 0.0001112 $(\mathrm{m^2/mJ})^2$, respectively for the three surfaces; the corresponding γ_{sv} values are 11.78, 17.85, and 35.22 $\mathrm{mJ/m^2}$, respectively. Since the β values do not show any dependence on the solid surface tensions, a weighted mean β was calculated: a value of 0.001247 $(\mathrm{m^2/mJ})^2$ was obtained [15]. Calculations of γ_{sv} values with slightly different β values, i.e. in the above range, have very little effect on the outcome.

The γ_{sv} values calculated from Eq. (17) are given in Table 2 for the solid surfaces shown in Fig. 8: FC-721-coated mica, FC-722-coated mica and silicon wafer, heat-pressed teflon (FEP), hexatriacontane, cholesteryl acetate, poly(propene-*alt*-*N*-(*n*-hexyl)maleimide), poly(ethylene terephthalate) (PET), poly(propene-*alt*-*N*-(*n*-propyl) maleimide), and poly(propene-*alt*-*N*-methylmaleimide). It can be seen that the γ_{sv} values calculated from the equation-of-state approach [10] are quite constant, essentially independent of the choice of the liquids. We conclude that this indeed confirms the validity of the approach to determine solid surface tensions from contact angles.

Table 2 Summary of contact angles of various solids by different techniques. The γ_{sv} values (mJ/m^2) were calculated from the equation-of-state approach [10]

Solid/technique	Liquid	γ_{lv} [mJ/m^2]	θ [deg]	γ_{sv} Equation-of-state, Eq. (17)
FC-721-coated mica/mica/ADSA-P/static angles [15, 16]	Pentane	15.56	38.89	12.41
	Hexane	18.13	50.48	12.24
	Methanol	22.30	62.39	12.24
	Decane	23.43	65.97	11.98
	Methyl acetate	25.10	68.28	12.27
	Dodecane	25.44	69.82	12.03
	Tetradecane	26.55	73.31	11.63
	Hexadecane	27.76	75.32	11.63
	trans-Decalin	29.50	76.71	12.03
	cis-Decalin	31.65	79.87	12.04
	Ethyl cinnamate	38.37	88.20	12.11
	Dibenzylamine	40.63	92.06	11.64
	DMSO	43.58	94.47	11.90
	1-Bromonaphthalene	44.01	95.29	11.75
	Diethylene glycol	45.04	96.84	11.56
	Ethylene glycol	47.99	99.03	11.82
	Thiodiethanol	54.13	103.73	12.20
	Formamide	57.49	107.32	11.90
	Glycerol	63.11	111.38	12.17
	Water	72.75	119.05	12.06
FC-721-coated mica/capillary rise/dynamic angles [33]	Dodecane	25.03	70.4	11.67
	2-Octanol	26.00	73.5	11.31
	Tetradecane	26.50	73.5	11.55
	1-Octanol	27.28	75.1	11.47
	Hexadecane	27.31	75.6	11.34
	1-Hexadecene	27.75	74.0	12.01
	1-Decanol	28.29	76.6	11.51
	1-Dodecanol	29.53	79.2	11.31
FC-722-coated mica/ADSA-P/dynamic angles [26]	Decane	23.88	67.36	11.87
	1-Pentanol	26.01	72.95	11.46
	trans-Decalin	27.19	73.38	11.92
	Hexadecane	27.62	75.94	11.39
	1-Decanol	28.99	78.84	11.18
	cis-Decalin	32.32	79.56	12.44
	Ethyl cinnamate	37.17	86.54	12.21
	Dibenzylamine	40.80	90.70	12.21
	DMSO	42.68	90.95	12.88
	1-Bromonaphthalene	44.31	93.81	12.44
	Diethylene glycol	44.68	94.22	12.43
	Ethylene glycol	47.55	97.87	12.11
	Diiodomethane	49.98	101.18	11.70
	Thiodiethanol	56.26	104.56	12.67
	Formamide	59.08	108.49	11.98
	Glycerol	65.02	111.73	12.75
	Water	72.70	118.69	12.23
FC-722-coated silicon wafer/ADSA-P/dynamic angles [27]	Hexane	18.50	50.83	12.43
	2-Octanol	26.42	74.74	11.17
	Hexadecane	27.62	75.64	11.48
	Glycerol	65.02	111.89	12.67
Teflon (FEP)/ADSA-P/static angles [15]	Decane	23.43	43.70	17.54
	Dodecane	25.44	47.96	17.98
	Tetradecane	26.55	52.51	17.53
	Hexadecane	27.76	53.75	17.99
	trans-Decalin	29.50	58.14	17.81
	cis-Decalin	31.65	62.60	17.71
	Dimethylformamide	35.57	68.52	17.94
	Ethyl cinnamate	38.37	72.61	17.96
	Dibenzylamine	40.63	75.99	17.84

Table 2 (*Continued*)

Solid/technique	Liquid	γ_{lv} [mJ/m^2]	θ [deg]	γ_{sv} Equation-of-state, Eq. (17)
	DMSO	43.58	80.35	17.58
	1-Bromonaphthalene	44.01	79.70	18.08
	Diethylene glycol	45.04	81.48	17.85
	Ethylene glycol	47.99	85.56	17.54
	Formamide	57.49	95.38	17.57
	Glycerol	63.11	100.63	17.60
	Water	72.75	111.59	16.16
Teflon (FEP)/capillary rise/dynamic angles [33]	Dodecane	25.03	47.8	17.72
	2-Octanol	26.00	52.3	17.21
	Tetradecane	26.50	52.6	17.47
	1-Octanol	27.28	54.4	17.48
	Hexadecane	27.31	53.9	17.68
	1-Hexadecene	27.75	54.2	17.86
	1-Dodecanol	29.53	55.7	18.59
	Dimethylformamide	35.21	68.6	17.70
	Methyl salicylate	38.85	72.2	18.38
Hexatriacontane/capillary rise/dynamic angles [1, 34]	Ethylene glycol	47.7	79.2	20.28
	Thiodiethanol	54.0	86.3	20.31
	Glycerol	63.4	95.4	20.56
	Water	72.8	104.6	20.27
Cholesteryl acetate/capillary rise/dynamic angles [1, 35]	Ethylene glycol	47.7	77.0	21.29
	Thiodiethanol	54.0	84.3	21.30
	Glycerol	63.4	94.0	21.33
	Water	72.8	103.3	21.05
Poly(propene-*alt*-*N*-(*n*-hexyl) maleimide)/ADSA-P/dynamic angles [25, 28]	*cis*-Decalin	32.32	28.81	28.54
	Diethylene glycol	44.68	61.04	26.68
	Glycerol	65.02	82.83	28.62
	Water	72.70	92.26	27.76
PET/ADSA-P/static angles [15]	Diethylene glycol	45.04	41.19	35.40
	Ethylene glycol	47.99	47.52	35.10
	Thiodiethanol	54.13	55.57	35.99
	Formamide	57.49	61.50	35.42
	Glycerol	63.11	68.10	35.82
	Water	72.75	79.09	36.02
Poly(propene-*alt*-*N*-(*n*-propyl) maleimide)/ADSA-P/dynamic angles [25, 28]	Iodonaphthalene	42.92	35.19	35.88
	1-Bromonaphthalene	44.31	30.75	38.61
	Diiodopropane	46.51	39.98	37.09
	Thiodiethanol	53.77	54.04	36.49
	Glycerol	65.02	70.67	35.69
	Water	72.70	77.51	36.97
Poly(propene-*alt*-*N*-methyl maleimide)/ADSA-P/dynamic angles [27]	Diiodomethane	49.98	30.71	43.65
	Glycerol	65.02	60.25	41.69
	Water	72.70	69.81	41.76

Outlook

A look at the likely future of research on contact angles and surface energetics will benefit from a view back, particularly at persistent obstacles to progress. There are three key attitudes which have hampered progress immensely:

(1) Contact angles are simple quantities which can be readily measured and interpreted by anybody, requiring no particular skill, methodology, or knowledge.
(2) Contact angle measurements show hysteresis. Hence, experimental contact angles are not equilibrium values, and therefore nothing useful can be learned from contact angle studies.

(3) Contact angles contain readily accessible information about intermolecular forces.

All of these propositions are false. With respect to the first proposition, the present paper shows how misleading conventional contact angle measurements on sessile drops can be.

The second proposition, while false, is the most difficult to refute. Operationally, the key question is not whether a contact angle is an equilibrium contact angle, but whether it can be used in conjunction with Young's equation. For instance, contact angle hysteresis caused by a modest degree of surface heterogeneity does not make Young's equation inapplicable. Furthermore, it is clear from the systems examined here that contact with a liquid will often affect the solid surface. This does not preclude the possibility that the advancing contact angle is the equilibrium angle; rather this equilibrium cannot be reached from "the other side", i.e., under receding conditions, because the solid surface is now different. Finally, it is difficult to see how the smooth curves of, e.g., Figs. 1, 7 and 8 could arise if the contact angle data did not contain surface energetic meaning.

Finally, the widely held view expressed in the third proposition is demonstrably false, from curves like those in Figs. 1, 7 and 8. For a given solid, the contact angle depends only on the liquid surface tension, not directly on the intermolecular forces which give rise to these surface tensions. Referring to the curve for FC-721 in Fig. 1, the fact that methanol falls on the same smooth curves as hexane and decane would imply that the polar component of surface tension of methanol is zero, i.e. methanol would have to be classified as non-polar, which is clearly absurd.

The present state of knowledge is well summarized in the equation-of-state approach outlined above. The approach can be used to good advantage in numerous applications, including interfacial phenomena of powders [41, 42]. Nevertheless, there remain many unsolved questions relating to the measurements and interpretation of contact angles.

(1) While vapour adsorption does not appear to play a significant role in the contact angle data presented here, it is quite possible that it could enter the picture for more hydrophilic surfaces, particularly, if $\gamma_{lv} \approx \gamma_{sv}$ or $\gamma_{lv} < \gamma_{sv}$, i.e. cases which were excluded in the present study. While adsorption may produce contact angle patterns different from those shown in Fig. 8, this would not necessarily imply that Young's equation is not valid or that contact angle approaches might not be feasible. The answer to these questions would require general criteria to distinguish the effect of vapour adsorption on the contact angles from all other effects.
(2) Recently, self-assembled monolayers (SAMs) have been widely used to synthesize surfaces of different chemical compositions and wettabilities However, it is expected that penetrations of liquid into the SAMs are inevitable. Given that two solids having the same solid surface tensions and that a specific liquid may penetrate into one solid but not the other, the contact angles on the two chemically identical surfaces can be different: one would be due to pure energetic effects, and the other to the effects of the changed energetics and penetrations. In the latter case, attempts [43] to interpet surface energetics from contact angle approaches naively, e.g. by means of the above equation-of-state approach, could be misleading.
(3) It is well known that roughness affects the experimentally observed contact angles. However, there are as yet no general criteria to quantify roughness and at what level of smoothness surface topography has no longer an effect on the contact angle. The answer to such questions will have to involve considerations of line tension [5].
(4) It is still an open question of whether or not β in Eq. (17) is a universal constant, i.e. independent of the solid surface. Such a question can be addressed only after a large body of accurate contact angle data on various solids has been generated.
(5) In the experimental patterns of Fig. 8, it was found that liquid properties, i.e. intermolecular forces, do not act independently on the contact angles. Indeed, this is supported by a recent study [44] using van der Waals theory. Further development or modeling of this molecular theory or any other sound theory should allow us to gain a better understanding of contact angles and to contribute to the present debate in terms of contact angle interpretation.
(6) While the slip/stick contact angle patterns, e.g., of that shown in Fig. 4c and similar patterns cannot be used to determine surface energetics, they are still very interesting and worth investigating.
(7) Routine production of systems which are free of contact angle hysteresis might be considered an important goal. We are firmly convinced that all it would produce is further confirmation of the contact angle story presented here. We expect more benefits from the much simple systematic study of receding contact angles, of available systems, using concepts and procedures described above.

Acknowledgements This research was supported by the Natural Science and Engineering Research Council of Canada (Grants: No. A8278 and No. EQP173469), Ontario Graduate Scholarships (D.Y.K.), and University of Toronto Open Fellowships (D.Y.K.).

References

1. Neumann AW (1974) Adv Colloid Interface Sci 4:105
2. Marmur A (1996) Colloids Surf A 116:25
3. Li D, Neumann AW (1996) Theromodynamic status of contact angles. In: Neumann AW, Spelt JK (eds) Applied Surface Thermodynamics. Marcel Dekker, New York, pp 109–168
4. Grundke K, Bogumil T, Gietzelt T, Jacobasch H-J, Kwok DY, Neuman AW (1996) Progr Colloid Polym Sci 101:58
5. Gaydos J, Neumann AW (1996) Line tension in multiphase equilibrium systems. In: Neumann AW, Spelt JK (eds) Applied Surface Thermodynamics. Marcel Dekker, New York, pp 169–238
6. Sedev RV, Petrov JG, Neumann AW (1996) J Colloid Interface Sci 180:36
7. Fowkes FM (1964) Ind Eng Chem 12:40
8. Driedger O, Neumann AW, Sell PJ (1965) Kolloid-ZZ Polym 201:52
9. Neumann AW, Good RJ, Hope CJ, Seipal M (1974) J Colloid Interface Sci 49:291
10. Spelt JK, Li D (1996) The equation of state approach to interfacial tensions. In: Neumann AW, Spelt JK (eds) Applied Surface Thermodynamics. Marcel Dekker, New York, pp 239–292
11. Owens DK, Wendt RC (1969) J Appl Polym Sci 13:1741
12. van Oss CJ, Chaudhury MK, Good RJ (1988) Chem Revs 88:927
13. van Oss CJ, Ju L, Chaudhury MK, Good RJ (1989) J Colloid Interface Sci 128:313
14. Good RJ, van Oss CJ (1992) The modern theory of contact angles and the hydrogen bond components of surface energies. In: Schrader M, Loeb G (eds) Modern Approaches to Wettability: Theory and Applications. Plenum Press, New York, pp 1–27
15. Li D, Neumann AW (1992) J Colloid Interface Sci 148:190
16. Li D, Xie M, Neumann AW (1993) Colloid Polym Sci 271:573
17. Rotenberg Y, Boruvka L, Neumann AW (1983) J Colloid Interface Sci 93:169
18. Cheng P, Li D, Boruvka L, Rotenberg Y, Neumman AW (1983) Colloids Surf 93:169
19. Kwok DY, Li D, Neumann AW (1994) Colloids Surf A 89:181
20. Ward CA, Neumann AW (1974) J Colloid Interface Sci 49:286
21. Defay R (1934) Etude Thermodynamique de la Tension Superficielle. Gauthier-Villars, Paris
22. Defay R, Prigogine I (1954) Surface Tension and Adsorption. Bellemans A (collab.), Everett DH (trans), Longmans & Green, London, p 222
23. Li D, Gaydos J, Neumann AW (1989) Langmuir 5:1133
24. Li D, Neumann AW (1994) Adv Colloids Interface Sci 49:147
25. Kwok DY, Gietzelt T, Grundke K, Jacobasch H-J, Neumann AW (1997) Langmuir 13:2880
26. Kwok DY, Lin R, Mui M, Neumann AW (1996) Colloids Surf A 116:63
27. del Rio OI, Kwok DY, Wu R, Alvarez JM, Neumann AW. Contact angle measurements by axisymmetric drop shape analysis and an automated polynomial fit program. Colloid Surf A, accepted for publication
28. Kwok DY, Lam CNC, Li A, Leung A, Neumann AW (1998) Langmuir 14:2221
29. Neumann AW (1964) Z Phys Chem (Frankfurt) 41:339
30. Kwok DY, Li D, Neumann AW (1996) Capillary rise at a vertical plate as a contact angle technique. In: Neumann AW, Spelt JK (eds) Applied Surface Thermodynamics. Marcel Dekker, New York, pp 413–440
31. Kloubek J, Neumann AW (1969) Tenside 6:4
32. Neumann AW, Good RJ (1970) Techniques of Measuring contact angles. In: Good RJ, Stromberg R (eds) Experimental Methods in Surface and Colloid Science Series, Vol 11. Plenum Press, New York, pp 31–91
33. Kwok DY, Budziak CJ, Neumann AW (1995) J Colloid Interface Sci 173:143
34. Hellwig GHE, Neumann AW (1968) 5th Int Congr on Surface Activity, Section B, p 687
35. Hellwig GHE, Neumann AW (1969) Kolloid-ZZ Polym 40:229
36. Berthelot D (1898) Compt rend 126, 1703:1857
37. Dupré A, (1969) Théorie Mécanique de la Chaleur. Gauthier-Villars, Paris
38. Maitland GC, Rigby M, Smith EB, Wakeham WA (1981) Intermolecular Forces: Their Origin and Determination. Clarendon Press, Oxford
39. Kestin J, Mason EA (1973) AIP Conf Proc 11:137
40. Israelachvili JN (1972) Proc R Soc London A 331:39
41. Li D, Neumann AW (1996) Wettability and surface tension of particles. In: Neumann AW, Spelt JK (eds) Applied Surface Thermodynamics. Marcel Dekker, New York, pp 509–556
42. Li D, Neumann AW (1996) Behaviour of Particles at Solidificatioin Fronts. In: Neumann AW, Spelt JK (eds) Applied Surface Thermodynamics. Marcel Dekker, New York, pp 557–628
43. Drelich J, Miller JD (1994) J Colloid Interface Sci 167:217
44. van Giessen AE, Bukman DJ, Widom B (1997) J Colloid Interface Sci 192:257

Progr Colloid Polym Sci (1998) 109:185–191
DISPERSIONS

I. ul Haq
E. Matijević

Preparation and properties of uniform coated inorganic colloidal particles. 12. Tin and its compounds on hematite

Received: 29 May 1997
Accepted: 3 June 1997

Supported by the NSF grant CHE-9423163

I. ul Haq[1] · Prof. Dr. E. Matijević (✉)
Center for Advanced Materials Processing
Clarkson University
Potsdam, New York 13699-5814, USA

[1]On leave from:
National Center for Excellence
in Physical Chemistry
University of Peshawar
N.W.F.P. Pakistan

Abstract Colloidal spherical particles of tin hydroxide of narrow size distribution were obtained by heating acidic solutions of stannous sulfate and urea at elevated temperatures over a range of experimental conditions. The original particles were amorphous, which on calcination at 750 °C converted into crystalline SnO_2, without a change in their morphology. The same tin compound was also precipitated in the form of uniform coatings on ellipsoidal hematite particles by heating dispersions of the latter in acidified solutions of stannous sulfate and urea at 80 °C for 1 h. The size of the shell could be altered over a broad range by adjusting several experimental parameters. The amorphous coating was converted into crystalline SnO_2 by heating the composite powders at 750 °C, without interaction with the cores. On heating at 350 °C in the flow of hydrogen, the calcined coated particles were reduced to metals, composed of Fe, Sn, and Fe_3Sn_2. In contrast, under similar conditions SnO_2 particles could not be changed to metallic tin. A mechanism involved in the reduction process is offered.

Key words Coated hematite – hematite, monodispersed – particle coating – tin films – tin oxide – tin oxide coating

Introduction

Over recent years much attention has been paid to the study of tin oxides and their composites, because of their importance in catalysis [1], adsorption [2], ceramics [3], as gas sensors [4], etc. In some of the studies it was noted that the performance of these materials in different applications could be improved, if their morphological properties and chemical compositions were carefully controlled.

Several procedures have been reported in the literature for the production of tin oxides in the form of powders [5–11] and films [12–16]. These techniques include gas-phase condensation [5], sol–gel precipitation [6], hydrolysis of tin salts in solutions [7–9] and aerosols [10], chemical vapor deposition [11–14], physical vapor deposition [14], spray pyrolysis [14, 15] and d.c. magnetron reactive sputtering [16].

In addition, dispersions of tin particles were obtained by γ-irradiation of $SnCl_2$ solutions [17] and pure tin films on glass were prepared by d.c. magnetron sputtering [18]. While SnO_2 gas sensors were prepared by oxidation of Sn thin films [19], apparently there is no information available on the opposite process, i.e. formation of tin layers by reduction of SnO_2.

Recently, a number of studies have dealt with coating of uniform particles with shells of different chemical compositions [20]. In principle, the method consists of precipitating a shell on preformed cores, suspended in the solutions of metal salts of interest. For example, yttrium basic carbonate [21], and silica [22] layers of different thickness were obtained on hematite by this procedure. It was also shown that the composition of the coatings could be altered by post-treatments, such as by converting yttrium basic carbonate shells to yttrium oxide by calcination of coated hematite particles [21]. Metallic nickel shells on manganese oxide cores were produced by reducing a NiO layer with hydrogen at elevated temperatures [23]. Many other systems of such composite particles have been investigated as reviewed elsewhere [20].

In this study the coating of colloidal hematite (α-Fe_2O_3) particles with tin hydroxide, tin oxide, and metallic tin layers is described. The ellipsoidal shape of hematite was chosen in order to more readily distinguish precipitated spherical colloidal tin hydroxide, should the latter appear in the presence of coated particles. Initially, tin hydroxide was precipitated onto ellipsoidal iron hydroxide cores in acidic $SnSO_4$–urea solutions by aging the latter suspensions at elevated temperatures. The shells could be converted to SnO_2 by calcination and to pure tin, when hematite particles coated with SnO_2 were reduced in H_2 at higher temperatures. Under the same conditions the α-Fe_2O_3 cores were reduced to iron. It was shown that in this latter process the presence of iron was essential for the reduction of SnO_2 to metallic tin. It would appear that the described system is the first, showing the feasibility of producing tin films from SnO_2.

It is noteworthy that uniform colloidal tin hydroxide dispersions, precipitated in the $SnSO_4$–urea solutions, could be converted to SnO_2, but it would not be transformed into metallic tin particles by reaction with hydrogen.

Experimental section

Reagents

All chemicals were reagent grade and were used as received. Their stock solutions were filtered through 0.2 μm pore size membranes in order to remove any possible contaminants and were never kept for longer than one week.

To prepare the stock solution of stannous sulfate, a desired amount of this salt was suspended in 50 cm^3 of concentrated sulfuric acid and heated on a hot plate until a clear liquid was obtained. The latter was transferred into a 250 cm^3 volumetric flask and the volume was made up to the mark with doubly distilled water.

Preparation of tin hydroxide particles

In principle, the procedure for the preparation of colloidal tin hydroxide particles consisted of heating acidified stannous sulfate solutions in the presence of urea. The properties of the resulting solids and their yield depended on the concentration of the reactants, pH, the reaction temperature, and time. For this reason the effects of the above experimental parameters were investigated by aging for 10–120 min in a preheated water bath at 60–85 °C, solutions containing 1×10^{-3}–6×10^{-3} mol dm^{-3} $SnSO_4$, 0.5–2 mol dm^{-3} urea (CON_2H_4), and 0.05–0.5 mol dm^{-3} HCl, using Pyrex tubes tightly closed with Teflon-lined caps.

Colloidal dispersions of essentially spherical particles of narrow size distribution were obtained from solutions of 2.5×10^{-3}–4.6×10^{-3} mol dm^{-3} $SnSO_4$, 1.0–1.85 mol dm^{-3} urea and 0.1–0.4 mol dm^{-3} HCl, kept at 70–80 °C for 10–60 min. In all these experiments the molar ratio $[SnSO_4]/[CON_2H_4]$ in the reactant mixture was maintained at 2.5×10^{-3}.

Expect for the above conditions, either gelatinous or agglomerated solids were obtained.

Preparation of hematite particles

Hematite (α-Fe_2O_3) particles were prepared as described elsewhere [24] by keeping at 100 $\pm$ 2 °C for 48 h a solution containing 0.02 mol dm^{-3} $FeCl_3$ and 3×10^{-4} mol dm^{-3} NaH_2PO_4 in 500 cm^3 glass stoppered Pyrex bottles, using a preheated air convection oven. The resulting solids were washed with water and either dried in a desiccator, or redispersed in 500 cm^3 of doubly distilled water. The scanning electron micrograph (SEM) in Fig. 1 illustrates such particles, which were used as the core material in

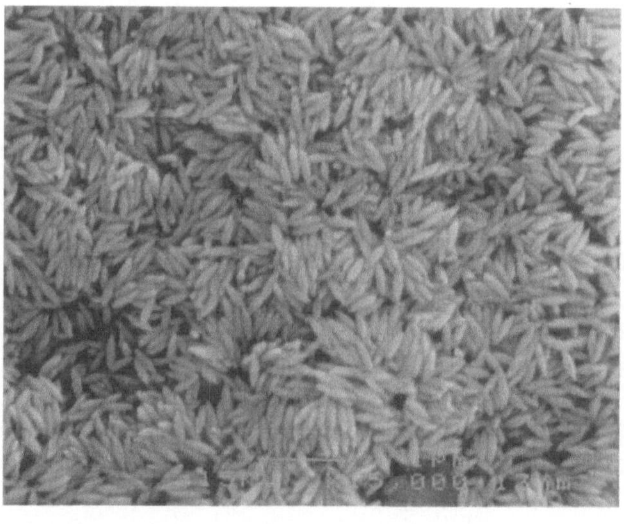

Fig. 1 Scanning electron micrograph (SEM) of hematite particles obtained by heating a solution containing 0.02 mol dm^{-3} $FeCl_3$ and 3×10^{-4} mol dm^{-3} NaH_2PO_4 at 100 $\pm$ 2 °C for 48 h in air

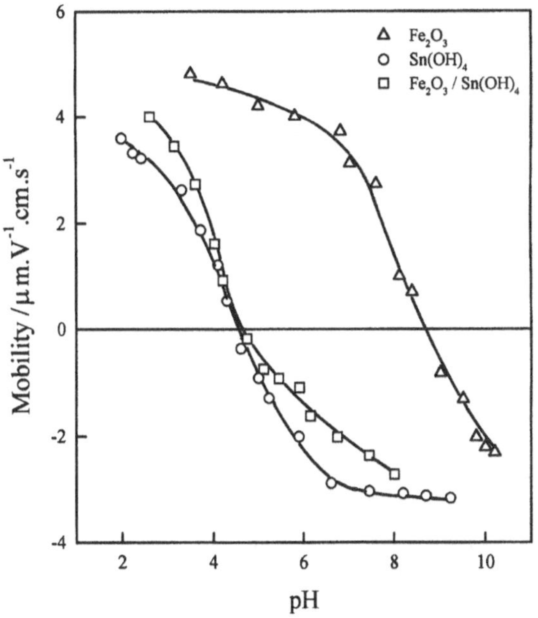

Fig. 2 Electrophoretic mobilities as a function of the pH of hematite particles shown in Fig. 1 (△); of tin hydroxide particles shown in Fig. 3A (○); and of coated (tin hydroxide on hematite) particles shown in Fig. 7A (□)

studies described below. The α-Fe_2O_3 ellipsoids were $\sim$0.4 μm long and $\sim$0.1 μm wide. The electrophoretic mobility (Fig. 2) of these particles yielded an isoelectric point (i.e.p) at pH $\sim$8, in agreement with the value reported earlier [25].

Coating of hematite particles

Uniform shells of tin hydroxide on hematite particles were obtained under limited set of conditions, since the process of forming them was sensitive to the experimental parameters and, especially, to the ratio of the amount of the precursor solid to the concentration of the reactants. In cases other than described below, mixtures of coated hematite and independent tin hydroxide colloids were formed. Thus, disperisons consisting of α-Fe_2O_3 cores with $Sn(OH)_4$ shells were only produced when aqueous systems of 0.15–0.3 g dm^{-3} hematite, 2.5×10^{-3}–4.6×10^{-3} mol dm^{-3} $SnSO_4$, 1.0–1.85 mol dm^{-3} urea, and 0.31 mol dm^{-3} HCl, were heated for 15–60 min at 80 °C with constant agitation using a magnetic stirrer. Stirring the same system was essential for the process, since under the same conditions a mixture of coated particles and colloidal tin hydroxide was formed when aging was carried out without agitation. The so-prepared solids were filtered using 0.2 μm Nuclepore membranes, washed several times with distilled water, and dried in a desiccator before characterization.

Calcination

Calcination of the desired solids was carried out in a tube furnace, equipped with a programmable controller, at 750 °C for 1 h at the heating rate of 5 °C min^{-1}.

Reduction

A known amount (200–400 mg) of the selected solids was transferred into a quartz microreactor and heated at 350 °C for 4–7 h in a stream of hydrogen at the flow rates of 80–150 cm^3 min^{-1}. The solids were recovered from the reactor and weighed again, in order to determine the loss during the reduction process.

Characterizations

All solids were inspected by scanning electron microscopy (SEM) and analyzed by X-ray diffractometry (XRD), and thermogravimetric (TGA) and differential thermal (DTA) analyses. The content of tin in these solids was determined by inductively coupled plasma-optical emission spectroscopy (ICP-OES). The electrophoretic mobilities of some of the dispersions were determined as a function of the pH at the constant ionic strength of 1×10^{-3} mol dm^{-3} with the PenKem 3000 instrument.

Results

Particles of tin compounds

Scanning electron micrograph in Fig. 3A illustrates typical particles obtained under the conditions given in the legend, the average diameter of which is $\sim$0.2 μm. Such dispersed matter first appeared in the reactant solution after $\sim$6 min of aging and the particles formation continued for an hour, yielding eventually 0.75 g dm^{-3}. The final amount of the solids decreased from 0.75 to 0.52 g dm^{-3} by reducing the concentration of $SnSO_4$ in the reactant mixture from 3.7×10^{-3} to 2.5×10^{-3} mol dm^{-3}, while the shape of the particles remained the same. An increase in the particle size to $\sim$0.3 μm (Fig. 3B) was observed, if the concentration of HCl was increased from 0.18 to 0.4 mol dm^{-3} under otherwise the same conditions as that of Fig. 3A. In the latter case, the initial particle appearance was delayed to $\sim$9 min, while the yield was not affected after 1 h of aging.

The electrophoretic mobility as a function of the pH (Fig. 2) of tin hydroxide particles [Fig. 3A] gave the i.e.p.

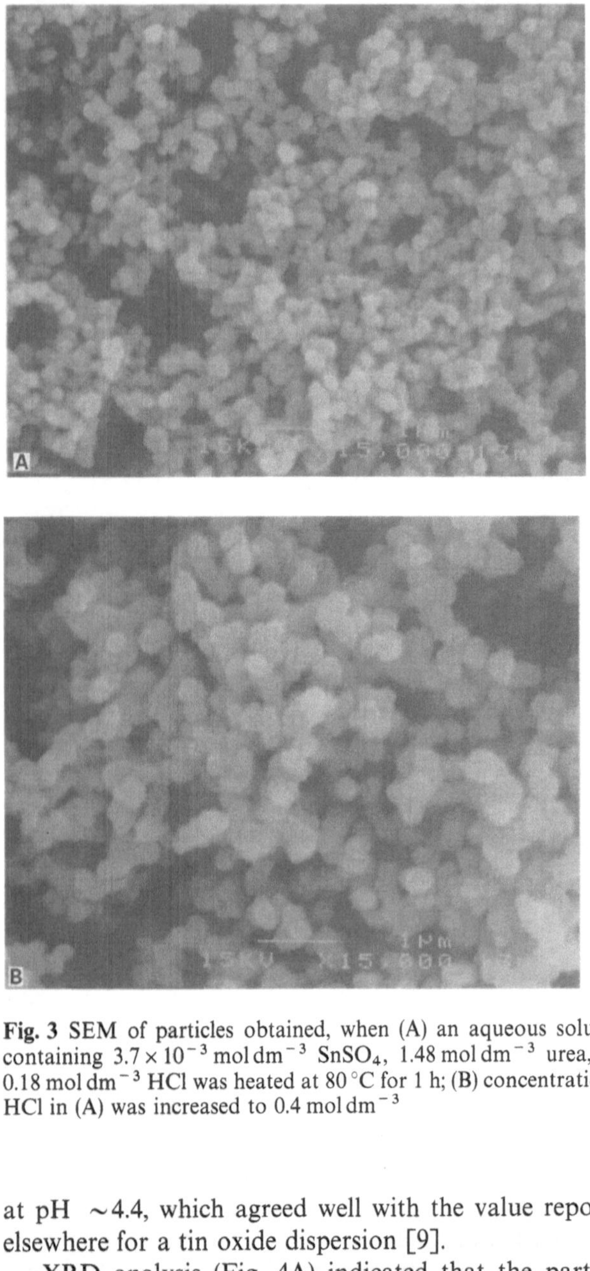

Fig. 3 SEM of particles obtained, when (A) an aqueous solution, containing 3.7×10^{-3} mol dm^{-3} $SnSO_4$, 1.48 mol dm^{-3} urea, and 0.18 mol dm^{-3} HCl was heated at 80 °C for 1 h; (B) concentration of HCl in (A) was increased to 0.4 mol dm^{-3}

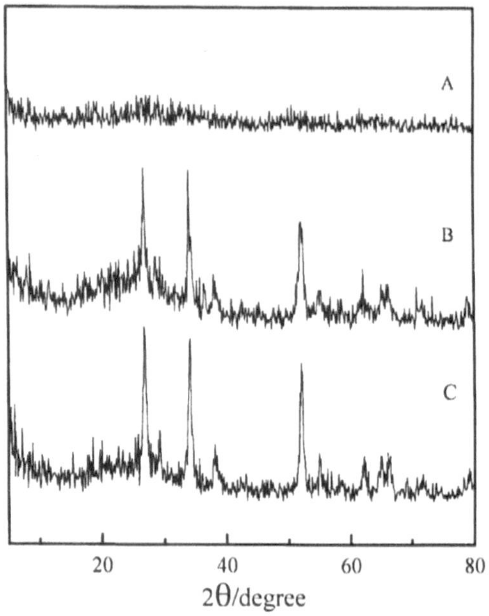

Fig. 4 X-ray diffraction patterns (XRD) of particles shown in Fig. 3A (A); in Fig. 6 (B); when particles shown in Fig. 6B are heated at 350 °C for 6 h in the flow of hydrogen (C)

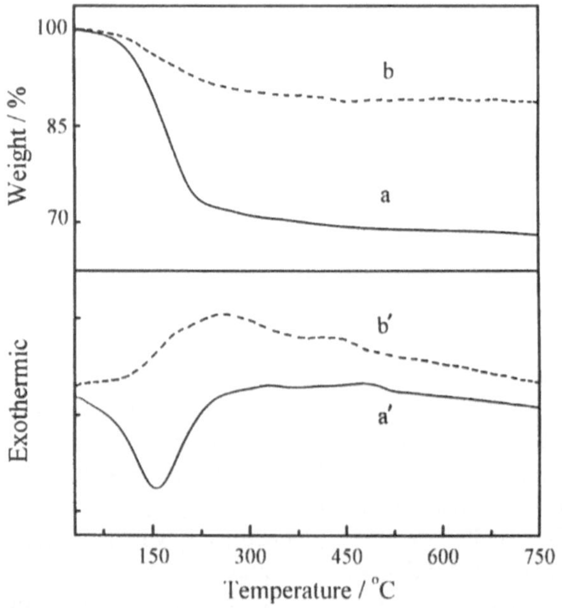

Fig. 5 (a, b) TGA and (a′, b′) DTA curves obtained with particles shown in Fig. 3A (—) and Fig. 7A (---)

at pH ~4.4, which agreed well with the value reported elsewhere for a tin oxide dispersion [9].

XRD analysis (Fig. 4A) indicated that the particles shown in Fig. 3A were amorphous. TGA and DTA traces (Fig. 5) showed a significant weight loss region and an endothermic peak over the temperature range 100–250 °C, respectively, which may be assigned to the loss of bound water and to the conversion of hydroxide to oxide groups. The weight loss of 33.1% indicates the chemical composition of the original particles to be $Sn(OH)_4 \cdot 2H_2O$. Assuming the reaction

$$Sn(OH)_4 \cdot 2H_2O \rightarrow SnO_2 + 4H_2O \ , \qquad (1)$$

the calculated weight loss should be 32.2%. Independent chemical analysis of dehydrated amorphous particles gave 66.8% Sn, which corresponds reasonably well to the content of tin in $Sn(OH)_4$ of 63.6 wt%.

On calcination in air, the resulting SnO_2 first remained amorphous, then slowly crystallized and it was finally converted to crystalline SnO_2 at 750 °C (XRD, Fig. 4B),

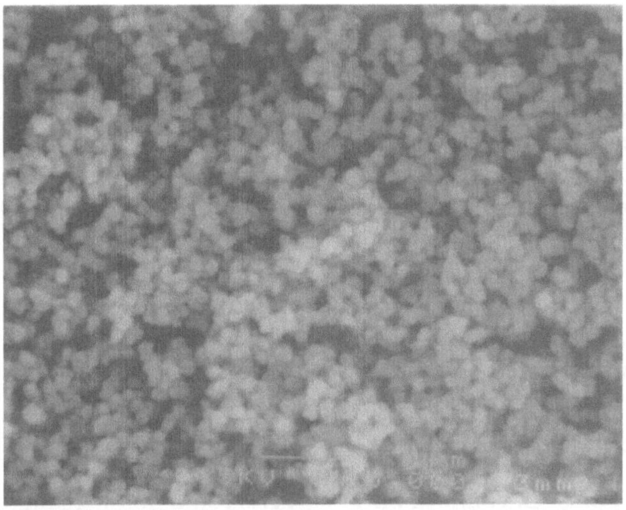

Fig. 6 SEM of particles obtained when the powder shown in Fig. 3A was calcined at 750 °C for 1 h in air

maintaining the same particle size and shape (SEM, Fig. 6). Because this process was slow and continuous, no peaks are observed in the DTA curve >280 °C.

The crystalline powder was readily redispersible in aqueous solutions.

Coated particles

Figure 7A displays a typical example of coated particles produced under the conditions described in the legend. XRD analysis (Fig. 8A) of these solids indicated that the shell was amorphous tin hydroxide, since only peaks characteristic of hematite were detected. The experimentally determined weight increase due to the coating was 89.5% as compared to the weight of hematite cores, which is equal to 47.2 wt% of $Sn(OH)_4$ in the final particles, yielding a molar ratio [Sn]/[Fe] = 0.32.

The amount of coating was found to decrease proportionally with increasing quantity of hematite in the reaction system. For example, for the sample illustrated in Fig. 7A the weight percent of tin hydroxide was reduced from 47.2 to 32.7% when the quantity of hematite was changed from 0.2 to 0.3 $g\,dm^{-3}$ in the coating solution. The shell also increased with the aging time; thus, under otherwise the same conditions the weight fraction of the coating was only 23.7% after 15 min as compared to 47.2% after 1 h of reaction.

The total weight loss of the coated particles on heating to ≥300 °C was ~9.5% (Fig. 5b), which agreed well with the expected loss of 9%, taking into consideration the weight fraction of the shell consisting of $Sn(OH)_4$.

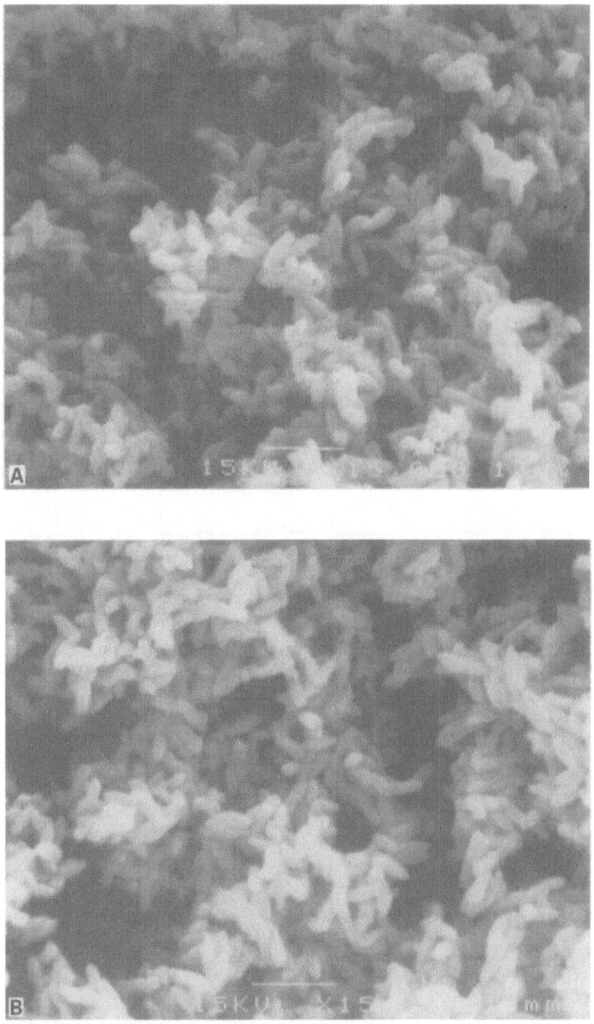

Fig. 7 SEM of particles obtained when (A) an aqueous dispersion containing 0.2 $g\,dm^{-3}$ hematite cores, 3.7×10^{-3} $mol\,dm^{-3}$ $SnSO_4$, 1.48 $mol\,dm^{-3}$ urea, and 0.34 $mol\,dm^{-3}$ HCl was heated for 1 h with constant stirring; (B) particles shown in (A) were calcined at 750 °C for 1 h in air

When calcined at 750 °C, the coating of these particles (Fig. 7A) was converted to SnO_2 (XRD, Fig. 8B) with no change in their morphology (Fig. 7B), and they remained fully redispersible in aqueous solutions.

The powder shown in Fig. 7B, calcined in a stream of hydrogen at 350 °C for 6 h did not change in terms of its dispersion properties. However, the particles became ferromagnetic and their XRD pattern was characteristic of Sn, Fe, and Fe_3Sn_2 (Fig. 9). The weight loss on reduction of ~48% was close to the calculated value of 46.6%.

An attempt to reduce tin oxide particles shown in Fig. 6, under the same conditions as above, failed. Indeed, no effect of this treatment on their chemical composition (XRD, Fig. 4C) or other properties could be detected.

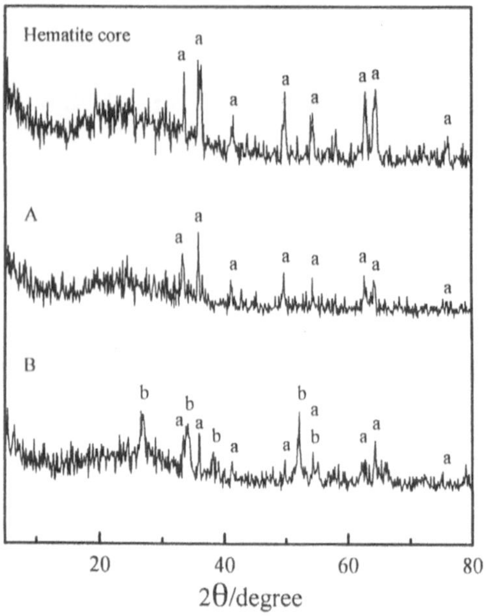

Fig. 8 XRD patterns of particles shown in Fig. 1 (hematite cores); of the coated particles shown in Fig. 7A (A); and Fig. 7B (B). Symbols: a, Fe_2O_3; b, SnO_2

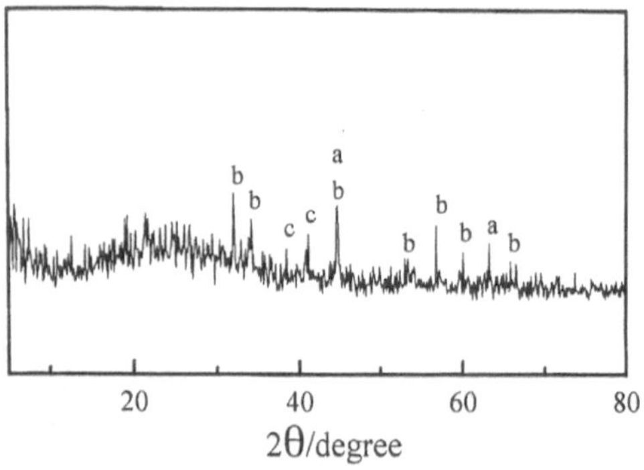

Fig. 9 XRD pattern of particles obtained when the particles shown in Fig. 7B were reduced at 350 °C in a flow of hydrogen for 6 h at the flow rate of 120 $cm^3\,min^{-1}$. Symbols: a, Fe; b, Sn; c, Fe_3Sn_2

Discussion

As mentioned above, tin hydroxide precipitated as spherical colloidal particles of narrow size distribution using the method above. With the same reactants and properly adjusted conditions, uniform shells of $Sn(OH)_4$ were formed on hematite cores as confirmed by the SEM, XRD, and chemical analysis. On calcination tin hydroxide particles and shells were converted to SnO_2. This finding indicated, that cores remained inert during the coating and calcination processes. Such behavior of coated particles was reported on other systems before [23, 25, 26].

By heating the coated particles at 350 °C in the flow of hydrogen, both the cores and shells are transformed into metallic particles, composed, respectively, of Fe, Sn, and Fe_3Sn_2. In contrast, SnO_2 particles could not be reduced under the same conditions to tin. In order to account for these observations, the following reaction scheme is proposed for the reduction of the coated particles:

$$Fe_2O_3 + 3H_2 \rightarrow 2Fe + 3H_2O \,, \quad (2)$$

$$4Fe + 3SnO_2 \rightarrow 2Fe_2O_3 + 3Sn \,, \quad (3)$$

$$Fe_2O_3 + 3H_2 \rightarrow 2Fe + 3H_2O \,. \quad (4)$$

Following this mechanism, hydrogen permeates through the tin oxide coating and reduces Fe_2O_3 cores to Fe (Eq. (2)), which in turn reduces SnO_2 coating to Sn. Hematite produced in reaction (3) is again converted to Fe by hydrogen according to reaction 4. The calculated free energy for the reaction given by Eq. (3) was found to decrease from 70 $kJ\,mol^{-1}$ at 25 °C to 50 $kJ\,mol^{-1}$ at 350 °C. While the trend is favorable, the results would indicate that the reduction should still not take place spontaneously at 350 °C. However, the calculation neglects the possible role of hydrogen in the process.

The XRD analysis (Fig. 9) also indicates the presence of a bimetallic Fe–Sn alloy as a result of the reduction

Fig. 10 Schematic presentation of various phases in the reduction of α-Fe_2O_3 coated particles with SnO_2 to pure metals

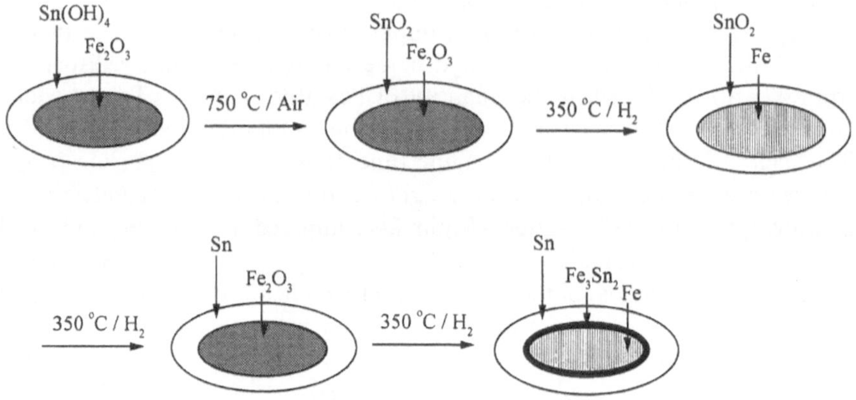

processes. However, due to the weak signals, it was not possible to identify with certainty the composition of this fraction of the shell. The Fe–Sn phase diagram gives Fe_3Sn_2 as the alloy over the 500–750 °C range [27]. While the reduction in the present case is carried out at a somewhat lower temperature, it may be feasible that such a bimetallic compound is formed at the iron/tin interface. The described reaction path is schematically displayed in Fig. 10.

It is noteworthy, that under analogous conditions, SnO_2 particles could not be reduced to metallic tin. This finding showed that close contact of core and coating was necessary for the reduction processes. It has been reported that SnO_2 is used as sensor for hydrogen gas [14] and can be reduced to metallic Sn only in an r.f. excited hydrogen plasma [28]. Thus, it would appear that the procedure described in this study represents the first reaction path to reduce SnO_2 to Sn by chemical processing.

References

1. Corma A, Martinez A, Martinez C (1996) Appl Catal A 144:249
2. Zubina TS, Dobrovolskii YA (1995) Russ J Electrochem (1995) 31:1280
3. Cerri JA, Leite ER, Gouvea D, Longo E, Verela JA (1996) J Am Ceram Soc 79:799
4. Mishra VN, Agarwal RP (1994) Sensors & Actuators B 21:209
5. Jimenez VM, Gonalezelipe AR, Espinos JP, Justo A, Fernandez A (1996) Sensors Actuators B 31:29
6. Lu E, Chen SY, Peng SY (1996) Catal Today 30:183
7. Berger F, Beche E, Berjoan R, Klein D, Chambaudet A (1996) Appl Surf Sci 93:9
8. Ocaña M, Serna CJ, Matijević E (1991) Mater Lett 12:32
9. Ocaña M, Matijević E (1990) J Mater Res 5:1083
10. Ocaña M, Matijević E (1990) J Aerosol Sci 21:811
11. Liu Y, Zhu W, Tan OK, Yao X, Shen Y (1996) J Mater Sci Mater Electron 7:279
12. Shirakata S, Yokyama A, Isomura S (1996) Jpn J Appl Phys Part 2 Lett 35:L722
13. Liu Y, Zhu W, Tse MS (1995) J Mater Sci Lett 14:1185
14. Ansari SG, Gosavi SW, Gangal SA, Karekar RN, Aiyer RC (1997) J Mater Sci Mater Electron 8:23
15. Murakami K, Yagi I, Kaneko S (1996) J Am Ceram Soc 79:2557
16. Huang JL, Kuo DW, Shew BY (1996) Surf Coat Technol 79:263
17. Henglein A, Giersig M (1994) J Phys Chem 98:6931
18. Musil J, Matous J, Vlcek J, Koydl L (1994) Czech J Phys 44: 565
19. Yoo KS, Cho NW, Song HS, Jung HJ (1995) Sensors Actuators B 25:474
20. Matijević E (1996) In: Pelizzetti E (ed) NATO ASI Ser 3, Kluwer Academic Publishers, Vol 12, p 1
21. Aiken B, Matijević E (1988) J Colloid Interface Sci 126:645
22. Ohmori M, Matijević E (1992) J Colloid Interface Sci 150:594
23. Haq I, Matijević E, Akhtar K (1997) Chem Mater (submitted)
24. Ozaki M, Kratohvil S, Matijević E (1984) J Colloid Interface Sci 102: 146
25. Haq I, Matijević E (1997) J Colloid Interface Sci, in press
26. Haq I, Matijević E (1993) Colloids Surf A81:153
27. Binary Alloy Phase Diagrams (1990) NSRDS ASM Int 2nd Ed, Vol 2, p 1774
28. Thomas JH (1983) Appl Phys Lett 42: 794

Progr Colloid Polym Sci (1998) 109:192–201

A. Pohlmeier
M. Ilic

Binding reactions at the solid–liquid interface analyzed by the concept of kinetic and affinity spectra: Cd^{2+} in montmorillonites

Received: 8 August 1997
Accepted: 5 September 1997

Dr. Andreas Pohlmeier (✉)
Forschungzentrum Jülich ICG-7
52425 Jülich
Germany

Dr. M. Ilic
Inst. Phys. Chem. Univ. Beograd
Yu-11000 Beograd
Yugoslavia

Abstract The influence of heterogeneity on the ion exchange of Cd^{2+} in three important clay minerals, Mg-, Na-montmorillonite and Na-illite is investigated by analyzing the exchange isotherms, and the kinetics. For Mg-montmorillonite the desorption by the cationic surfactant DTAB is also examined. By means of the regularization technique CONTIN affinity and kinetic spectra are obtained. In the montmorillonites two processes can be distinguished: the fast sorption and desorption at the outer surface, controlled by the ion exchange process itself and a slower intercalation step, whose rate determining step is the activated diffusion inside the interlayer space and in case of Na-montmorillonite a simultaneous aggregation of the clay platelets. At illite only the exchange at the outer surface takes place, no intercalation can be observed. All processes are influenced by the heterogeneity of the surface, the affinity spectra are considerably broadened.

Key words Montmorillonite – illite – Cd^{2+} – heterogeneity – CONTIN – kinetics – ion exchange

Introduction

The mobility of environmental contaminants like heavy metal ions in soils and waters is controlled by complexation and ion exchange processes at the surfaces of colloidal minerals and organic polymers. These substances are characterized by a considerable degree of heterogeneity [1], that may influence their binding behaviour. This heterogeneity must therefore be taken into account for a quantitative description of the binding kinetics and affinities of toxic metal ions and also for the description of the transport. In this article the sorption of Cd^{2+} at important clay minerals and its mobilization by a cationic surfactant are chosen as an example to demonstrate the applicability of the concept of kinetic and affinity spectra for the sorption as well as for the desorption reaction.

Layer silicates

Important minerals of the clay fraction of soils are layer silicates like illite and montmorillonite. These 1:2 layer silicates consist of one layer of Al^{3+}-ions, octahedrally, coordinated with oxygen (*O-layer*), and two adjacent layers of tetrahedrically coordinated Si^{4+}-ions (*T-layer*, see also Fig. 1). Some of these ions are substituted, e.g., by Mg^{2+} or Fe^{3+}, thus a constant negative charge is created ranging from 0 to -2 per cent unit cell. For example, montmorillonites carry about -0.4 charges and illites about -0.8 charges per unit cell [2]. Due to this isomorphic substitution inside the crystal lattice ion exchange sites are created that may bind counterions like Na^+, K^+; Ca^{2+}, Mg^{2+}, or Cd^{2+} on the outer surface of these platelets. Depending on the layer charge and the type of charge-compensating counterions the layer silicates may exist in

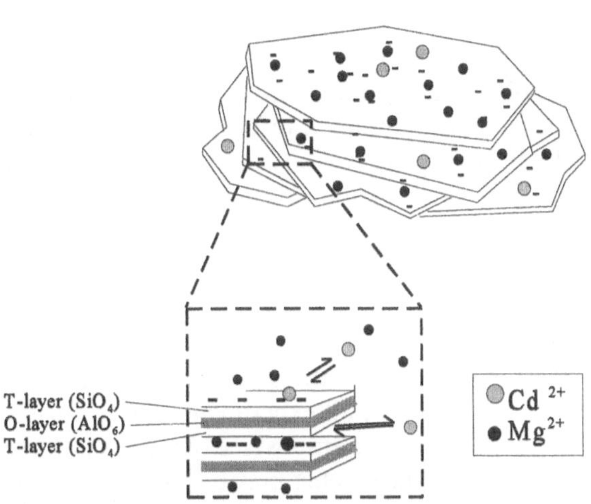

Fig. 1 Structure of Mg-montmorillonite with exchanged Cd^{2+} (schematic). Upper part: irregular shape and heterogeneous charge distribution. Lower part: exchange of Cd^{2+} in the outer surface and in the interlayer space

aqueous dispersion as aggregated stacks or as single platelets. The aggregated state is favored by a high layer charge and small counterions like K^+, well known minerals are illite-type layer silicates. An example of layer silicate favoring the dispersed state is Na-montmorillonite, which is nearly completely dispersed due to its low layer charge and the big hydrated counterion Na^+. Montmorillonites possessing 2-valent counterions like Ca^{2+}-, or Mg^{2+} exist in an aggregated state but their interlayer space is still accessible for other ions like Cd^{2+}.

The charge density at the surface of clay minerals that creates the binding sites varies over the surface of layer silicates [3]. Furthermore, these minerals carry at their edges pH-dependent charges due to protonation/deprotonation reactions at -AlOH residues. Also the size of the platelets may vary from few nanometers to several microns, and their shape is irregular, and different types of binding sites may exist at the outer surface and in the interlayer space as it is schematically shown in Fig. 1 [2]. All these properties contribute to a heterogeneity of the surface influencing their binding properties with respect to metal ions. So the classical description of the exchange process by single, discrete coefficients like rate constants and ion exchange constants is not satisfactory since the heterogeneity must be combined with reaction models. In this work it will be demonstrated how the introduction of *distribution functions* of the Gibbs free enthalpy (ΔG_j) and of the rate coefficients k_j leads to affinity and kinetic spectra. Both procedures will be demonstrated in this article for the binding of Cd^{2+} at two montmorillonites and one illite.

Theory

Distribution functions

General distribution functions of the affinity or ion exchange constants $S(K)$, and of the rate constant $S(k)$ are defined. K and k, respectively, are local affinity and rate coefficients, and the distribution functions describe the relative contribution of a local binding site to the total observable reaction. A plot of $S(K)$ vs. $\log(K)$ or of $S(k)$ vs. $\log k$, respectively, are termed in the following as *affinity* or *kinetic spectrum*. The relation between these distribution functions and the experimentally observed binding isotherm or time curve, respectively, are derived in the following:

The reactive surface of the minerals is considered as an ensemble of independently reacting types of binding sites Z_j^{2-}, originally occupied with an alkali or earth alkali ion. At each type of site the following *local reaction* takes place:

$$Cd^{2+} + M_mZ_j \underset{k_{j,b}}{\overset{k_{j,f}}{\rightleftarrows}} CdZ_j + mM^{x+} , \quad (1)$$

where M^{x+} represents a mono- of divalent metal ion, $k_{j,f}$ and $k_{j,b}$ are the local forward and backward rate constants. All sites of the type Z_j^{2-} differ from each other in their local affinity and their kinetics of the exchanging metal ions.

Isotherm

The local equilibrium is described by Eq. (2):

$$K_{j,ex} = \frac{x_{Z_jCd}\, c_M^m}{x_{Z_jM_m}\, c_{Cd^{2+}}} \frac{f_M^m}{f_{Cd}} = \frac{k_{j,f}}{k_{j,b}} , \quad (2)$$

with the local ion exchange coefficient $K_{j,ex}$, and the mole fractions of occupied binding sites, χ. "c" are the concentrations of the free metal ions, "f" are the activity coefficients, and $k_{j,f}$ or $k_{j,b}$ are the forward or backward rate coefficients, respectively. The index "j" denotes a binding site of one type. Introducing the mass balance equations for Cd^{2+}, M^{x+}, and binding sites Z^{2-} one obtains the ion exchange equation for divalent exchange:

$$c_{Z_jCd} = -\frac{1}{2}P \pm \sqrt{\frac{P^2}{4} - Q} \quad (3)$$

with $P = (c_{Z_j}^0 - c_M^0 - K_{j,ex}(c_{Cd}^0 + c_{Z_j}^0))/(K_{j,ex} - 1)$, $Q = (K_{j,ex} c_{Z_j}^0 c_{Cd}^0)/(K_{j,ex} - 1)$ and $c_{Z_jCd} = \chi_j c_{Z_j}^0$. $c_{Z_j}^0$, c_M^0 and c_{Cd}^0 are the total concentrations of the binding sites of type "j", of M^{2+} and of Cd^{2+}, respectively. It should be noted at this place that Eq. (3) is not defined for $K_{j,ex} = 1$, due to the denominator in the definitions of P and Q.

In many cases it can be arranged by the experimental conditions that the total concentrations of Cd^{2+} and binding sites Z^{2-} are much smaller than the total concentration of M^{x+}. Then the latter quantity may be regarded approximately constant over the whole range of the isotherm and it may be combined with $K_{j,ex}$ to give a pseudo-Langmuir coefficient $K_j = K_{j,ex} f_{Cd}/c_M^m f_M^m$. Additionally, the mole fractions in Eq. (2) may be multiplied with the total concentration of binding sites, c_Z^0, without changing the numerical value of K_j. By this way we get

$$K_j = \frac{K_{j,ex} f_{Cd}}{c_M^m f_M^m} = \frac{c_{Z_jCd}}{c_{Z_jM_m} c_{Cd}} \tag{4}$$

with

$c_{Z_jCd} = \chi_{Z_jCd} \cdot c_{Z_j}^0$ and $c_{Z_jM_m} = \chi_{Z_jM_m} \cdot c_{Z_j}^0$.

Rearranging of Eq. (4) yields Langmuir's equation:

$$c_{Z_jCd} = c_{Z_jCd}^0 \frac{K_j c_{Cd}}{1 + K_j c_{Cd}} , \tag{5}$$

where c_{Z_jCd} is the concentration of Cd^{2+} bound at the sites of type "j" and $c_{Z_jCd}^0$ is the maximum concentration of Cd^{2+} that can be bound at these sites. Eqs. (3) and (5) are termed as the local binding isotherms, and their relation to the experimentally obtainable isotherms is given in the following.

The only measurable quantity is the total concentration of bound Cd^{2+}-ions, c_{ZCd}, which is the sum over the ensemble of binding sites "Z_j":

$$c_{ZCd} = \sum_j^n c_{Z_jCd} = \sum_j^n S_j G_j(K_j, c_{Cd}) , \tag{6}$$

where S_j is a weighting factor and $G(K_j, c_{Cd})$ represents Eqs. (3) and (5) for the ion exchange and the Langmuir isotherm. As a limiting case the sum in Eq. (6) may be written as an integral:

$$c_{ZCd} = \int_0^\infty S(K) G(K, c_{Cd}) \, dK + \varphi \tag{7}$$

with the distribution function $S(K)$ and the kernel $G(K, c_{Cd})$, which is identical to Eqs. (3) and (5), respectively. It must be emphasized that this way of regarding the processes at the local binding sites implies no assumptions besides the validity of the mass action law, Eq. (2), and the independence of the binding sites. A plot of $S(K)$ versus $\log K$ or $\log K_{ex}$, respectively, is termed an affinity spectrum. In the classical case, a binding process is described by a single Langmuir function, it consists of a single delta function. If a system is very heterogeneous, a broad distribution function is expected. In this work the distribution function is obtained by direct inversion of the experimental data c_{ZCd} without any a priori assumptions about its shape by means of the regularization technique CONTIN, in the section *Solution techniques*.

Kinetics

For monitoring the kinetics the stopped flow method is employed taking care that all experiments are performed under pseudo first order conditions (see the section *Material and methods*). Also the kinetics may be influenced by the heterogeneity of the surface binding sites, so the rate equation describing the exchange at this single type of site reads according to Eq. (1):

$$\frac{dc_{Z_jCd}}{dt} = k_{j,f}(c_{Z_jM} c_{Cd}) - k_{j,b}(c_{Z_jCd} c_M^m) , \tag{8}$$

where all concentrations depend on the time. $k_{j,f}$ and $k_{j,b}$ are the local forward and backward rate coefficients. In case of pseudo first order conditions [4] the corresponding rate law reads:

$$\Delta c_{Z_jCd}(t) = \Delta c_{Z_jCd}(0) \exp(-k_j t) , \tag{9}$$

where $\Delta c_{Z_jCd}(t)$ is the difference of the concentration of Z_jCd at time "t" and at equilibrium, and $\Delta c_{Z_jCd}(0)$ is the difference of the concentration of Z_jCd at time "0" and at equilibrium.

In Eq. (9) an observable local rate coefficient k_j is introduced, which is related to $k_{j,f}$, $k_{j,b}$, and the concentration species for M^{x+} as a monovalent alkali metal ion (e.g. Na^+):

$$k_j = k_{j,b}(c_M^2 + 4c_{Z_jCd} c_M) + k_{j,f}(c_{Cd} + c_{Z_jM_2}) . \tag{10}$$

If M^{x+} is a divalent metal ion like Mg^{2+}, the expression for k_j reads

$$k_j = k_{j,f}(c_{ZM} + c_{Cd}) + k_{j,b}(c_{ZCd} + c_M) . \tag{11}$$

In Eq. (9), "$\Delta c_{Z_jCd}(t)$" is not measurable, since its absolute concentration is very small. Analogously to Eq. (7), Eq. (12) is derived by integrating over the whole ensemble of binding sites:

$$\Delta c_{ZCd}(t) = \int_0^\infty S(k) \exp(-kt) \, dk + \varphi , \tag{12}$$

where $S(k)$ is the distribution function of the rate constant k_j for the reaction described by Eq. (1), and φ is a noise term. Equation (12) has the same mathematical structure like the integral binding equation, Eq. (7), with the exponential decaying function as the kernel. A plot of $S(k)$ versus $\log k$ is termed the *kinetic spectrum*. In case of a single exponential decay, i.e. if all types of sites are identical (homogenic limiting case), $S(k)$ is a delta function. If the system is very heterogeneous, the kinetic spectrum shows that a broad distribution over several orders of magnitudes of k.

Solution techniques

The distribution functions $S(K, c_{Cd})$ for affinity and $S(k, t)$ for the kinetics may be obtained from the experimental data $f(x)$ by inversion of the integral equations: Eq. (7) including Eqs. (3) or (5), and Eq. (12) for the kinetics, respectively. The solution of these ill-posed equations is not trivial, since due to the noise term an infinite number of solution exists. Three different families of solution techniques are commonly employed to overcome this problem: (i) analytical techniques, (ii) semianalytical techniques and (iii) direct numerical procedures. Some commonly used methods should be described here in brief:

(i) the most widespread methods are of analytical character, i.e. they base upon isotherm equations like the general Freundlich isotherm containing parameters that can be easily fitted to experimental data by Marquardt's algorithm [5]. In this case Eq. (7) including Eq. (5) reads:

$$c_{ZCd} = c^0_{ZCd} \frac{(Kc_{Cd})^\beta}{(1 + (Kc_{Cd})^\beta)}, \tag{13}$$

where β is a heterogeneity parameter. To these isotherms belong corresponding distribution functions of a well defined shape which are calculable with the fitted parameters. For the above example this is Sips' distribution function [6]:

$$S(K) = \frac{\sin(\pi\beta)}{\pi(2\cos(\pi\beta) + (K/K_m)^\beta + (K_m/K)^\beta}. \tag{14}$$

Analogous equations exist for the analysis of the kinetics like the Elovich equation, but it could be shown that this equation is not convenient for the description of adsorption kinetics [7, 8]. These techniques are easy to handle but have the disadvantage that the distribution functions are restricted to a discrete number of peaks (1 or 2) of a well defined shape. The ensemble average value and the width of the peaks are the only variable parameters.

(ii) From a theoretical point of view it is not necessary to constrain the shape and number of peaks of the kinetic and the affinity spectra as in the fitting procedures. The semianalytical procedures solve the integral equation with the assumption of zero noise and derive an approximate analytical solution. One type of convenient procedures for this purpose are the local isotherm approximations of Koopal and van Riemsdijk (LIAs, [9]). Examples are the condensation approximation (CA) and the logarithmic symmetric approximation (LOGA). The characteristic feature is that analytical approximations for the distribution functions are obtained containing the 1st or higher orders of derivatives of the experimental data. Due to this, these techniques are sensible to oscillations and a smoothing step must preceed the actual analysis. But in contrast to the aforementioned fitting procedures, the LIAs imply no a priori assumption about the shape of the distribution functions and the number of peaks. These techniques have also been applied successfully to the analysis of the kinetics of metal ion desorption from natural humic substances [10]. Also the procedure applied by Olson and Shuman [11] to the kinetics of copper dissociation from humic acids belongs to this family of solving techniques.

(iii) The third family of methods for solving Eqs. (7) and (12) are numerical techniques like the expectation maximization method (EM, [12]), the cross validation (CV, [13]), singular value decomposition (SVD, [14]) and the regularization techniques (e.g. INTEG of von Szombathely [15], REMEDI of Koopal [16], and the program developed by Cernik et al. [17]).

As mentioned above the calculations of the distribution functions $S(g)$ in Eqs. (7) and (12) constitute ill-posed problems. Thus there exists a large number of possible solutions, of which each fits the data well within experimental error. Even worse, the errors in the solution are unbounded. To stabilize the optimal solution of such kind of problems the regularization is one of the best known methods [15–23]. Its basic principle consists in expanding the minimizing least-squares condition by an additional term called the regularizer favoring a certain type of solution by implying any prior knowledge like the smoothness and parsimony of the solution:

$$V(\alpha) = ||\mathbf{M}_\varphi^{-1/2}(\mathbf{y} - \mathbf{Ax})||^2 + \alpha\mathbf{R}^2, \tag{15}$$

where $V(\alpha)$ is the regularized variance, α is a parameter determining the weight of the regularizer, $\mathbf{M}_\varphi$ is the covariance matrix of the experimental data, the vector $\mathbf{y}$ contains the experimental data, the matrix $\mathbf{A}$ is composed of the kernel ($G(K, c_{Cd})$ or $G(k, t)$), $\mathbf{x}$ are the discretized values of the independent variable c_{Cd} or t, respectively, and $\mathbf{R}^2$ is the regularizer. The latter can be composed of convenient functions like the second derivative of the distribution function, which leads to a smooth distribution.

A further principle employs certain constraints of the solution range for the distribution function on the basis of known a priori information. In comparison to the fitting techniques this method possesses the important advantage that it makes no assumptions about the shape of the distribution function and the number of peaks. In the present work the regularized, constrained least-squares algorithm CONTIN, originally developed by Provencher, is applied, its basic principles are described exactly in the literature [19–21]. The characteristic feature of CONTIN is the determination of the strength of the regularization. The program uses the F-test for choosing the optimal solution. This tests if the increase of the variance of the regularized solution compared to that of the unregularized is caused at random, a parameter probl(α) is calculated that should lie in the region around 0.5. For all computations in this article CONTIN is used in the version of Ruf [22, 23]. The kernel is left unchanged for the calculation of the kinetic spectra and for the evaluations of the isotherms Eqs. (3) and (5), respectively, are introduced as kernels.

Materials and methods

Na- and Mg-montmorillonites were prepared from bentonite (Südchemie AG, Germany) by a threefold ion exchange with 1 M $NaNO_3$ or $Mg(NO_3)_2$ followed by dialysis for 6 weeks. Na-illite was obtained from Erbslöh GmbH, Germany and prepared in the same way. The $<2\ \mu m$ fractions were separated by sedimentation (60 cm, 48 h). The c.e.c was determined as 9.8×10^{-4}, 9.0×10^{-4}, and 2.7×10^{-4} mol g^{-1} for Na-, Mg-montmorillonite and Na-illite, respectively and the BET surface areas were 89 m^2 g^{-1}, 70 and 43 m^2 g^{-1}, respectively. For all ion exchange experiments $Cd(NO_3)_2$ (Fluka AG, Switzerland, pA-grade) was used and the solutions were buffered with 0.005 mol l^{-1} PIPES (pA grade, Fluka AG, Switzerland). For the mobilization experiments the cationic surfactant dodecyl-trimethyl-ammoniumbromide (DTAB) was used (Merck, Germany). The kinetics were monitored as described in [18] by means of the heavy metal indicator PAR (pA-grade, from Janssen Chimica, Geel, Belgium). All solutions were buffered at pH = 7.0 with 2×10^{-4} mol l^{-1} PIPES.

Ion exchange isotherms was recorded electrochemically in situ by detecting the concentrations of free Cd^{2+} by the DPP or DPASV technique as described in [8, 18]. The kinetics of the ion exchange were measured with the stopped flow technique with optical density detection [18]. The dead time of the instrument is 20 ms and absorbance changes of less than 0.0002 can be reliably detected. The instrument is thermostated with an accuracy of $\pm$ 0.1°C, its temperature ranges from 0°C to 45°C. The displacement of Cd^{2+} by the cationic surfactant was measured by adding a solution of DTAB to a suspension of Cd-containing clay minerals, the released amount of Cd^{2+} was detected by the indicator PAR.

Results and discussion

Sorption isotherms

The sorption isotherms of Cd^{2+} are measured at pH = 6.9 with a suspension concentration of 0.05 g l^{-1}. Figure 2 shows the sorption isotherm and the fits with the affinity spectrum (——) according to Eq. (7) including Eq. (3) and with a single discrete ion exchange constant $K_{ex} = 1.6$ (----) according to Eq. (2) with $j = 1$. It is obvious that the affinity spectrum describes the course of the ion exchange isotherm much better than the classical description, however the latter is still valid within the experimental error bars. Figure 3 shows the corresponding affinity spectrum and for comparison the affinity spectra obtained for the adsorption of Cd^{2+} on Na-montmorillonite (data taken from [18]) and Na-illite (data taken from [8]). The spectra at the montmorillonites are characterized by a similar half width at half height of $\pm 0.5 \log(K_{ex})$ units, the spectrum of Na-illite is narrower (HWHH = $0.3 \log(K_{ex})$). The mean exchange coefficients are $K_{ex} = 1.0$ for Mg-montmorillonite, and $K_{ex} =$ 3.0 l mol^{-1} for Na-montmorillonite and Na-illite. These values agree also very well with data obtained for ion exchange of other clay minerals [24, 25]. The affinities at both the Na-clays are identical, this means that the nature of the clay is less dominant than the nature of the exchanging ions. The differences lie in the heterogeneity of the surfaces: the montmorillonites appear more heterogeneous than the illite, the reason is presumely that montmorillonites may intercalate Cd^{2+}, whereas the interlayer space of illite is not accessible for Cd^{2+}. The characteristic feature of all affinity spectra is the pronounced "tailing" to the low affinity side, a consequence of the fact that the isotherms are not monitored till a constant sorbed amount is obtained. This is experimentally extremely difficult, since the errors of c_{ZCd} are proportional to c_{Cd}, and they increase strongly above $c_{Cd} = 0.05$ mmol l^{-1}, as indicated by the error bars in Fig. 2. More insight into the sorption reaction allows the analysis of the reaction rate.

Sorption kinetics

Mg-montmorillonite. As discussed in *Material and methods* section, the kinetics are fast and therefore are measured by the stopped flow technique, as described in detail elsewhere [18]. The concentration of free Cd^{2+} is monitored by the indicator PAR (pyridyl-azo-resorcinol)

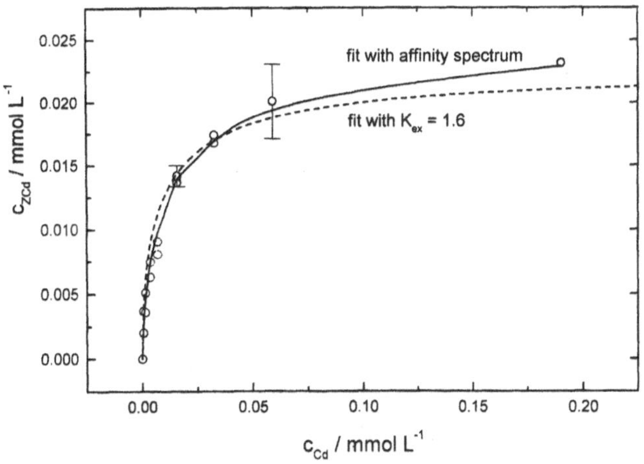

Fig. 2 Ion exchange isotherm of Cd^{2+} in Mg-montmorillonite at pH = 6.9 and $T = 25\,°C$ at a suspension concentration of 0.05 g/l. (——) fit with the affinity spectrum according to Eq. (7) including Eq. (3), (----) fit with $K_{ex} = 1.6$ according to Eq. (2) with $j = 1$

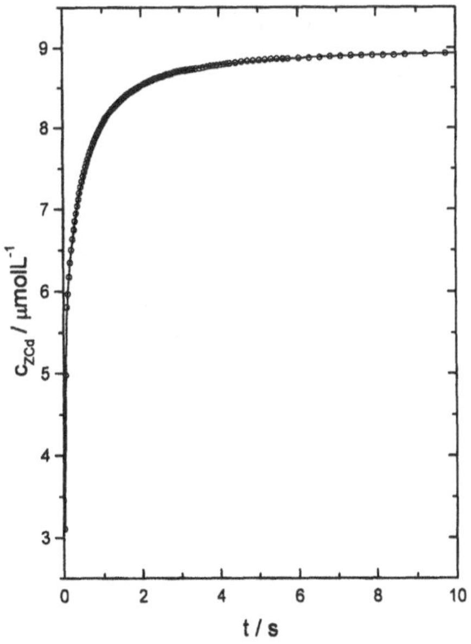

Fig. 4 Stopped flow experiment: concentration of bound Cd^{2+} (c_{ZCd}) as a function of time. $c^0_{Cd} = 15\,\mu M$, 0.05 g l^{-1} Mg-montmorillonite, 0.005 M $NaNO_3$, pH = 6.6, $T = 25\,°C$

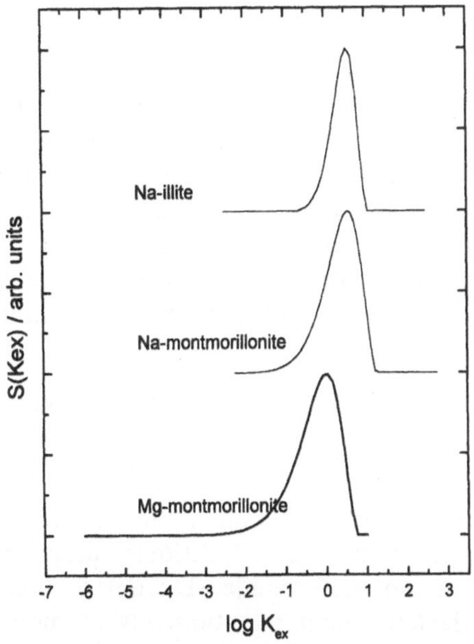

Fig. 3 Affinity spectrum $S(K_{ex})$, for the ion exchange of Cd^{2+} in Mg-montmorillonite, calculated from the isotherm data from Fig. 2. For comparison the corresponding affinity spectra for the ion exchange at Na-montmorillonite (from [18]) and Na-illite (from [8]) are also shown

which forms at pH-values around 7 only weak complexes with Cd^{2+}. By means of a calibration curve the concentration of free Cd^{2+} can be computed and from this the concentration of bound Cd^{2+}. Figure 4 shows c_{ZCd} as a function of time when 15 µM solution of $Cd(NO_3)_2$ reacts with 0.05 g l^{-1} Mg-montmorillonite at pH = 6.6 and $I = 0.005$ mM $NaNO_3$. The reaction is complete after about 10 s, the line shows the fit with the kinetic spectrum according to Eq. (12). In Fig. 5 is included the corresponding kinetic spectrum together with those obtained for different temperatures. Bimodal reactions are observed: the fast process with mean rate coefficients of 30–40 s^{-1} lies in the order of magnitude of the decomplexation reaction [18], and the main, slow process is characterized by a rate coefficient in the order of magnitude of 1 s^{-1}. This agrees with the results obtained earlier [7] and corresponds to the intercalation of Cd^{2+} into the interlayer of Mg-montmorillonite. The mean activation energy obtained from an Arrhenius plot (Fig. 6, lower part), is 20 ± 5 kJ mol^{-1}, this value also confirms the results obtained in [7]. Pure diffusion controlled reactions of bivalent metal ions are characterized by an activation energy smaller than 14 kJ mol^{-1} [4], and for chemically controlled complexation reactions of Cd^{2+} it should be in the order of magnitude of 32 kJ mol^{-1} [18]. The conclusion may be drawn that the intercalation is controlled by an activated diffusion, a reasonable assumption since the inter-layer space is very narrow with 10 Å. An evaluation of mean rate constants obtained for different total concentrations of Cd^{2+} according to Eq. (9) is not possible, since at least two processes are participating in the whole reaction.

The fast reaction is also characterized by a mean activation energy of about 20 kJ mol^{-1} (Fig. 6), however, this value is very uncertain, since the corresponding rate coefficients are at the edge of the observation window which is

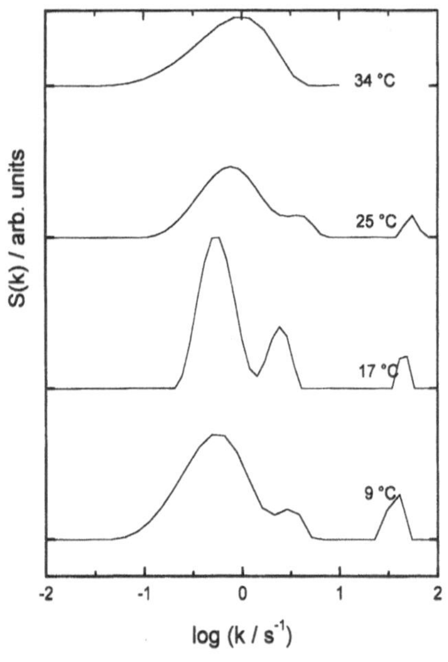

Fig. 5 Kinetic spectra for the ion exchange reaction of Cd^{2+} in Mg-montmorillonite at different temperatures. $S(k)$ is the normalized distribution function with respect to the highest peak. Equal area representation: the area under each peak in this half-logarithmic scale corresponds to the relative contribution of the respective process

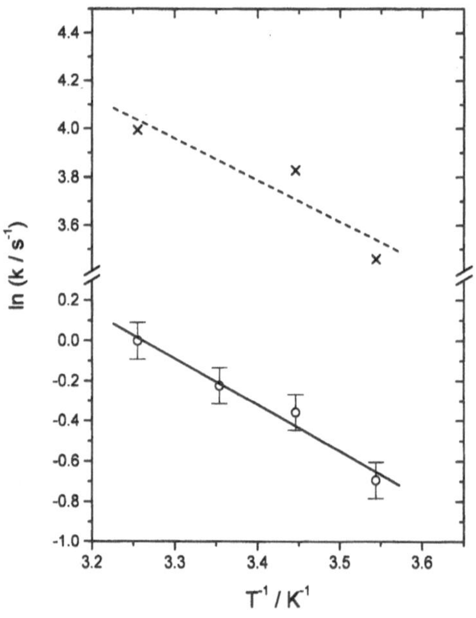

Fig. 6 Arrhenius plot of the mean rate coefficients k_{slow} and k_{fast} for the slow, main process (lower part) and the fast process (upper part), respectively, found for the ion exchange kinetic spectra in Fig. 5

delimited by the rate of the indicator reaction, and should not be discussed further.

Na-montmorillonite. The kinetics of ion exchange at Na-montmorillonite is investigated in detail elsewhere [18], only the main results are reviewed here. Figure 7 shows the kinetic spectra obtained for different temperatures at an suspension concentration of $0.05\ g l^{-1}$ and a concentration of Cd^{2+} of 50 μM. Bimodal distribution functions are obtained, of which the main process with $k_{mean} = 11\ s^{-1}$ at 25 °C corresponds to the ion exchange at the outer surface. Its activation energy of $28\ kJ mol^{-1}$ proves that the ion exchange reaction itself is the rate determining step, not the diffusion to the surface. The slow process is caused by aggregation of the clay platelets and the intercalation of Cd^{2+} into the newly formed interlayer space.

Na-illite. The results of the kinetics are also investigated in detail earlier [8]. In this case only one process is found, characterized by a half-width at half-height (HWHH) of $0.2 \log(k(s^{-1}))$ units. It could be assigned to the ion exchange at the outer surface of illite, no intercalation is observed. The activation energy of $35\ kJ mol^{-1}$ proves that diffusion is not the rate determining step. From the concentration dependence of the mean rate coefficients, the mean ion exchange coefficient could be calculated from a plot according to Eq. (10), with $k_j = k_{mean}$. Figure 8 shows the results. A straight line is obtained, from which $k_{f,mean} = 1.2 \times 10^6\ l mol^{-1} s^{-1}$ and $k_{b,mean} = 0.6 \times 10^6\ l^2 mol^{-2} s^{-1}$ are computed. The mean ion exchange coefficient calculated from these data, $K_{ex} = 2\ mol l^{-1}$ agrees well within the experimental error with that obtained from the affinity spectrum (see Fig. 3). This agreement clearly proves the validity of the evaluation of the kinetics and the isotherms by distribution functions.

Desorption

The kinetics of the desorption reaction are investigated in this work. As mobilization agent the cationic surfactant dodecyltrimethyl ammoniumbromide (DTAB) is used, since cationic surfactants bind quantitatively at montmorillonites and perform a stoichiometric ion exchange [26, 27]. Also this process is experimentally observed by the stopped flow technique, since the mobilization reaction turned out to be fast. A suspension of $0.05\ g l^{-1}$ Mg-montmorillonite containing 10 μM Cd^{2+} reacts with a solution of 50 μM DTAB in buffered solution at pH = 6.5. The concentration of DTAB is same as the total cation exchange capacity of the Mg-montmorillonite suspension, therefore a quantitative adsorption of DTAB and mobilization of Cd^{2+} takes place. Figure 9 shows the results for $T = 25$ °C. Like the sorption reaction, this desorption is characterized by a bimodal distribution function. The fast reaction is slower than that of the sorption reaction, but

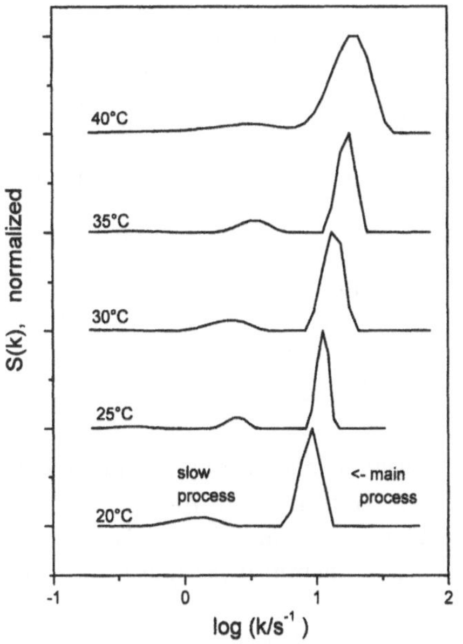

Fig. 7 Kinetic spectra for the ion exchange reaction of Cd^{2+} in Na-montmorillonite ($c^0_{Cd} = 10\ \mu M$, $0.05\ g\,l^{-1}$ Mg-montmorillonite, pH = 7.0) at different temperatures (from [18]). $S(k)$ is the normalized distribution function with respect to the highest peak. Equal area representation: the area under each peak in this half-logarithmic scale corresponds to the relative contribution of the respective process

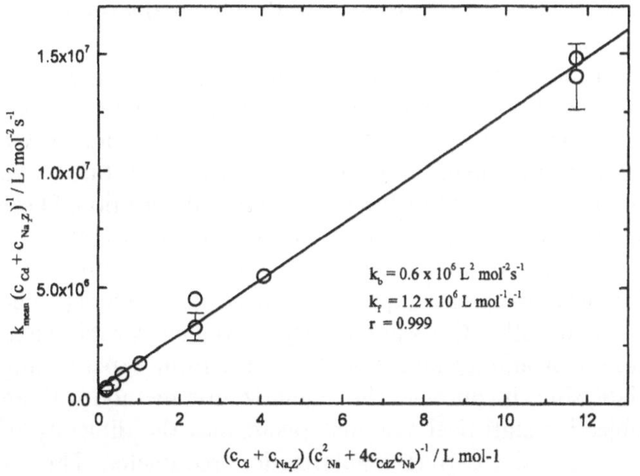

Fig. 8 Ion exchange kinetics of Cd^{2+} in Na-illite (from [8]) plot of k_{mean} for different concentrations of Cd^{2+} at an illite concentration of $0.05\ g\,l^{-1}$ at $T = 25\ °C$ and pH = 7.1 according to Eq. (10)

shows no temperature dependence. It is probably the mobilization of Cd^{2+} from the outer surface of the montmorillonite, controlled by the diffusion of DTAB to the surface.

Figure 10 summarizes the temperature dependence of the mean rate coefficients of the slow process by means of an Arrhenius plot. The activation energy is with

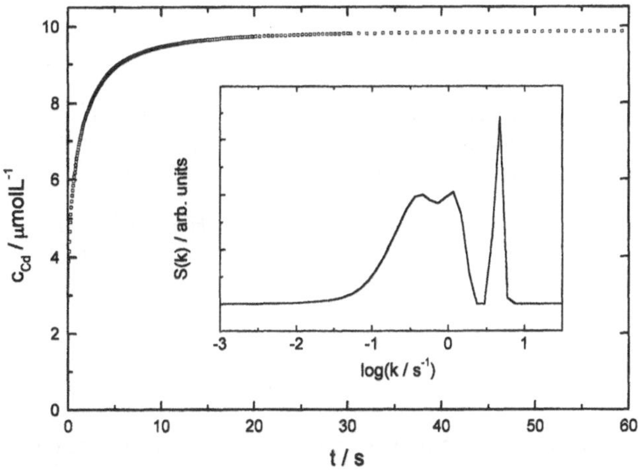

Fig. 9 Mobilization of $10\ \mu M\ Cd^{2+}$ from Mg-montmorillonite by the cationic surfactant DTAB ($50\ \mu M$) at pH = 6.5 and $T = 25\ °C$. Inset: corresponding kinetic spectrum according to Eq. (12)

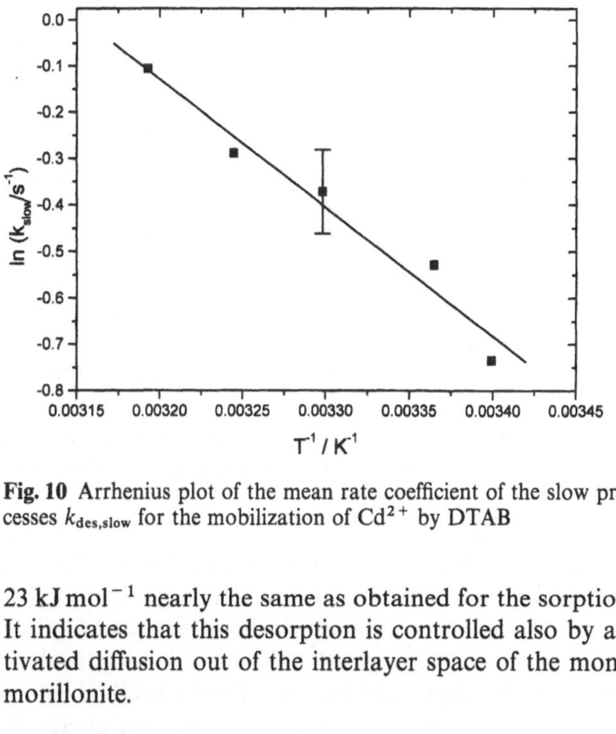

Fig. 10 Arrhenius plot of the mean rate coefficient of the slow processes $k_{des,slow}$ for the mobilization of Cd^{2+} by DTAB

$23\ kJ\,mol^{-1}$ nearly the same as obtained for the sorption. It indicates that this desorption is controlled also by activated diffusion out of the interlayer space of the montmorillonite.

General discussion and outlook

This article shows the usefulness of the kinetic and affinity spectra for the analysis of ion exchange processes of Cd^{2+} at the clay minerals montmorillonite and illite in dilute suspensions. Distribution functions of the rate and the ion exchange coefficients were obtained that characterize the observable processes much better than the classical description by discrete coefficients. All results are summarized in Table 1.

Table 1 Compilation of kinetic- and thermodynamical data on the ion exchange of Cd^{2+} in different clay minerals (HWWH = half-width at half-height of the kinetic and affinity spectra)

	Mg-montmorillonite	Na-montmorillonite	Na-illite
$k_{fast,mean}$ (s^{-1})	> 30	10	5 to 30
HWHH/log$k$ (s^{-1})	Not det.	0.1	0.2
$\Delta H^{\#}$ (kJ mol^{-1})	20	28	35
$\Delta S^{\#}$ (J mol^{-1} K^{-1})	Not det.	− 130	− 105
$k_{slow,mean}$ (s^{-1})	1	2.4	—
HWHH/log$k$ (s^{-1})	0.5	0.2	—
$\Delta H^{\#}$ (kJ mol^{-1})	20	37	—
$\Delta S^{\#}$ (J mol^{-1} K^{-1})		− 110	—
$k_{ex,mean}$	1.0	3.0 mol l^{-1}	3.0 mol l^{-1}
HWHH/log(K_{ex})	0.5	0.5	0.3
Ref.	This work	[18]	[8]

The ion exchange processes are fast, they take place in the order of magnitude of 0.1–10 s. At first an ion exchange at the outer surface is observed for all investigated minerals. It is characterized by mean rate coefficients of about 10 s^{-1} and a considerable degree of heterogeneity, as indicated by the values of the half-width at half-height (HWHH) of the distribution functions of 0.2 log(k(s^{-1})) units. This inhomogeneity is probably caused by the heterogeneity of the charge density at the outer surface of the clay minerals [3]. The activation energy of this process is 28 kJ mol^{-1} for Na-montmorillonite and 35 kJ mol^{-1} for Na-illite. These values are higher than expected for the diffusion controlled reaction, proving that the ion exchange itself are the rate determining steps. These results confirm the mechanism obtained for the ion exchange of alkali metal ions at other montmorillonites [28, 29]. The large negative data of the activation entropies (Table 1) show the associative nature of the transition complex during the ion exchange. Also for the desorption of Cd^{2+} by the cationic surfactant DTAB a fast mobilization from the outer surface, characterized by mean rate coefficient of 6 s^{-1} is observed.

At Mg- and Na-montmorillonite a slow process is found in contrast to illite. This proves that at illite no intercalation takes place. This intercalation is most obvious at Mg-montmorillonite, which exists even in dilute suspension in an aggregated state. It is characterized by a considerable degree of heterogeneity, the HWHH values are between 0.2 and 0.5. Also the activation energy is greater than expected for diffusion controlled reactions, the data indicate an activated diffusion into the interlayer space. The very high value found for Na-montmorillonite is probably caused by simultaneous intercalation and aggregation, since in that case a monovalent ion is exchanged by a divalent ion, which causes a change of the aggregation state, whereas at Mg-montmorillonite the degree of aggregation should not change significantly. The heterogeneity may be due to the irregular shape of the particles, since the intercalation is diffusion controlled. The desorption of Cd^{2+} by DTAB from Mg-montmorillonite also takes place from the interlayer space with mean rate coefficients of 0.5 s^{-1}. The activation energy at 23 kJ mol^{-1} is as high as for the sorption process, and shows also for this case that the diffusion out of the interlayer space is the rate determining step.

The exchange isotherms are also investigated by distribution functions and affinity spectra are obtained. The experimental data are much better described by these spectra than by classical discrete ion exchange coefficients. The mean values of the ion exchange affinity spectra lie in the same order of magnitude as found in the literature for other ion exchange process in clay minerals. The widths of the distribution functions (0.3 < HWWH < 0.5) indicate the heterogeneity of the minerals surface, most probably caused by the inhomogeneity of the charge distribution at the surfaces.

Heterogeneity phenomena may occur with many sorption reactions at natural minerals and organic polymers. The analysis of the kinetics and the equilibrium by the concept of kinetic and affinity spectra turns out to be a very convenient tool for a quantitative description. These distribution functions are connected to the classical description by discrete coefficients by their mean value, making these analyses comparable to the classical models. The used algorithm CONTIN is very convenient for the calculation of affinity and kinetic spectra from experimental data, since it combines the regularization method with an objective statistical test that overcomes the difficulty of choosing the correct regularization parameter. Thus it prevents from oversmoothing or introducing artefacts. In future this method will be applied to binding and mobilization reactions of contaminants at other important soil components, so a better understanding of the influence of heterogeneity is expectable.

Acknowledgements The authors wish to thank H. Ruf, (MPI for Biophysics, Frankfurt, Germany) and A. Topp (University of Cologne, Germany) for the latest version of CONTIN, S. Haber-Pohlmeier (University of Graz, Austria) for the modifications of CONTIN necessary for the affinity spectra and H. Rützel (Research Center Jülich, Germany) for the determination of the isotherms.

References

1. Buffle J (1988) Complexation Reactions in Aquatic Systems: An Analytical Approach. Ellis Horwood, Chichester
2. Jasmund K, Lagaly G (1993) Tonminerale und Tone. Steinkopf Verlag, Darmstadt
3. Malla PB, Robert M, Douglas LA, Tessier D, Komarneni S (1993) Clays Clay Min 41:412
4. Strehlow H, Knoche W (1977) Fundamentals of Chemical Relaxation. VCH, Weinheim
5. Bevington PR (1992) Data Reduction and Error Analysis, 2nd ed. McGraw-Hill, New York
6. Sips R (1948) J Chem Phys 16:490
7. Pohlmeier A (1994) Progr Colloid Polym Sci 95:113
8. Pohlmeier A, Rützel H (1996) J Colloid Interface Sci 181:297
9. Nederlof MM, van Riemsdijk WH, Koopal LK (1990) J Colloid Interface Sci 135:410
10. Nederlof MM, van Riemsdijk WH, Koopal LK (1994) Environ Sci Technol 28:1048
11. Olson DL, Shuman MS (1983) Anal Chem 55:1103
12. Stanley BJ, Guiochon G (1993) J Phys Chem 97:8098
13. Craven P, Wahba G (1979) Numer Math 31:377
14. Vos CHW, Koopal LK (1985) J Colloid Interface Sci 105:183
15. Von Szombathely M, Brauer P, Jaroniec M (1992) J Comput Chem 13:17
16. Koopal LK, Vos CHW (1993) Langmuir 9:2593
17. Cernik M, Borkovec M, Westall JC (1995) Environ Sci Technol 29:413
18. Haber-Pohlmeier S, Pohlmeier A (1997) J Colloid Interface Sci 188:377
19. Provencher SW (1982) Comput Phys Commun 27:213
20. Provencher SW (1982) Comput Phys Commun 27:229
21. Provencher SW (1984) CONTIN Users Manual EMBL Technical Report, Heidelberg
22. Ruf H (1993) Adv Colloid Interface Sci 46:333
23. Ruf H (1989) Biophys J 56:67
24. Thellier C, Sposito G (1988) Soil Sci Soc Am J 52:979
25. Laudelout H (1987) In: Newman ACD (ed) Chemistry of Clays and Clay Minerals. Mineralogical Society, Monograph No. 6, Longman, London, pp 225–235
26. Röhl W, von Rybinski W, Schwuger MJ (1991) Progr Colloid Polym Sci 84:206
27. Zhang ZZ, Sparks DL, Scrivner NC (1993) Environ Sci Technol 27:1625
28. Crooks JE, El-Daly H, El-Sheik MY, Habib AFM, Zaki A (1993) Int J Chem Kinet 25:161
29. Tang L, Sparks DL (1993) Soil Sci Soc Am J 57:42

Progr Colloid Polym Sci (1998) 109:202–213

E. Klumpp
H. Lewandowski
J.-M. Séquaris

Surface modifications of soil minerals by amphiphilic substances (surfactants, synthetic and natural macromolecules)

Received: 23 October 1997
Accepted: 27 October 1997

Dr. E. Klumpp (✉) · H. Lewandowski
J.-M. Séquaris
Institute of Applied Chemistry
Research Center Jülich
D-52425 Jülich
Germany

Abstract Surface modifications of major inorganic soil components such as metal oxides and clay minerals by amphiphilic substances such as surfactants, synthetic polymers and natural macromolecules are reviewed. Some resulting features concerning the colloidal behaviour of soil components and the immobilization of nonionic organic contaminants are discussed.

Key words Adsorption – soil minerals surfactants – macromolecules – organic contaminants – surface modification

Introduction

Synthetic surfactants and water-soluble ionic and nonionic polymers are passed into the environment. Studies on the behaviour of these amphiphilic substances in soils, i.e. on their mobility, degradability and damaging effects, are only available to a limited extent [1]. Because of the complexity of the processes involved and the soil diversity, unique statements on the influence of these synthetic amphiphilic substances are hardly possible. Similarly, the soil behaviours of natural amphiphilic macromolecules such as humic acids are far from being comprehensively described due to the great heterogeneity of their structures characterized, e.g., by a large molecular weight distribution and the presence of various functional groups [2]. It can be generally stated that this universal character of humic acids sums up the physicochemical properties of synthetic well-defined substances like surfactants and polyelectrolytes which are often used as chemical models in the study of the environmental behaviour of humic substances.

The amphiphilic character of these substances can be expressed in various ways in the soil properties. According to their surface active properties with soil particles, they can modify the capillary properties of the soil by reducing the water surface tension in the porous structures. A possible dispersing effect on the original soil aggregate structure not only affects the soil particle distribution in the soil subsurface but also promotes their transport into deeper soil layers by percolating waters. The modification of the hydrophilic properties of the soil particles by the amphiphilic substances also renders their surfaces more hydrophobic, which enhances the adsorption of poorly water-soluble pollutants and thus affects their transport.

In this short review, fundamental aspects of the soil behaviour of these substances are considered: their adsorption at major inorganic soil particle surfaces and the possible consequences of this surface modification on soil processes. Thus, recent results obtained in our institute (under the direction of Prof. Schwuger) will be briefly reviewed in the case of surfactants and synthetic polymers as well as other statements from the literature. The effect of natural macromolecules on soil aggregation and stability is shown together with preliminary results obtained with humic acids.

Surface modification of soil minerals by surfactants

Sorption of surfactant

Cationic surfactants adsorb strongly onto soil and sediments because of favourable electrostatic interactions with the predominantly negatively charged surfaces of natural materials. Despite the importance of these compounds, there has not been any extensive systematic investigation of their adsorption onto natural materials such as soils and sediments [3].

Recently, the factors have been studied that control the sorption of the cationic surfactant dodecylpyridinium bromide (DP) on environmental and pristine surfaces [4]. The adsorption isotherms are nonlinear, even at very low surface coverages of DP. This behaviour can be described by assuming a heterogeneous mixture of adsorption sites. The sorption depends primarily on the CEC (cation exchange capacity) of the sorbent. The effects of electrolyte type and concentration on adsorption also indicate that the primary adsorption mechanism is a cation exchange. A judgement about any correlation with organic carbon is hardly warranted.

The adsorption of cationic surfactants on other sorbents, which are important constituents or potential analogues of environmental matrices, has been studied in more detail [5–7]. The cationic surfactant adsorption on layer silicates is of particular importance because cationic surfactant/clay mineral complexes are used for landfill sealing.

The cationic surfactant isotherm on the clay mineral is of the high-affinity type (Fig. 1) [8–10]. The slope of the isotherm is very steep because of the electrostatic attraction between cationic surfactants and negatively charged clay minerals. However, almost complete ion exchange up to CEC does not take place, but the isotherm already changes into a flattened shape at about 60–70% of the CEC. In the plateau region, cationic surfactant is adsorbed by hydrophobic interactions between the alkyl chains of the surfactant molecules even beyond the CEC. A surfactant bilayer is formed on the surface of the clay mineral and the isotherm exhibits a plateau.

In the case of the so-called organoclays, the quantitative exchange with the cationic surfactant corresponds to the CEC. If this takes place on a swelling layer silicate, such as montmorillonite, cationic surfactants are intercalated in the interlayers, and the layer interspaces thus become organophilic. Consequently, the adsorption properties of the layer silicates change fundamentally [11].

The adsorption isotherms of the cationic surfactant HTAC (hexadecyl trimethylammonium chloride) on the oxide minerals SiO_2 and Al_2O_3 are S-shaped. The course of the isotherms can be partitioned into three segments (Fig. 3), as is frequently the case for the adsorption of ionic surfactants on oppositely charged surfaces. It is to be assumed that the adsorption in the first region is electrostatically controlled. With increasing surfactant concentration (stage II) the slope of the isotherms changes distinctly, which shows that the rising degree of coverage increases the affinity of the surfactants to the surface. The adsorption is mainly determined by the interactions between the alkyl chains of the surfactants. Surfactant aggregates can be formed on the surface, the so-called hemimicelles, due to the analogy of micelle formation in solution. In stage III, a further adsorption takes place by supplementing the adsorbate layer to bilayers [12].

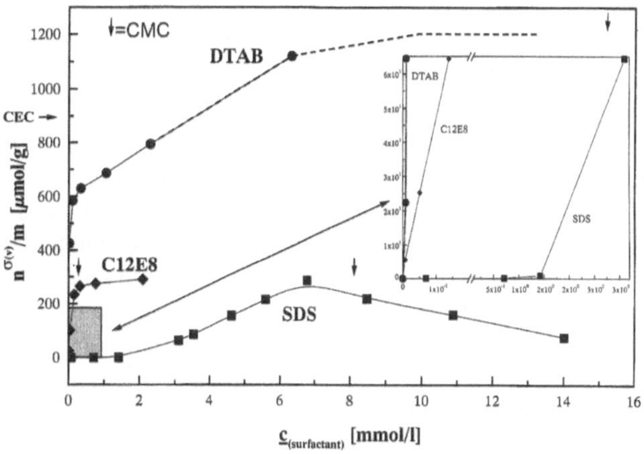

Fig. 1 Adsorption isotherms of different surfactants on Ca^{2+}-montmorillonite

With respect to their effect on the soil environment, the most studied surfactants are the nonionics [13]. For example, the sorption of four nonionic surfactants with a soil containing an organic content of 0.96% has been evaluated [14]. Sorption was observed to follow the Freundlich isotherm at concentrations below the CMC. The isotherms showed a plateau above the CMC. However, little has been reported on the sorption mechanism, as is frequently the case in studies with natural soil samples.

Recently, the adsorption of dodecyl octaethylene glycolether ($C_{12}E_8$) on different clay minerals has been compared [10]. The adsorption isotherms are of the L2 type (Fig. 1) and thus mostly based on the physisorption. The adsorption of C_nE_m type surfactants on clays probably takes place predominantly via the hydrated ethylene glycol groups due to ion-dipole interactions and hydrogen bridge bonding. The extent of adsorption strongly depends on the surface area of the respective layer silicate. For montmorillonites, it is confirmed by X-ray measurements that the nonionic surfactant penetrates into the layer interspace displacing water and forming bilayers [10].

Little is known about anionic surfactant adsorption, such as linear alkylbenzene sulphonate (LAS) in soil. It is assumed that this so-called "soft" detergent is easily degraded limiting its accumulation in soil and groundwater compartments [15, 16]. However, interaction studies between anionic surfactants and soil at high concentrations are also relevant since these are of significance in soil washing. Thus, a precipitation is assumed for the adsorption of anionic surfactants on the clay mineral surface. This mechanism was also proposed for the interactions of dodecyl sulphate ions (DS^-) with Ca^{2+}-kaolinite and Ca^{2+}-illite [17, 18]. The adsorption isotherm of DS^- on Ca^{2+}-montmorillonite is S-shaped (Fig. 1) [19]. Surface precipitation takes place after exceeding the solubility product. X-ray diffraction measurements have shown that precipitation only takes place on the outer montmorillonite surfaces.

While the organo-mineral complexes in relation to the soil aggregation and structure are discussed in considerable detail [20], only little has been published about surfactant effects on soil stability [13]. On the other hand, because of its practical importance for aspects such as flotation, the influence of surfactants on the dispersion stability of minerals has been more intensively studied [21, 22].

Adsorption of organic contaminants on surfactant-modified soil minerals

The uptake of nonionic organic contaminants (NOCs) by soils is strongly correlated with the soil organic matter content and is a partition process. The bonding of surfactants on soil has consequences similar to the immobilization of humic matter. The hydrophobization of the surface fundamentally changes the adsorption properties of the soil components, so that the adsorption of NOCs is favoured. This surface modification is therefore used in the application of surfactants in soil sealing for soil protection, but it also occurs due to their accumulation in soils and sediments after passing into the environment in different ways (e.g. as pesticide dispersing agents).

Recently, cationic surfactants have been used in order to enhance the sorption capacity of soil with respect to NOCs. The sorption characteristics of soils were modified by the addition of HDTMA (hexadecyl trimethylammonium) cations in an amount equivalent to the CEC of the soil [23, 24]. The adsorption enhancement is hundredfold and independent of the clay content in the respective soil. Comparing the organic matter normalized sorption coefficients K_{om}, it was found that the adsorbed surfactant is 10- to 30-fold more effective than natural soil organic matter and that the log K_{om} values of the HDTMA-treated soils closely agree with the corresponding log K_{ow} values (octanol-water partition coefficient).

As already discussed, the originally hydrophilic aluminosilicate surface becomes organophilic by replacing hydrated cations such as Ca^{2+} or Na^+ by cationic surfactants. The extent of enhanced NOCs adsorption depends on the degree of hydophobization of the layer silicate surface. It therefore increases with a growing surfactant load, an increasing surfactant chain length and the number of alkyl chains per surfactant molecule [25–29].

Recently, a systematic study was carried out with nitrophenol on alkyl and dialkyl ammonium/bentonite complexes (the bentonite consisted of 95% montmorillonite) [30]. The adsorption isotherms reveal that the enhanced adsorption of this relatively readily water-soluble substance only takes place from an exchanged amount of 32% of the CEC ("32% CEC") upwards in the case of alkyltrimethylammonium (C_n) derivatives. If the hydrophilic layer silicate surface is covered by a dialkyldimethylammonium ($2C_n$) derivative, this effect already occurs at "16% CEC" as expected.

The kinetic studies carried out with these systems have also confirmed this finding. It can be established that in addition to the exchanged amount of alkyl chains their orientation in the interlayer space also has an influence. On "32% CEC" $2C_{12}$ bentonite, the adsorption of 4-nitrophenol proceeds much more rapidly (half-life = 2.94 min) than for the corresponding "64% CEC" C_{12} bentonite (half-life = 8.57 min), although both bentonites are covered by the same amount of alkyl chain. On the one hand, the closeness of neighbouring alkyl chains seems to be important for adsorption enhancement. Furthermore, the sorption process is an intercalation in the interlayers, i.e. in addition to the structure of the surfactant layer other parameters also play a role, such as basal spacing and intercalated amount of water.

Thus, it was found that the adsorption of 4-nitrophenol on organo-bentonites leads to additional layer expansion and that this does not only depend on the volume of the adsorbed contaminant. Depending on the surfactant, nitrophenol intercalation produces a change in the structure of the adsorbed layer [30]. Figure 2 shows, as a function of the adsorbed nitrophenol amount, from which surfactant or alkyl chain amount onwards an enhanced layer expansion is to be observed. It can be clearly seen that at least 64% of the exchangeable ions must be replaced by C_{12} ions in C_{12} bentonite in order to observe an additional layer expansion due to the 4-nitrophenol adsorption. With a double-chained $2C_{12}$ only half of this amount is required, i.e. 32% of the CEC, for the same effect. The sudden change in basal spacing of the $2C_{12}$ bentonite,

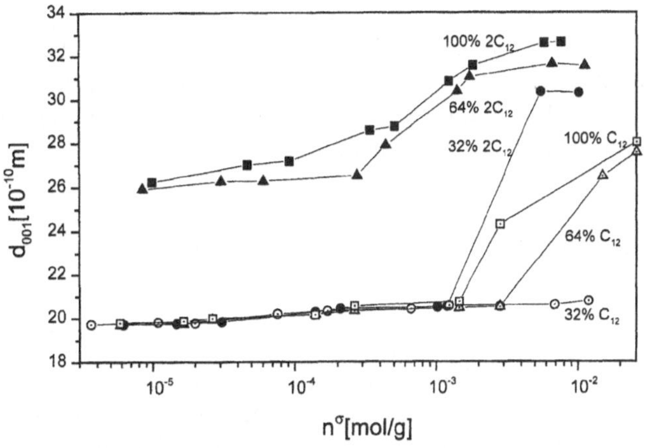

Fig. 2 Basal spacings of C_{12} and $2C_{12}$ bentonite complexes with different amounts of exchanged cationic surfactant as a function of adsorbed 4-nitrophenol amount [30]

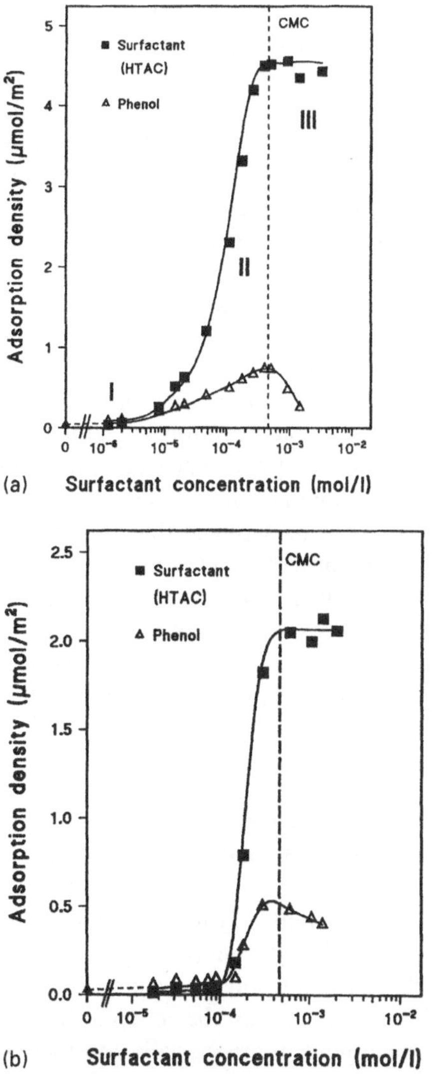

Fig. 3 (a) Mixed adsorption of hexadecyl trimethylammonium chloride (HTAC) and phenol on SiO_2 (pH = 6.3). (b) Mixed adsorption of HTAC and phenol on Al_2O_3 (pH = 9.6) [12]

which is overproportional to the contaminant adsorption, suggests a restructuring effect.

A recent study compared the influence of the cationic surfactant DTAB and the nonionic surfactant $C_{12}E_8$ on the sorption of biphenyl on four different layer silicates [10, 31]. Although the adsorption mechanism of the two surfactants is completely different at small surface coverages, the plateau values of the adsorption isotherms are of the same order of magnitude and the same layer expansion is to be observed in the case of swelling layer silicates. For the biphenyl adsorption, it was found that the organic carbon adsorption coefficient K_{oc} (the Henry coefficient is related to the organic carbon content of the adsorbent) for DTA^+ layer silicates is hardly influenced by the type of layer silicate. On the other hand, the K_{oc} values of $C_{12}E_8$ layer silicate complexes are generally different and higher. The reason is assumed to be the different surfactant layer structure based on the different adsorption behaviour of the various Na^+ clay minerals. The sorption mechanisms of $C_{12}E_8$ remain to be clarified. It is to be assumed that no classical monolayer formation takes place in this case and that a formation of surfactant aggregates is possible.

As can be seen from the above examples, knowledge of the surfactant adsorption plays a crucial role in understanding the possible interactions between the adsorbates in the surfactant/nonionic contaminant/soil (soil component) system. The mechanisms of surfactant adsorption seem to be better clarified for oxides (see the previous subsection).

The adsorption behaviour of mixtures of cationic surfactant HTAC and phenol on SiO_2 and α-Al_2O_3 is shown in Fig. 3 [12]. In these experiments, the initial phenol concentration was kept constant. The general shapes of surfactant isotherms (see the previous subsection) are not affected by the addition of phenol, but the isotherms are shifted to lower concentrations and the CMC is lowered.

The course of the isotherms is very similar in both systems (Fig. 3a and b). At low surfactant concentrations, only a weak increase in surfactant adsorption was observed with increasing equilibrium concentration. The influence of HTAC on phenol adsorption is not significant either. The phenol molecules can either be adsorbed on the mineral surface or on the surfactant anchors. It must be mentioned here that in the case of alcohols a clearly enhanced alcohol adsorption was also found in this concentration range [32].

With growing surfactant concentration (region II, c_T > hmc), the large increase in surfactant adsorption leads to a strongly enhanced phenol adsorption. Phenols are co-adsorbed in the surfactant surface aggregates, which can be present both as monolayers and as bilayers. This process can be interpreted as adsolubilization [12]. In region III (Fig. 3a and b), above the CMC, the surfactant adsorption density reaches a plateau and there is a decrease in phenol adsorption. The growth of the number of surfactant micelles leads to an increased solubilization of phenol in the micelles. This implies that the decrease in phenol adsorption may be considered to be a result of the competition between the solubilization of phenol in the adsorbed surfactant layer and in the micelles [12, 28, 29, 31].

On the other hand, in the case of the anionic surfactant SDS an opposite effect on the sorption of NOCs on clay minerals can be observed [33]. Indeed, at low SDS surface coverages on Ca-montmorillonite, the sorption of s-triazines is reduced. It can be assumed that this occurs due to the relatively high concentration of SDS molecules in the bulk, which specifically interact with the s-triazines in solution. At higher surface coverages of SDS, a relative enhancement of NOC adsorption is found, which is proportional to the assumed surface precipitation of $Ca(DS)_2$ (see the previous subsection).

Different features of the surface modification of soil inorganic components by water-soluble synthetic polymers

Some features of the water-soluble synthetic polymers at water/soil interfaces are reviewed from two aspects. Thus, the parameters which control the polymer adsorption itself and the impacts of their presence on the colloidal behaviour of soil inorganic components are specially considered.

In a general way, the following points are of interest in the study of the water-soluble synthetic polymers behaviour in the soil environment:

(1) The description of the *adsorption behaviour* of synthetic water-soluble polymers at soil inorganic components is a prerequisite for the study of the polymer transport mechanism into the soil. This investigation allows the retarding factors to be appreciated which impede a free polymer flow into the groundwater. *Structural conformations* adopted by the adsorbed polymer influence its *desorption* and may also drive its *biodegradation.*

(2) Polymer effects on the *transport of other soil substances* are also envisaged in the case of an *adsorption concurrence* for the same adsorption sites at soil inorganic components. On the other hand, this modified surface soil components by polymers can also enhance an "*absorption*" of other soil substances in the *adsorbed polymer layer*. In addition to this direct effect in the transport feature of soil substances, other indirect effects resulting from the modification of the *soil hydraulic conductivity* also must be considered. Indeed, water-soluble polymers can change the *aggregate state* of the soil. This modification of the soil colloid stability affects the water transport or air exchanges leading to the flows of soluble or suspending soil substances being influenced.

Adsorption of synthetic polymers at soil inorganic components

The interactions of synthetic polymers with soil inorganic particles are mainly based on their adsorption behaviours. Depending on the extent and the nature of the ionic character of the polymer, various interaction modes can be described. In a rapid overview extending from more general colloidal properties in the aqueous phase to specific chemical interactions at the inorganic particle surface, the different forces driving the polymer adsorption can be thus listed [34–37].

In a general way, depending on the pH and the electrolyte composition, an aqueous phase with a low solubility property favours polymer accumulation at the particle interface. Long-range interactions between polymers and particle surfaces have been intensively treated in the DLVO theory in the case of Van der Waals attraction forces counterbalanced by double-layer electrostatic repulsion forces emerging from ionic polymer and inorganic particle surfaces. Short-range interactions of a more specific type, like H-bridging and inner-metal surface complexation, require a dual compatibility between functional groups from polymers and the particle surface. The net result of the adsorption forces makes a polymer anchoring at the solid surface possible through a multi-site binding process. This cooperativity factor can give a high affinity or "irreversible" character to the interaction by shifting the equilibrium polymer concentration to very low values.

Depending on the polymer conformation at the surface, this polymer adsorption more or less modifies the chemical and physical surface properties of the inorganic soil colloids. Indeed, it can be stated that a close-contact adsorption between the the polymer and the particle surface through train segments is a prerequisite for a possible chemical modification of surface functional groups while loops and end segments in the adsorbed layer located at some distance from the particle surface would modify its long-range interactions of a more physical nature.

Adsorption behaviour of nonionic and anionic polymers at clay minerals

The relative importance of the electrical properties of clay mineral surfaces can be illustrated by the adsorption behaviour of two synthetic polymers carrying different functional groups under electrolyte conditions close to soil solution ($I = 0.01$ M). Polyvinylpyrrolidone (PVP) and acrylic–maleic acid (PAA-PMLA) copolymer nonionic and anionic polymers, respectively, have found several industrial applications, for example as ingredients in newly developed detergents. Though these polymers are nontoxic, their low biodegradability [38–40] has required investigations on potential mobility in soils, especially in terms of adsorption- desorption processes at clay minerals, [41–45].

In Fig. 4, the adsorption isotherms of PVP and PAA–PMLA copolymer at montmorillonite and kaolinite, two different clay minerals, are reported [42–45]. It can be seen that the maximal adsorbed amount of PVP is always the highest. However, the difference in the adsorption level between the two polymers is lower at the kaolinite surface than at montmorillonite. Taking into account the electrical surface properties of both basal and edge surfaces of clay minerals as well as their respective surface area, this adsorption behaviour can be explained as follows. First, it can be stated that PVP, a nonionic polymer, adsorbs at the total surface area of both clay minerals and thus reflects, a priori, their maximal binding capacity independently of the presence of superficial charges. In the case of PAA–PMLA copolymer, it has been demonstrated that the adsorption only takes place at edge surface sites where positively charged aluminol groups favour the adsorption of the polymer with an anionic character. Indeed, in neutral and weak acid pH regions, an electrostatic repulsion due to the ionization of the carboxylic groups impedes a binding to the negatively charged basal planes. The ratio of the edge surface to the total surface area is thus a determining factor for the difference in the binding extent of PAA–PMLA copolymers at kaolinite and montmorillonite. Thus, it can be shown that the ratios of adsorbed PAA–PMLA copolymers to PVP polymer amounts at the two clay minerals give satisfactory results for the ratios of edge surface to the total surface area, i.e. $\approx$ 1–2% and 15–20% for montmorillonite and kaolinite respectively.

Fig. 4 Adsorption isotherms of PAA–PMLA copolymer and PVP on two different clay minerals at pH 4.5–5.5 in 0.01 M NaCl. A: montmorillonite; B: kaolinite; clay minerals 5 g/l. □–□, PVP; ▲–▲, PAA–PMLA copolymer

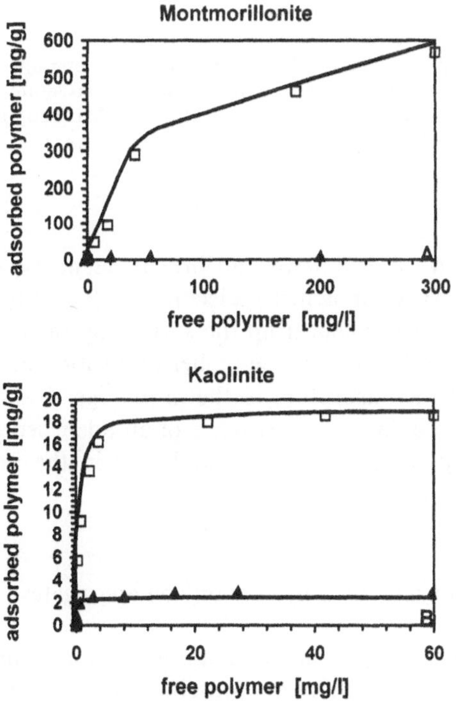

Conformation of the polymer adsorbed layer

In a general way, the adsorption behaviour of polymers has been modelled in terms of loop-train-tail conformations at the surface. At low surface coverage, the polymer chains lie flat. The fraction of train segments in the adsorbed layer is very high. At high surface coverage, the relatively large increase of the hydrodynamic thickness of the polymer layer (δ_H) in comparison with the adsorbed amount is assumed to be determined by the fraction of segments in loops and more particularly in extended tails [35, 36]. In Fig. 5, the δ_H data obtained with different PVP molecular weights [44] at kaolinite are plotted against the corresponding adsorbed amounts. As predicted in the mean-field theory treatment of Scheutjens and Fleer, the results show no dependence of δ_H in a molecular weight

Fig. 5 Effect of the adsorbed amount on the hydrodynamic layer thickness (δ_H) for different PVP molecular weights. ■–■, M_w 5000 g/mol; □–□, M_w 10 000 g/mol; ○–○, M_w 24 000 g/mol; ●–●, M_w 44 000 g/mol; ▲–▲, M_w 245 000 g/mol; ◆–◆, M_w 400 000 g/mol; △–△, M_w 600 000 g/mol [44]

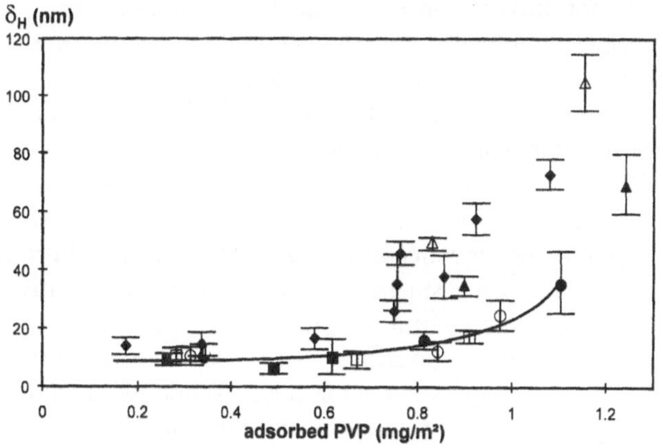

range from 5000 g/mol to 600 000 g/mol, at least up to an adsorbed amount of 0.6 mg/m^2. Furthermore, this observation also holds for molecular weights $\leq$ 44 000 g/mol up to an adsorbed amount of about 1 mg/m^2. A "universal" character regardless of the molecular weight is found for the dependence of δ_H on the adsorbed amount which is adjusted by varying the PVP molecular weight and/or the concentration. However, increasing deviations from the "universal" relationship at high adsorbed amounts are observed for the highest PVP molecular weights. This behaviour could be related to an increase of the solvency of PVP samples. From the experimental results [44] for the molecular weight dependence of δ_H, measured here by a microelectrophoretic method, it can also be added that the fraction of segments in coil structure or loops at the surface would principally determine the thickness of the PVP adsorbed layer.

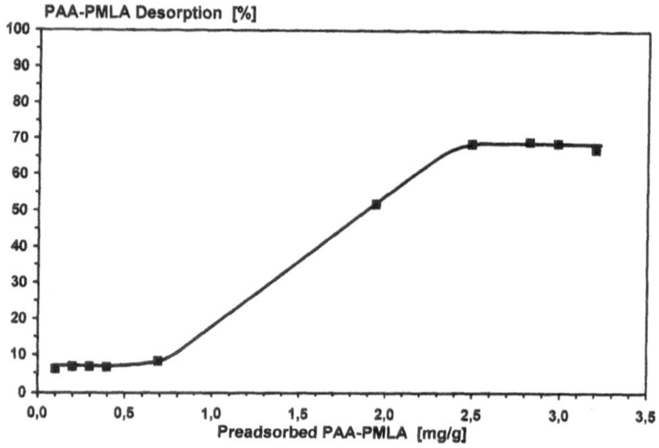

Fig. 6 Desorption percentage of PAA–PMLA copolymer in the presence of 150 mM phosphate equivalent STP as a function of initial preadsorbed PAA–PMLA copolymer amount on kaolinite at pH 4.5–5 in 0.01 M NaCl. Kaolinite 5 g/l, [43]

Desorption behaviour of polymer

In the case of the PAA–PMLA copolymer, it can be established that the previously described general conformation for the adsorbed polymer also determines its desorption behaviour. Taking into account the chemical binding of the polymer carboxylic groups through a ligand exchange mechanism at the edge sites of kaolinite clay mineral, a desorption of the PAA–PMLA copolymer is only possible in the presence of displacer molecules such as phosphate derivatives which react in a similar way with the surface [42, 43].

In Fig. 6, desorption experiments with a fixed maximum of sodium tripolyphosphate (STP) concentration of 50 mM or 150 meq/l, but with various adsorbed amounts of PAA–PMLA copolymer along the entire adsorption isotherm (Figs. 4 and 7), also indicate the influence of the *adsorbed polymer conformation* on the desorption. The observed percentages of PAA–PMLA copolymer desorption are plotted against the amount of copolymer initially preadsorbed. The percentage of desorption is found to be lowest for surface concentrations lower than 0.7 mg/g of adsorbed copolymer. This is followed by a rise in the percentage of desorption up to a constant value of about 70% under adsorption saturation conditions. The initial almost nondesorbing region of up to 0.7 mg PAA–PMLA/g kaolinite indicates that PAA–PMLA copolymer is strongly bound to the surface. Thus, at very low PAA–PMLA copolymer concentration in the interfacial solution region, it has been suggested that a rather quantitative saturation of the aluminol binding sites can be achieved through a flat adsorption by train segments of single PAA–PMLA copolymer molecules. While with more concentrated polymer solution, the high desorption can be related to the fewer adsorbed segments per one PAA– PMLA copolymer molecule thus facilating a more effective displacement from the surface by the competing STP. In the latter case, the higher desorption in the final region of the adsorption isotherm in Figs. 4 and 7 thus indirectly confirms the existence of an adsorbed polymer layer containing more loops and tails in the interfacial region.

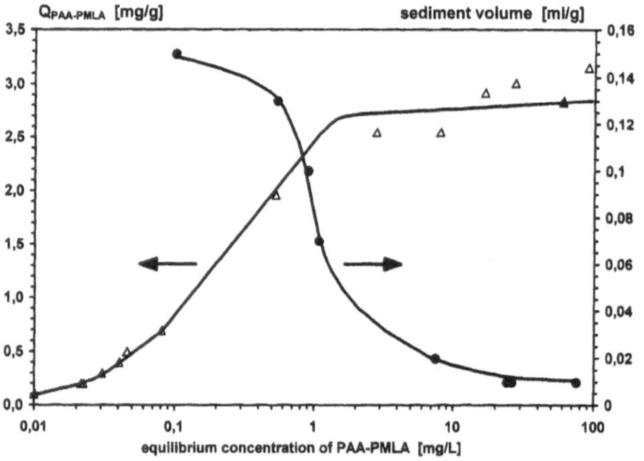

Fig. 7 Semi-logarithmic plot of the adsorbed amount of PAA–PMLA copolymer and the sediment volume of kaolinite as a function of the equilibrium concentration of PAA–PMLA copolymer (see other conditions in Fig. 4). △–△, adsorption isotherm $Q_{PAA-PMLA}$; ●–●, sediment volume [43]

Effects of synthetic polymers on soil aggregates

Porous structures consist of inorganic soil components which are stabilized by the bridging action of multivalent cations, clay materials and organic macromolecules such as polysacharride and humic acids. This air- and water- filled soil matrix is essential for the biological activity of the soils. Indeed, the variously sized capillary, channel or slit structures in the soil aggregates determine the remaining or the further flow of dissolved and suspended essential nutritients for the microbiological and vegetal lives. The formation of these structures is primarily based on physicochemical interactions which are controlled by the colloidal properties of the soil components.

Thus, processes like aggregation or flocculation are mainly driven by the variation of the electrical and chemical properties of the contacting particle surfaces. Aggregation phenomena are generally defined as the result of a screening of the double layer electrostatic repulsion by increasing the salt concentration or its multivalent nature. Flocculation phenomena are more restrictedly assigned to the "aggregation" in the presence of organic macromolecules. It results in more or less loosely stabilized structures which are designated as the soil aggregates.

The importance of this colloidal behaviour with clay minerals in a soil porous medium may be described by pore clogging driven by swelling and dispersing or deflocculation processes.

Colloidal behaviour of polymer-modified inorganic soil components

The effects of surface modification by polymers can be roughly described as bringing inorganic soil particles in close contact -flocculation- or on the contrary as maintaining them apart in solution–dispersion, [34–36, 46–48]. In a general way, these phenomena are first determined by the ionic state and the colloidal size of the polymer. The dosage ratio of polymer to surface area particles is another parameter which conditions the heterogeneous character of the interaction as well as the screening dimension of the adsorbed surface modification.

In the case of interactions between polymers and soil colloid surfaces of unlike charge, see for example anionic polycarboxylate and positively charged hematite or cationic hydroxy-Al, -Fe polymers and negatively charged clay minerals, a rather quantitative surface charge neutralization at low polymer concentrations screens the electrostatic stability of the system and induces a flocculation. This effect is further enhanced by a bridging mechanism between particle surfaces through the binding of large-sized polymers. Thus, high-molecular-weight polymers or an extended configuration of the ionized polymer under low electrolyte conditions can favour this process. However, at high polymer concentrations, a redispersion of the particles is often observed, which can be attributed to a newly gained electrostatic stabilization due to adsorbed polymer layer. Indeed, an excess of ionized groups in the adsorbed layer which overcompensates the surface charge density can be related to a more coiled structure adopted by the adsorbed polymer at high polymer/binding site surface concentration ratios. A steric stabilization due to an unfavourable entropic term in the intraminglement energy between interacting adsorbed polymer layer is also found with dispersing polymers when the nonionic character of the polymer is pronounced.

In the case of clay minerals in dilute electrolyte solution, the well-established presence of a dual charge, i.e. the permanent negative charge in the face surface and the pH-dependent positive charge located at the broken edge surface, is responsible for edge-face aggregations between clay minerals particles as usually illustrated by the "house-of-cards" structure. They also play a determinant role in the interactions with ionic polymers. In the case of polycarboxylic acids, such as the PAA–PMLA copolymer, ligand exchange reactions of carboxylate groups take place at the positively charged aluminol groups of the edge surface. A consequence of this localized adsorption is the dispersion of the "house-of-cards" aggregates. In Fig. 7, a clear concomitant decrease of the sediment volumes due to the edge-face aggregates with the adsorption of PAA–PMLA copolymer supports a superequivalent adsorption of carboxylate groups at aluminol sites on the edges [43]. An increase of the negative overall charge of clay minerals leads to the dispersion of the clay suspension. Thus, more parallelly oriented negatively charged kaolinite particles give less voluminous sediment volumes.

In the case of interactions between similarly charged polymers and soil colloid surfaces, it can be generally stated that interactions are rather weak under low electrolyte conditions where according to the DLVO theory the electrostatic repulsion forces are well developed. However, under soil conditions, an important contribution of polyvalent cations such as Ca, Al and Fe must be considered in the adsorption of organic polyanions at negatively charged soil colloid surfaces. Thus, depending on the metal speciation in soil electrolyte solution, an aggregation mechanism must be envisaged where polyvalent cations play bridging effects between similarly charged polymers and colloid surfaces.

Significance of natural macromolecules for soil processes

Natural macromolecules in the soil

Natural macromolecules important with regard to soil processes are substances produced by animals and plants or their by-products. Among others, polynucleotides (nucleic acids), polypeptides (proteins), polysaccharides, lignin, and polyisoprenes belong to this group of polymers, which are also called biopolymers. The main important degradation products in the soil are the humic substances. They differ from biopolymers because they are not built up of uniform, recurring monomers.

The natural macromolecules are interesting for soil chemistry, because e.g. of their surface active behaviour. This holds for polysaccharides and humic acids in particular. Thus, hydrophobic areas (carbohydrate framework) and hydrophilic areas (phenolic and carboxylic functional groups) are linked in one molecule. The macromolecules adsorb at the solid/liquid interfaces in the soil, e.g. at the mineral components (layer silicates, oxides) and change the surface properties of these compounds. Here, polysaccharides and humic acids play a crucial role because of their importance for the aggregation and stability of the soil [49].

Polysaccharides

Polysaccharides are macromolecules built up by linking small sugar units together through glycosidic bonds. The basic modules, the monosaccharides, are either polyhydroxy aldehydes or polyhydroxy ketones with mainly five or six carbon atoms in a linear carbon skeleton. These five- and six-membered skeletons can form cyclic hemiacetals by intramolecular linking of the carboxylic group with the hydroxylic group of the same sugar molecule. Polysaccharides can be composed of one or more kinds of monosaccharides and these modules can be arranged in either a linear or a branched fashion. Because they are not genetically determined like proteins, they are usually not monodisperse, which means that molecules of a given sort of polysaccharide can differ in the degree of polymerization, in proportions of monosaccharides, in proportion of linkage types, or in distribution of side chains [50]. Carbohydrates and above all polysaccharides are important substances in soil. Depending on the soil type, they represent between 5 and 25% of the organic matter.

Humic substances

A definition of "humic substances" is not possible in the normal chemical or biological terminology. These substances do not correspond to a unique chemical entity and, accordingly, cannot be described in clear structural terms, but they can be defined operationally [51]:

"Humic substances" are a general category of naturally occurring, biogenic, heterogeneous organic substances that can generally be characterized as being yellow to black in colour, of high molecular weight, and refractory. Three fractions of humic substances can be defined in terms of their solubility. The "humin" fraction is not soluble at any pH value, "humic acids" are not soluble in water under acidic conditions (<pH 2), but become soluble at greater pH, and "fulvic acids" are soluble under all pH conditions. This classification in terms of pH-dependent solubility is only one possibility of defining different fractions of humic substances [52], but it is a relatively effective criterion because it depends on important chemical characteristics, like acidic functional groups, molecular weight, aromaticity, and so on. Thus, the molecular weight of fulvic acid is very much smaller than that of humic acid (fulvic acid: 500–2000 g/mol; humic acid: 50 000–100 000 g/mol, with a few molecules up to 250 000 g/mol) [2].

The formation cannot be described by one concept, nor is any common chemical structure known for fulvic and humic acids. However, it is possible to define some typical functional groups which characterize the properties of humic substances. For example, the reactivity is due to a large extent to their high content of oxygen-containing groups, including carboxyl groups, phenolic- and/or enolic-OH, alcoholic OH, and the C=O double bond of quinones, hydroxyquinones, and α,β-unsaturated ketones. Absolute values for the various groups in different types of humic substances cannot be given quantitatively, but there are some tendencies. The content of oxygen-containing functional groups in fulvic acids is higher than for any other naturally occurring organic polymer, the total acidity (COOH and phenolic OH) of fulvic acids is higher than for humic acids and more oxygen in the humic acids occurs in ether linkages [2].

Natural macromolecules and soil structure

The influence of natural macromolecules on the soil structure shows the importance of surface modification for living systems. Plant growth depends on the arrangement of soil particles into stable secondary particles or aggregates, because the quality of the aggregates directly influences such important soil properties as aeration, water penetration and retention, mechanical impedance to roots, and emergence of shoots. It has been known for many decades that the organic matter in the soils is responsible for the formation and stabilization of soil aggregates. Above all soil polysaccharides and humic substances are important [53–63].

The starting point for investigations on soil structure was the assessment of the degree of aggregation in many different types of soils. A correlation between the total organic matter content of the soils and the degree of aggregation was found. It was also shown that microorganisms incubated in previously sterilized soil have an increasing effect on the aggregation [59]. These experiments and those that followed led to the conclusion that extracellular polysaccharides produced by many different organisms have binding properties for soil particles. However, the correlation between the content of carbohydrates and the aggregation can also be interpreted in the opposite way: good aggregation may encourage the growth of polysaccharide-producing organisms [64]. So a direct chemical treatment of soil aggregates was needed to test the involvement of polysaccharides. The main technique for modifying soil hydrocarbons is oxidation with periodate. Oxidized polysaccharides are unstable under alkaline conditions. Comparing the stability of the soil structure of untreated and oxidized material shows the significance of polysaccharides for aggregation. It was shown that there are differences between synthetic aggregates made with soil polysaccharides and natural aggregates. Synthetic aggregates were completely destroyed but natural aggregates remained stable [64]. Periodate oxidation and other chemical reactions with natural and synthetic aggregates and the finding that cultivation influences the effects of the chemical treatments [65] showed that only a part of the structural stability of soil aggregates depends on polysaccharides. Other substances identified as responsible for stable aggregates are fungal mycelia, mucilage and mucigel of roots, humic substances, and polyvalent cations [61]. Two stages in the formation of a stable soil structure have been postulated. In the first stage, microbial gums (polysaccharides) act as aggregating agents, and in the second stage, more complex material (mucigel and humic substances) stabilizes these aggregates. Incubation experiments show the comparative effect of polysaccharides and humic substances [66]. The addition of glucose and two microbially produced exocellular polysaccharides, xanthan gum and alginate, to a soil in which the natural structure had been destroyed by mechanical crushing produced stabilized, reformed aggregates. However, there are differences in the mechanism. The stabilizing effects of the polysaccharides are directly due to the binding action of these compounds the mineral soil particles. On the other hand, the glucose does not act directly as a binding agent. The results obtained from the glucose treatment are due to the production of exocellular polysaccharides by microorganisms as a result of metabolizing the added glucose. This in situ production of polysaccharides caused more stable aggregates than by merely adding polysaccharides.

For the investigations with humic substances monoionic (Ca^{2+}-)soils were produced, in which the aggregates had been destroyed by ion-washing procedures. No stable aggregates could be made by adding carbohydrates to these soils. The effect of humic substances depends on the incubation technique. The addition of humic substances as a simple physical mixture is also ineffective, but the addition of humic substances through carefully controlled surface adsorption processes [63] produces stable aggregates and the effect is more stable than the effect of the carbohydrates with the crushed samples. The best aggregating effect was achieved with a combination of adsorbed humic substances together with glucose.

These experiments [62, 63, 66] show that polysaccharides are capable of producing stable aggregates, but that the effect is transitory and declines as the polysaccharides are decomposed. Adsorbed humic acid also produces stable aggregates and the effect is more persistent. The increased persistence is due to the relative resistance of humic acid to biological decomposition in the soil compared to polysaccharides. The most stable and persistent reformed aggregates are produced by the combined action of polysaccharides and humic acid. In all instances the involvement of soil microorganisms is an essential ingredient of the aggregate formation and stabilization process.

Humic acid effects on the colloidal behaviour of inorganic soil compounds

As also reported in the case of synthetic polycarboxylate, the surface charge of the colloidal particles characterizes their structural behaviour. In the soil, mainly humic acids affect the surface charge of the inorganic compounds. For example, the colloidal stability of aluminium oxide can be influenced by the adsorption of humic acid [67]. During adsorption the surface charge of the oxide is changed from positive to negative (Fig. 8b). The amount of humic acid that is needed to balance the surface charge of the oxide also causes the greatest sediment volume (Fig. 8a). If the surface is charged (negatively or positively) there are repulsive forces between the particles and no solvent will be entrapped. The sediment is closely packed. The situation is different for a surface that is electrically neutral. There are no repulsive forces and the particles can already start to aggregate in dispersion. The preformed flakes deposit and built a loosely packed sediment with solvent molecules in the cavity between the particles.

Besides the organic and biological compounds, inorganic polyvalent cations have an important influence on the aggregation of soil particles [68–70]. The ions act as bridges between the anionic groups of the polymeric organic matter and the negatively charged surfaces of clay particles. Arias et al. examined the effects of associations between humic acids and iron or aluminium on the

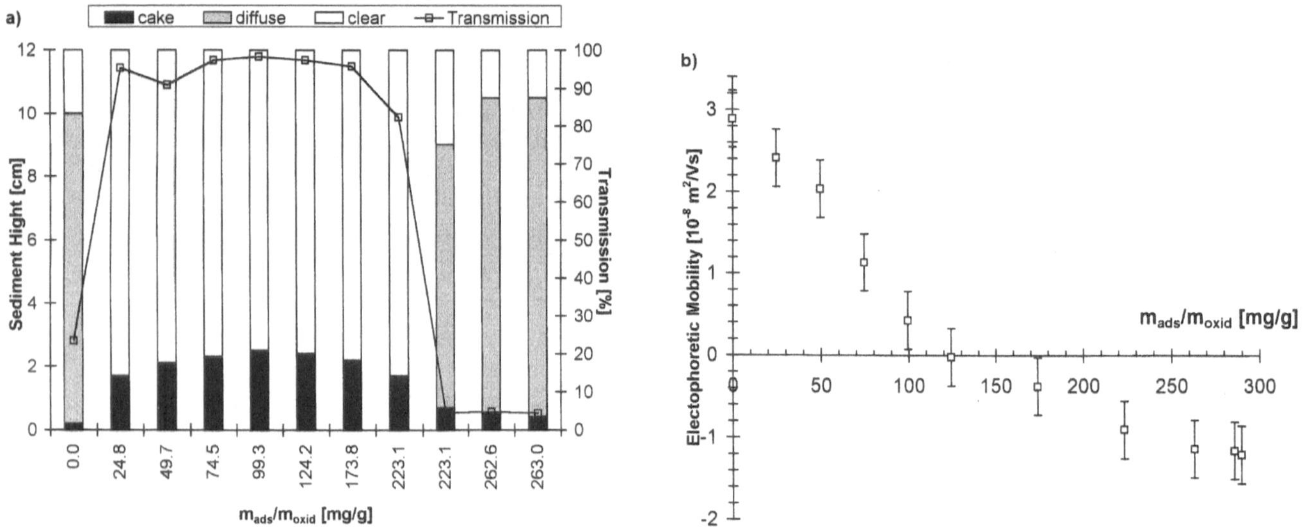

Fig. 8 Sediment height (a) and electrophoretic mobility (b) of Al_2O_3 with different amounts of humic acid adsorbed (pH ≈ 5; electrolyte: 0.1 M NaCl)

flocculation and aggregation of kaolin and quartz [71]. They treated suspensions of kaolin or quartz with metal or humic acid or both. In the absence of added metal ions the proportion of humic acids retained by the substrate decreased as the humic acid dose increased. In the presence of Fe or Al, the proportion of humic acids removed from the solution increased and was always >90%. Mainly the aluminium influences the aggregation effect of humic acid. Compared with untreated kaolin, humic acid alone narrowed the pH-interval where the kaolin flocculates. Associations between humic acid and aluminium widened the flocculation interval and induced the formation of aggregates greater than 100 μm. Associations with iron barely affected the pH-interval.

References

1. Kuhnt G, Knief K (1990) Ökologisches Verhalten von Tensiden in Böden, UBA-Report, Kiel
2. Stevenson FJ (1994) Humus Chemistry: Genesis, Composition, Reactions. Wiley, New York
3. Larson RJ, Vashon RD (1983) Dev Ind Microbiol 24:425–434
4. Brownawell BJ, Chen H, Collier JM, Westall JC (1990) Environ Sci Technol 24(8):1234–1241
5. Law JP, Kunze GW (1966) Soil Sci Soc Am Proc 30:321–327
6. Schwarz R, Heckmann K, Strnad J (1988) J Colloid Interface Sci 124:50–56
7. Bijsterbosch BH (1974) J Colloid Interface Sci 47:186–198
8. de Keizer A (1990) Progr Colloid Polym Sci 83:118–126
9. Mehrian T (1992) PhD Dissertation, University of Wageningen
10. Rheinländer T (1993) PhD Dissertation, University of Düsseldorf
11. Boyd SA, Jaynes WF, Ross BS (1991) In: Baker RA (ed) Organic Substances and Sediments in Water, Vol 1, Lewis, Chelsea
12. Schieder D, Dobias B, Klumpp E, Schwuger MJ (1994) Colloids and Surfaces 88:103–111
13. Piccolo A, Mbagwu JSC (1989) Soil Sci 147(1):47–54
14. Liu Z, Edwards DA, Luthy RG (1992) Water Res 26(10):1337–1345
15. Holt MS, Bernstein SL (1992) Wat Res 26(5):613–624
16. Larson RJ, Federle TW, Shimp RJ, Ventullo RM (1989) Tenside Surfactants Detergents 26(2):116–121
17. Hanna H, Somasundaran P (1979) J Colloid Interface Sci 70:181–187
18. Siffert B, Zundel J (1985) Clay Min 20:189–194
19. Ilic M, Gonzalez J, Pohlmeier A, Narres HD, Schwuger MJ (1996) Colloid and Polymer Sci 274(10):966–973
20. Emerson WW, Foster RC, Oades JM (1986) In: Huang PM and Schnitzer M (eds) Interactions of Soil Minerals with Natural Organics and Microbes, Soil Sci Soc America, Madison
21. Moudgil BM, Soto H, Somasundaran P (1987) In: Somasundaran P, Moudgil BM (eds) Reagents in Mineral Technology, Marcel Dekker, New York
22. Dobias B (1993) In: Dobias B (ed) Coagulation and Flocculation, Theory and Applications. Marcel Dekker, New York
23. Lee JF, Crum JR, Boyd SA (1989) Environ Sci Technol 23:1365–1372
24. Boyd SA, Lee JF, Mortland MM (1988) Nature 333:345–347
25. Boyd SA, Mortland MM, Chiou CT (1988) Soil Sci Soc Am J 52:652–657
26. Boyd SA, Shaobai S, Lee JF, Mortland MM (1988) Clays and Clay Minerals 36(2):125–130

27. Smith JA, Jaffè PR, Chiou CT (1990) Environ Sci Technol 24(8):1167–1172
28. Klumpp E, Heitmann H, Lewandowski H, Schwuger MJ (1992) Progr Colloid Polym Sci 89:181–185
29. Klumpp E, Heitmann H, Schwuger MJ (1993) Colloids and Surfaces 78:93–98
30. Heitmann H (1996) PhD Dissertation, University of Dortmund
31. Rheinländer T, Klumpp E, Schwuger MJ (1998) J Dispersion Sci Techn 19(2–3):379–398
32. Monticone V, Treiner C (1994) J Colloid Interface Sci 166:394–403
33. Herwig U, Klumpp E, Narres HD, (1998) submitted
34. Theng BKG (1979) In: Formation and Properties of Clay–Polymer Complexes. Elsevier, Amsterdam
35. Cohen Stuart MA, Fleer GJ, Lyklema J, Norde W, Scheutjens JMHM (1991) Adv Colloid Interface Sci 34:477–535
36. Fleer GJ, Cohen Stuart MA, Scheutjens JMHM, Cosgrove T, Vincent B (1993) In: Polymers at Interfaces. Chapman & Hall, London
37. Séquaris J-M (1997) In: Schwuger MJ (ed) Detergents in the Environment, Surfactants Sciences Series, Vol 65, Ch 7. Marcel Dekker, New York
38. Opgenorth H-J (1992) In: Hutzinger O (ed) The Handbook of Environmental Chemistry,Vol 3, Part F. Springer, Berlin, pp 337–350
39. Langbein I (1997) In: Schwuger MJ (ed) Detergents in the Environment, Surfactant Science Series, Vol 65, Ch 8. Marcel Dekker, New York
40. Hamilton JD, Reinert KH, Freeman MB (1994) Environ Sci Technol 28: 187A–192A
41. Sastry NV, Séquaris J-M, Schwuger MJ (1995) J Colloid Interface Sci 171: 224–233
42. Blockhaus F (1996) PhD Dissertation, University of Düsseldorf
43. Blockhaus F, Séquaris J-M, Narres HD, Schwuger MJ (1997) J Colloid Interface Sci 186:234–247
44. Hild A, Séquaris J-M, Narres HD, Schwuger MJ (1997) Colloids Surfaces A 123–124:515–522
45. Hild A (1998) PhD Dissertation, University of Düsseldorf
46. Gu B, Doner HE (1993) Soil Sci Soc Am J 57:709–716
47. Ben-Hur M, Malik M, Letey J, Mingelgrin U (1992) Soil Sci 153:349–356
48. De Boodt MF (1990) In: De Boodt MF, Hayes MHB, Herbillon A (eds) Soil Colloids and Their Associations in Aggregates, Ch 8. Plenum Press, New York
49. Haynes RJ (1992) Soil Biochemistry. In: Nierenberg WA (ed) Encyclopedia of Earth System Science. Academic Press, San Diego, pp 209–218
50. Lehmann J, Rapp K (1986) Carbohydrates. In: Ullmann's Encyclopedia of Industrial Chemistry, Vol A5. Verlag Chemie, Weinheim
51. Aiken GR, McKnight DM, Wershaw RL (1985) Humic Substances in Soil, Sediment, and Water. Wiley, New York
52. Ziechmann W (1980) Huminstoffe. Verlag Chemie, Weinheim
53. Browning GM (1938) Soil Sci Soc Am Proc 2:85–96
54. Browning GM, Milam FM (1944) Soil Sci 57:91–106
55. Woodruff CM (1940) Soil Sci Soc Am Proc 4:13–18
56. Alderfer RB, Merkle FG (1941) Soil Sci 51:201–212
57. McCalla TM (1942) Soil Sci Soc Am Proc 7: 209–214
58. McHenry JR, Russel MB (1944) Soil Sci 57:351–357
59. Martin JP (1945) Soil Sci 59:163–174
60. Martin JP (1946) Soil Sci 61:157–166
61. Cheshire MV (1979) Nature and Origin of Carbohydrates in Soils. Academic Press, London
62. Chaney K, Swift RS (1986) J Soil Sci 37: 329–335
63. Chaney K, Swift RS (1986) J Soil Sci 37: 337–343
64. Mehta NC, Sreuli H, Muller M, Deuel H (1960) J Sci Food Agric 11:40–47
65. Greenland DJ, Lindstrom GR, Quirk JP (1962) Soil Sci Soc Am Proc 26: 366–371
66. Swift RS (1991) Effects of humic substances and polysaccharides on soil aggregation. In: Wilson WS (ed) Advances in Soil Organic Matter Research: The Impact on Agriculture and the Environment. Redwood Press, Melksham, pp 153–162
67. Vermöhlen K (1998) PhD Dissertation, University of Düsseldorf
68. Edwards AP, Bremmer JM (1967) J Soil Sci 18:64–73
69. Mortland MM (1970) Adv Agron 22: 75–117
70. Tisdall JM, Oades JM (1982) J Soil Sci 33:141–163
71. Arias M, Barral MT, Diaz-Fierros F (1996) Eur J Soil Sci 47:335–343

Progr Colloid Polym Sci (1998) 109:214–220

I. Regdon
Z. Király
I. Dékány
G. Lagaly

Microcalorimetric studies of S/L interfacial layers: thermodynamic parameters of the adsorption of butanol–water on hydrophobized clay minerals

Received: 2 September 1997
Accepted: 15 September 1997

I. Regdon · Z. · Király · I. Dékány (✉)
Deparment of Colloid Chemistry
Attila József University
Aradi v.t. 1
H-6720 Szeged
Hungary
E-mail: i.dekany@chem.u-szeged.hu

G. Lagaly
Institute of Inorganic Chemistry
University of Kiel
D-24098 Kiel
Germany

Abstract Interaction of hydrophobized surfaces with mixtures of water and organic solvents is important in many practical applications. Alkylammonium clay minerals are excellent models for studying such processes. An example is the adsorption of *n*-butanol–water on vermiculite primed with dodecyl-, octadedcyl- and dodecyldiammonium ions. The thermodynamic functions $\Delta_{21}G^{s}$, $\Delta_{21}H^{s}$, $\Delta_{21}S^{s}$ for the adsorption of butanol from water at the interfaces were obtained from the reduced specific surface excess values and the calorimetric heat of displacement. They reveal the strong influence of the alkyl chain length. Two contributions are important: the enthalpy change due to the rearrangement of the alkyl chains from flat to paraffine-type orientation and the gain of entropy due to conformational changes of the chains.

Key words Adsorption from solution – alkylammonium ions – clay minerals – microcalorimetry – swelling – thermodynamics – verimiculite

Introduction

Water-based solvents and dispersion media are increasingly used due to the need of improved environmental control. This requires a profound knowledge of adsorption processes in binary or even multicomponent liquid mixtures. In many cases the dispersed particles adsorb both components of the solvent, e.g. water and alcohols. Adsorption measurements yield the composite isotherms (surface excess isotherms) from which the true adsorption isotherms must be derived [1–4]. To understand these adsorption processes more clearly, the surface excess isotherms were combined with calorimetric measurements [5–12]. In evaluating the data, the heat estimated in diluted solutions should be corrected by a dilution term [8–12].

A fascinating group of adsorbing materials are the 2 : 1 clay minerals which are used in an immense number of applications [13]. In many uses they are applied in hydrophobized form; the exchangeable cations on the external and internal surfaces are replaced by alkylammonium ions. The material becomes hydrophobic, i.e. it is difficult to be wetted by water. In spite of this behavior organic derivatives adsorb water in the interlayer spaces with expansion of the layer-to-layer distance [14].

A particular case is the adsorption from partially miscible liquids. The adsorption of n-butanol–water on graphite was previously described [8, 9, 12]. When alkylammonium clay minerals are dispersed in butanol–water, increasing amounts of butanol are adsorbed with increasing butanol concentration (Fig. 1). Alkylammonium clay minerals as adsorbents provide the advantage that the volume of the adsorption phase can be derived from the basal spacing, the dimensions of the unit cell, the interlayer charge density, and the molecular volume of the alkylammonium ions. The layer-to-layer separation of many organophilic clay minerals changes with the composition of the liquid mixture [15]. In other cases it remains almost constant but larger than in the absence of the liquid

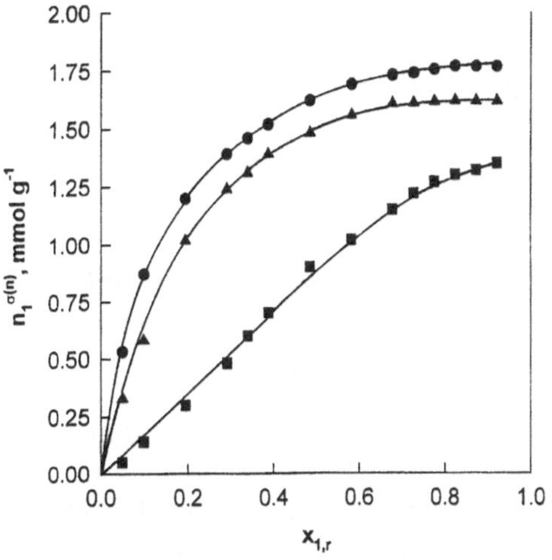

Fig. 1 Adsorption excess isotherms of *n*-butanol–water on dodecylammonium (●), dodecyldiammonium (■) and octadecylammonium vermiculite (▲)

mixture. Dodecylammonium and octadecylammonium vermiculites were used as examples with varying basal spacings. In contrast, bridging the layers by dodecyldiammonium ions makes the interlayer separation independent on the composition of the butanol–water mixture.

Theoretical background

The reduced specific adsorption excess, $n_1^{\sigma(n)}$, for adsorption from liquid mixtures is

$$n_i^{\sigma(n)} = n^0(x_i^0 - x_i) = n^0 \Delta x_i \ . \tag{1a}$$

The Ostwald–Izaguirre equation gives the material balance of adsorption for a binary liquid mixture [5, 6]:

$$n_1^{\sigma(n)} = n^0 \Delta x_1 = n_1^s - n^s x_1 \ , \tag{1b}$$

where n^0 (mmol g^{-1}) is the total amount of the liquid mixture per unit mass of adsorbent; x_i^0 and x_i are the mole fractions of component i before and after adsorption; n_i^s is the amount of component i actually present in the adsorption layer and $n^s = n_1^s + n_2^s$.

When the adsorption excess isotherm $n_1^{\sigma(n)} = f(x_1)$ and the activities of the bulk liquid phase are known, the free enthalpy of displacement in the adsorption layer, $\Delta_{21}G$, is obtained from the Gibbs adsorption equation [5, 6, 11, 12]:

$$\Delta_{21}G = -RT\int_0^{x_1} \frac{n_1^{\sigma(n)}}{(1-x_1)x_1}\left(1 + \frac{\mathrm{d}\ln\gamma_1}{\mathrm{d}\ln x_1}\right)\mathrm{d}x_1 \ . \tag{2}$$

The activity coefficient term in the brackets vanishes for ideal or ideally diluted solutions (like *n*-butanol in water).

In a flow replacement experiment, the integral enthalpy of displacement, by definition, is [7]:

$$\Delta_{21}H = \sum_{x_1=0}^{x_1} (\Delta_d H - \Delta \mathrm{H_{mix}}) = f(x) \ , \tag{3}$$

where $\Delta_d H$ is the enthalpy difference when Δn_2^s mole of component 2 at the S/L interface is displaced by $\Delta n_1^s = -\Delta n_2^s/r$ mole of component 1 during a concentration step of Δx_1 [8–10, 12]:

$$\Delta_d H = \Delta n_1^s(h_1^s - h_1^l) + \Delta n_2^s(h_2^s - h_2^l) \ , \tag{4}$$

where h_i^s and h_i^l are the partial molar enthalpies of component i in the adsorption layer and in the bulk liquid phase, respectively. It may be noted that if Δx_1 is small enough, the enthalpy of mixing (ΔH_{mix} in Eq. (3)) at the interface between the replacing and replaced solutions becomes negligible [16–18].

The integral enthalpy ($\Delta_{21}H^1$), entropy ($\Delta_{21}S^1$), and free enthalpy ($\Delta_{21}G^1$) of bulk dilution provide the link between the corresponding displacement and adsorption quantities [10, 12]:

$$\Delta_{21}G = \Delta_{21}G^s + \Delta_{21}G^1 \ , \tag{5}$$

$$\Delta_{21}H = \Delta_{21}H^s + \Delta_{21}H^1 \ , \tag{6}$$

$$\Delta_{21}S = \Delta_{21}S^s + \Delta_{21}S^1 \ , \tag{7}$$

$\Delta_{21}H^s$, $\Delta_{21}S^s$ and $\Delta_{21}G^s$ are the integral enthalpy, entropy, and free enthalpy of adsorption, and $\Delta_{21}H$, $\Delta_{21}S$ and $\Delta_{21}G$ are the corresponding displacement quantities.

When the solute is preferentially adsorbed from ideally diluted solution, $n_1^s \approx n_1^{\sigma(n)}$ and the bulk dilution terms are [10]:

$$\Delta_{21}G^1 = -n_1^{\sigma(n)}(\Delta_{sol}\mu_1^\infty + RT\ln x_1^l) \ , \tag{8}$$

$$\Delta_{21}H^l = -n_1^{\sigma(n)}\Delta_{sol}h_1^\infty , \tag{9}$$

$$\Delta_{21}S^l = -n_1^{\sigma(n)}(\Delta_{sol}s_1^\infty - R\ln x_1^l) \ . \tag{10}$$

The relationship between the immersion and surface excess thermodynamic quantities has been further discussed by Schay [5], Everett [6] and Woodbury and Noll [17, 18]. The standard enthalpy $\Delta_{sol}h_1^\infty$, entropy $\Delta_{sol}s_1^\infty$, and free enthalpy $\Delta_{sol}\mu_1^\infty$ of component 1 in component 2 at infinite dilution are listed in tables of solution thermodynamics [19].

Combining Eq. (6) with Eq. (9) and Eq. (5) with Eq. (8) leads to the integral enthalpy of adsorption and the free energy of adsorption [8, 10]:

$$\Delta_{21}H^s = \Delta_{21}H + n_1^{\sigma(n)}\Delta_{sol}h_1^\infty \ , \tag{11}$$

$$\Delta_{21}G^s = \Delta_{21}G + n_1^{\sigma(n)}RT\ln x_1 + \Delta_{sol}\mu_1^\infty n_1^{\sigma(n)} \ . \tag{12}$$

The entropy term is obtained from the enthalpy and free enthalpy of adsorption:

$$T\Delta_{21}S^s = \Delta_{21}H^s - \Delta_{21}G^s \,. \tag{13}$$

The experimental values of $\Delta_{21}H$, are converted into $\Delta_{21}H^s$, and $\Delta_{21}H^s$ is plotted vs. x_1. On the other hand, $\Delta_{21}G^s$ is obtained from Eq. (12) after numerical integration of Eq. (2). Finally, $T\Delta_{21}S^s$ is calculated by Eq. (13).

The differential molar enthalpy of displacement, $\Delta_{21}h_1$, is obtained from Eq. (4) [10, 12]:

$$\Delta_{21}h_1 = \frac{\Delta(\Delta_d H)}{\Delta n_1^s} = (h_1^s - h_1^l) - r(h_2^s - h_2^l) \,. \tag{14}$$

For ideally dilute solutions $h_1^l = \Delta_{sol}h_1^\infty$ and $h_2^l = 0$. Therefore, the differential molar enthalpy of adsorption, $\Delta_{21}h_1^s$, is [10, 12]:

$$\Delta_{21}h_1^s = \Delta_{21}h_1 + \Delta_{sol}h_1^\infty = h_1^s - rh_2^s \,. \tag{15}$$

Materials and methods

Alkylammonium vermiculite

Vermiculite from South Africa was made hydrophobic by cation exchange with dodecylammonium, octadecylammonium and dodecyldiammonium chloride (340 K, 1 month, 0.1 M surfactant in 1.5-fold excess related to the CEC, pH = 4.5) [15, 20–22]. The derivatives were washed several times with alcohol–water mixtures (1:1). This purification procedure was followed by a Soxhlet extraction with a mixture of isopropanol–water (1:1) for 48 h. Before the adsorption and microcalorimetric measurements, these organovermiculites were dried in a vacuum dessicator over night at 340 K. The total number of alkylammonium ions per formula unit was $\xi = 0.71$ mole/mole, determined from the C-content [23].

Adsorption experiments

n-Butanol of p. a. purity (Reanal, Hungary) was used without further purification. Twice distilled water was saturated with *n*-butanol in a separating funnel maintained at 298 ± 0.1 K. The butanol–water solutions were prepared by diluting the lower, butanol saturated phase with water. At 298 K the mole fraction of *n*-butanol in water at saturation is $x_1^{sat} = 1.877 \times 10^{-2}$ [12].

The adsorption excess isotherms at 298 ± 0.1 K were determined by dispersing samples of alkylammonium vermiculite in butanol–water mixtures as described earlier [20]. Because of the partial miscibility, the mole fractions x_1^0 and x_1 in Eq. (1) were replaced by $x_{1,r}^0$ and $x_{1,r}$: also $x_{1,r}^0 = x_1^0/x_1^{sat}$ and $x_{1,r} = x_1/x_1^{sat}$.

Flow microcalorimetry

The integral enthalpy of displacement of water by butanol was determined in an LKB 2107 flow sorption microcalorimeter (Bromma, Sweden) at 298 ± 0.01 K. The sorption cell was loaded with 0.3–0.5 g alkylammonium vermiculite. Initially, pure water was circulated at a constant flow rate of 24 $cm^3 h^{-1}$. The cumulative heats of displacement were measured by increasing the alcohol concentration in small steps [7–9, 12, 20]. Heats of mixing were not detected in blank runs when teflon powder was used as an inert solid. Thus, the heat effects measured step-by-step could be directly assigned to the displacement process [7, 9, 12].

X-ray diffractometry

The basal spacing of the alkylammonium vermiculites, d_L, was determined with a Philips PW-1830 diffractometer (Cu K_α radiation, $\lambda = 1.54$ nm) at $1° \leq 2\theta \leq 20°$. The sample holder containing suspensions of alkylammonium vermiculite in butanol–water was covered by Mylar foil (25 μm thickness) to prevent evaporation of the dispersion medium. The basal spacings were reproduced within ±0.02 nm.

Results and discussion

The choice of the system, alkylammonium vermiculites in butanol–water, was guided by earlier results of adsorption and calorimetric studies on graphite, which demonstrated the importance of the dilution term in interpreting the calorimetric measurements [8, 12].

The surface excess isotherms for dodecylammonium, dodecyldiammonium and octadecylammonium vermiculite (Fig. 1) reveal that the amount of butanol adsorbed increases with $x_{1,r}$. Since the mole fraction of butanol is very small, $n_1^{\sigma(n)}$ is almost identical with the true amount of butanol adsorbed, $n_1^s \approx n_1^{\sigma(n)}$. The isotherms reflect the different interlamellar structure formed in the presence of butanol–water. The interlayer space containing dodecyl- and octadecylammonium ions widens with increasing butanol concentration (Figs. 2a and 3) because more and more liquid molecules penetrate between the layers. This is illustrated in Fig. 4 by the quantity $V_{int} - V_{alk}$ which expresses the volume of liquid taken up between the silicate

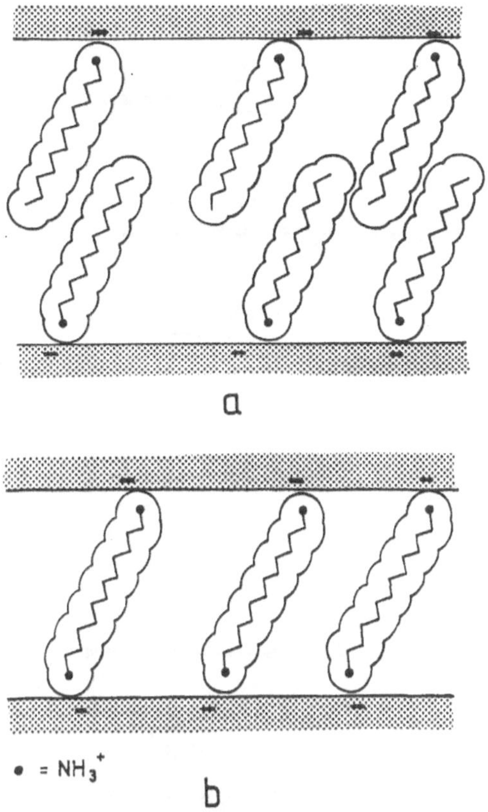

Fig. 2 Orientation of the alkyl chains (in ideallized all-trans conformation) between the silicate layers in butanol–water. (a) Dodecylammonium vermiculite and (b) dodecyldiammonium vermiculite

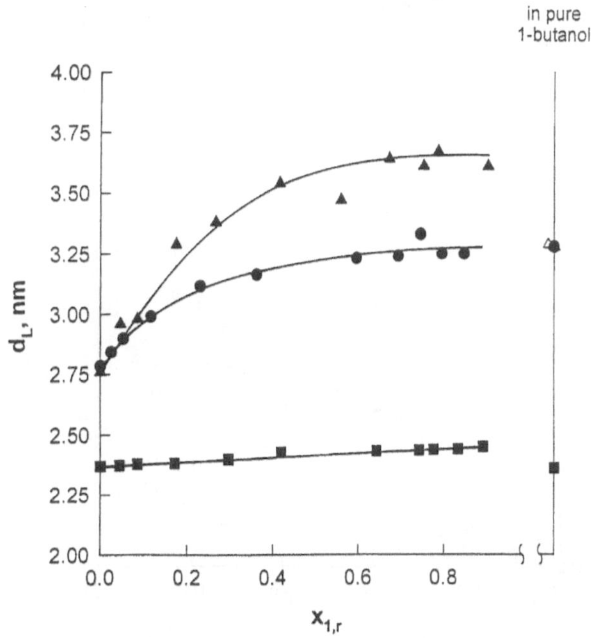

Fig. 3 Basal spacing of dodecylammonium (●), dodecyldiammonium (■) and octadecylammonium vermiculite (▲) in butanol–water

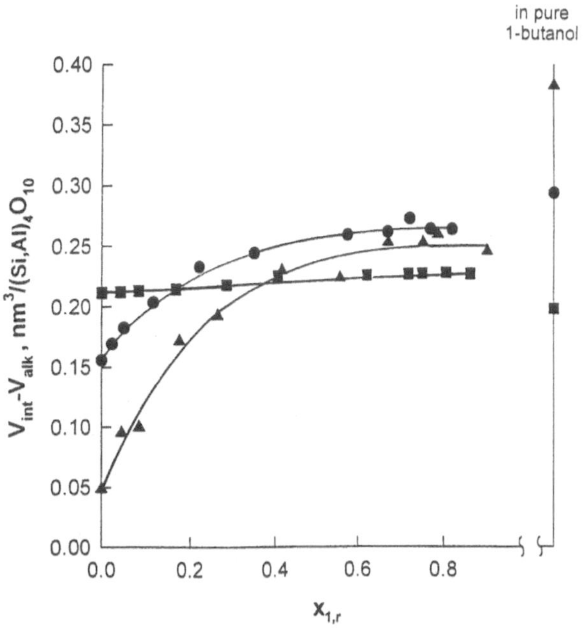

Fig. 4 Interlamellar free volume of dodecylammonium (●), dodecyldiammonium (■) and octadecylamonium vermiculite (▲) in butanol–water

layers and the alkylammonium ions (V_{int} = volume of the interlayer space, calculated from the basal spacing and the dimensions of the unit cell, V_{alk} = volume of the alkylammonium ions, see [3, 4, 20]). Water expands in the interlayer space of the octadecylammonium vermiculite less significantly than that of the dodecylammonium derivative as long as the butanol content is below $x_{1,r} \approx 0.6$. When $x_{1,r}$ approaches 1, the volume increments $V_{int} - V_{alk}$ approximate the same value (Fig. 4).

The swelling of the dodecyldiammonium vermiculite is restricted by the bridging alkyl chains (Fig. 2b). The constancy of the basal spacing indicates that the chain orientation (in an angle of 56° to the silicate layer) is independent on the composition of the liquid mixture.

The free enthalpy of displacement of water by butanol, $\Delta_{21}G$, was calculated from the surface excess isotherms (Eq. (2)). It decreases with increasing $x_{1,r}$ and reaches smaller values for the alkylammonium vermiculites (-10 and -8.4 J g^{-1}) than for the dodecyldiammonium derivative (-4 J g^{-1}) (Fig. 5).

The enthalpy of the displacement process, $\Delta_{21}H$, was measured in the flow sorption microcalorimeter (Fig. 6). It is endothermic and increases to a plateau at 6 J/g (dodecyldiammonium vermiculite) and 14 J/g (octadecylammonium vermiculite). The reaction of the dodecylammonium derivative is nearly thermoneutral.

When the enthalpy of displacement is corrected by the dilution term ($\Delta_{sol}h_1^{\infty} = -9.30$ kJ/mol for butanol–water,

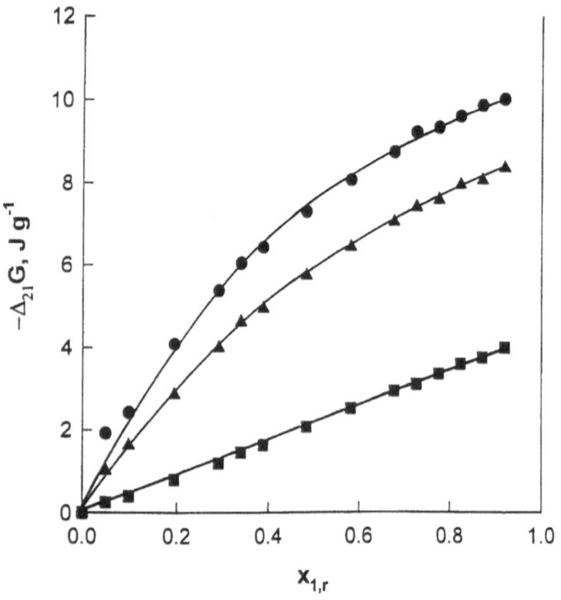

Fig. 5 Free enthalpy of displacement of water by butanol on dodecylammonium (●), dodecyldiammonium (■) and octadecylammonium vermiculite (▲)

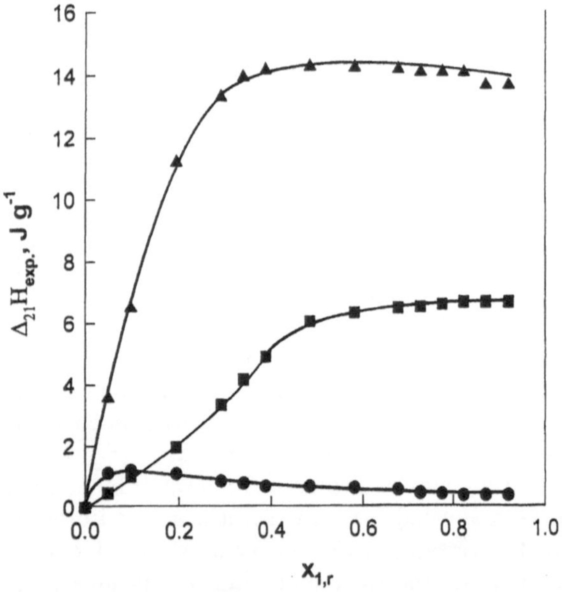

Fig. 6 Integral enthalpy isotherms of displacement of water by butanol (microcalorimetric measurements) on dodecylammonium (●), dodecyldiammonium (■) and octadecylammonium vermiculite (▲)

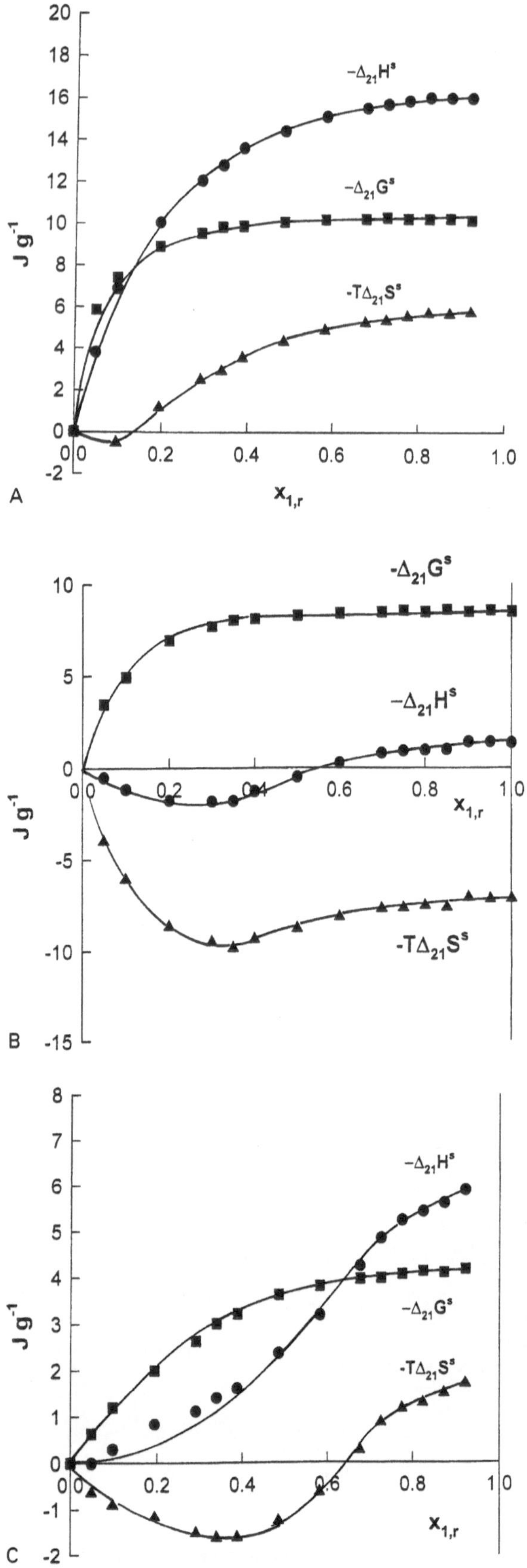

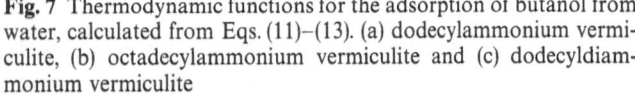

Fig. 7 Thermodynamic functions for the adsorption of butanol from water, calculated from Eqs. (11)–(13). (a) dodecylammonium vermiculite, (b) octadecylammonium vermiculite and (c) dodecyldiammonium vermiculite

Table 1 Basal spacings and thermodynamic parameters for *n*-butanol–water/alkylammonium vermiculites

	Dodecylammonium vermiculite	Octadecylammonium vermiculite	Dodecyldiammonium vermiculite
Molecular mass (g)[1)]	510	570	450
Basal spacing (nm)			
Dried	2.03	2.76	1.53
In water	2.78	2.76	2.37
In butanol	3.28	3.28	2.36
$\Delta^{21}G^{s\,x\to 1}_{1,r}$	− 10.5	− 4.2	− 8.5
$\Delta_{21}H^{s}$ ($J\,g^{-1}$)	− 16.2	− 6.0	− 1.3
$T\Delta_{21}S^{s}$	− 5.7	− 1.8	7.2
$\Delta_{21}h_1^{s}$ ($kJ\,mol^{-1}$)	− 9.0	1.8	− 4.2

[1)] Calculated from the molecular mass of the silicate layer (≈ 378) and the molecular mass of 0.71 alkylammonium ions or 0.355 alkyldiammonium ions.

Eq. (11),) the enthalpy of adsorption, $\Delta_{21}H^{s}$, becomes negative when dodecylammonium ions are interlayer cations. It approximates $-16.2\,J\,g^{-1}$ at $x_{1,r}\to 1$ (Fig. 7a, Table 1). While the displacement of the first water molecules by butanol is accompanied by small positive values of $\Delta_{21}S^{s}$, at $x_{1,r} > 0.1$, $\Delta_{21}S^{s}$ becomes negative. With longer alkylammonium ions in the interlayer space, $\Delta_{21}H^{s}$ is very small and changes from positive to negative at $x_{1,r}\approx 0.6$ (Fig. 7b). When the swelling is limited by bridging alkyl chains, $\Delta_{21}H^{s}$ decreases to $-6\,J\,g^{-1}$ whereas the entropy changes sign from positive to negative at $x_{1,r} = 0.67$ (Fig. 7c).

The integral enthalpies of adsorption of butanol from water are plotted against the amount of butanol adsorbed in Fig. 8. Good straight lines are obtained in this representation suggesting that the behaviour of the adsorption layer is close to ideal. The differential molar enthalpies of adsorption are listed in Table 1 together with some other thermodynamic parameters. The adsorption process is more exothermic on dodecylammonium vermiculite ($-9.0\,kJ\,mol^{-1}$) than on dodecyldiammonium vermiculite ($-4.2\,kJ\,mol^{-1}$), the alkyl chain density for the latter being smaller by a factor of two. The enthalpy of adsorption is slightly endothermic on octadecylammonium vermiculite ($1.8\,kJ\,mol^{-1}$) which, in part, may be related to the superposition of the swelling process on the adsorption process.

The most interesting aspect concerns the entropy of adsorption, $\Delta_{21}S^{s}$. In interlayer spaces with dodecyldiammonium ions water is replaced by butanol at constant interlamellar volume (Fig. 4). The mixing of water and butanol makes $\Delta_{21}S^{s}$ positive (Fig. 7c) as expected for a mixing process. At higher proportions of butanol, at $x_{1,r} > 0.67$, an interlamellar process of ordering occurs which makes the entropy of adsorption negative.

When one end of the alkyl chain is free (dodecyl- and octadecylammonium ions), the conformational entropy of

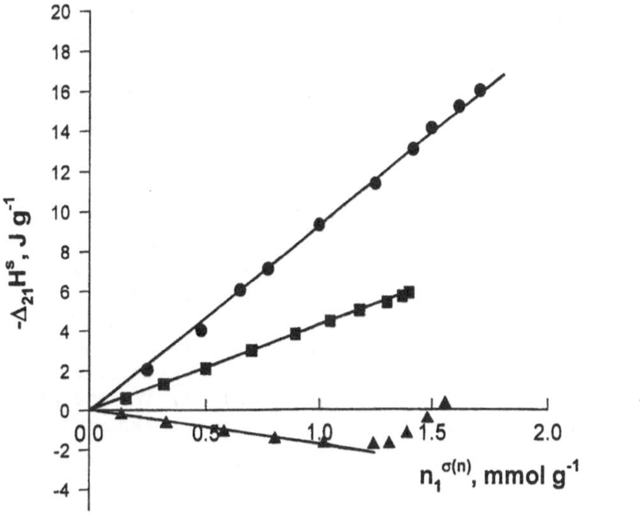

Fig. 8 Integral enthalpies of adsorption as functions of the amount of butanol adsorbed. Dodecylammonium (●), dodecyldiammonium (■) and octadecylammonium vermiculite (▲)

the chains plays an important role. The entropy change, $\Delta_{21}S^{s}$, for the dodecylammonium derivative is negative at $x_{1,s} > 0.1$ (Fig. 7a). As the alkyl chains have considerable conformational freedom, the interlamellar structure assumes a higher degree of ordering when water is replaced by butanol. The amount of butanol in the interlayer space also increases with $x_{1,r}$ (Fig. 4). As the reaction is highly exothermic ($\Delta_{21}H^{s} = -16.2\,J\,g^{-1}$), $\Delta_{21}G^{s}$ is <0 in spite of the negative entropy term. The extent of swelling and the energy required to change the alkyl chain orientation increase with increasing alkyl chain length, and $\Delta_{21}H^{s}$ is reduced; $\Delta_{21}H^{s}\approx 0$ for octadecylammonium ions. As the alkyl chains pointing into the interlayer space (Fig. 2b) gain more conformational entropy with increasing chain length, $\Delta_{21}S^{s}$ increases and becomes positive for the octadecylammonium vermiculite (Fig. 7b).

Conclusion

Evaluation of the thermodynamic functions reveals the strong influence of the alkyl chain length on the adsorption process. An important contribution is the enthalpy required to move the alkyl chain from the flat position (with strong van der Waals contacts to the surface oxygen atoms) into the paraffine-type orientation. Therefore, $\Delta_{21}H^s$ is strongly exothermic for dodecylammonium vermiculite but small (changing from positive to negative at $x_{1,r} \approx 0.6$, Fig. 7b) for the octadecylammonium derivative. A further important contribution arises from the entropy term due to the conformational changes of the alkyl chains. When the alkyl chains point to the interlayer space (Fig. 2b), they gain more conformational freedom compared to the flat position. The resulting gain in entropy increases with the alkyl chain length: $T\Delta_{21}S^s$ is negative for dodecylammonium vermiculite but positive for the octadecylammonium derivative.

References

1. Dékány I, Szántó F, Nagy LG, Fóti G (1975) J Colloid Interface Sci 50:265
2. Dékány I, Szántó F, Nagy LG (1986) J Colloid Interface Sci 109:376
3. Dékány I, Szántó F, Weiss A, Lagaly G (1985) Ber Bunsenges Phys Chem 89:62
4. Dékány I, Szántó F, Weiss A, Lagaly G (1986) Ber Bunsenges Phys Chem 90:422, 427
5. Schay G (1976) Pure Appl Chem 48:373
6. Everett DH (1981) Pure Appl Chem 53:2181
7. Dékány I, Zsednai Á, Király Z, László K, Nagy LG (1986) Colloid Surfaces 19:47
8. Király Z, Dékány I (1990) Colloid Surfaces 49:95
9. Király Z, Dékány I (1990) Prog Colloid Polym Sci 83:68
10. Király Z, Dékány I, Nagy LG (1993) Colloid Surfaces A 71:287
11. Király Z, Dékány I (1988) Colloid Surfaces 34:1
12. Király Z, Dékány I (1989) Chem Soc Faraday Trans 85:3373
13. Jasmund K, Lagaly G (eds) (1993) Tonminerale und Tone. Struktur, Eigenschaften, Anwendungen und Einsatz in Industrie und Umwelt. Steinkopff Verlag, Darmstadt
14. Weiss A (1963) Clays Clay Min X (10th Natl Conf on Clays Clay Min) 191–224
15. Dékány I, Szántó F, Nagy LG (1978) Colloid Polym Sci 256:150
16. Groszek AJ (1970) Proc Roy Soc London, Ser A 314:473
17. Woodbury GW Jr, Noll LA (1983) Colloid Surfaces 8:1
18. Woodbury GW Jr (1987) Colloid Surfaces 28:233
19. Abraham MH (1984) J Chem Soc, Faraday Trans 1, 80:153
20. Regdon I, Király Z, Dékány I, Lagaly G (1994) Colloid Polym Sci 272:1129
21. Dékány I, Szántó F, Nagy LG (1978) J Colloid Interface Sci 103:321
22. Szántó F, Dékány I, Patzkó Á, Várkonyi B (1986) Colloid Surfaces 18:359
23. Lagaly G (1994) Layer charge determination by alkylammonium ions. In: Mermut A (ed) Layer Charge Characteristics of 2:1 Silicate Clay Minerals, CMS Workshop Lectures, Vol 6. The Clay Mineral Society, Boulder, pp 1–46

Progr Colloid Polym Sci (1998) 109:221–226

N. Kallay
A. Čop
D. Kovačević
A. Pohlmeier

The use of electrokinetic potential in the interpretation of adsorption phenomena. Adsorption of salicylic acid on titanium dioxide

Received: 23 July 1997
Accepted: 5 September 1997

Prof. Nikola Kallay (✉) · A. Čop
D. Kovačević
Laboratory of Physical Chemistry
Faculty of Science
University of Zagreb
Marulićev trg 19//POB 163
10001 Zagreb
Croatia

A. Pohlmeier
Institute of Applied Physical Chemistry
Research Center Jülich
52425 Jülich
Germany

Abstract The adsorption of salicylic acid on titanium dioxide was measured as a function of pH and interpreted using a modified adsorption isotherm based on the surface complexation model, which takes into account dissociation of the acid in the bulk of the solution and the electrostatic effect on the adsorbed ionic species. It was concluded that both singly charged and neutral species are adsorbable, and both adsorption equilibrium constants were determined.

Key words titanium dioxide – salicylic acid – adsorption – electrokinetics

Introduction

In the last decades, with the development of the ecological science, the adsorption processes have become a popular subject of many investigations, especially the adsorption of organic acids on metal oxides [1].

In the past, a phenomenological approach was applied to the interpretation of adsorption equilibria, i.e. experimental results were analyzed using more or less semi-empirical adsorption isotherms. The agreement of experimental data with a particular isotherm (e.g. Langmuir or Freundlich) was used as the basis for the understanding of different adsorption mechanisms. From the mid-1970s, the important role of analyzing the surface processes, and particularly the adsorption phenomena, was taken over by the surface complexation model (SCM) [2]. When interpreting the adsorption equilibria on the basis of this model, it is necessary to define or postulate surface active groups and the mechanism of the corresponding surface reactions. Accordingly, in order to use SCM for adsorption of organic acids, one has to postulate or assume the structure of the surface complex and also the specific structure of the electrical interfacial layer, i.e. planes in which the surface species are located. In some cases, such assumptions incorporated in the interpretation procedure may lead to unrealistic or doubtful results.

One of the ways of overriding these problems is based on the interpretation of simultaneous adsorption and electrokinetic measurements. As demonstrated in previous reports [3–6], if the electrostatic interactions and dissociation of the acid are taken into account, the interpretation results in the equilibrium constant of adsorption, and also in the charge and area occupied by the adsorbed species.

In such an approach, electrokinetic measurements provide information on surface electrostatic interactions. Dissociation of the acid in the bulk is considered through the corresponding equilibrium constants so that several possibilities regarding the charge of adsorbable species could be examined.

In the first version of this approach (adsorption of oxalic, citric [3, 4] and iminodiacetic [5] acid on hematite), any assumption regarding the electrical interfacial layer structure was avoided by considering only adsorption data at the isoelectric point at which no electrostatic effects were expected. Other data points, outside the isoelectric conditions, were used to calculate the surface potential, which then enabled comparison with the measured electrokinetic ζ-potential. The comparison showed that ζ-potential was always lower than the surface potential (at the onset of the diffuse layer), which led to the introduction of the Gouy–Chapman theory and to the concept of electrokinetic slipping plane separation.

A more advanced approach, used for adsorption of salicylic acid on hematite [6], introduced the slipping plane separation as an adjustable parameter. Consequently, all data points were used in the calculation procedure, which reduced the number of necessary experimental runs. This approach enables also analysis of the case when more than one kind of ionic species are simultaneously adsorbed on the surface, which is the subject of this report.

Adsorption isotherm

The surface complexation model could be also used to derive the classical Langmuir isotherm

$$\frac{1}{\Gamma} = \frac{1}{\Gamma_{\max}} + \frac{1}{\Gamma_{\max} K c_{\mathrm{eq}}}, \qquad (1)$$

where Γ is the equilibrium surface concentration, $\Gamma_{\max}$ the maximum surface concentration, c_{eq} the equilibrium concentration of the adsorbable species, and K the adsorption "equilibrium constant".

For adsorption of ionizable species (like salicylic acid) one should refine the simple Langmuir equation. Since salicylic acid dissociates in the bulk of the solution, three different forms (H_2L, HL^-, L^{2-}) are present and one should not introduce the total analytical equilibrium concentration of the acid into the above equation. Instead, one needs to assume which species actually adsorb and use the respective equilibrium concentration. In addition, the adsorption of charged species is markedly influenced by the electrostatic potential in the interfacial layer, i.e. by the potential at the plane in which the adsorbed species (ions) are located (surface potential).

Effect of acid dissociation

The proposed interpretation assumes several possibilities regarding the species being adsorbed:

(i) In the classical approach, all species are adsorbable with the same adsorption affinity (same K value):

$$c_{\mathrm{eq}} = c_{\mathrm{eq,tot}}, \quad \Gamma = \Gamma(\mathrm{H_2L}) + \Gamma(\mathrm{HL^-}) + \Gamma(\mathrm{L^{2-}}) .$$

(ii) Only neutral species are adsorbed:

$$c_{\mathrm{eq}} = [\mathrm{H_2L}], \quad \Gamma = \Gamma(\mathrm{H_2L}) .$$

(iii) Only singly charged species are adsorbed:

$$c_{\mathrm{eq}} = [\mathrm{HL^-}], \quad \Gamma = \Gamma(\mathrm{HL^-}) .$$

(iv) Both neutral and singly charged species are adsorbed with different equilibrium constants:

$$c_{\mathrm{eq}} = [\mathrm{HL^-}] + [\mathrm{H_2L}], \quad \Gamma = \Gamma(\mathrm{HL^-}) + \Gamma(\mathrm{H_2L}) .$$

Adsorption of doubly charged species (L^{2-}) is not considered here because their concentration is negligible in the examined pH range.

Electrostatic effect

The effect of surface charge (electrostatic potential) on the adsorption affinity, i.e. on the value of the "equilibrium constant" K, is taken into account through the interaction parameter representing the electrostatic interaction energy:

$$K = K^0 \exp(-zF\varphi_{\mathrm{a}}/RT), \qquad (2)$$

where K^0 is the standard (intrinsic) equilibrium constant and φ_{a} the surface potential affecting the state of the adsorbed species of charge number z, while F, R and T have their usual meaning.

In such an interpretation, K^0 is the real equilibrium constant representing the "chemical part of the free energy". It is equal to K at the isoelectric point where ζ and φ_{a} approach zero. The surface potential, φ_{a}, can be evaluated from the measured electrokinetic ζ-potential via the Gouy–Chapman theory, assuming a certain distance for the slipping plane separation (e.g. 5–20 Å):

$$\varphi_{\mathrm{a}} = (2RT/F)\ln\left(\frac{\exp(-s\kappa) + \tanh(F\zeta/4RT)}{\exp(-s\kappa) - \tanh(F\zeta/4RT)}\right), \qquad (3)$$

where s is the slipping plane separation (the distance between the slipping plane and the plane in which the adsorbed species are located), κ is the Debye–Hückel reciprocal distance, and ζ is the measured electrokinetic potential.

Combining Eqs. (1) and (2) leads to

$$\frac{1}{\Gamma} = \frac{1}{\Gamma_{max}} + \frac{1}{\Gamma_{max} K^0 \exp(-zF\varphi_a/RT) c_{eq}} , \quad (4)$$

where c_{eq} is the bulk equilibrium concentration of adsorbable species expressed in moles per litre. The above equation may be used for interpretation of adsorption equilibrium if only one kind of ionic species is adsorbed.

Simultaneous adsorption of different ionic species

For the treatment of simultaneous adsorption of different ionic species with specific equilibrium constants some other modifications of the original expression are needed. In the case of salicyclic acid, the presence of doubly charged anions (L^{2-}) in the examined pH region is negligible, so simultaneous adsorption of neutral (H_2L) and singly charged (HL^-) species is the only realistic assumption.

The binding constants of neutral ($z = 0$, $K_{ads,0}$) and singly charged ($z = -1$, $K_{ads,1}$) species on the oxide surface are defined as

$$K_{ads,0} = K^0_{ads,0} = \frac{\Gamma(H_2L)}{[H_2L](\Gamma_{max} - \Gamma)} , \quad (5)$$

$$K_{ads,1} = K^0_{ads,1} \exp(F\varphi_a/RT) = \frac{\Gamma(HL^-)}{[HL^-)(\Gamma_{max} - \Gamma)} . \quad (6)$$

The bulk concentrations of H_2L and HL^- (expressed in mol dm^{-3}) species are calculated from the total acid concentration at the equilibrium (c_{eq}) and the measured pH using the literature values [7] of the dissociation equilibrium constants (K_1 and K_2):

$$[H_2L] = \frac{c_{eq}[H^+]^2}{[H^+]^2 + K_1[H^+] + K_1K_2} , \quad (7)$$

$$[HL^-] = \frac{c_{eq}K_1[H^+]}{[H^+]^2 + K_1[H^+] + K_1K_2} . \quad (8)$$

In the pH range $3 < pH < 6$, the concentration of L^{2-} species is negligible, so one cannot expect their adsorption. Consequently,

$$\Gamma = \Gamma(HL^-) + \Gamma(H_2L) . \quad (9)$$

By combining Eqs. (5)–(9), the following expression is obtained:

$$\Gamma = \frac{K^0_{ads,0} c_{eq}[H^+]^2(\Gamma_{max} - \Gamma)}{([H^+]^2 + K_1[H^+] + K_1K_2)} + \frac{K_1 c_{eq} K^0_{ads,1}[H^+](\Gamma_{max} - \Gamma)\exp(F\varphi_a/RT)}{[H^+]^2 + K_1[H^+] + K_1K_2} . \quad (10)$$

The above relationship could be linearized in the following way:

$$\frac{\Gamma([H^+]^2 + K_1[H^+] + K_1K_2)}{[H^+]^2 c_{eq}(\Gamma_{max} - \Gamma)} = K^0_{ads,0} + K_1 K^0_{ads,1} \frac{\exp(F\varphi_a/RT)}{[H^+]} . \quad (11)$$

According to Eq. (11) the slope of the plot of

$$\frac{\Gamma([H^+]^2 + K_1[H^+] + K_1K_2)}{[H^+]^2 c_{eq}(\Gamma_{max} - \Gamma)} \text{ vs } \frac{K_1 \exp(F\varphi_a/RT)}{[H^+]}$$

is equal to $K^0_{ads,1}$ while the intercept yields $K^0_{ads,0}$.

Experimental

Chemicals

All chemicals (HCl, NaOH, NaCl and salicylic acid) used in this study were of analytical purity grade. Titanium(IV) oxide (Alfa, Johnson Matthey GmbH, Karlsruhe, Germany) was purified by dialysis. Specific surface area of powder, measured by the BET method, was obtained as 2.0 $m^2 g^{-1}$.

Adsorption measurements

Adsorption measurements were performed at 25 °C in the following way: 10 g of TiO_2 powder was dispersed in 20 cm^3 of salicylic acid aqueous solution so that mass concentration, γ, was always 500 g dm^{-3}. Sodium hydroxide or hydrochloric acid solution was used to adjust the pH. In one run, the total concentration of salicylic acid was kept constant. The systems were constantly mixed for 180 min. The ionic strength was controlled by sodium chloride. The supernatant was separated by filtration. The concentration of salicylic acid in the supernatant solution was determined spectrophotometrically (UV-Vis-NIR spectrophotometer Cary 05, Varian, USA). The absorbance was measured at 296 nm, which is the characteristic band of the salicylic acid. No significant effect of the pH on the spectra of salicylic acid was observed over the pH range of interest.

Electrokinetic measurements

The electrokinetic data, at 25 °C, were obtained with the Otsuka ELS-800 electrophoretic light scattering instrument. In these measurements, the dilute dispersion remainings after sedimentation were used, containing a sufficient amount of titanium(IV) oxide particles for the electrophoretic measurements. This procedure ensured that the equilibrium concentration of salicylic acid and the pH were the same as in the corresponding adsorption experiments.

Results

Figure 1 presents the effect of pH on the adsorption and electrokinetic data for titanium(IV) oxide/salicyclic acid system. Contrary to the behavior of other similar systems [6], the pH dependency of both surface concentration and electrokinetic potential is not a smooth continuous function. This finding is not due to experimental inaccuracy, since the repeated runs showed the same patterns. Accordingly, one may not expect a simple interpretation of these results.

The experimental data were interpreted on the basis of different assumptions, like in the previous report [6]:

(i) All species are adsorbable by the same and constant adsorption affinity:

$$c_{eq} = c_{eq,tot}, \qquad \Gamma = \Gamma(H_2L) + \Gamma(HL^-) + \Gamma(L^{2-}) .$$

(ii) Only neutral species are adsorbed:

$$c_{eq} = [H_2L], \quad z = 0, \qquad \Gamma = \Gamma(H_2L) .$$

(iii) Only singly charged species are adsorbed:

$$c_{eq} = [HL^-], \quad z = -1, \quad \Gamma = \Gamma(HL^-) .$$

Additionally, the following assumption was also applied:

(iv) Both neutral and singly charged species are adsorbed:

$$c_{eq} = [HL^-] + [H_2L], \quad \Gamma = \Gamma(HL^-) + \Gamma(H_2L) .$$

Assumption (i) was tested using the original Langmuir isotherm (1), assumptions (ii) and (iii) were tested using the plot based on Eq. (4), while Eq. (11) was applied for the last assumption (iv). The latter interpretation included the adsorption parameters (Γ_{max} and s) as obtained previously [6].

In calculating surface potentials from the measured electrokinetic ones the slipping plane separations of 10, 15 and 20 Å were used for case (iii).

Case (i) is presented in Fig. 2. Assumption (i) disagrees markedly with the experiment since no linearity of the plotted function was found. Moreover, Γ^{-1} decreases with c^{-1}, contrary to the prediction of the classical Langmuir model. As shown in Fig. 3, the disagreement with expectation was also found when adsorption of neutral species only was assumed (ii). When adsorption of HL^- species

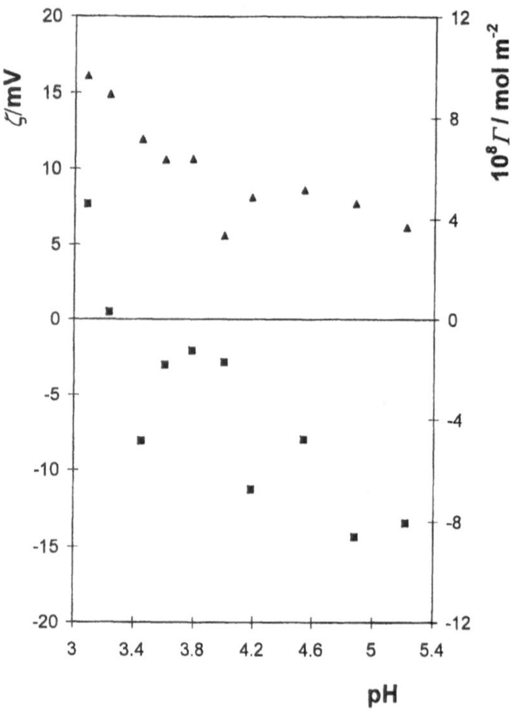

Fig. 1 The effect of pH on the surface concentration of salicylic acid on titanium(IV) oxide (▲) and on the electrokinetic potential of titanium(IV) oxide particles (■), $I_c = 0.006$ mol dm^{-3}; $\gamma = 500$ g dm^{-3}; $c_{in} = 6.4 \times 10^{-4}$ mol dm^{-3}; $t = 25\,°C$. (Note that for electrokinetic experiments a low mass concentration of particles was used)

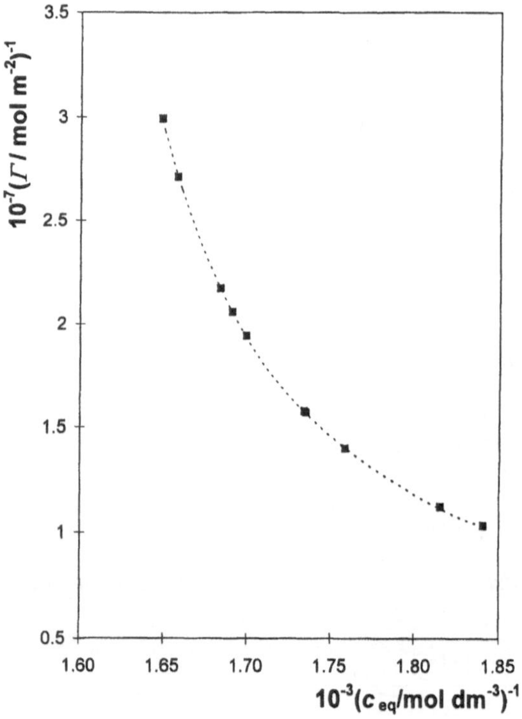

Fig. 2 Data from the experiment displayed in Fig. 1, as interpreted by Eq. (1), assuming that all species are adsorbable by the same and constant adsorption affinity (i)

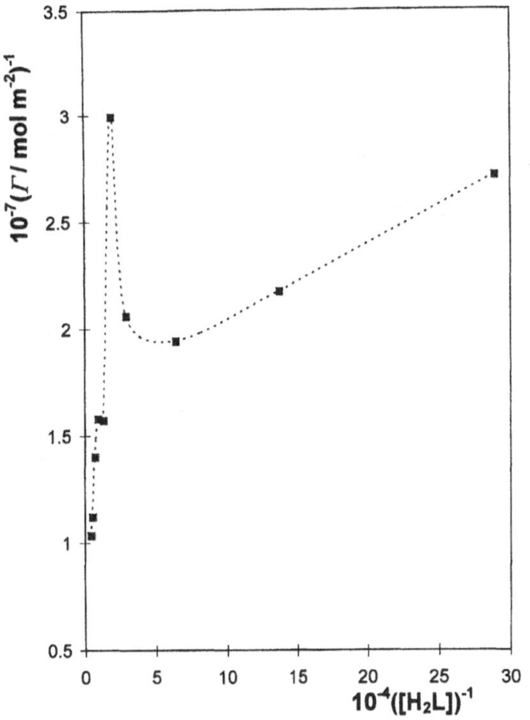

Fig. 3 Data from the experiment displayed in Fig. 1, as interpreted by Eq. (4), assuming that only the neutral H_2L species are adsorbed (ii); $c_{eq} = [H_2L]$; $z = 0$

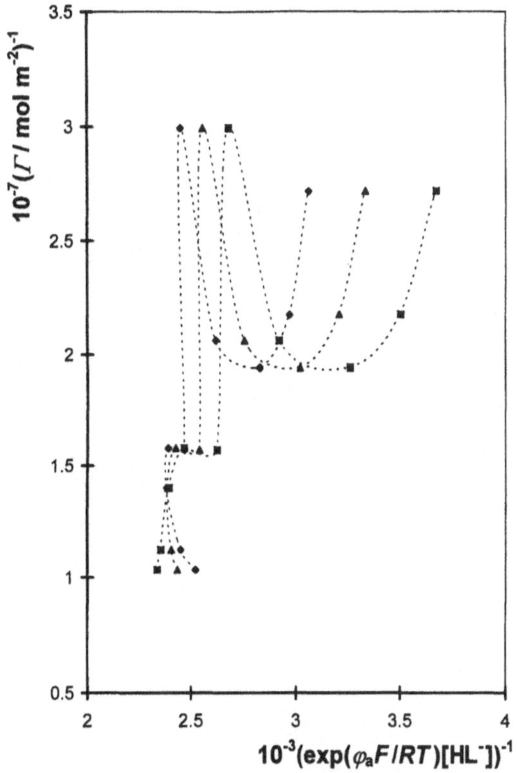

Fig. 4 Data from the experiment displayed in Fig. 1, as interpreted by Eq. (4), assuming that only HL^- species adsorb (iii); $c_{eq} = [HL^-]$; $z = -1$. In the Gouy–Chapman equation (3), the slipping plane separations of 10 Å (◆), 15 Å (▲) and 20 Å (■) were assumed

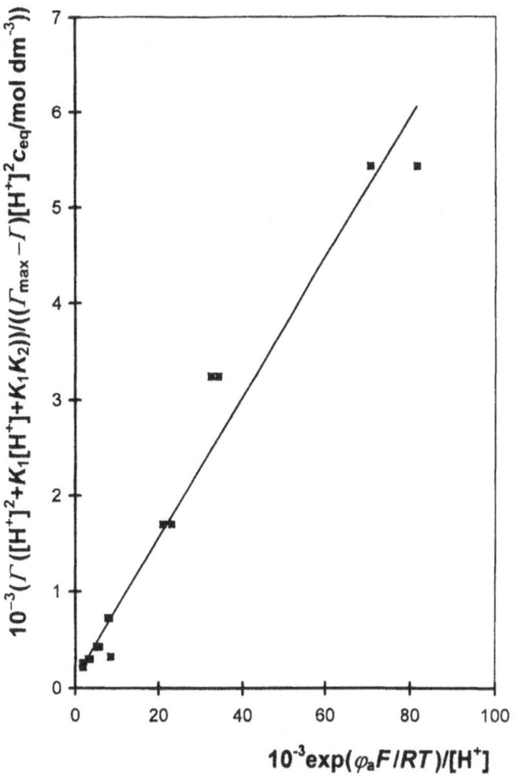

Fig. 5 Data from the experiment displayed in Fig. 1, as interpreted by Eq. (11), assuming that both H_2L and HL^- species adsorb (iv); $\Gamma = \Gamma(HL^-) + \Gamma(H_2L)$. The following values of parameters were used [7]: $s = 15$ Å, $\Gamma_{max} = 2 \times 10^{-6}$ mol m^{-2}

was assumed (iii), regardless of the introduced value for the slipping plane separation, no straight line was obtained (Fig. 4).

All the above assumptions were found to be inapplicable to the adsorption of salicylic acid on titanium(IV) oxide. Therefore, a more complex assumption was introduced, i.e. the adsorption of both neutral and singly charged species. For this assumption, the plot based on Eq. (11) yielded a straight line (Fig. 4). The best linearity was obtained for $s \approx 15$ Å.

The above tests were applied also to the second experimental run and similar results were obtained. The interpretation of data based on assumption (iv) from two experiments with the same initial concentration of salicylic acid ($c_{in} = 6.4 \times 10^{-4}$ mol dm^{-3}) yielded the following adsorption equilibrium constants:

$$K^0_{ads,0} = 150 \pm 50, \qquad K^0_{ads,1} = 60 \pm 10 .$$

Discussion

This study deals with the problem of adsorption of an organic acid at the metal oxide/aqueous interface. The proposed interpretation of adsorption data considers dissociation of adsorbate in the bulk of solution and the electrostatic effect evaluated from the electrokinetic data. The interpretation is very sensitive to the charge of the species that actually adsorb which enables their accurate detection. The interpretation avoids any assumption regarding the structure of the inner part of the interfacial layer, and therefore it is less speculative in comparison with the common approach based on the surface complexation model. However, in doing so one cannot obtain any specific information on the mechanism of the surface reaction and the structure of the surface complex; the only information that could be obtained are the charge number of the surface complex, equilibrium constant and surface area per adsorbed ion or molecule. The problems related to the interpretation are twofold. First, competition and mutual relations with other surface reactions are neglected, which may be solved by performing measurements at a low surface coverage. The second problem arises from the use of the Gouy–Chapman theory and introduction of the electrokinetic slipping plane separation concept. In discussing this problem, one should start with some findings that are not doubtful. It was demonstrated that the 0-plane potential (φ_0) is of the Nernstian type with the slope lower than the theoretical one [8–14]. The interpretation of the adsorption equilibria requires the potential to which bound counterions are exposed (φ_β) to be lower than φ_0 but significantly higher than the measured electrokinetic potential (ζ) [15]. This knowledge makes two different assumptions regarding the location of the onset of the diffuse layer (d-plane) possible. First, one may assume that d-plane is close to the electrokinetic slipping plane so that $\varphi_0 > \varphi_\beta > \varphi_d = \zeta$. The other choice is that β-plane is close to d-plane, and for such an assumption $\varphi_0 > \varphi_\beta = \varphi_d > \zeta$. Coagulation measurements [15, 16] support the latter choice since the potential at the onset of diffuse layer (φ_d), which governs the aggregation process, should be higher than ζ, but lower than the Nernstian one.

In this study, the adsorption of salicylic acid on titania was investigated. Contrary to other similar systems [3–6], the interpretation showed clearly that more than one kind of ionic species are adsorbable with relatively low affinities. For example, the binding equilibrium constant for salicylic acid is about 30 times higher for hematite with respect to titania. This finding can be compared with the complexation of relevant cations in the bulk of solution. The salicylic acid complexes with Ti^{4+} are significantly less stable [7] than for Fe^{3+}. This phenomenon is of a general nature; all organic ligands form more stable complexes with Fe than with Ti-ions. Accordingly, one may conclude that adsorption of organic acids will be more pronounced on hematite and other ion oxides than on titania, and especially silica.

References

1. Cornell RM, Schwertemann U (1996) The Iron Oxides, VCH Verlagsgesellschaft, Weinheim, Germany
2. Yates DE, Levine S, Healy TW (1974) J Chem Soc Faraday Trans I 70:1807
3. Kallay N, Matijević E (1985) Langmuir 1:195
4. Zhang Y, Kallay N, Matijević E (1985) Langmuir 1:201
5. Torres R, Kallay N, Matijević E (1988) Langmuir 4:706
6. Kovačević D, Kallay N, Antol I, Pohlmeier A, Lewandowski H, Narres HD Colloids Surf, in press
7. Sillen G, Martell AE (1964) Stability constants of Metal-Ion Complexes, The Chemical Society, London
8. Penners NHG, Koopal LK, Lyklema J (1986) Colloids Surf 21:457
9. Bousse L, De Roij NF, Bergveld P (1983) Surf Sci 135: 479
10. Ardizzone S, Siviglia P, Trasatti S (1981) J Electroanal Chem 122:395
11. Kallay N, Babić D, Matijević E (1986) Colloids Surf 19:375
12. Blesa MA, Kallay N (1988) Adv Colloid Interface Sci 28:111
13. Avena MJ, Camara OR, De Pauli CP (1993) Colloids Surf 69:217
14. Kallay N, Sprycha R, Tomić M, Žalac S, Torbić Ž (1990) Croat Chem Acta 63:467
15. Čolić M, Fuerstenau DW, Kallay N, Matijević E (1991) Colloids Surf 59:169
16. Kallay N, Čolić M, Fuerstenau DW, Jang HM, Matijević E (1994) Colloid Polym Sci 272:554

Progr Colloid Polym Sci (1998) 109:227–231

K. Tertsch
H. Versmold

Electro-osmosis in electrophoretic light scattering experiments

Received: 25 June 1997
Accepted: 1 July 1997

K. Tertsch · Prof. Dr. H. Versmold (✉)
Technische Hochschule
Lehrstuhl für Physikalische Chemie II
Templergraben 59
D-52062 Aachen
Germany

Abstract Electrophoretic/electro-osmotic light scattering experiments are reported. Particular emphasis is attributed to the various effects resulting from (a) the electro-osmotic mobility μ_{eo} due to added indifferent electrolyte, (b) the dependence of μ_{eo} on the pH, (c) the influence of coating the cell walls and particles with the macromolecule poly-DADMAC, (d) contracting the double layers, (e) the duration of the coating process and (f) the switching frequency of the external electric field.

Key words Light scattering – electrophoretic – electro-osmotic

Introduction

Charged colloid particles in dispersion perform electrophoretic motion when an electric field E is applied to the cell containing the dispersion. Such a field, however, does not only induce electrophoretic motion of the particles but also causes electro-osmotic flow due to the double layers close to the cell walls [1]. If the field is instantaneously switched on, the system passes through a transient period with a complicated time dependent flow pattern in the cell. After the transient period, a stationary electro-osmotic flow field of the solvent is built up in which the electrophoretic movement of the particles takes place. Thus, apart from being an interesting scientific topic by itself, the knowledge of the electro-osmosis in a cell is prerequisite for a reliable determination of the electrophoretic mobility.

To be more specific, we consider electrophoretic light scattering (ELS) experiments. These are usually carried out in scattering cells or capillaries made from quartz glass. Due to the dissociation of surface OH-groups a negatively charged surface results which is in contact with a positively charged diffuse double layer. If an electric field E is applied, the H^+-ions in the diffuse double layer migrate towards the cathode. The drag exerted by the migrating ions on the solvent molecules results in a laminar flow of the solvent parallel to the cell walls, i.e. an electro-osmotic flow. Since the cells are usually closed, a back-flow in the center of the cell results. Under stationary conditions the general form of the hydrodynamic flow profiles for different cell geometries are well known [1]. For cylindrical and rectangular cells parabolic velocity profiles occur which are characterized by only one parameter, the electro-osmotic velocity v_{eo} or the electro-osmotic mobility $\mu_{eo} = v_{eo}/E$.

As pointed out the electro-osmotic flow in a cell is due to the double layer close to the cell walls. Thus, the cell wall material, the solvent, and surface dissociation reactions are of primary importance. However, other physicochemical quantities like the ionic strength, pH, and adsorption of macromolecules also play an important role for the double layer close to the cell wall and thus influence the electro-osmosis. It is important to note that these latter parameters may also affect the double layers of the colloidal particles and, therefore, their electrophoretic mobility. In fact, often ELS experiments are carried out in order to investigate the dependence of the electrophoretic mobility on these parameters. Thus, for a successful and reproducible determination of electrophoretic data from ELS experiments the influence of the abovementioned parameters on the electro-osmosis must be known.

In this paper we investigate the influence of the parameters, ionic strength, pH, and coating of the cell walls with a polycation, on the electro-osmotic mobility μ_{eo}. Further, in order to avoid electrode polarization ELS experiments are usually carried out by applying an alternating E field. We show that too high switching frequencies of the field cause significant deviations from the commonly used stationary parabolic electro-osmotic flow profile. The caveat from this latter investigation is that the commonly used geometric position of the stationary levels [1] is no longer valid under this condition.

Experimental

Electrokinetic light scattering experiments were carried out with a Zetasizer III, Malvern Instruments, England. In contrast to earlier heterodyne ELS experiments performed by our group [2] this instrument applies Laser-Doppler-Anemometry (LDS) [3, 4]. Details are given in the manual describing the instrument [5]. All experiments were carried out with a Malvern capillary cell Type AZ4 [5].

In order to make the electro-osmotic velocity profile measurable tracer particles which scatter light are usually used in an ELS experiment. Uncharged particles or particles at their isoelectric point (i.e.p.) would directly reflect the electro-osmotic flow in a cell. However, such dispersions are highly instable and coagulate rapidly. In order to avoid such complications charge stabilized particles were used in the present investigation at the expense that their electrophoretic velocity v_E must be taken into account.

The measurements were performed with two species of standard latex particles, IDC 306 (diameter $\sigma = 306$ nm) and IDC 686 ($\sigma = 686$ nm), purchased from Interfacial Dynamics Corporation (IDC), Portland, USA. These particles are rather monodisperse concerning both size and charge. In order to avoid particle adsorption on the cell walls low particle concentrations of the order $n = 4 \times 10^8\ \mathrm{cm}^{-3}$ were used. The dispersions were prepared from stock dispersions by dilution with salt solutions of definite ionic strength. All measurements were performed at $25 \pm 0.2\ ^\circ$C.

The electrophoretic/electro-osmotic particle velocity v_P in a closed cylindrical capillary cell is given by [1]

$$v_P(x) = (v_E + v_{eo}) - 8v_{eo}(x - x^2)\ , \tag{1}$$

where v_E is the electrophoretic velocity of the particles, v_{eo} is the electro-osmotic velocity, x is the relative distance measured from the center of the cylindrical capillary as a fraction of the diameter $2r$, i.e. $x = (1 - y/r)/2$ [1]. At the capillary axis ($y = 0$) the particle velocity is $v_P = v_E - v_{eo}$, at the wall ($y = r$) the particle velocity is $v_P = v_E + v_{eo}$. From Eq. (1) it is apparent that the particle velocity is parabolic and symmetric about the capillary axis.

For each dispersion the electrophoretic/electro-osmotic velocity profile was determined by measuring the mobility at nine different depth in the cell. The data were least square fit to the parabolic velocity profiles, Eq. (1), from which the electro-osmotic mobility, $\mu_{eo} = v_{eo}/E$, and the electrophoretic mobility, $\mu_E = v_E/E$, were obtained. Typical fits to experimental data are given below. The error of these two quantities is estimated to be 3%.

Results and discussion

As mentioned above in this paper the influence of (a) the ionic strength of the dispersion, (b) the pH, (c) adsorbed cationic macromolecules on the cell walls, and (d) the switching frequency of the applied electric field on the electro-osmotic mobility μ_{eo} and the flow profile in a cylindrical light scattering cell is investigated.

In Fig. 1 the electrophoretic/electro-osmotic mobility profiles for the IDC 306 particles at three NaCl ionic strengths 10^{-5}, 10^{-4}, and $10^{-2}\ \mathrm{mol\,l^{-1}}$ at low particle number density are shown. The solidly drawn curves, fitted according to Eq. (1), almost perfectly reproduce the experimental data points. Crossing of the three parabola at the stationary level indicates that under the conditions chosen, μ_E is almost identical so that the profiles can be compared regarding the electro-osmotic flow.

In Fig. 2 the resulting electro-osmotic mobility is shown as a function of the NaCl ionic strength, which ranges from 10^{-5} to $10^{-2}\ \mathrm{mol\,l^{-1}}$. Results for the two particle species IDC 306 and IDC 686 at particle number densities $n = 3.6 \times 10^8\ \mathrm{cm}^{-3}$ and $n = 7.2 \times 10^8\ \mathrm{cm}^{-3}$, respectively, are shown. Obviously, there is a significant

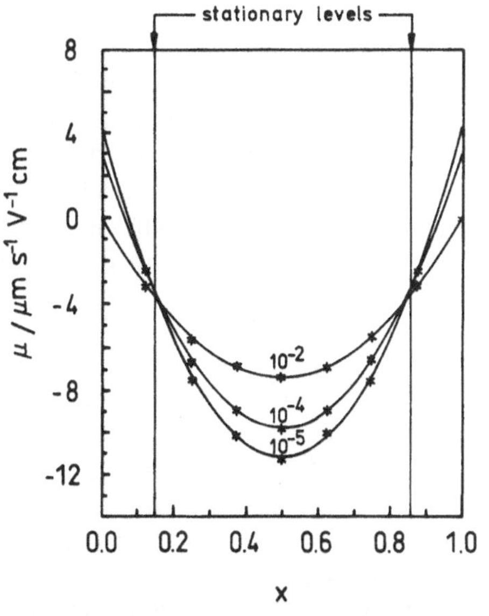

Fig. 1 Electrophoretic/electro-osmotic mobility profiles of dispersions containing IDC 306 particles ($n = 3.6 \times 10^8$ particles cm^{-3}) at the NaCl ionic strengths 10^{-5}, 10^{-4}, and 10^{-2} $mol\,l^{-1}$

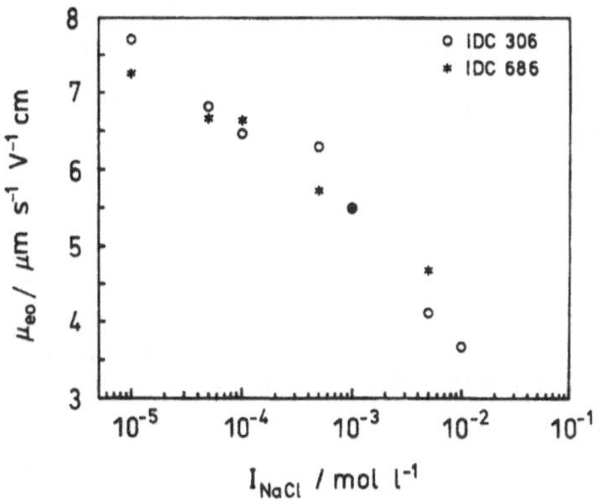

Fig. 2 Electro-osmotic mobility $\mu_{eo} = v_{eo}/E$ as a function of the NaCl ionic strength

decrease of the electro-osmotic mobility μ_{eo} with increasing ionic strength. Further, although the two particle species have different electrophoretic mobilities μ_E, within the experimental accuracy the electro-osmotic mobility μ_{eo} does not depend on the particle species used for the determination. By applying a different light scattering method, AWPS, Schätzel et al. [6] also investigated the electro-osmotic velocities v_{eo} of uncoated quartz cells. Their results are in good agreement with those shown in Fig. 1.

A quantitative interpretation of the observed behavior does not seem to be possible at present [7]. On the other hand, a qualitative explanation of the decrease of μ_{eo} with increasing ionic strength can be given. We recall that μ_{eo} is related to the ζ-potential of the cell wall as

$$\mu_{eo} = -\frac{\varepsilon_0 \varepsilon_r \zeta}{\eta} , \qquad (2)$$

where, $\varepsilon_0\varepsilon_r$ is the permittivity and η the viscosity of the medium. The dependence of the ζ-potential on the ionic strength and the pH for various surface materials has been discussed by Healy et al. [7] in terms of a site dissociation model which reproduces at least the broad features. For a negative surface like SiO_2 at a given pH the effect of an indifferent electrolyte is that the ζ-potential goes from a high negative value at low salt concentration to values close to zero at high salt concentration. According to Eq. (2) this leads to a high electro-osmotic mobility μ_{eo} at low NaCl ionic strength which decreases with increasing salt concentration. This is the experimentally observed behavior shown in Fig. 2.

Next, we consider the pH dependence of μ_{eo} which is shown in Fig. 3. The dispersions were prepared from IDC 306 particles at a concentration $n = 3.6 \times 10^8$ particles cm^{-3} and NaCl ionic strength $I = 1 \times 10^{-3}$ $mol\,l^{-1}$. The pH was adjusted by adding 0.01 *n* HCl or 0.01 NaOH. At low pH there is a strong increase of μ_{eo} with increasing pH, which saturates at about pH = 7 and eventually decreases again for pH > 10.

Starting from low pH it is known that at constant ionic strength the ζ-potential of a SiO_2 surface first assumes higher negative values due to the stronger dissociation of the OH surface groups. Saturation occurs as all groups are dissociated. In order to obtain pH = 10 a NaOH concentration of 10^{-4} $mol\,l^{-1}$ is required, which is 0.1 times the background NaCl ionic strength in the dispersions. At this and higher NaOH addition the overall ionic strength of the dispersion noticeably increases, which in

Fig. 3 pH dependence of μ_{eo} for dispersions of IDC 306 particles at $n = 3.6 \times 10^8$ particles cm^{-3} and NaCl ionic strength $I = 1 \times 10^{-3}$ $mol\,l^{-1}$

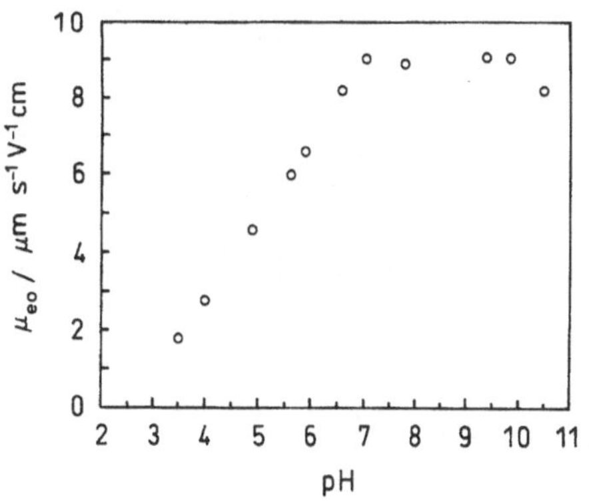

turn leads to a decrease of the ζ-potential and according to Eq. (2) to a decrease of μ_{eo}. This explains the initial increase, the saturation and the decrease of μ_{eo} shown in Fig. 3.

It is well known that coating of the cell walls with polyelectrolytes can have a strong influence on the double layer of the cell walls and thus on the electro-osmotic mobility μ_{eo} and the flow profile in a cell. Small electro-osmosis is desirable if μ_E is to be measured at the stationary levels. In this case the resulting low slope of the electro-osmotic profile at the stationary levels makes focusing to the stationary level uncritical. If the electro-osmotic mobility could be suppressed completely, μ_E could be measured at any depth in the cell. Accordingly, various materials as agarose gel, polyethyleneglycol, methylcellulose etc. have been used as coatings for cell walls [8–10]. One disadvantage is that none of the coatings is stable over extended periods of time. Further, usually the electro-osmosis is only partly removed.

Here, we report experiments in which the poly-cation p-DADMAC (polydiallyldimethylammoniumchloride) was used as coating material. After careful cleaning the cell and extended washing with distilled water, the cell was coated with a p-DADMAC solution. After the coating the cell was rinsed extensively with distilled water again.

Figure 4 shows that at least in principle the electro-osmosis can be removed completely with p-DADMAC (○ ○ ○). Also included in Fig. 4 is the electro-osmotic profile for the uncoated cell. The two experimental curves intersect at the stationary levels of the cell, i.e. at 14.7% and 85.3% of the capillary diameter. It turns out, however, that this coating is stable only for rather short periods of time. Also the preparation of the electro-osmosis free coating was not reproducible. On the other hand lower coverage of the cell walls with p-DADMAC can be carried out reproducible. For example a coverage which reduces the electro-osmosis by 70% is stable for more than 8 h. Coating of the cell is easily accomplished and requires little time and material. We therefore consider it as an interesting technique.

The amount of adsorbed polymer material depends on the concentration of the polymer solution and on the contact time of the solution with the walls. In Fig. 5 the influence of the concentration of the p-DADMAC solution is shown for $c_{p\text{-DADMAC}} = 1.25 \times 10^{-4}$ g/(100 g solution) and $c_{p\text{-DADMAC}} = 2.50 \times 10^{-4}$ g/(100 g solution). In both cases the contact time of the solution with the cell walls was 5 min. Also included in Fig. 5 is the electro-osmotic profile of the uncoated cell. Figure 6 shows the effect of different contact times $t = 5$ and 15 min for coating with a $c_{p\text{-DADMAC}} = 5 \times 10^{-4}$ g/(100 g solution). For the longer coating time a suppression of the electro-osmotic flow by 74% has been achieved.

In order to avoid electrode polarization ELS experiments are usually carried out by applying an alternating E field. Little attention has been paid to the fact that the commonly applied Eq. (1) which is valid for stationary electro-osmotic flow in a cell may loose its validity at too

Fig. 5 Influence of the concentration of p-DADMAC solutions for $c_{p\text{-DADMAC}} = 1.25 \times 10^{-4}$ g/(100 g solution) and $c_{p\text{-DADMAC}} = 2.50 \times 10^{-4}$ g/(100 g solution). The contact time was 5 min. Also shown is the electro-osmotic profile of the uncoated cell

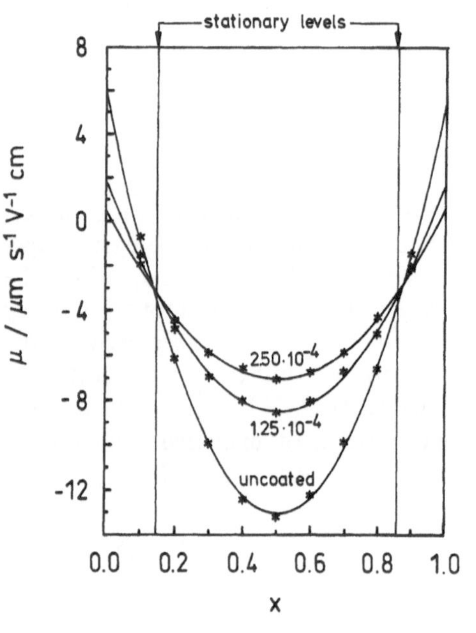

Fig. 4 Electrokinetic mobility profile for the uncoated (∗ ∗ ∗) and the p-DADMAC coated cell (○○○)

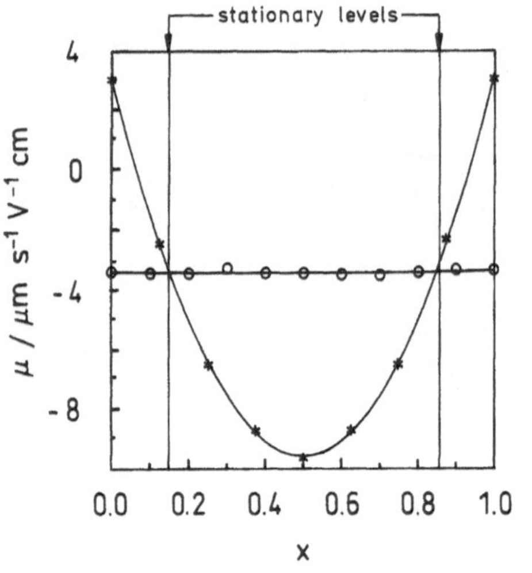

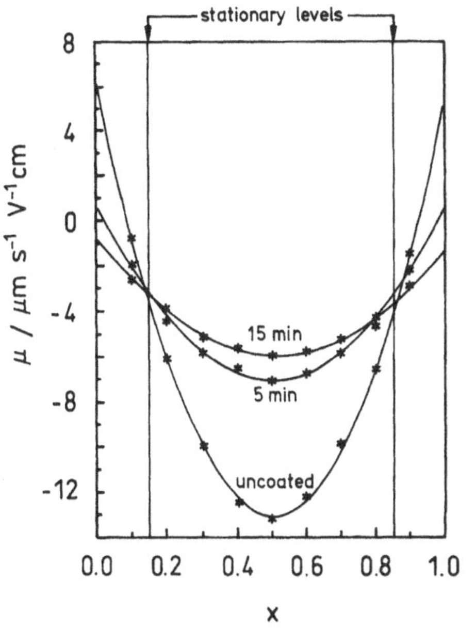

Fig. 6 The effect of contact time for coating a cell with a solution of $c_{\text{p-DADMAC}} = 5 \times 10^{-4}$ g/(100 g solution)

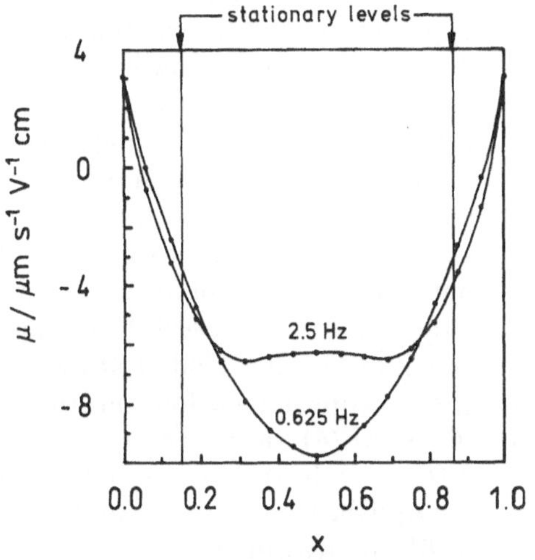

Fig. 7 Electro-osmotic profiles for the switching frequencies 0.625 and 2.5 Hz. For details see text

high switching frequencies of the *E* field. If the electric field applied to a closed electrophoresis cell is switched on, it takes about 100 ms until the electro-osmotic velocity profile reaches its stationary parabolic profile. For faster switching frequencies the electro-osmotic profile does not achieve the stationary profile. Under such conditions an electro-osmotic flow field is set up, which significantly deviates from Eq. (1).

In Fig. 7 electro-osmotic profiles for the two switching frequencies 0.625 and 2.5 Hz are shown. The measurements were carried out with IDC 306 particles at a particle number density $n = 3.6 \times 10^{-8}\,\text{cm}^{-3}$. At the lower frequency 0.625 Hz the profile can be perfectly be fit to a parabola, i.e. the flow profile is well described by Eq. (1). On the other hand at 2.5 Hz strong deviations from the low frequency profile occur. Although the deviations are largest in the central part of the cell, it is important to note that the stationary levels are shifted from the commonly used 14.7% and 86.3% of the cell diameter towards the cell walls. Schätzel et al. [11] calculated flow profiles for different frequencies by solving the Navier–Stokes equations. Qualitatively their calculations are in good agreement with our profiles shown in Fig. 7. Quantitatively the calculations predict deviations already at lower frequencies (0.625 Hz), which we were unable to confirm. Commercial ELS instruments allow measurements at frequencies for which the mentioned distortions of the electro-osmotic profile occur. Measurements of the electrophoretic mobility μ_E merely at the stationary levels do not allow to check the flow profile and can therefore be erroneous.

In conclusion, in this paper we have shown that several sources of error may occur in ELS experiments. However, we have also shown that usually these errors can be avoided by measuring the full flow profile.

Acknowledgements The authors are indebted to Dr. Fischer, Hoechst AG, Frankfurt, for supplying them with poly-DADMAC material. Financial support of the Deutsche Forschungsgemeinschaft and the Fonds der Chemischen Industrie is gratefully acknowledged.

References

1. Hunter RJ (1981) Zeta Potential in Colloid Science. Academic Press, New York
2. Palberg T, Versmold H (1989) J Phys Chem 93:5296
3. Pike ER (1977) In: Photon Correlation Spectroscopy and Velocimetry. Plenum Press, New York
4. Durst F, Melling A, Whitelaw S (1981) Principles and Practice of Laser-Doppler-Anemometry. Academic Press, London
5. Malvern Instruments (1989) Betriebshandbuch, ZetaSizer III, München
6. Miller JF, Schätzel K, Vincent B (1991) J Colloid Interface Sci 143:532
7. Healy W, White LR (1978) Adv Colloid Interface Sci 9:303
8. Herren BJ, Shafer SG, van Alstine J, Harris J, Snyder R (1987) J Colloid Interface Sci 115:46
9. Doren A, Lemaitre J, Rouxhet PG (1989) J Colloid Interface Sci 130:146
10. Van Oss CJ, Fike RM, Good RJ, Reinig JM (1974) Anal Biochem 60:242
11. Schätzel K, Weise W, Sobotta A, Drewel M (1991) J Colloid Interface Sci 143:287

Progr Colloid Polym Sci (1998) 109:232–243

E. Koglin
S.M. Kreisig
T. Copitzky

Adsorption of organic molecules on metal nanostructures: State of the art in SERS spectroscopy

Received: 29 September 1997
Accepted: 13 October 1997

Dr. E. Koglin (✉) · S.M. Kreisig
T. Copitzky
Institute of Applied Physical Chemistry
Research Center Juelich
D-52425 Juelich
Germany

Abstract Surface-enhanced Raman spectroscopy (VIS-SERS, FT-SERS) and surface enhanced Raman microprobe spectroscopy (micro-SERS) of molecules adsorbed on metal nanostructures is reviewed. Advantages and applications of these surface vibrational spectroscopic methods particularly for in situ study of the chemical identity, structure, orientation, chemical and electrochemical reaction of anions, surfactants, environmental pollutants, biomolecules and dye molecules adsorbed on charged Ag metal surfaces are discussed. Different applications show that these spectroscopic techniques are a powerful in situ method to study the liquid/solid interface and characterize self-assembled monolayers (SAMs) at high resolution. Moreover, the high enhancement factor of the surface Raman scattering intensity creates a new technique for obtaining vibrational spectra of organic and inorganic molecules from dilute aqueous solutions down to 10^{-14} mol dm^{-3}. Micro-SERS spectroscopy permits the acquisition of surface Raman spectra from substance spots down to 1 μm in size or other forms of microsamples approaching the femtogram level.

In addition it is shown that colloidal metal nanoparticles can be used as probe marker to study the location and identification of adsorbed molecules on different substrates (nano-TLC, oxides, clay minerals) by means of SERS spectroscopy. In order to reveal the optimal conditions for this kind of SERS spectroscopy, atomic force microscopy (AFM) was applied to investigate the surface morphology of the Ag labelled substrates (overlayer SERS spectroscopy).

Key words SERS spectroscopy – monolayer adsorption – AFM trace analysis – ab initio calculations

Introduction

The discovery that Raman vibrational signals from molecules adsorbed on nanometer scale metal particle structures are enhanced by 10^6–10^9 has caused extraordinary interest and excitement [1–6]. This Raman technique, known as Surface-enhanced Raman scattering (SERS), offers new possibilities as a spectroscopic probe of surface chemistry and dynamics:

(i) The ability to determine, in situ, the chemical identity, structure, orientation, conformation and dynamics of adsorbed molecules in the first monolayer.

(ii) Extraordinary sensitivity and selectivity to the interfacial species.

(iii) Competitive adsorption reactions, displacement kinetics and coadsorption effects.

(iv) Operation at solid/gas, solid/liquid, and even solid/solid interfaces.

At present, both experimental investigations and theoretical calculations have reached an interesting level of maturity and demonstrated the applicability of the technique to problems of genuine surface analytical significance. Combining SERS spectroscopy and laser Raman microprobe spectroscopy has enabled in situ analysis of surface spots containing quantities of material down to subpicogram levels [7–9].

Characteristics of SERS

The major theoretical concepts which are concerned with the origin of the surface enhancement effect can be classified within two categories: the long-range classical electromagnetic enhancement (CEME) and a short-range "chemical" or "electronic" contribution to SERS. Consequently, from the CEME point of view, nanometer scale structures of metal particles are ideal Raman intensity enhancers, causing the adsorbates to experience an intensity enhancement of ca. 10^6. The CEME predicts long-range enhancement from the resonance of localized surface-plasmon polaritons and is not sensitive to the chemical nature of the adsorbed molecule.

One of the main problems in attempting to interpret SERS within the CEME framework of this mechanism is the high sensitivity of SERS in the first monolayer: short-range effect, or "first-layer" SERS. Theories which invoke a molecular mechanism and consider the molecular interaction between the molecule and the metal particle are presented in detail by Otto [5] and Birke et al. [3]. The mechanism of Otto proceeds by increased electron–photon coupling at an atomically rough metal surface and by temporary charge transfer to orbitals of the adsorbates. The comprehensive development of the charge-transfer theory of SERS by Lombardi et al. incorporates the Herzberg–Teller mixing of zero-order Born–Oppenheimer electronic states by means of vibronic interaction terms in the Hamiltonian.

The intensity of SERS includes therefore a variety of contributions such as CEME enhancement (G_{EM}), enhancement due to charge-transfer interaction (G_{CT}), the surface coverage (θ) and orientation (O) of the adsorbates, and some extra enhancement (G_{ex}). The SERS intensity can be written as a product of these various contributions:

$$I \sim G_{EM}(\omega_0, \omega_s) G_{CT}(\omega_0, \omega_s, V) \theta(V)\, O(V) G_{ex}(\omega_0, \omega_s, V) ,$$

where the parameters in parentheses indicate the frequencies of the incident laser (ω_0), on the frequencies of the scattered photons (ω_s) and on the surface charge V.

Experimental investigations of biomolecules and long-chain surfactants have indicated that the sensitivity of the SERS enhancement is limited by the short-range enhancement for distances smaller than 10 Å from the surface [10]. Thus, SERS spectroscopy is a very sensitive method to detect parts lying close (first monolayer) to a charged surface of an adsorbed macromolecule.

Topography of the nanostructures

A simple experimental verification of SERS spectroscopy will be realized by molecules on colloidal dispersions of nanometer-sized metallic particles, i.e. by adsorption on the so-called metal hydrosols. Silver hydrosols which are prepared by reduction of silver nitrate with $NaBH_4$ (Creighton sols) showed a single adsorption band at $\lambda = 390$ nm, characteristic of silver particles with a diameter of ca. 15 nm. These structures have been well characterized by STM microscopy.

Figure 1 shows, as an example, the STM photomicrograph of the Ag-Creighton colloid. Observations on silver electrodes have demonstrated that no Raman signals are found for a freshly prepared nonroughned surface immersed into an electrolyte, containing low concentrated adsorbates. In order to "see" an enhancement, e.g. to increase the surface Raman scattering above about 10^2, an oxidation–reduction cycle (ORC) must be applied to the electrode surface. The ex situ ORC surface roughness of the electrode surface used for our SERS experiments is topographically characterized using AFM microscopy in elucidating the quality of the SERS spectra.

Figure 2 shows the AFM images of the freshly prepared silver electrode surface after an ORC. It was found that the intensities of the acquired SERS signals run

Fig. 1 STM photomicrograph of Ag-Creighton colloids on a HOPG substrate. The imagined area: $X = Y = 124$ nm

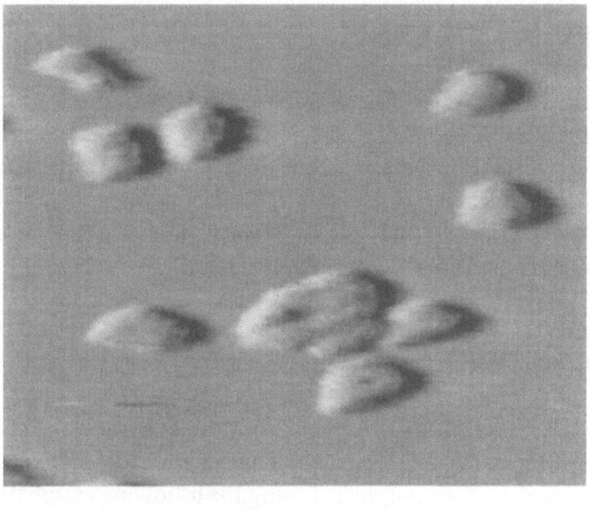

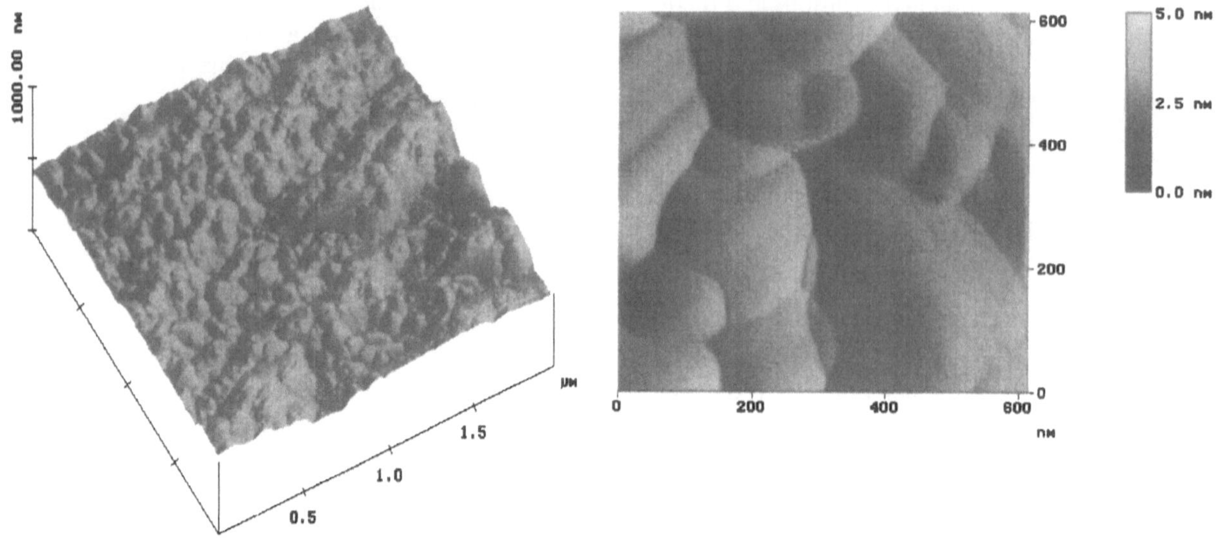

Fig. 2. AFM images of the freshly prepared silver electrode surface after an ORC in the μm and nm scale

through a maximum at a charge of $Q = 130\,\text{mC/cm}^2$ so that in all following experiments this optimized ORC could be applied. The surface image of such an optimized roughened surface in a nano-meter scale is also illustrated in Fig. 2.

The basic aim of our analytical investigations on SERS was the combination of chromatography as a separation technique and SERS as an ultra-sensitive detection method as previously reported [11], e.g. a metal colloid can be sprayed on to a thin-layer chromatographic plate (HPTLC) as a probe particle to analyse in situ the chromatogram spot. The deposited nanoparticles act as antennae. Meanwhile, this SERS depositing technique is widely used to provide information on non-SERS-active surfaces.

Experimental SERS methods

An SERS apparatus basically consists of a laser excitation source (visible spectral range, VIS or near infrared range, NIR), the liquid Raman unite or a potential-controlled electrochemical cell, the optics for collecting the molecular surface scattering, and a Raman spectrometer (dispersive monochromator or a FT-interferometer).

Opto-electrochemical preparation

The detection of SERS signals from the electrode surface requires specific surface roughening of the electrode surface which was accomplished using the following roughening procedure. The surface is pre-treated by running the ORC in the electrolyte solution without the molecule studied (ex situ SERS). The voltage is stepped to $+0.2\,\text{V}$ vs. SCE in 0.1 M KCl solution in order to pass between 100 and $150\,\text{mC/cm}^2$ and subsequently stepped back to $-0.6\,\text{V}$. For intensity standardisation the procedure was put under the digital Coulomb meter control ($130\,\text{mC/cm}^2$ for the multiple and repeating measurements) and yielded a very reproducible SERS-active electrode surface. This quantity of surface charge yields the most reproducible SERS signals.

In order to avoid conformation changes of the molecule in the solution we chose for an ex situ roughening, i.e. ORC in the electrolyte solution in the absence of the substance and the molecular solution under investigation is then added to the opto-electrochemical cell.

The SERS spectra are obtained at a negative potential in the electrochemical cell. In this region from 0.0 to $-1.4\,\text{V}$, the non-Faradaic region, no electrochemical change occurs for all investigated substances.

Electrochemical equipment

The opto-electrochemical cell consisted of a Teflon cylinder with a diameter of ca. 0.8 mm and a capacity of ca. 0.5 ml of solution. The working electrode was a polycrystalline Ag metal disk, ca. 2 mm in diameter, enclosed in a Teflon holder. This metal electrode was prepared before each experiment by mechanical polishing with an Al_2O_3 suspension, ultrasonic treatment and electrochemical cleaning by H_2 evolution. The electrochemical equipment consisted of a potentiostat (PAR, model 173) using a three-electrode system and a function generator (PAR, model 175) as programmer for the oxidation–reduction cycle (ORC). A digital Coulomb meter was used to measure the charge transfer during the ORC, which indicates the degree of metal dissolution and recrystallization of the electrode. All potential measurements were made with respect to the saturated calomel electrode (SCE).

SERS microprobe set up

Raman spectra in the visible spectral range were obtained with an Instruments S.A. MOLE-S3000 triple, computerised, spectrometer equipped with multichannel (E-IRY 1024) data acquisition. The excitation was performed with an argon-ion laser (Spectra Physics, model 2020-03) operating at 457, 488 and 514 nm or via a Dye laser system at 645 nm. The microscope attached to the MOLE-S3000 provides a way of precisely focusing the laser light on the electrode surface (diameter of 1 μm, 8 mW of power). The collection of one Raman spectrum takes about 5–10 s. This implies that during one ORC (5 mV s^{-1}) 20–40 spectra can be recorded. With 600 grooves mm^{-1} holographic gratings the spectral coverage was 1600 cm^{-1}.

NIR-FT-Raman spectra

The instrument used was a Bruker FT-Raman system RFS 100 and a cooled Ge diode detector. A Nd : YAG laser was used as the exciting source (λ_{ex} = 1064 nm; 250 mW of power at the sample) in the backscattering configuration. All spectra were recorded at 4 cm^{-1} resolution.

Atomic force microscopy

The surface morphology images of the roughened electrode surface were recorded using a Nanoscope III microscope equipped with a four-segment laser beam deflection sensor (Digital Instruments). All images were acquired in air (static contact/variable deflection mode) with Si_3N_4 cantilever (force constant 0.12 N m^{-1}) having pyramidal tips with 70° cone angle and 20–80 nm effective radii of curvature.

Quantum chemical ab initio calculations

In order to utilise SERS spectra to study adsorption geomtries of molecules and their mixtures on charged surfaces we need precise assignments of vibrational wavenumbers. This was achieved by comparison of observed band positions and intensities in IR and Raman spectra with wavenumbers and intensities from quantum chemical ab initio calculations of different molecules. Information on the adsorption geometry in the first monolayer at the silver electrode surface can be obtained on the basis of these assignments because the magnitude and direction of the dipole moment fluctuations in the CEME and CT model are directly related to the character of the normal modes.

All calculations were performed with the Gaussian 94 suite of programs [12] on a Cray supercomputer (J90, T90). The geometry was fully optimised without imposing external symmetry constraints using the HF/6-31G* basis set. Thereafter a force constant calculation was performed to obtain the vibrational frequencies and the corresponding IR and Raman intensities. The resulting wavenumbers from this type of calculations are on average 10% larger than the experimental data. For comparison with the results of FTIR and FT-Raman spectroscopy, the calculated wavenumbers were therefore multiplied by a constant scaling factor of 0.89.

Density functional theory (DFT) methods have recently become available for molecular structure optimizations and frequency calculations. These methods promise to provide calculations at a very high level of theory [13].

Coadsorbed halide ions are very important for the binding of cationic molecules in the first monolayer. The positively charged parts of the molecule are surrounded by halide ions and this molecule/halide ion/surface complex is strongly adsorbed even on a positively charged surface. Therefore, in order to determine the adsorption geometry in this surface complex DFT ab initio calculations are carried out. In this calculation, the silver surface was modelled by constructing an FCC Ag_{10} cluster and the electron distribution and the vibrations of the adsorbed halide ion on this silver cluster was calculated.

Applications

Adsorption of anions

In order to elucidate the nature of the *molecule/electrolyte ion/surface complex* in the first adsorption layer, the effect of the counterions on the adsorption process has to be studied. In general, the following vibration modes are expected in the molecule/metal surface system: (i) the enhanced internal modes of the components of the adsorbed organic molecule, (ii) the interfacial vibrations of the adsorbed molecule interacting with the metal surface, (iii) the vibrations of counterions interacting with the charged metal surface, i.e. Ag/Cl, Ag/Br, Ag/NO_3 vibrations. As an

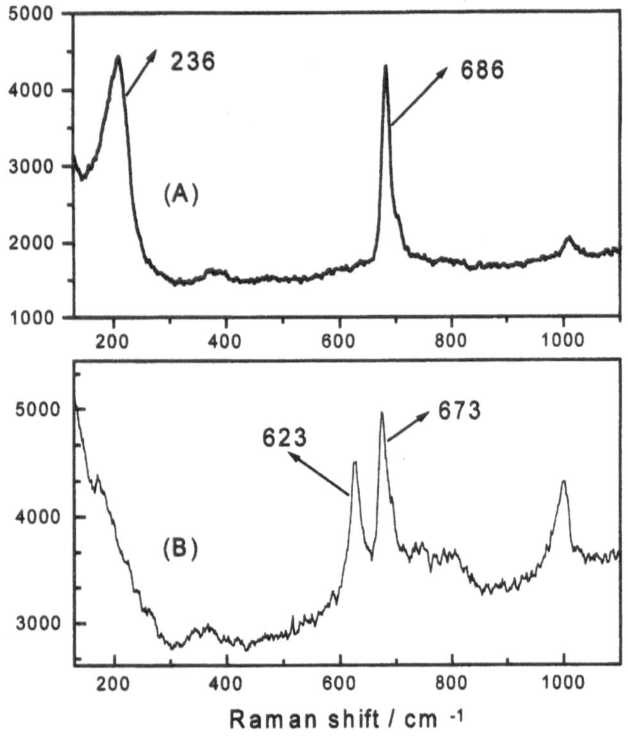

Fig. 3 Micro-SERS spectra of melamine: Ag electrode; ex situ roughening; E_s of -0.2 V vs. SCE; concentration of melamine, 10^{-2} M; excitation 514.5 nm; laser power, 4 mW; laser spot focus, 1 μm; (A) 1 M KCl electrolyte concentration; (B) 10^{-3} M KCl electrolyte concentration

example the adsorption process of melamine (1,3,5-tri-amino-s-triazine) on a roughened silver electrode was investigated. Figure 3 shows the micro-SERS spectra in the wavenumber range between 150 and 1100 cm^{-1} at a potential of -0.2 V vs. SCE. The potential of -0.2 V corresponds to a positively charged surface. There are two intensity maxima of the SERS vibrational modes at this potential. The first maximum is at 236 cm^{-1} and the second intense maximum at 686 cm^{-1}. The broad low lying surface micro-SERS band at 236 cm^{-1}at the electrode potential of $E_s = -0.2$ V appears just after the complete reduction of the adherent AgCl and Ag-melamine layer and can be attributed to the Ag^+/Cl^- stretching mode. The intensity of the band at 236 cm^{-1} strongly decreases upon a shift of the potential from -0.2 to -0.8 V. Since chloride ions are completely desorbed at $E_s = -0.8$ V the Ag^+/Cl^- stretching mode is missing at the neutral or negative charged surface.

The intense SERS band in the whole potential region at 686 cm^{-1} is the characteristic in-plane "ring breathing 2" mode of the malamine molecule of the triazine ring in the trisubstituted triazine molecule.

Decreasing the Cl^- concentration from 1 M KCl to 0.001 M KCl affects the adsorption process of the active melamine surface complex and markedly changes the SERS spectrum appearance. Especially, the relative intensity of the out-of-plane vibration at 623 cm^{-1} dramatically depended on the bulk chloride ion concentration (cf. Fig. 3B). The intensity of the Ag^+/Cl^- vibration is strongly decreased in the 0.001 M KCl electrolyte solution.

The DFT calculated $Ag_{10}Cl$ cluster model vibration of the Ag–Cl interfacial band is illustrated in Fig. 4a. The unscaled DFT calculated frequency is at 231 cm^{-1}. The observed frequency in the NO_3^- electrolyte solution is observed at 244 cm^{-1} and the calculated frequency is at 228 cm^{-1} (cf. Fig. 4b).

Surfactants

Advantages and applications of the SERS vibrational spectroscopic method are discussed, particularly for the in situ study of the organization and distribution of surfactants at the liquid/solid interface. Some fundamental features of the mechanism of adsorption of surface active agents on metal nano-particle surfaces at the solid/liquid interface are extremely important in a variety of fields such as detergency, adhesion, lubrication, dewetting and corrosion inhibition. The synergistic coadsorption of surfactants and organic pollutants with soils or other natural adsorbents in aqueous environment is a mechanism of fundamental importance in determining the fate and distribution of these amphiphilic substances in the environment. In addition, even small amounts of surfactants produce major effects with respect to the transport and desorption of pollutants in soils and sediments. The in situ investigation of the adsorption kinetics of organic pollutants, its interaction with surfactant molecules, and the removal of contaminants by means of surfactants is an open field, still largely unexplored, mainly because of a lack of an appropriate technique for determining adsorption processes in the first solid/liquid adsorption layer.

SERS spectroscopy has been used to study, in situ, the adsorption behavior of different kinds of surfactants on charged metal surfaces [14–18]. The investigations have shown that this surface spectroscopy is an extremely powerful technique for monitoring the interfacial behavior of surfactants, for assessing adsorbate orientation, adsorbate conformation, investigations of coadsorbed surfactants, competitive and displacement adsorption, characterising self-assembled monolayers and for probing the effects of various environmental factors on the adsorbate/substrate interaction.

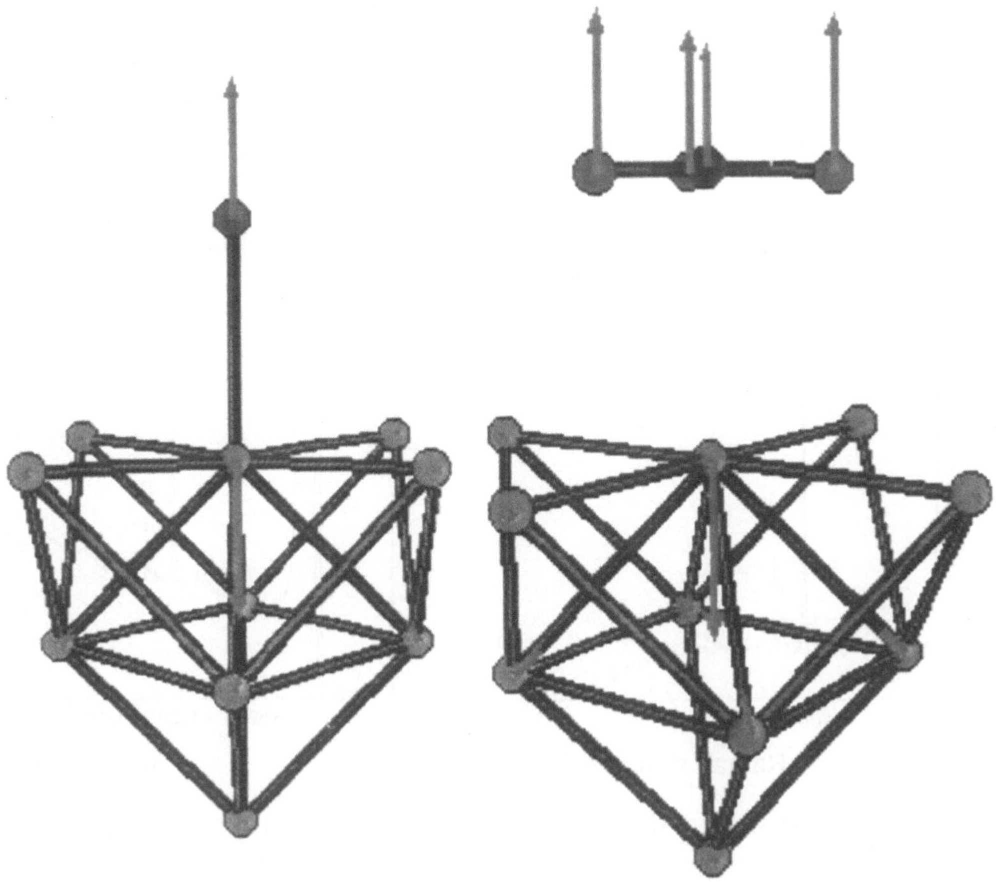

Fig. 4 DFT calculated vibrations of the Ag–Cl and Ag–NO_3 interfacial bands

The importance of the determination of the adsorption geometry of surfactants in the first monolayer is stressed in Fig. 5a which shows the adsorption behavior of cetylpyridinium bromide at the positively charged Ag colloid surface. The same results are obtained after an SERS-optimized ex situ roughening by means of an ORC in a 0.1 M KBr solution. In this case 1 μl of a 10^{-3} M CPB solution was applied to the roughened and dried electrode surface (cf. Fig. 5b; SAMs).

There are two intensity maxima of the SERS vibrational modes of the CPB molecule at this electrode potential. The first maximum is located at 1026 cm^{-1} and this SERS band can be attributed to the enhanced ring breathing mode (ab initio calculated mode ν_{24}) of the pyridinium head group. The second intense band at 164 cm^{-1} is attributed to the Ag–Br interfacial band. Comparing the relative intensities of the NIR-FT-SERS spectrum and the normal FT-Raman spectrum, we can clearly see the strong enhancement of the pyridinium ring breathing mode of the pyridinium head group. In addition, we can recognize that the enhancement of the pyridinium ring vibrations and the hydrocarbon modes are quite different. The characteristic tail vibration at 1301 cm^{-1} (CH_2 twisting vibration of hydrocarbon tail) is completely absent in the adsorbed state. The CH_2-scissoring vibration at 1452 cm^{-1} shows a very low intensity. On the basis of the short-range sensitivity of the SERS enhancement, we can conclude that the pyridinium head group is attached to the surface, leaving the hydrocarbon chain directed away from the surface. Because vibrations with atomic motion perpendicular to the surface couple more effectively to the surface electromagnetic waves, the strong intensity of the enhanced ring breathing mode (1026 cm^{-1}) suggests that the pyridinium head group is in an upright configuration (edgewise orientation with coadsorption of Br^- ions).

The strong SERS signals of the cationic CPB surfactant head group on highly positively charged silver surfaces and the intense Ag/Br^- SERS interfacial band at 164 cm^{-1} suggest that the strong adsorption process is a result of the formation of a pyridinium/bromide/surface complex. The positively charged nitrogen (N^+) atom of the CPB molecule can adsorb on the electron cloud (calculated in the $Ag_{10}Br$ cluster surface model) surrounded on the adsorbed Br^- ions in the interfacial region. Therefore, it is possible that the adsorbed bromide ion acts as a bridge for the surfactant adsorption on a positively charged surface. In addition, an electron transfer between the metal surface and the CPB molecule through the

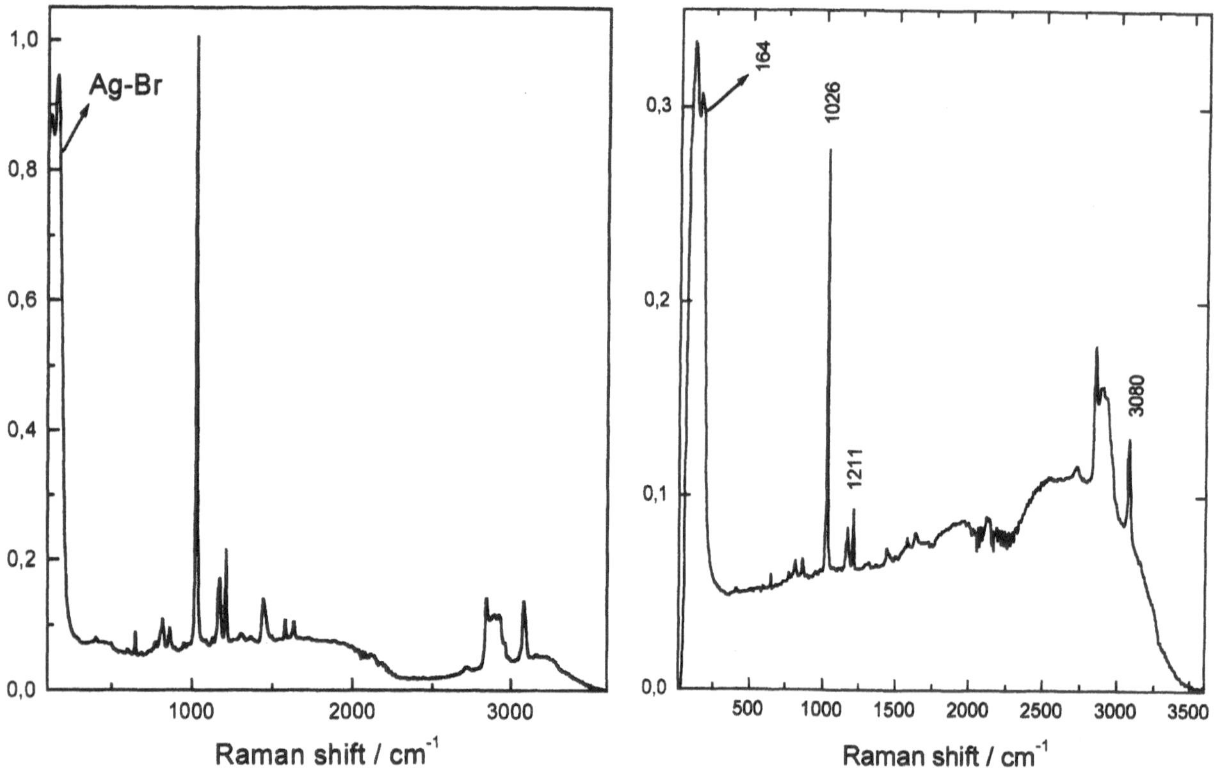

Fig. 5 Colloid NIR-FT-SERS of cetylpyridium bromide (CPB) adsorbed on Ag colloids (A) and adsorbed (SAMs) on a dried Ag electrode surface (B)

adsorbed counterions can explain the strong SERS intensity by an extra charge transfer type resonance SERS contribution to the overall surface enhancement factor at the positively charged surface (B and C term resonance from the Herzberg–Teller theory [3]).

Environmental pollutants

The assessment of environmental pollution by trace-level organics requires increasingly sophisticated measurement methods. Because SERS spectroscopy provides both rich vibrational spectroscopic information and sensitivity as a result of the large surface enhancement effect, in principle it is an ideal tool for trace analysis, interfacial (solid/liquid) investigations and chromatographic detection. The micro-SERS spectra of PCDDs and PCDFs adsorbed on Ag electrode surfaces, Ag hydrosols and SERS-activated high performance thin-layer plates (HPTLC) have been recorded and the results are compared, attempting to clarify the molecular adsorption geometry in the first monolayer [8]. The synergistic coadsorption of surfactants and organic pollutants with soils or other natural adsorbents in aqueous environment is a mechanism of fundamental importance in determining the fate and distribution of these substances in the environment. SERS spectroscopy has been used to study, in situ, the adsorption process of coadsorbed cationic surfactants (binary mixtures) and displacement adsorption of cetylpyridinium chloride (CPC) of preadsorbed nitrophenols on charged silver nanoparticle surfaces [15, 18]. These investigations have shown that this surface spectroscopy is an extremely powerful technique for monitoring the intefacial behavior of surfactants an organic pollutants, for assessing adsorbate orientation, displacement kinetics, competitive adsorption and for probing the effect of various environmental factors on the adsorbate/substrate interaction.

NIR-FT-SERS spectra of 18 polyaromatic hydrocarbons (PAHs) were measured on silver surface which had been previously roughened by an oxidation–reduction procedure in an opto-electrochemical cell [19]. Detection limits were around 10^{-9} mol for most compounds. The combination of micro-SERS spectroscopy with dried roughened electrode substrates results in an extremely low detection limit. As an example Fig. 6 shows the micro-SERS spectrum of pyrene adsorbed on an electrochemical roughened and dried Ag surface in comparison with the micro-Raman spectrum of a microparticle of pyrene.

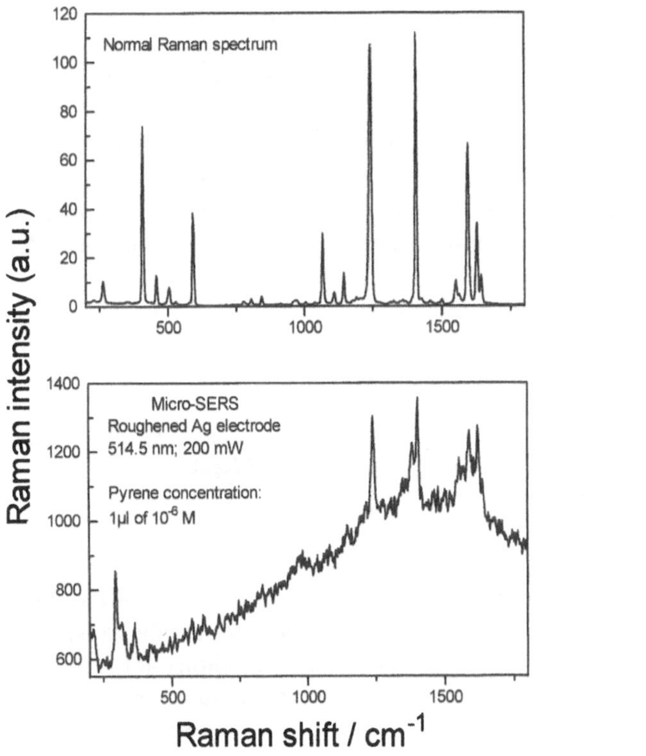

Fig. 6 Raman microprobe analysis of pyrene on roughened and dried Ag surface

Biomolecules

The molecular structure and dynamics of chemisorbed biomolecules are of great importance in order to elucidate the behavior of these molecules at the solid/liquid interface. About 200 original papers devoted to the study of different classes of biological molecules (nucleic acid basis, amino acids, water-soluble membranes, photosensitive proteins, DNAs, drug-DAN complexes, extracts of ocular lenses, living cells, chromosomes, cell virus interactions, cancer cells) have been published in the last 15 years [20–22]. Some of these currently being investigated are: the selective study of cell membrane components, the determination of the distribution of drugs within a living cell and on the cellular membrane, the analysis of biomedical mixtures and extracts [23].

Since the discovery of left-handed Z-DNAs, Raman and SERS spectroscopy has become a powerful tool to identify the Z-DNA conformation under different environmental conditions [24]. One main advantage for biochemistry is that surface Raman spectra are obtainable from highly dilute aqueous solutions and that surface Raman spectroscopy at solid/liquid interfaces becomes a general tool for the study of the physicochemcical phenomena that take place in such environments. Figure 7 shows the

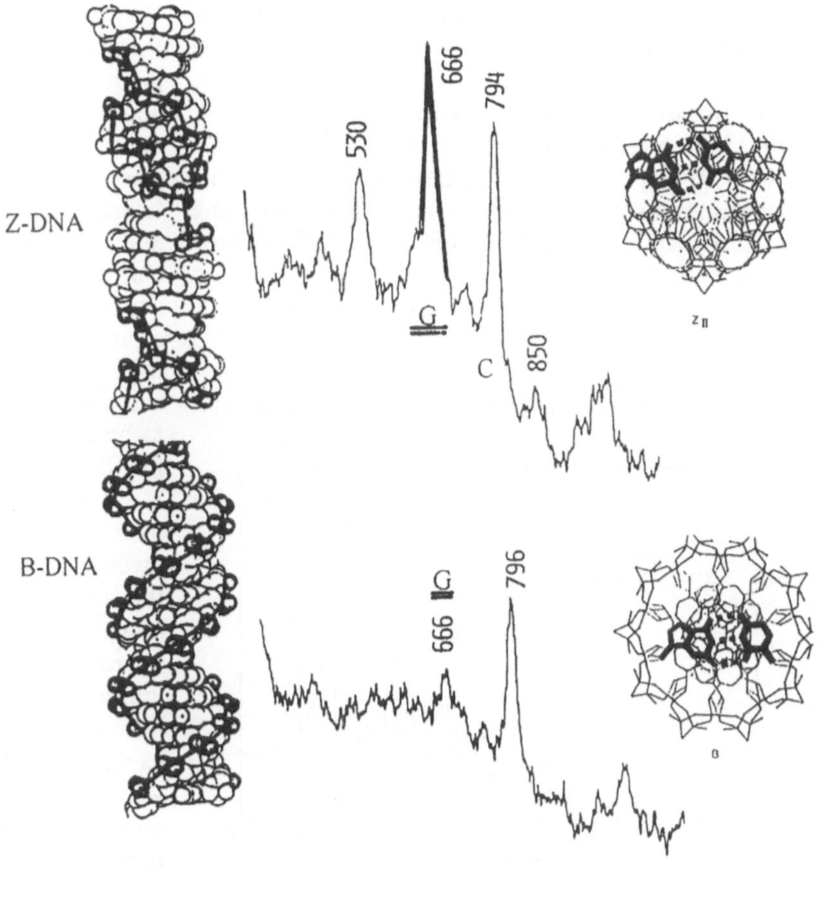

Fig. 7 Ag electrode SERS spectra of poly (dG-dC) × poly(dG-dC), B and Z(Eu^{+3}) structure. Concentration of the nucleotide units 2×10^{-4} M, Eu^{+3} 10^{-4}, 0.15 M NaCl; laser excitation line 514.5 nm

electrode SERS spectra for B and Z structures of poly(dG-dC)* poly(dG-dC) adsorbed at a potential near the potential of zero charge (pzc). Keeping in mind that the SERS effect is a short range phenomenon results in a strong increase in intensity of the peripheral syn-position of the guanine moieties in the Z structure. Therefore, these results given not only general insight on the physicochemical aspects of the recognition of B and Z structures of DAN in a more complex biological interfacial system but provide also a method to monitor, in situ, the B–Z conversion.

Dye molecules

The Raman scattering cross-section of an adsorbed molecule can be increased by utilizing a laser excitation frequency which is in resonance with an electronic transition in that molecule. This molecular resonance Raman scattering and the SERS effect can combine to give SERRS (surface-enhanced resonance Raman scattering) so that the limit of detection is further increased. Therefore, detecting resonant molecules in solutions at very low concentrations in many fields of research and development, such as environmental science, biochemistry, medicine, pharmacology and analytical chemistry was achieved [8, 15, 22]. Another striking feature of SERRS spectroscopy is that fluorescence of the adsorbate can be completely quenched by the metal surface which generates a high-quality surface Raman spectrum.

The lowest detected concentration levels were 4×10^{-15} M on Ag microelectrode for a cyanine day and colloid SERRS spectra of rhodamine 6G (R6G) at concentrations as low as 8×10^{-16} M in colloidal silver solutions activated by NaCl ions [25]. The colloid SERRS spectroscopy of R6G shows the possibility of using SERRS for single molecule detection [26]. Moreover, the high enhancement factor of the SERRS effect combined with the Raman microprobe spectroscopy can create a new technique for obtaining very high-resolution surface vibrational spectra.

Fig. 8 SERRS spectrum of 10^{-12} M rhodamine 6G in silver colloidal solution

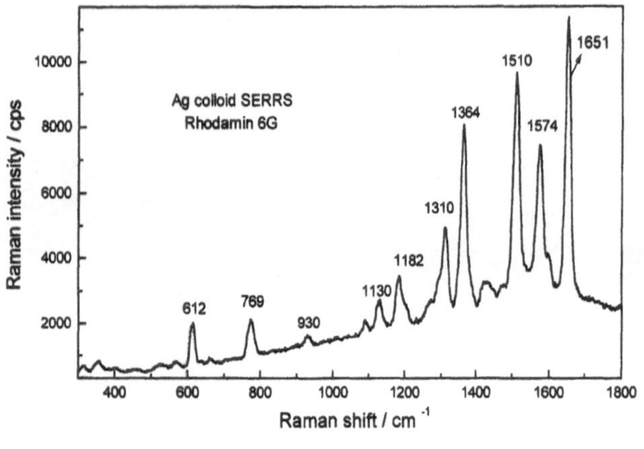

Another common application of SERRS spectroscopy is the study of adsorption behaviour and conformation of chromophore-containing biomolecules (haem proteins, flavoproteins, phytochrome) of the solid/solution interface.

To show the high sensitivity of SERRS spectroscopy Fig. 8 demonstrates the adsorption of R6G on Ag hydrosols at a concentration as low as 10^{-12} M.

Colloids as probe particles (overlayer SERS)

Direct analysis of HPTLC spots

High-performance thin-layer chromatography (HPTLC) is a useful, inexpensive and easy to handle separation technique and hence the method of choice for numerous applications. Identification of the separated species can be performed via R_f-values, chemical methods (spray reactions), and spectrometry (UV/VIS, IR, Raman), although those techniques are rather non-specific or sometimes limited due to their lack of sensitivity.

One of the significant limitations on the applications of Raman spectroscopy in the field of HPTLC chromatogram spot characterization is the problem of fluorescence and reflected background scattering. An important advance in methodology was made by SERS spectroscopy. It was shown that excellent Raman spectra could be obtained for nanogram to picogram levels of nonresonant and fluorescent substances on filter paper, paper chromatographic supports, or high-performance thin layer chromatographic plates using SERS or SERRS spectroscopy [27–30]. For HPTLC this SERS effect is accomplished by spraying chromatograms with colloidal silver solutions. As an example of this general SERS overlayer technique Fig. 9 shows the topography of a silver colloidal SERS activated silica gel (KG60) HPTLC plate.

The laser Raman microprobe has been used in combination with HPTLC-SERS spectrometry for the in situ investigation of spots on these nano-TLC plates. It permits the acquisition of Raman spectra from HPTLC spots down to 1 μm in size or other forms of microsamples approaching the subpicogram level. In order to further improve the sensitivity of the HPTLC/SERS method we have carried out experiments in the field of well defined vacuum-deposited Ag films onto the separated and developed HPTLC plates. The HPTLC plate was mounted on a holder inside a vacuum chamber where silver was thermally evaporated onto the plate. The evaporation rate

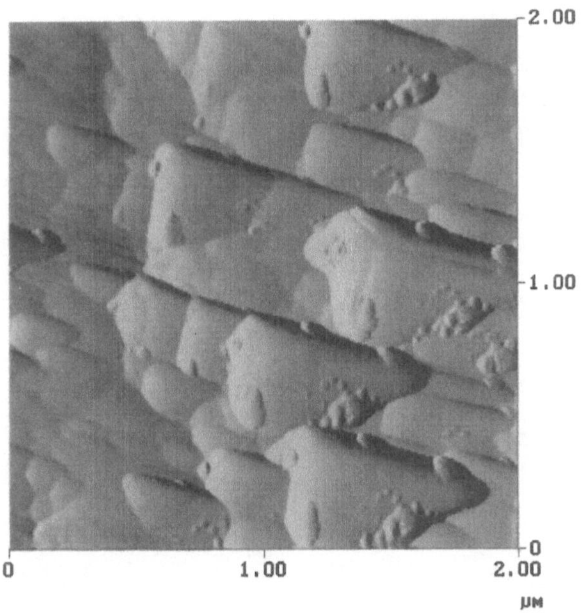

Fig. 9 AFM (tapping mode) surface plot of a silver colloidal SERS activated silica gel plate (KG60)

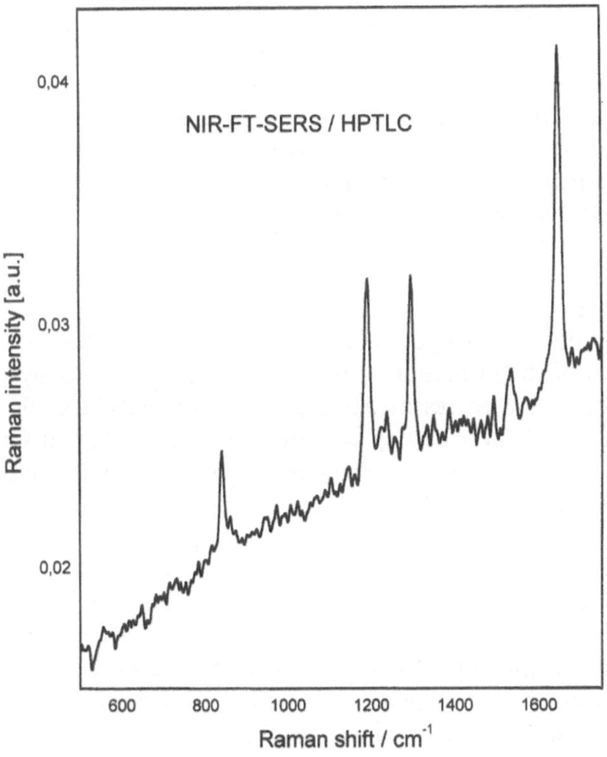

Fig. 10 HPTLC/FT-SERS analysis of paraquat. Silver-coated nano-TLC plate (silca gel 60, KG60)

and the silver thickness were controlled in order to find out the most intense Raman signals. To show the sensitivity of NIR-FT-SERS in combination with the Ag coating technique on HPTLC plates Fig. 10 demonstrates the detection of the herbicide paraquat on a KG60 plate spot at a concentration as low 5 ng.

In order to interpret the HPTLC/FT-SERS spectra it is first necessary to assign accurate frequencies of the Raman bands. Therefore, to complement the HPTLC/FT-SERS studies, we carried out quantum mechanical ab initio calculations at the Hartree–Fock level. The corresponding calculated vibrations of the four strong bands of paraquat at 848, 1204, 1308 and 1651 cm^{-1} in the HPTLC/FT-SERS spectrum are shown in Fig. 11.

Abdsorption on clay minerals

The adsorption of organic pollutants in soils and sediments take place primarily on colloidal layer silicates (clay minerals such as kaolinite, montmorillonite and illite). The organic molecules alter the adsorption properties of layer silicates, depending on type and structure, by means of ion exchange, physisorption and intercalation of the adsorbed molecules between silicate layers. Therefore, the interaction between organic pollutants and clay minerals is of great importance for environmental, waste disposal, and technical research. New techniques that appear to be capable of answering some basic questions regarding organic pollutants adsorption on clay minerals is Fourier transform surface-enhanced Raman scattering and micro-SERS. Utilization of the Colloid-FT-SERS techniques for obtaining detailed information on the adsorption mechanism of cationic surfactants on kaolin was achieved [31, 32]. Since the positively charged Ag marker colloids bind non-specifically to the negatively charged basal planes of the silicate kaolin, the FT-SERS spectra can give information about the location of the cationic surfactants on the clay mineral. On the basis of the short range sensitivity of the enhancement, it was concluded that in the case of the cetylpyridinium surfactant (CPC) adsorption that the pyridinium head group in the kaolin/CPC/Ag-colloid complex is attached to the basal planes of kaolin, leaving the hydrocarbon chain away from the surface. These investigations have shown that the Colloid-SERS spectroscopy can give information about the direct location of molecules in the surrounding of the adsorbed silver nono-particles in a clay suspension. In contrast the colloid electron microscopy gives only information about the location of the colloid electron microscopy gives only information about the location of the colloids at the clay minerals [33].

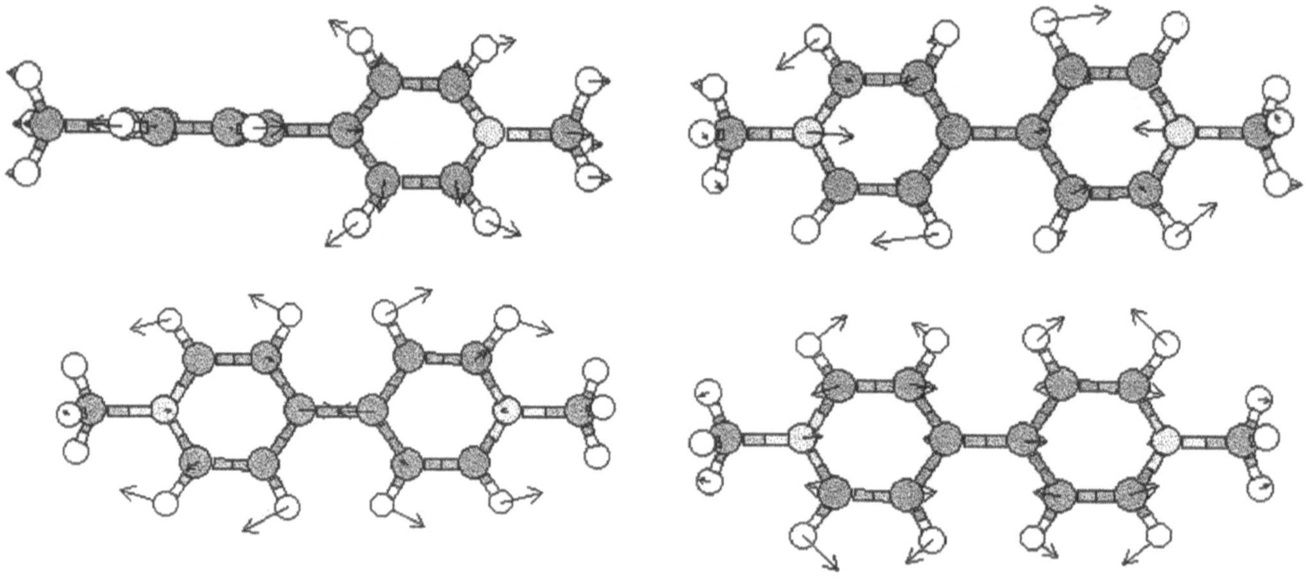

Fig. 11 Calculated normal mode displacement of the 848, 1204, 1308 and 1651 cm^{-1} vibrations of the paraquat molecule

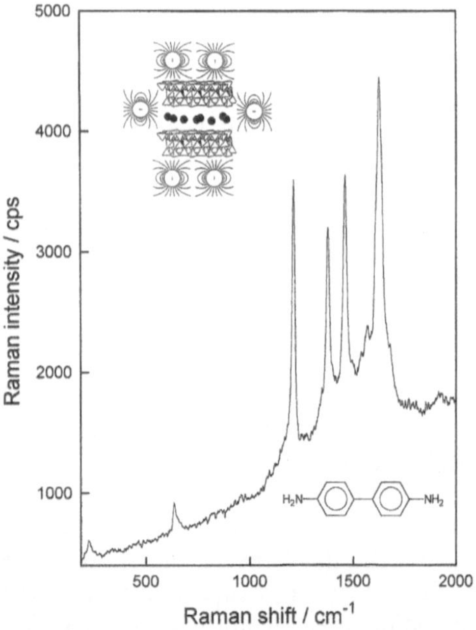

Fig. 12 SERRS microprobe analysis of adsorbed aniline on Cu(II)-montmorillonite: Benzidine di-cation is formed from aniline in the interlayer. MOLE-S 3000 Micro-Raman equipment. Excitation wavelength 457.9 nm, laser power at sample 8 mW, integration time 2 s, number of reads 60

A wide variety of organic molecules are known to interact with clay minerals, and the formation of polymers was observed in many cases. Among the clay minerals, montmorillonite shows interesting behaviour in these clay-organic systems. In the study of adsorption where several kinds of adsorbed species exist, resonance Raman spectroscopy is often useful in offering selective information on the relevant species in the interlayer of transition-metal (Cu^{2+}, Fe^{2+}, Ru^{3+}) ion-exchanged montmorillonite [34, 35]. By means of Colloid-SERRS spectroscopy the Raman scattering intensity of species in the interlayer of these montmorillonite clay minerals can be enhanced by many orders of magnitude [31, 32]. Application of the Colloid-Micro-SERS spectroscopy for obtaining detailed information on the interlayer reactions of aniline on Cu(II)-montmorillonite is such an example. These measurements are made of the adsorption of aniline from aqueous solution by Cu(II)-montmorillonite and the formation of species in the interlayer were investigated after SERS activation with Ag hydrosols. The high quality SERRS spectrum of the formation of benzidine in the clay interlayer from aniline, represented in Fig. 12, shows that the Colloid-SERRS spectroscopy can provide a new potential for the investigation of interactions of organic compounds and clay minerals in the interlayer.

Conclusion

The examples of VIS-SERS, FT-SERS, and Micro-SERS measurements in different research fields reviewed in this article were selected to illustrate the sensitivity, molecular specificity of adsorption processes, accuracy, easy SERS sample preparation, and significant manifold applications of Raman analysis in the adsorbed state. One main advantage is that Raman spectra are obtainable from highly diluted solutions, indicating that SERS and SERRS have great potential for detecting extremely small concentrations of organic materials. Another advantage of the SERS (SERRS) spectroscopy is to obtain, in situ, vibrational spectroscopic information from molecules in the first molecular adsorption layer of the solid/liquid interface.

References

1. Chang R, Laube B (1984) CRC Crit Rev Solid State Mater Sci 12:1
2. Creighton J (1988) In: Clark R, Hester R (eds) Spectroscopy of Surfaces, Ch 2, p 37
3. Birke RL, Lombardi JR (1988) In: Gale RJ (ed) Spectroelectrochemistry:Theory and Practice, Ch 6, Plenum Press, New York, p 263
4. Garrel R (1989) Anal Chem 61:401A
5. Otto A, Mrozek I, Grabhorn H, Akemann W (1992) J Phys Condens Matter 4:1143
6. Angel M, Myrick L (1992) In: Burns DA, Ciurczak EW (eds) Handbook of Near-Infrared Analysis – Practical Spectroscopy, Series 13. Marcel Dekker, New York, p 225
7. Van Duyne RP, Haller KL, Altkorn RI (1986) 126:190
8. Koglin E, Schwuger MJ (1992) Faraday Discuss 94:213
9. Sharonov S, Nabiev I, Chourpa I, Feofanov A, Valisa P, Manfait M (1994) J Raman Spectrosc 25:699
10. Koglin E, Sequaris J-M (1983) J Phys C 10:487
11. Koglin E (1990) J Planar Chromatogr 3:117
12. Gaussian 94 (1992) Revision A, Frisch MJ, Trucks GW, Head-Gordon M, Gill PMW, Wong MW, Foresman JB, Johnson BG, Schlegel HB, Robb MA, Replogle ES, Gomperts R, Andres JL, Raghavachari K, Binkley JS, Gonzalez C, Martin RL, Fox DJ, Defrees DJ, Baker J, Stewart JJP, Pople JA, Gaussian Inc, Pittsburgh PA
13. DGauss 3.0/UC-3.0 Cray Research Inc
14. Sun S, Birke R, Lombardi J (1990) J Phys Chem 94:2005
15. Koglin E, Laumen B (1994) Borgarello E, Prog Colloid Polym Sci 95:143
16. Kreisig SM, Tarazona A, Koglin E, Schwuger MJ (1996) Langmuir 12:5279
17. Tarazona A, Kreisig S, Koglin E, Schwuger MJ (1997) Prog Colloid Polym Sci 103:181
18. Koglin E, Tarazona A, Kreisig S, Schwuger MJ (1997) Colloids Surfaces A 123–124:523
19. Steward SD, Fredericks PM (1995) J Raman Spectrosc 26:629
20. Koglin E, Séquaris J-M (1986) In: Dewar M et al (eds) Topics Current Chemistry, Vol 134, Springer, Berlin 1
21. Chumanov GD, Efremov RG, Nabiev IR (1990) J Raman Spectrosc 21:43, 49, 333
22. Cotton TM, Kim JH, Chumanow GD (1991) J Raman Spectrosc 22:729
23. Nabiev I, Chourpa I, Manfait M (1994) J Raman Spectrosc 25:13
24. Sequaris JM, Koglin E (1985) In: Alix A, Bernard L, Manfait M (eds) Spectroscopy of Biological Molecules, Wiley, New York, p 237
25. Kneipp K, Wang Y, Dasari R, Feld M (1995) Appl Spectrosc 49:780
26. Nie S, Emory SR (1997) Science 275:1102
27. Sequaris J-M, Koglin E (1987) Anal Chem 59:525
28. Soper S, Ratzlaff K, Kuwana T (1990) Anal Chem 62:1438
29. Rau A (1993) J Raman Spectrosc 24:251
30. Koglin E (1993) J Planar Chromatography 6:88
31. Koglin E, Schwuger MJ (1994) In: Yu N, Li X (eds) Raman Spectroscopy, INCORS Wiley, New York, p 628
32. Koglin E, Tarazona A, Narres HD (1995) In: Elsen A, Grobet P, Keung M, Leeman H, Schoonheydt R, Toufar H (eds) Euroclay, Interfasechemie, p 124
33. Thiese PA (1942) Z Elektochem 48: 675
34. Soma Y, Soma M, Harada I (1983) Chem Phys Lett 99:153
35. Soma Y, Soma M, Harada I (1985) Chem Phys Lett 89:738

Progr Colloid Polym Sci (1998) 109:244–253

M. Zizlsperger
W. Knoll

Multispot parallel on-line monitoring of interfacial binding reactions by surface plasmon microscopy

Received: 13 October 1997
Accepted: 20 October 1997

M. Zizlsperger · Prof. Dr. W. Knoll (✉)
Max-Planck-Institut für Polymerforschung
Postfach 3148
D-55021 Mainz
Germany

Abstract Surface plasmon microscopy combined with image analysis software is shown to be well suited to monitor in situ the binding of analytes from the aqueous phase to the elements of a sensor chip arranged in a 2D matrix, each being individually functionalized with a specific binding site. We used an ink jet technique to generate a 4 × 4 matrix of sensor spots functionalized by a self-assembled monolayer of binary thiol mixtures exposing biotin as the "ligand" for the specific binding of streptavidin. Kinetic binding curves as well as the amount of analyte adsorbed to the surface of each sensor element can thus be monitored for all 16 channels in parallel.

Key words Biosensing – surface plasmon microscopy – multispot recording – surface functionalization – biorecognition

Introduction

The label-free detection of recognition and binding events between analytes from solution (ligands, antigens, DNAs, etc.) and their reaction partners (receptor proteins, antibodies, capture probe DNA, etc.) functionally fixed to the solid/liquid interface has been in the center of biosensor research for many years. Various detection schemes have been proposed based on different interfacial physical properties that change upon the binding reaction, the most prominent ones being related to (1) the mass coverage change of a quartz crystal microbalance (QCM) modifying its shear mode frequencies [1, 2], (2) changes in the electrical capacity and resistivity of the interface detected by electrochemical techniques [3, 4] and (3) optical properties that are monitored by a whole battery of techniques, including ellipsometry and evanescent wave techniques like surface acoustic waves (SAW) [5] surface plasmon (PSP) [6, 7] and optical waveguide spectroscopies [8]. The latter two principles have already been implemented in commercially available instruments that combine the sensitivity to even subtle changes of the optical thickness of the interface upon binding with smart molecular architectures of the functional biosensor surface layer aiming at optimizing the specific recognition while simultaneously minimizing the non-specific binding [9].

A major drawback of all these methods is the fact that they are integrating the sensor signal over relatively large sample areas. This is either intrinsically linked to the way the method works, e.g. in QCM or SAW device, or is a consequence of the mode of operation, e.g. given by the spot-size of the laser beam used to read out the optical thickness change.

However, with the invention of the surface plasmon microscope [10] (which was later extended to the optical waveguide microscope [11]), we could show that the information about the build-up of an interfacial layer by binding reactions obtained from the Fresnel analysis of the angular shift of the surface plasmon resonance measured, e.g. in the Kretschmann configuration averaged over the laser spot, can be derived with the same quality and sensitivity, however, with a lateral resolution that can reach the few microns level [12].

The reason for this lies in the physical principles of surface plasmon optics: A PSP mode that is launched by the resonant coupling to the plane waves of a laser beam, has a spatial "lifetime" given by the propagation length

L_x which in turn depends on all the loss mechanisms that contribute to the damping of the evanescent wave: the intrinsic losses giving rise to dissipation in the metal layer that guides the mode which are described by the imaginary part ε'' of the complex dielectric function of the metal $\tilde{\varepsilon} = \varepsilon' + i\varepsilon''$, the radiative losses through the thin metal layer by surface plasmons that couple-out again through the prism, and re-radiation by surface roughness giving rise to radiative decay channels for non-ideal thin stab configurations [13]. Any thin film coating of higher optical density than the environment increases the intrinsic losses by pushing the electromagnetic field of the mode deeper into the (lossy) metal, and, of course, can contribute through own electronic excitations that cause dissipation of PSP energy.

As has been described in detail, this leads to propagation lengths L_x that vary greatly for different metals, e.g. Ag and Au, and for different laser wavelengths used in the experiment: operating in the red means more ideal metals, i.e., lower losses by ε'' and hence longer propagation lengths (e.g., $L_x = 250\,\mu m$ for a Ag/air-interface at $\lambda = 1152\,nm$). Short propagation lengths and hence high lateral resolution in surface plasmon microscopy requires operation in the blue for Ag-substrates ($L_x = 5\,\mu m$ at $\lambda = 456\,nm$) or can be done even with a HeNe laser if Au-substrates are employed ($L_x = 4\,\mu m$ at $\lambda = 633\,nm$) [14].

In any case, the minimum lateral dimension of the sample area that is required to allow for the coherent excitation of a surface plasmon mode with a well-defined dispersion behavior, i.e., is observed – for a given laser wavelength – at a well-determined angle of incidence, is given by the propagation length L_x. This means that two neighboring areas on a sample surface can be distinguished as individual sites with, e.g., different coating thicknesses, if they are each about L_x in size (or, more precisely, if their homogeneous lateral extension along the propagation direction is at least L_x).

Once this was worked out it was then an obvious step to subdivide a sensor area of, say, 1 cm^2 into many small spots of a sensor array and read them individually by surface plasmon microscopy. If one would push the integration limits to, e.g. 10 μm^2 spot size (absolutely compatible with the physics that govern SPM) one then would deal with a sensor array of 10^6 reaction elements. To make life easier we prepared and describe in the following the read-out of only 4×4 elements, each being ca. 200 μm in diameter, and being about 100 μm apart from each other.

However, a major problem that had to be solved was the individual functionalization of each of the single sensor elements. Our strategy was to use endgroup-derivatized thiolates and their ability to self-assemble on Au-surfaces to generate such a multi-functional matrix sensor [15]. Obviously, the quick and easy rubber-stamp concept [16] for the generation of patterned self-assembled functional monolayers does not work in this case because it yields only identical SAM domains. A possible strategy would be a step- and repeat approach, by which a homogeneous monolayer of a particular functionality covering at first the whole sensor surface is then locally re-functionalized by UV-desorbing on a small spot the chemisorbed thiolates and replace them by exposing the chip (and thereby the generated small area of bare gold) to a solution containing thiols with another endgroup functionality [17, 18]. This might work for a few steps but certainly not for thousands of re-functionalizations because each time all spots are exposed to the thiol solution and hence some "cross-talk" would be unavoidable.

We describe here another strategy based on a drop-on-demand or ink-jet principle which allows for a fully automated, computer-controlled pattern generation for a sensor matrix of 4×4 elements each of which can, in principle, be individually functionalized. Again, for simplicity, we limited our investigation to 4 different functionalities each with a 4-fold redundancy which allows for reproducibility tests.

Experimental

Principle of chip preparation

On a high refractive index glass slide (LaSFN9/Schott) a thin gold film ($d = 47\,nm$) is deposited by thermal evaporation at a pressure of about 5×10^{-7} mbar. After preparation the gold-coated glass substrate was allowed to cool under vacuum for 30 min. The fresh gold layer was then immediately fixed on the working platform of the chip preparation system. This system consists of a *xyz*-positioning system with a spatial resolution of 10 μm in *xy*-direction and a silicon piezoelectric micropump (GeSiM/Dresden), which is able to eject microdroplets of a few nanoliters volume. Both parts are controlled by a computer program, based on a software package of GeSiM. Next to the gold-coated slide a microtiter-plate is fixed on the platform, filled with all needed solutions and also with pure solvent for the washing processes. This is schematically depicted in Fig. 1.

To start the chip preparation, the piezo pump, which is fixed on the *z*-arm of the positioning system, is driven to the first bin with a certain thiol solution. The pump absorbs approximately 100 μl solution and is then driven to one or more positions above the gold layer. The pump ejects now each time one drop of about 1 nl volume from a height of 1 mm above the gold surface.

The drops usually have an average diameter on the surface of about 200 μm. Their separation distance is between

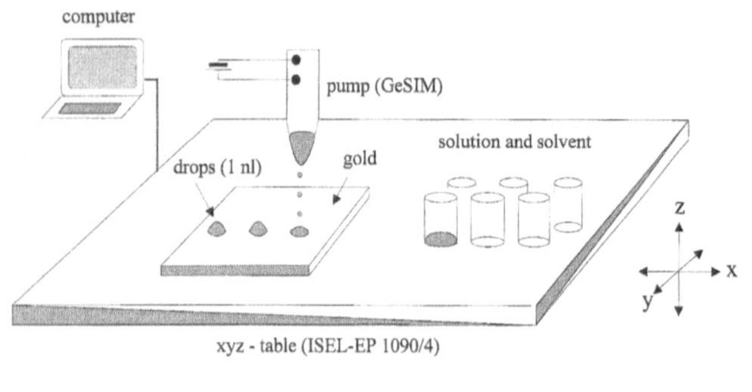

Fig. 1 Principle of the chip preparation

200 and 400 μm. This large distance was chosen in order to prevent in the final rinsing step cross contaminations between different monolayers.

After placing all drops of the first solution, the pump is washed by rinsing automatically in a solvent bin. This is done by absorbing and depositing solvent fluid several times and by dipping the pump deeply into the solvent. After the washing process the system goes on with absorbing the same or a different thiol solution and again with setting down several drops. These steps are repeated automatically until all drops are placed. This takes between 10 and 20 min, depending on the number of different thiol solutions employed.

At a normal relative humidity (70%) the drops evaporate within a few seconds. So the system was placed under 100% humidity conditions at room temperature (22 °C), which allows to reach life times of the drops of more than 2 h. During this time the solved molecules, in our case alkanethiols, have enough time to self assemble on the gold surface and to form a homogeneous monolayer. After this, the sample is removed from the preparation system, immediately pressed on a Teflon cuvette with a volume of 60 μl and rinsed by injecting a few milliliters of solvent into the cell to remove unspecifically bound thiol molecules. This working step was far from being optimized, and sometime resulted in a sub-monolayer contamination of the Au-substrate between the single spots, but never led to a measurable cross-functionalization from one spot to another.

Principle of surface plasmon microscopy with image analysis

A system consisting of a normal surface plasmon microscopy setup with a linearly polarized HeNe laser at a wavelength of 632.8 nm ($P \sim 5$ mW) and an additional controlling and image analysis software was developed (cf. Fig. 2a). The setup of the SPM was constructed in the Kretschmann configuration. In this case the momentum of the exciting photon is increased by the refraction index of a 90° glass prism. This allows for the excitation of plasmon surface polaritons at the interface between the metal and the dielectric medium. By scanning the angles of incidence

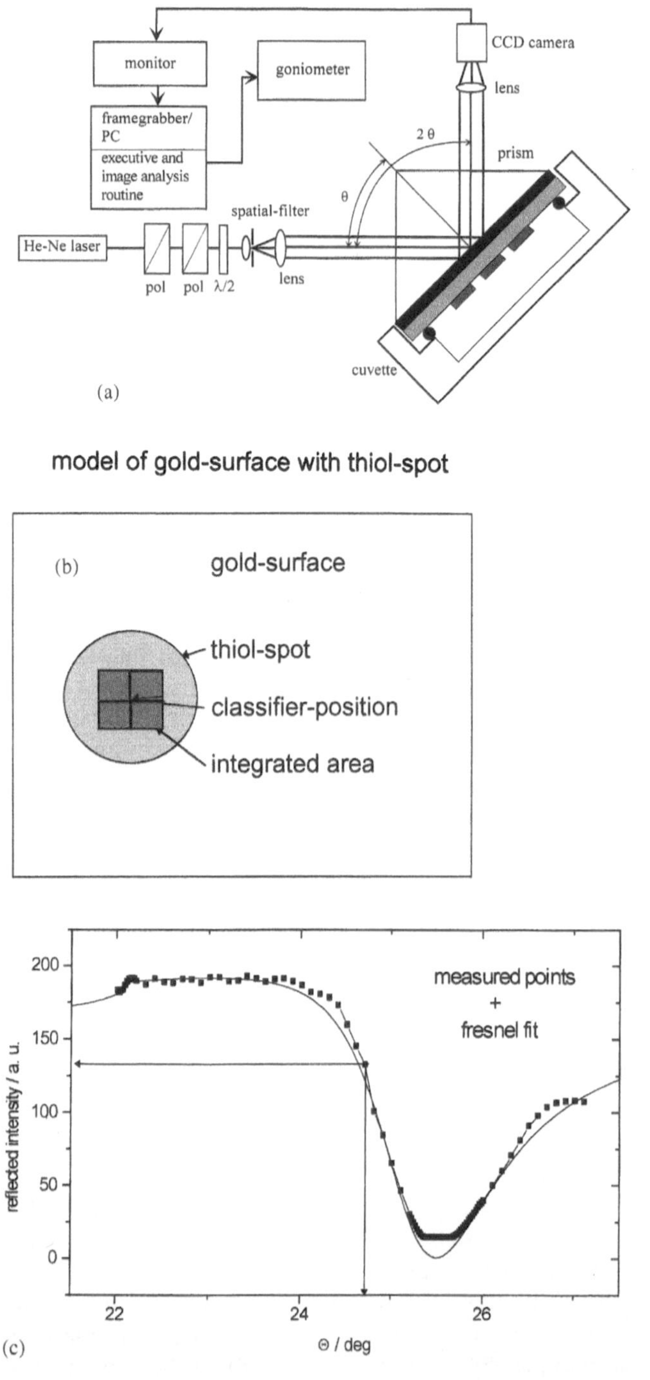

Fig. 2 (a) Schematic diagram and principle of surface plasmon microscopy with image analysis. (b) Schematic of a single sensor element and its actively analyzed area defined by a classifier routine. (c) Reflectivity curve (reflected average intensity versus angle of incidence) generated for one sensor element by the image analysis software.

of the laser beam, θ, the reflected light shows a sharp minimum at a certain resonance angle. This angle is defined by the energy and momentum matching conditions between the laser photons and the surface plasmons. With an additional thin dielectric coating on the surface generated, e.g., by a molecular binding event, a shift of this resonance angle to higher angles is obtained. This effect is the basic sensing mechanism, which is also used in surface plasmon microscopy.

An expanded laser beam illuminates the surface, in our case an area of 6 mm^2, where differences in the optical thickness result at a given angle of incidence in a contrast in the reflected plasmonic light reflected from different sensor elements. This reflected, diffracted and scattered light is then Fourier-backconverted by a lens and detected by a CCD camera. The images taken by the camera are digitized by a framegrabber computer card and are stored automatically as usual image files on the computer.

To reach an automatic measurement and data analysis, it was necessary to create a special software routine. For an experiment two different measurement modes are possible. The first is the angle scan measurement. Here, sample and camera position are rotated in $\theta/2\theta$ at an angular resolution of 1/1000°, controlled by a computer. At every programmed angle one image is acquired automatically.

The second mode is the kinetic measurement. In this mode at a fixed angle a time-dependent image acquisition program is activated. The time resolution is given by the transfer time of the camera and the framegrabber card ($\Delta t \sim 600$–1000 ms).

Both kinds of measurements were done on-line with the solvent being in contact with the surface layer and consisted usually of about 200 images.

One problem that had to be solved was the diameter of the illuminated area changing during an angle scan measurement, and slight shifts in the relative coordinates of the spots on the CCD image, both effects being the result of using a 90° prism. It is necessary to ensure in all images the analysis of the same position, relative to the center of each thiol spot. To this end, the software tool MINOS (Stemmer) was used to create classifier software routines. These are network structures with a searching algorithm, which can learn via example images to identify a special image object. In our case, one classifier is able to identify one corresponding monolayer spot in all taken images after learning with usually 1–5 images. It delivers for each image the center coordinates of this spot (cf. Fig. 2b).

The learning process must be done for as many classifiers as monolayer spots were created. One or more additional classifiers are responsible for the reference. They identify image positions between the thiol spots with only the bare gold surface. Another possibility, which was used sometimes, was to produce carefully a small scratch in the gold layer near the thiol matrix, in order to be able to read the bare glass surface as reference for the total reflection edge.

Now the image analysis software runs automatically in a loop through the whole sequence of images. The parallel use of all classifiers delivers the coordinates of all corresponding spots and the image analysis then calculates the average gray values out of 400 image pixels around these points. The number of pixels was chosen as a compromise between neglecting the thinner edge of the spots and having as much integration areas as possible, to reduce the noise-level.

In the case of analyzing a scan measurement these gray values are plotted now as a function of the angle of incidence. This way, one obtains simultaneously as many angle dependent reflectivity curves as there were spots created on the chip, together with one or more reference curves. The optical thicknesses of the layers can be calculated on the basis of Fresnel's equations given the angular shift of the resonance minima compared with the reference (cf. Fig. 2c).

In the case of a kinetic measurement the gray values are plotted versus the reaction time. This way one can monitor the binding events on each of the monolayer spots simultaneously in real time.

Materials

The spot functionalization was achieved with binary mixed monolayers prepared with two different alkanethiols dissolved in aqueous solution. Their structure formula is given in Fig. 3. The biotinylated thiol 1, HS-Prop-DADOO-X-Biotin ($M_w = 575.80$) and the hydroxyterminated thiol 2, HS-C2-Amidoethoxyethanol ($M_w = 193.27$) were synthesized by Boehringer Mannheim GmbH. Both thiols were dissolved in 0.5 M NaCl at a concentration of 5×10^{-4} M. In the experiments they were used as binary mixtures in different molar ratios. The biotin endgroup of thiol 1 serves as a model ligand fixed to the solid/liquid interface to which the protein streptavidin can bind from solution. Protein concentration was 5×10^{-6} M in 0.5 M NaCl.

The OH-terminated thiol 2 inside the self-assembled monolayers serves as a diluent molecule separating the biotin-moieties from each other. This had been shown to enhance the binding capacity [15]. In order to reach an even more efficient binding, the biotinylated thiol 1 contains a spacer arm, which elevates the biotin moiety off the surface allowing for a more facile access to the streptavidin binding cavity.

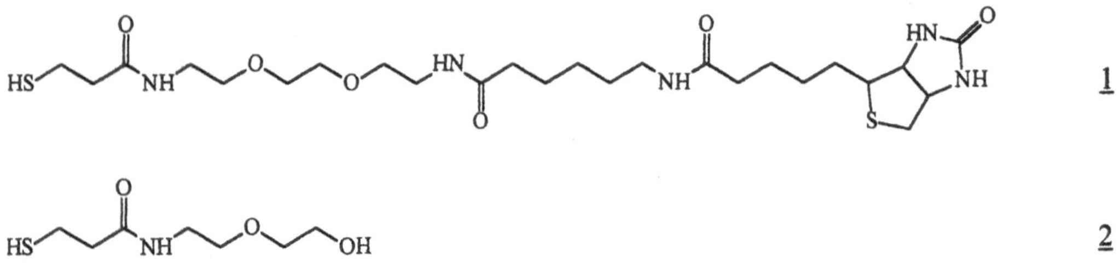

Fig. 3 Structure formula of the two employed water-soluble thiols

For the preparation of the aqueous solutions the water was obtained from an ion exchange purification train (MilliQ system, Millipore, 18.2 $M\Omega$) .

In the microscopic setup we used slides and prisms made from LaSFN9 glass ($n = 1.85$ at $\lambda = 633$ nm), which allows for measurements of the samples against solvent, owing to the lower resonance angles.

The images were recorded with a 8 bit TV camera (Hamamatsu, CCD) and were digitized by a framegrabber computer card (Imas, Imascan Precision).

For the calculation of the layer thicknesses the refractive indices of the thiols were assumed to be $n = 1.5$ and for the streptavidin to be $n = 1.45$.

Results and discussion

Identical functionalization on cell spots

The first set of experiments was designed so as to allow for a general test of (1) the multispot detection approach and (2) in particular, whether the reproducibility of the binding result is spot position independent. For this aim a plain Au substrate was functionalized with 12 identical thiol spots composed of a binary mixture of 10 mol% of the biotinylated compound 1 and 90 mol% of the OH-terminated diluent thiol 2.

Figure 4 shows a series of SPM images taken from that sample after washing and mounting it to the microscopic set-up but before filling the cuvette with 0.5 M NaCl. The images displayed are taken at different angles of incidence (and observation) as indicated.

Several features are noteworthy:

(1) The qualitative impression is that all 12 spots tune into SPS resonance at roughly the same angle, a few tenths of a degree above the bare Au surface surrounding the spot.

(2) The individual spots show definitely not the same light intensity corresponding to an obviously non-homogeneous illumination of the observed area. Also within each spot, there are slight intensity variations, although this is a minor effect. We should point out, however, that no attempts were made to spatially clean the illuminating laser beam.

(3) On top of these features there are "optical defects" arising from interference patterns of dust particles somewhere along the pathway of the laser beam. But no attempts have been made to work in a dust-free environment. However, as we will see these local structures on the images do not compromise their information content.

(4) Also evident in these images is the fact, that the spot position is slightly fluctuating around the theoretical coordinates that were chosen strictly equidistant. This "noise" is beyond the reproducibility level of the X, Y-control of the ink-jet and rather mirrors local fluctuations of the wettability of the Au-surface by the thiol-solution. This effect also leads to a slight variation of the spot diameter.

As it was described in the experimental section, images like the ones shown in Fig. 4 but recorded with much smaller angular increments are analyzed for all 12 spots simultaneously and the average gray value of each spot at each angle stored. For a single spot these gray values, when plotted as a function of the angle of incidence, correspond to a regular reflectivity curve as it could be monitored in the usual Kretschmann configuration (cf. Fig. 2c).

Examples of such curves taken for one arbitrary spot are shown in Fig. 5. Just for reference purposes also a total internal reflectivity curve of a bare glass spot on the otherwise gold-coated slide was imaged and used for the angular calibration of the set-up at the edge of the critical angle for total internal reflection. The obtained curves show the typical features, however, with a few specific differences.

(1) Obviously, these curves show a higher noise level than normal (good) reflectivity scans do. This is due to the fact, that the beam intensity of the HeNe laser per spot area, of course, is largely reduced compared to the total intensity. Moreover, only about 20% of the spot area is

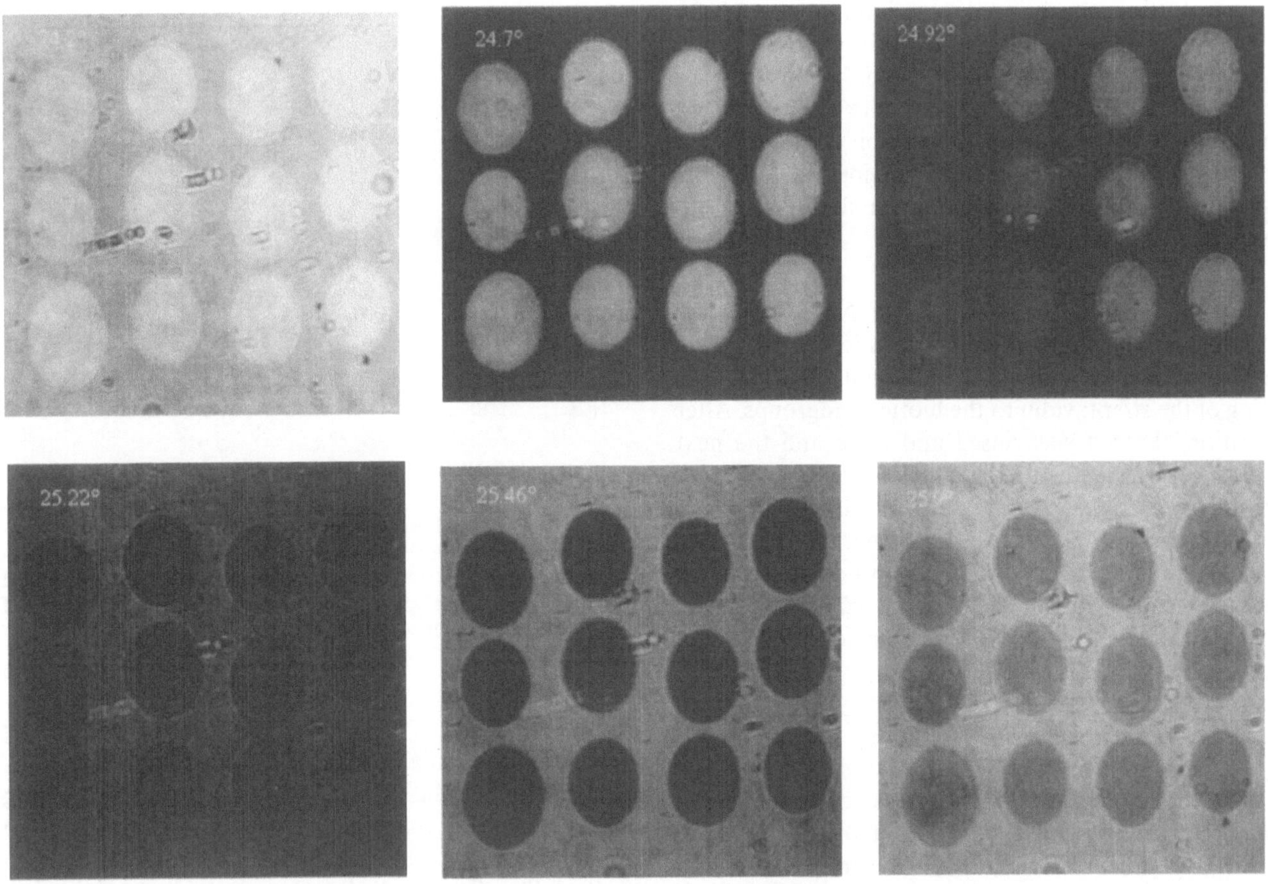

Fig. 4 Series of surface plasmon microscopy images taken from a 3 × 4 matrix of identically functionalized sensor elements at various angles of incidence, as indicated

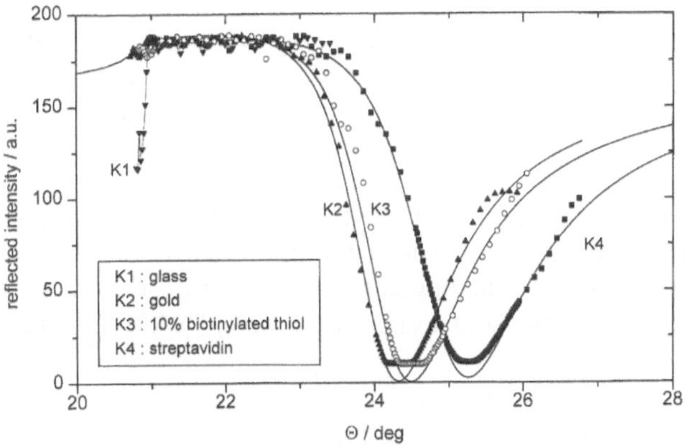

Fig. 5 Reflectivity scans taken on one single sensor matrix element before and after the binding of streptavidin. For comparison, the reflectivity curve for bare Au, as well as for the bare glass are given. All data were taken in air

currently used by our image analysis routine which causes an additional decrease of the integrated intensity. On the other hand, the employed laser was far from being stable as can be seen even better in the kinetic curves shown and discussed below (cf. Figs. 8 and 9). So, scaling all intensities to a separately monitored input intensity channel would reduce the noise level. Eventually, however, if one increases the integration density on the chip and hence the spot size will have to go down and so the number of pixels on the CCD camera which one integrates in order to get the mean gray value of the area will be further reduced, the noise will go up again, of course. The question as to how far this can be compensated by using a stronger (and more stable) light source and by using longer integration times, needs to be answered. One should keep in mind, however, that it is not the individual intensity which yields the desired thickness information, but rather the whole reflectivity curve whose weighted fit to a Fresnel calculation gives the relevant parameters, namely the resulting angular shift.

(2) The dynamical range of the employed camera imposes some distortions of the reflectivity curves. This can be clearly seen in the low reflectivity region of the curves where a cut-off of the real minimum is obvious (but it should be mentioned that also non-linearities of the camera-response function to different intensity levels lead to

a curve distortion, which is not so serious, however) [12]. The reason for this is that we purposefully choose a rather cheap multipurpose camera in order to demonstrate the potential of the method without expensive equipment, but then had to accept an intensity resolution of only 7 bit, at best. However this seems to be no major limitation because, once again, it is only the angular shift of the resonance that is analyzed.

After the cell was filled with 0.5 M NaCl buffer the 5×10^{-6} M streptavidin solution was injected, which resulted in the formation of a protein monolayer by specific binding of the streptavidin to the biotin headgroups. After completion, the cell was rinsed and dried and the next series of angle-dependent images was taken.

If the layer thicknesses obtained for all 12 spots from such angular shifts are plotted in a histogram (Fig. 6) one can see that, in fact, with reasonable reproducibility all 12 sensor elements give the same result: ca. 14 Å for the thickness of the mixed thiol SAM and an additional thickness increase of ca. 38 Å for the bound streptavidin layer. This clearly demonstrates for this model system that a reproducible surface functionalization by the employed drop-on-demand can be achieved on all sensor elements and that the protein binding to these recognition sites gives the (nearly) identical thickness information on all spots. This parallel read-out, therefore, can be considered to be equivalent to the sequential procedure, just is substantially faster.

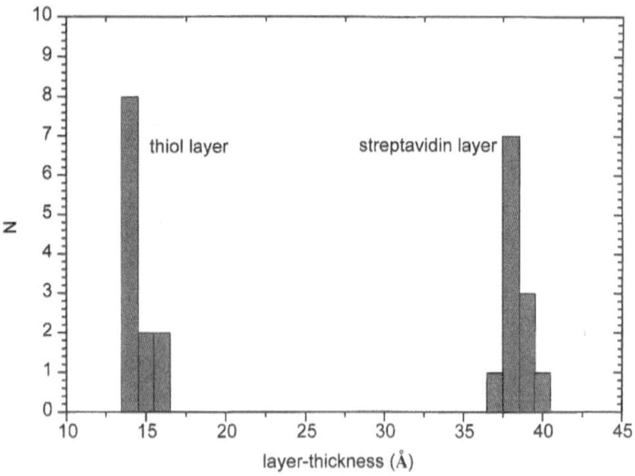

Fig. 6 Histograms of the layer thicknesses obtained for the 12 identically functionalized mixed thiol spots (containing 10% of the biotinylated thiol 1) and of the streptavidin layer thicknesses obtained by simultaneously analyzing the SPM images of the 3 × 4 matrix of sensor elements (cf. Fig. 4)

Parallel monitoring of binding reactions to sensor elements of different functionality

The next step was the read-out of the spot matrix modified with SAMs of different functionalities. As mentioned already, we choose for this purpose binary mixtures of the two thiols 1 and 2 in different molar ratios corresponding to 0, 1, 10, and 100% of biotin-labeled thiol. These mixtures were chosen because their binding capacity for streptavidin was shown to vary between zero and a full protein monolayer, corresponding to a thickness increment of 40 Å. The arrangement of the spots and their SAM pattern recorded in contact to buffer solutions are shown in Fig. 7a and b.

Many of the features seen and discussed in Fig. 4 can be found again: The heterogeneity of the illumination is even more pronounced and again many interference patterns from dust particles can be seen. One additional feature is the formation of dark streaks along the direction of the final rinsing of the sample. These originate from excess thiol deposited via the solution drop to the individual spots which is then washed away during the rinsing procedure and partially self-assembled to the free Au-surface surrounding the spots. This is only an artifact arising from our low-tech rinsing procedure but does not lead to any appreciable cross-contamination from one spot to another. The effect can be largely reduced by applying less thiol to the single drop, just enough to allow for a full monolayer deposition to the drop-covered Au spot, and by using more advanced liquid handling tools for the washing steps.

In principle, the darkness of the monolayer coated 4 × 4 spots which are on average in resonance, should vary in a systematic way, according to their respective small thickness differences and hence slight resonance angle shifts (cf. also Fig. 10). This effect, however, is totally obscured by the inhomogeneity of the sample illumination. But as we will see, this does not interfere with the analysis of the data (i.e., the angular scans constructed from such images, cf. Fig. 5) in terms of the actual SAM thickness determination.

Since the microscopic mode can constantly monitor images of the 4 × 4 spots in contact with the aqueous phase containing the analyte, one can follow the binding of the protein to the SAM-functionalized spots in real time by taking time lapse microscopy data at a fixed angle of incidence. Figure 8 shows an example of how the reflected intensity on 4 different selected spots varies as a function of time after injection of the 5×10^{-6} M streptavidin solution into the cell (see arrow). Given the frame transfer time of the CCD camera, which is ca. 50 Hz, the time resolution of this technique could be as high as $\Delta t = 20$ ms. However,

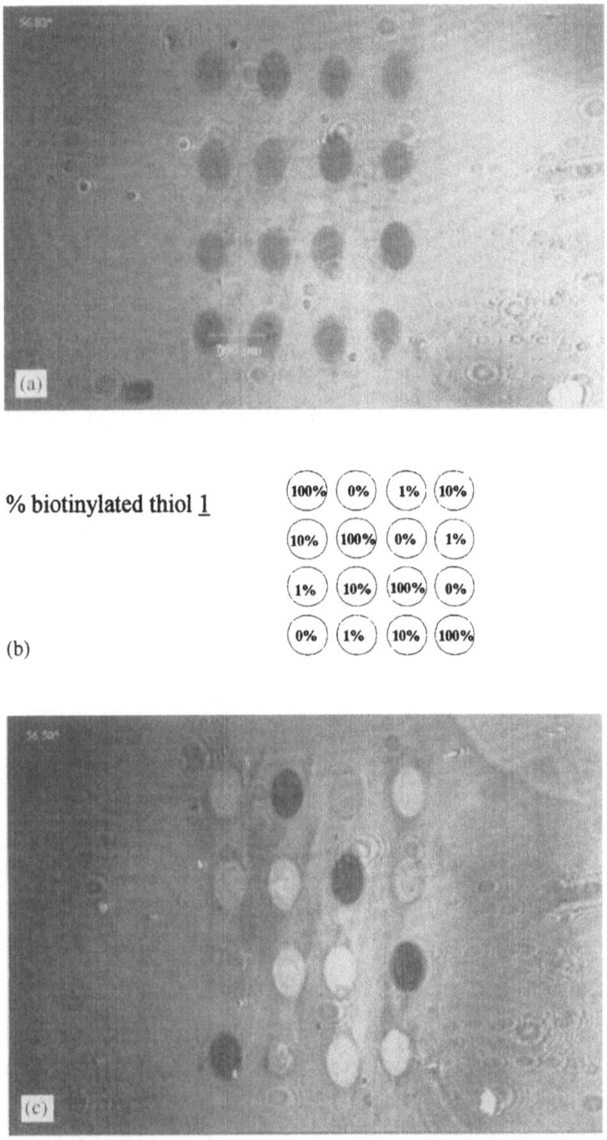

Fig. 7 (a) SPM image of a 4 × 4 sensor element matrix functionalized by binary thiol SAMs of 4 different compositions, containing 0, 1, 10, and 100% of the biotinylated compound 1. (b) Schematic of the functionalization pattern. (c) The same sensor matrix as in (a), but after binding of streptavidin to the thiol spots. All data are taken with the sensor surface in contact to the aqueous buffer solution

the framegrabber operation slows this time resolution down to about 0.6–1.0 s per image. For the diffusion-controlled adsorption of the protein to the interface a time resolution of $\Delta t = 1$ s in the early rapid change of the intensities and of $\Delta t = 10$ s in a later stage of the adsorption was considered to be sufficient. The angle of observation was chosen so as to observe a maximum change in reflected intensity upon binding of protein causing the resonance curve to shift to higher angles. An angular position corresponding to the lower part of the left slope of the resonance curve (cf. Fig. 5) was therefore selected, i.e. $\theta = 56.8°$.

The "binding" curves of Fig. 8 show the typical behavior found for the same type of experiments monitored in the usual spectroscopy mode: a rapid rise in the reflected intensity followed by a cross-over into a stable plateau. The intensity of this plateau depends on the specific functionalization of the corresponding Au spot by the SAM (see below).

Figure 9 compares the kinetic data taken during the same experiment but for 4 spots of the same functionalization, i.e., covered with a thiol monolayer containing 10 mol% of the biotin thiol 1. Slight differences in the early phase of the monolayer formation are probably due to subtle differences in the (convective) mixing of the streptavidin solution with the buffer in the cell and only mirrors a non optimized liquid handling operation. The final intensity increase, however, is nearly identical for all four displayed curves and proves the excellent reproducibility of this multispot recording.

The noise in the spectra shows components that are correlated intensity variations in all recorded reflectivities: as mentioned already, they are considered, therefore, to be due to intensity fluctuations of the incoming laser beam and could be easily scaled-out.

The full screen display of the 4 × 4 spots matrix mirrors the different plateau values of the four types of spots with different binding capacity (Fig. 7c) : The purely OH-functionalized spots show no intensity change at all compared to the image before injection of streptavidin (Fig. 7a) which indicates the passivation of the corresponding sensor-elements. For all the other spots it is again rather difficult to judge the degree of binding based on the relative intensity change because the inhomogeneous illumination and the impurity interference pattern overrule such a qualitative estimate.

Again, after the completion of the kinetic recording of the protein binding, which takes less than 5 min (cf. Figs. 8 and 9), a full angular scan is constructed for all 16 spots, and for each of the 4 identical ones their respective angular shifts are Fresnel-analyzed, and the obtained thickness values are averaged. Thus, a quantitative binding curve is obtained that is displayed in Fig. 10 (full symbols). The error bars give the rms-deviation from the average of these values. The full curves with the open symbols were the data obtained by normal surface plasmon spectroscopy, i.e., measured in a sequential way and, therefore, requiring substantially more measuring time [19]. The data obtained on a single shot by the microscopic mode are in quantitative agreement and hence prove the validity of the approach.

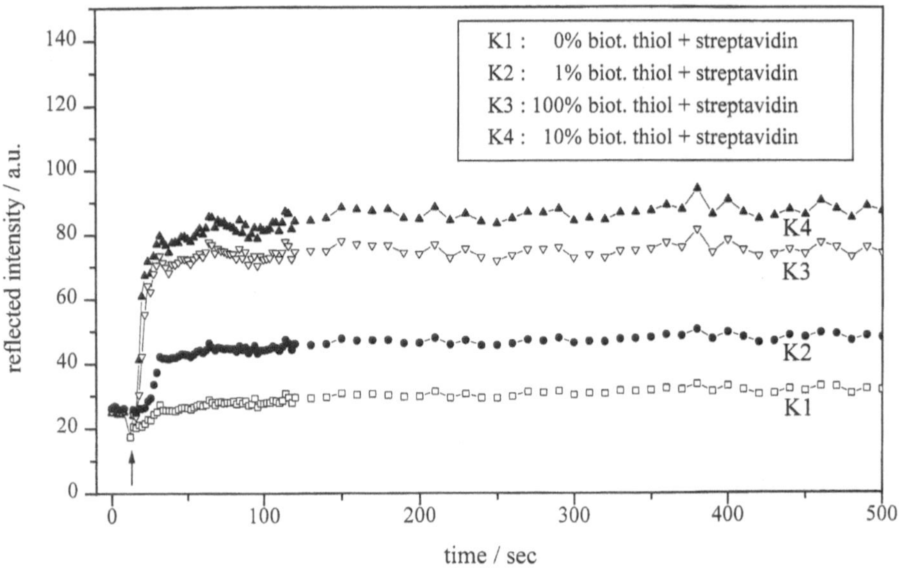

Fig. 8 Kinetic scan, i.e. reflected intensity recorded at a fixed angle of incidence as a function of time, for 4 different sensor elements, each functionalized with a different mixed thiol SAM, as indicated

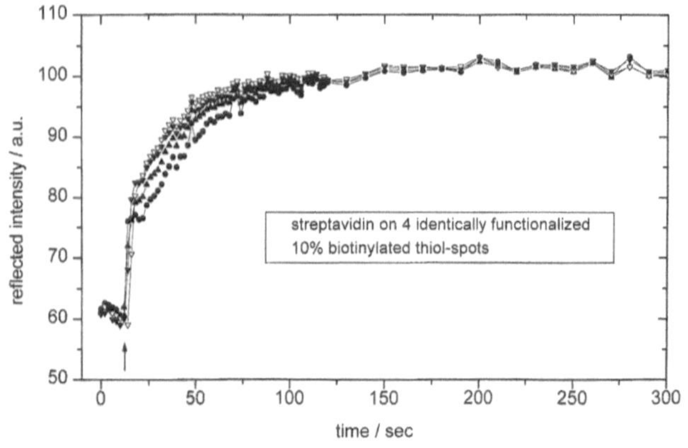

Fig. 9 Kinetic scan recorded for 4 different spots, but each with the same mixed thiol SAM (containing 10% of the biotinylated compound 1)

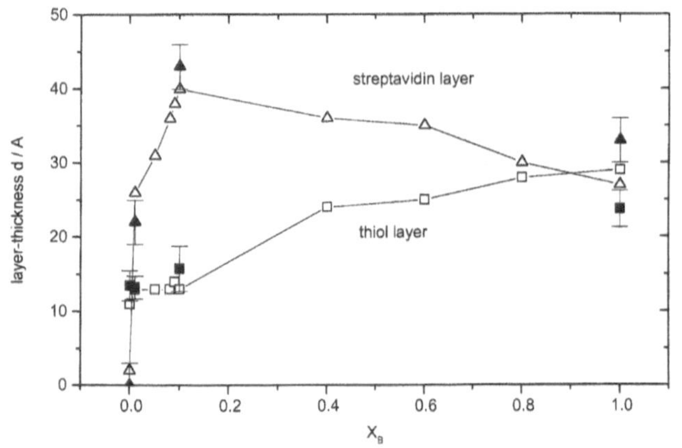

Fig. 10 Thickness data obtained from the parallel 4 × 4 multispot analysis of the binary mixed thiol SAMs of different composition (x_B denotes the mole fraction of the biotinylated compound 1 in the droplet solution), –■–, and of the thickness of the streptavidin layer bound to the individual sensor elements, –▲–. Given are the average thicknesses of 4 identical elements with their rms deviation. For comparison, thickness data obtained by the usual surface plasmon spectroscopy (i.e., in a serial mode, one sample after the other) are given by open symbols

Conclusions

The presented data demonstrate that surface plasmon microscopy combined with image analysis computer routines allow for a massive parallel on-line recording of binding events to individually functionalized sensor elements assembled in a 2D matrix on a sensor chip. The simultaneous recording and evaluation of 1000 sensor elements would be compatible with the proposed surface plasmon optical detection scheme. Kinetic data of the adsorption or desorption processes at the individual sensor spots could be monitored, as well as thicknesses equivalent to the amount of material adsorbed to the individual sensor element could be detected.

The employed ink-jet approach is not the only possible way to generate the multiplicity of surface functionalization: for small numbers of different functionalizations printing techniques are feasible, whereas the combination of lithographic patterning strategies coupled with combinatorial chemistry could result in a nearly unlimited number of individual sensor functionalities.

Acknowledgment This work was supported by Boehringer Mannheim. We are indebted to P. Sluka, R. Herrmann and G. Batz for numerous helpful discussions.

References

1. Okahata Y, Matsunobu Y, Ijiro K, Mukae M, Murakami A, Makiuo K (1992) J Am Chem Soc 114:8299
2. Ebato H, Gentry A, Herron JN, Müller W, Okahata Y, Ringsdorf H, Suci P (1994) Analyt Chem 66:1683
3. Lingler S, Rubinstein I, Knoll W, Offenhäusser A, Langmuir, submitted
4. Cooper JM, Morris DG, Ryder KS (1995) J Chem Soc Chem Commun 14:697
5. Roederer JE, Bastiaans GJ (1983) Anal Chem 55:2333
6. Schmitt FJ, Knoll W (1991) Biophys J 60:716
7. Knoll W (1991) MRS Bull XVI:29
8. Aust E, Ito S, Sawodny M, Knoll W (1994) TRIP 2:313
9. Loefaes S, Johnson B, Tegendal K, Roennberg J (1993) Coll Surf B1:83
10. Rothenhäusler B, Knoll W (1988) Nature 332:615
11. Hickel W, Knoll W (1990) Appl Phys Lett 57:1286
12. Hickel W, Knoll W (1990) J Appl Phys 67:3572
13. Raether H (1988) Surface Plasmons on Smooth and Rough Surfaces and on Gratings, Springer Tracts in Modern Physics, Vol 11. Springer, Berlin
14. Rothenhäusler B, Knoll W (1987) Surf Sci 191:585
15. Spinke J, Liley M, Guder HJ, Angermaier L, Knoll W (1993) Langmuir 9:1821
16. Jackman RJ, Wilbur JL, Whitesides GM (1995) Science 269:664
17. Tarlov MJ, Burgess DRF, Gillen G (1993) J Am Soc 115:5305
18. Piscevic D, Tarlov M, Knoll W (1995) Supramol Sci 2:99
19. Knoll W, Liley M, Piscevic D, Spinke J, Tarlov MJ (1997) Adv Biophys 34:231

Progr Colloid Polym Sci (1998) 109:254–259

K. Holmberg

Role of surfactants in water-borne coatings

Received: 2 April 1997
Accepted: 14 April 1997

Prof. K. Holmberg (✉)
Institute for Surface Chemistry
P.O. Box 5607
SE-11486 Stockholm
Sweden

Abstract The paper discusses the role of surfactants in latex polymerization and in post-emulsification of binders, such as alkyd resins. The advantage of polymerizable surfactants as emulsifier is pointed out. The paper further discusses competitive adsorption between surfactants and between surfactant and associative thickener in paint formulations.

Key words Paint – coating – alkyd emulsion – latex dispersion – surfactant – emulsifier – polymerizable surfactant – competitive adsorption

Introduction

Surfactants are key additives in water-borne coatings formulations. They are used as binder emulsifier and as pigment dispersant, they are needed to improve wetting on low energy substrates, to control foaming during application and processing, and to prevent film defects caused by surface tension gradients. In addition, surface active polymers, often referred to as associative thickeners, are widely used to optimize the rheological properties of the formulation, and anionic polyelectrolytes such as polyphosphates are commonly used as pigment dispersing agents. Taken together, a water-borne paint formulation is extremely complex with a plethora of low and high molecular weight compounds competing for available surfaces, such as binder droplets, pigment particles, and, although much smaller in surface area, the substrate to be painted. The situation is schematically illustrated in Fig. 1.

The majority of surfactants used in coatings formulations are standard anionic and nonionic amphiphiles, such as fatty alcohol sulfate, alkylaryl sulfonate and alcohol ethoxylate. Cationic and amphoteric surfactants are rarely used. A few types of speciality surfactants have found specific niches. Fluorosurfactants [1, 2] and silicone surfactants [3, 4] reduce surface tension to extremely low values. They are used in paint formulations to eliminate surface tension gradients that can form due to faster evaporation of the solvent from the coating edges than from the center. Acetylenic glycols, characterized by having two short, bulky hydrocarbon chains surrounding the polar group, are another type of niche surfactant. These non-micelle forming surfactants form expanded films on water surfaces which can withstand high surface pressures. They are widely used as antifoaming agents in coatings [5, 6].

The present paper focuses on recent developments in main stream paint surfactants used in latex dispersions and alkyd emulsions. The advantage of polymerizable surfactants as emulsifiers is pointed out and the important phenomenon of competitive adsorption is discussed.

Emulsifiers

The majority of water-borne paints are latex paints, i.e. aqueous dispersions of water insoluble polymers made by

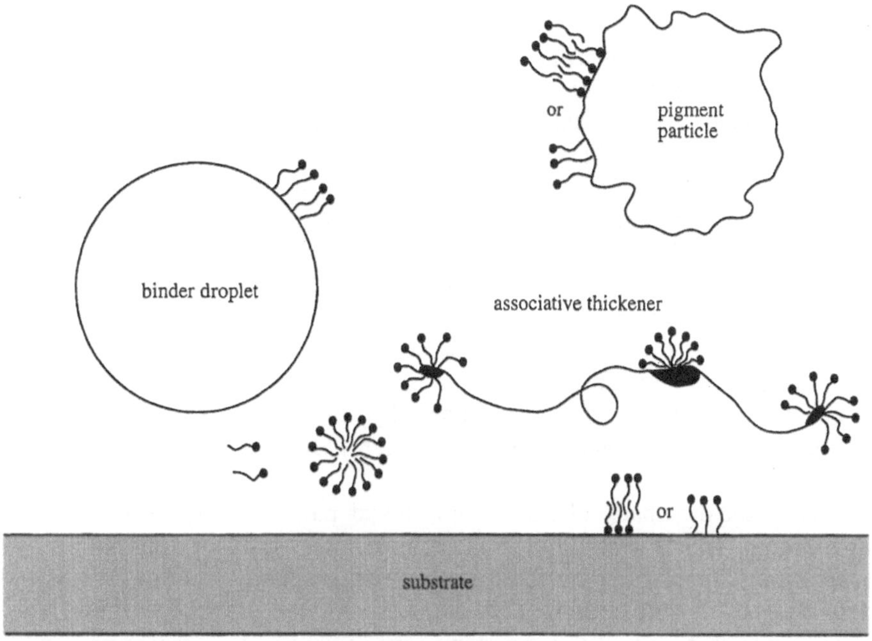

Fig. 1 A paint formulation containing emulsion droplets, pigment particles, associative thickener and surfactant

emulsion polymerization using free radical initiators. In the majority of cases the polymers are based on combinations of monomers, often with a high content of water insoluble entities such as methyl methacrylate, butyl acrylate and styrene and a much smaller fraction of water soluble monomers such as acrylic and methacrylic acid. The water soluble monomers give oligomeric acid segments at the latex particle surface which improves the colloidal stability of the formulation and adhesion and curing characteristics of the film. Most lattices have an average particle diameter in the range 100–500 nm.

The emulsifier used in latex preparation is often a combination of nonionic and anionic surfactants. The nonionic surfactant has traditionally been an alkylphenol ethoxylate, but environmental concern has caused a change over to other ethoxylated surfactants, such as fatty alcohol ethoxylate or fatty acid monoethanolamide ethoxylate. The ethoxylate is the surfactant mainly responsible for dispersion stabilization. The steric stabilization provided by surfactants with relatively long polyoxyethylene chains (>10 EO) is needed in order to retain stability at high solids content in the presence of electrolytes. Steric stabilization also gives proper shear stability to the latex. The presence of anionic surfactant, usually an alkyl sulfate or an alkylaryl sulfonate, during the latex synthesis is needed to compensate for the reverse temperature dependence of ethoxylated surfactants. For nonionics an increase in temperature leads to a decrease in water solubility and an increase in oil solubility. During the course of the emulsion polymerization there is an increase in temperature which would lead to formation of a water-in-oil emulsion if nonionic surfactants were the sole emulsifier. Upon depletion of the monomer phase, i.e. at high conversion, there would be a phase inversion into an oil-in water emulsion. Such a phase inversion leads to a very broad particle size distribution, since particle nucleation as well as reaction kinetics will be out of control [7, 8]. A way to circumvent the problem is to use a semi-continuous polymerization process with the monomer being slowly fed into the reactor during the polymerization.

Anionic surfactants provide electrostatic stabilization and such lattices exhibit good storage stability in formulations containing low or moderate salt concentration. However, lattices made with only anionic emulsifier are not stable at high electrolyte concentration. Evaporation of water during the drying process leads to a continuous raise in the ionic strength of the formulation. Often the stability limit is exceeded at a relatively early stage, leading to particle coagulation and consequent loss in film gloss.

Another common approach to water based coating formulations is post-emulsification of a polymer in water. Several condensation polymers, e.g. alkyds, i.e. fatty acid modified polyesters [9–12], polyurethanes [13, 14] and epoxy resins [15], have been made into dispersions by use of a suitable emulsifier and application of high shear. For instance, long oil alkyd resins of the type used in white spirit based formulations have been successfully emulsified using nonionic surfactants such as fatty alcohol ethoxylates, alkylphenol ethoxylates or fatty acid monoethanolamide ethoxylates [16–18]. Neutralization of alkyd carboxylic groups helps in producing small emulsion droplets and with the proper choice of surfactant, droplet diameters of less that 1 μm can be obtained. Such dispersions are sufficiently stable for most applications. It is interesting that whereas a nonionic surfactant needs to be added in an amount sufficient to give close packing of the emulsifier on the droplet surface, an anionic surfactant gives optimum effect in terms of stability already at concentrations that give very low packing density [19]. This is consistent with the different mechanisms by which nonionic and anionic surfactants exert stabilization, as discussed above. As for latex dispersions, alkyd emulsions stabilized only with anionic surfactants are highly sensitive to electrolytes.

The main drawback of water-borne alkyd paints is slow drying. This is partly due to the comparatively low evaporation rate of water but there is also strong evidence that catalysis of autoxidation does not work as well in water-borne as in solvent-borne systems. The distribution of the drier, in particular cobalt alkanoate, between the alkyd and water phases is believed to influence the early stages of drying of alkyd emulsions [11, 16].

Alkyd emulsions are also of interest in the industrial coatings market. The alkyds used for such applications, so called short oil alkyds, have a much higher viscosity and are most conveniently emulsified in a phase-inversion process. The emulsifier, which can either be a nonionic surfactant, an anionic surfactant or a combination, is dissolved in the alkyd at high temperature and water is added under low shear so that a water-in-oil emulsion is formed. For alkyds of very high viscosity the process must be performed in pressurized vessels to prevent boiling of the water. By adding more water and/or lowering the temperature, the emulsion is made to invert and form an oil-in-water emulsion [20]. During the inversion there is a rapid migration of the emulsifier from the alkyd phase to the water phase. The smallest droplet size, less than 1 μm, is obtained with emulsifiers that strongly favour distribution towards the water phase, and have relatively long polyoxyethylene chains to impart steric stabilization of the newly formed alkyd droplets [21].

Post-emulsification is simplified if the polymer itself is surface active. This can be achieved, e.g. by using polymers of high acid values or by using polymers with noncharged, hydrophilic segments such as polyoxyethylene chains. Such polymers can often be emulsified with considerably less surfactant, but the trade-off is water and chemical resistance of the paint film. If the polymer is sufficiently polar, no emulsifier at all is needed. However, such binders need to be crosslinked during curing in order to give acceptable film properties [22].

Polymerizable surfactants

In recent years there has been considerable interest in polymerizable surfactants as emulsifiers in emulsion and suspension polymerization [23, 24]. By using surfactants that become covalently bonded to the latex particle, many of the problems encountered with conventional surfactants can be avoided or at least minimized. Positive effects on the properties of both the dispersion itself and the dried film are often obtained [18, 25].

The surfactant-related problems in lattices, as well as in many other dispersions, arise from the fact that surfactants physically adsorbed on the particle surface may desorb into the bulk aqueous phase and that the equilibrium between surface and bulk surfactant concentration is governed by factors such as particle concentration, temperature, ionic strength and pH, all of which may be changed during storage, use and film formation. Since a certain surface concentration of surfactant is needed to give proper latex stabilization, a change in the adsorption–desorption equilibrium may severely affect rheology and stability of the dispersed system.

Formulations containing a latex in combination with another dispersion, such as a pigment slurry, constitute a particular problem from a stability point of view. The physically adsorbed latex surfactant may have higher affinity for the pigment than for the latex, a situation which often leads to latex instability. The surfactants used to stabilize the pigment are usually of a different type to those used for the latex. Hence, the two surfactants will then compete for both surfaces, the latex and the pigment, and the surface composition and coverage obtained in the equilibrium situation may be very different from that of the two components before mixing [26, 27]. This type of competitive adsorption may drastically affect rheology and stability of a formulation.

The presence of surfactant in the dried latex film may also impair film properties. During drying the surfactant is adsorbed on the latex particles. As the particles coalesce during the annealing process, the surfactant migrates out of the bulk phase and concentrates at the interfaces [28]. It has been shown that surfactant molecules preferably go to the film–air interface, where they align with their hydrophobic tails pointing towards the air. Calculations from ESCA spectra show that a lacquer film containing 1% surfactant may have an average surface surfactant concentration of around 50% [29]. Such a high concentration of a non-chemically incorporated, water-soluble component at the film surface will adversely affect adhesion properties and water resistance of the film.

Furthermore, atomic force microscopy (AFM) studies have shown that during the film forming process many conventional surfactants phase separate from the binder.

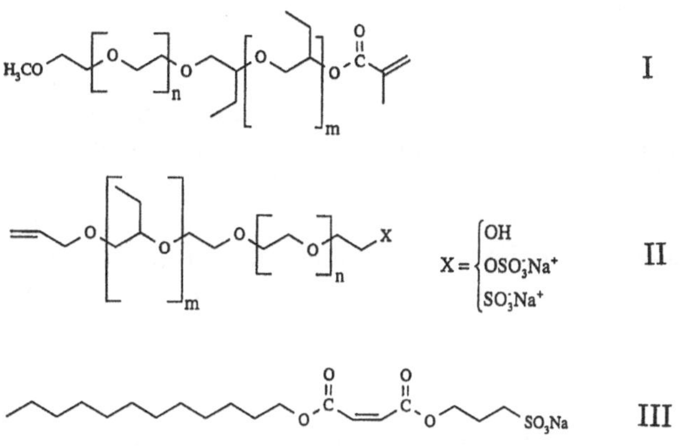

Fig. 2 Examples of polymerizable surfactants for lattices

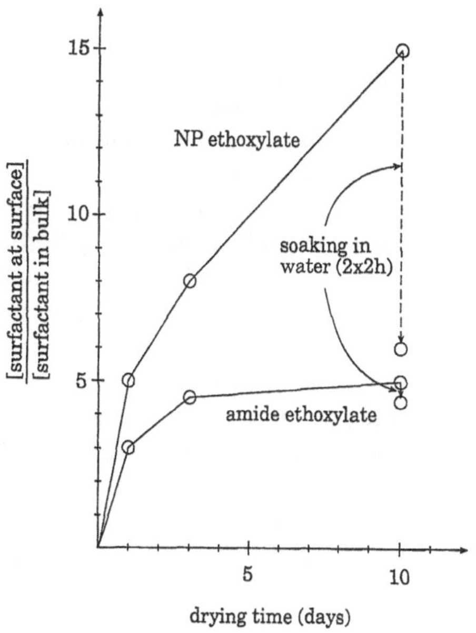

Fig. 3 Relative surface concentration of surfactant as a function of drying time as determined by ESCA. The NP ethoxylate is ethoxylated (12 EO) nonylphenol and the amide ethoxylate is ethoxylated (14 EO) monoethanolamide of linseed oil acid. The experimental conditions are given in Ref. [33]

When the surfactant has phase separated, the water flux may carry it to the film surface. Alternatively, it may accumulate in the interstices between the particles from where it will migrate to the film–air or film–substrate interface through a long term exudation process. Eventually the surfactant will be present in aggregates of considerable size, seen by AFM as "hills". After treatment with water, the surfactant aggregates are washed away and the "hills" are replaced by distinct "valleys". The rough surface will give rise to poor gloss.

It has been shown that many of the surfactant-related problems in latex paints can be minimized by the use of a polymerizable surfactant as emulsifier in the emulsion polymerization process [30–32]. Several types of reactive surfactants have been used for this purpose, some of which are shown in Fig. 2. Block copolymers of ethylene oxide and propylene or butylene oxide with a polymerizable group at one end (Structures I and II) have become popular due to ease of preparation. Of particular interest from performance point of view are surfactants which preferably undergo copolymerization rather than hompolymerization. A good example of such surfactants are maleate half esters of fatty alcohols, such as Structure III of Fig. 2 [31, 32]. However, even surfactants based on highly reactive groups such as maleate do not become quantitatively copolymerized during the emulsion polymerization [31]. The polymerizable emulsifiers are sometimes referred to as "Surfmers". Surface active polymerization initiators, "Inisurfs", and surface active chain transfer agents, Transurfs', have also been developed [23].

Surfactants capable of participating in autoxidative drying are of interest for the post-emulsification of alkyd resins [18, 25]. Ethoxylated monoethanolamides of unsaturated fatty acids are one such type of surfactant that can be chemically incorporated into the network during drying of an alkyd based coating film. Figure 3 illustrates the difference in surface composition with respect to surfactant for a polymerizable amide ethoxylate and a conventional nonionic surfactant of similar hydrophilic–lipophilic balance [33]. Surfactant concentrations at the film–air interface were measured by ESCA. As can be seen, both the conventional surfactant, a nonylphenol ethoxylate, and the amide ethoxylate accumulate at the surface and the concentration increases with time. Whereas the concentration vs. time curve is almost linear for the nonylphenol ethoxylate, it levels off for the amide ethoxylate. For the latter species, the distribution of surfactant in the film seems to be established within three days of drying.

The difference in behavior between the nonylphenol ethoxylate and the amide ethoxylate is probably due to the fact that the latter surfactant becomes immobilized through coupling to binder molecules during the drying process. Once covalently incorporated into the network, the migration process will cease. Another contributing factor for the low degree of migration of the amide ethoxylate could be that this surfactant is likely to be very compatible with the binder, a long oil alkyd resin. Surfactant–polymer compatibility has previously been found to be decisive in determining surfactant distribution in films [34, 35]. Surfactants are carried towards the surface by the flux of water during film drying and this process is particularly effective when there is poor compatibility between surfactant and polymer.

The effect on surface composition of soaking the dried film in water is also shown in Fig. 3. Whereas more than half of the nonylphenol ethoxylate disappears from the outermost surface layer (approximately 50 Å), the effect on the amide ethoxylate is small, in spite of the fact that both surfactants have about equal water solubility. This is a further indication of the amide ethoxylate being immobilized during the drying process, although one must keep in mind that the evidence shown in Fig. 3 is only indirect. The sensitivity of ESCA is not sufficient for monitoring disappearance of carbon–carbon double bonds, which would have been the most direct way of studying surfactant polymerization. However, studies on cobalt initiated autoxidation of ethyl esters of unsaturated fatty acids have shown that the oleate ester does not polymerize over 110 days, whereas the linoleate ester polymerises almost completely over three days [36]. These findings support the view that amide ethoxylates based on fatty acids with a high degree of unsaturation become covalently incorporated in the dried film.

Competitive adsorption

As discussed above, competitive adsorption of surface active agents is a common problem in paints. In fact, in a paint formulation, with its many different surface active species and its variety of interfaces, it is virtually impossible to maintain full control of the surface interactions. Uncontrolled desorption/adsorption of surfactants frequently gives rise to unexpected rheological effects and lack of dispersion stability. For instance, the nonionic surfactant needed as steric stabilizer of latex particles may preferentially adsorb on a hydrophobic pigment surface where it replaces the original dispersant. The net result will be that the latex particle will no longer be fully covered with nonionic surfactant, leading to reduced stability, particularly in formulations of high electrolyte concentration.

Fig. 4 Composition at the particle surface (■) and in the aqueous phase (●) as a function of total surfactant concentration in the aqueous phase. Adsorption from an aqueous solution of a 84:16 molar ratio mixture of sodium dodecyl sulfate and ethoxylated (10 EO) nonylphenol on a poly(butyl methacrylate) latex. The lines are predicted compositions (see text) (From Ref. [37])

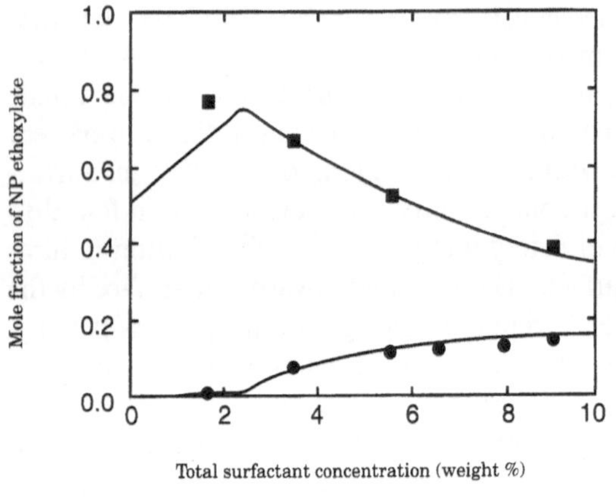

Figure 4 gives a good illustration of competitive adsorption of relevance to latex paints. A mixture of sodium dodecyl sulfate and ethoxylated (10 EO) nonylphenol is a common surfactant combination in latex preparation. Additional nonionic surfactant is often introduced into the formulation as mill base dispersant or as wetting agent. Additional anionic surfactant may be introduced as pigment or filler dispersant. As can be seen from the Figure, at low total surfactant concentration almost all nonionic surfactant is present on the latex particle surface. As the surfactant concentration is increased the ratio of nonionic to anionic surfactant at the surface approaches that in the bulk. The surfactant composition at the surface clearly varies with the total amount in the formulation and the preferential adsorption is greatest at the onset of micellization [37]. It has also been shown that the presence of polar solvents, which are commonly used to facilitate coalescence, affects the ratio of nonionic to anionic surfactant at the surface [38]. The experimentally determined surface composition from such surfactant mixtures agrees well with values predicted by treating the system as a pseudo three-phase system, consisting of a surface phase, a micellar phase, and a monomer phase, with only monomers adsorbing on the surface. The micelle–monomer equilibrium in the aqueous phase is given by the regular solution theory for mixed micelles [37].

The problem related to competitive adsorption has been accentuated with the incorporation of associative thickeners in the formation. These polymers, being highly surface active, have a strong driving force for hydrophobic surfaces such as latex particles. Adsorption behavior of hydrophobically modified polyurethanes, so-called HEUR thickeners, have been investigated in detail. If the latex surface is not fully covered by surfactant, these surface active polymers adsorb at the particle surface, often resulting in gels [39, 40]. Addition of nonionic surfactant often results in fluid, uniform dispersions, suggesting that the nonionic surfactant displaces the polymer from the latex surface.

In a systematic study on adsorption of mixtures of associative thickeners of the polyurethane type and nonylphenol ethoxylates on a hydrophobic latex it was demonstrated that polymers with hydrophobic side chains along the backbone could displace the nonionic surfactant provided the concentration of polymer hydrophobe units was high enough. For associative thickeners with only terminal hydrophobic chains it was found that the size of these chains is a decisive factor in competitive adsorption [41]. Also the distance between the hydrophobic end groups is of importance; the shorter the distance, i.e., the higher the concentration of hydrophobic chains, the more effective the polymer is in displacing the nonionic surfactant [42].

Competitive adsorption at solid surfaces is complex and it is therefore difficult to predict the outcome when new mixtures of amphiphiles or a new particle surface is introduced. For instance, the same mixture of nonionic surfactant and hydrophobically modified polyurethane showed a very different adsorption behavior on alpha olefin-maleic acid stabilized titanium dioxide than on the above-mentioned latex. On the coated pigment particle, the surfactant did not replace the associative thickener regardless of the structure of latter species [43].

Acknowledgement The author is grateful to Dr. Peter Weissenborn for valuable comments on the manuscript.

References

1. Dams R, Re S (1996) Polym Paint Colour J 186:13
2. Witte JD, Piessens G, Dams R (1994) Proc 22nd FATIPEC Congr, Vol 2, p 169
3. Easton T, Stephens D (1995) Verfkroniek 68:7
4. Adams JW, Heilen W (1995) Verfkroniek 68:43
5. Dougherty W, Medina S (1995) Eur Coat J 10:706
6. Derby R, Kleintjes HP (1990) The role of acetylenic glycols in the growth of water-based coatings. In: Karsa DR (ed) Additives for water-based coatings. Royal Society of Chemistry, Cambridge, UK
7. Jain M, Piirma I (1986) Polym Mater Sci Eng 54:358
8. Piirma I, Maw T (1984) Polym Bull 11:497
9. Bufkin BG, Grawe JR (1978) J Coatings Technol 50:65
10. Hofland A, Schaap F (1988) Polym Paint Colour J 178:620
11. Östberg G, Huldén M, Bergenståhl B, Holmberg K (1994) Progr Org Coatings 24:281
12. Bergenståhl B, Hofland A, Östberg G, Larsson A (1996) Alkyd emulsions. In: Salamone JC (ed) Polymeric Materials Encyclopedia. CRC Press, Boca Raton, FL
13. Dieterich D (1981) Progr Org Coatings 9:281
14. Rosthauser JW, Nacht-Kamp K (1986) J Coated Fabrics 16:39
15. Krishnamurti N (1983) Progr Org Coatings 11:167
16. Östberg G, Bergenståhl B, Sörensson K (1992) J Coatings Technol 64:33
17. Östberg G, Bergenståhl B, Huldén M (1994) J Coatings Technol 66:37
18. Holmberg K (1993) Surface Coatings Int 76:481
19. Östberg G, Bergenståhl B, Huldén M (1995) Colloids Surfaces A 94:161
20. Östberg G, Bergenståhl B (1996) J Coatings Technol 68:39
21. Weissenborn P, Bergenståhl B, Östberg G. To be published
22. Padget JC (1990) Additives for water-based coatings – a polymer chemist's view. In: Karsa DR (ed) Additives for water-based coatings. Royal Society of Chemistry, Cambridge, UK
23. Guyot A, Tauer K (1994) Adv Polym Sci 111:45
24. Guyot A (1996) Current Opinion Colloid Interface Sci 1:580
25. Holmberg K (1992) Progr Org Coatings 20:325
26. Huldén M, Sjöblom E (1990) Progr Colloid Polymer Sci 82:28
27. Kronberg B, Kuortti J, Stenius P (1986) Colloids Surfaces 18:411
28. Kientz E, Holl Y (1993) Colloids Surfaces A 78:255
29. Torstensson M, Rånby B, Hult A (1990) Macromolecules 23:126
30. Tauer K, Goebel K-H, Kosmella S, Stähler K, Neelsen J (1990) Makromol Chem Makromol Symp 31:107
31. Lam S, Hellgren AC, Sjöberg M, Holmberg K, Schoonbrood HAS, Unzué MJ, Asua JM, Tauer K, Sherrington DC, Montoya Goni A (1997) J Appl Polym Sci 66:187
32. Unzué MJ, Schoonbrood HAS, Asua JM, Montoya Goni A, Sherrington DC, Stähler K, Goebel K-H, Tauer K, Sjöberg M, Holmberg K (1997) J Appl Polym Sci 66:1803
33. Holmberg K (1996) Progr Colloid Polym Sci 101:69
34. Zhao CL, Dobler F, Pith T, Holl Y, Lambia M (1989) J Colloid Interface Sci 128:437
35. Evanson KW, Urban MW (1991) J Appl Polym Sci 42:2287
36. Muizebelt WJ, Hubert JC, Venderbosch RAM (1994) Progr Org Coat 24:263
37. Huldén M, Kronberg B (1994) J Coatings Technol 66:67
38. Kronberg B, Lindström M, Stenius P (1986) ACS Symp Ser 311, Ch 17
39. Karunasena A, Glass JE (1989) Progr Org Coatings 17:301
40. Karunasena A, Glass JE (1992) Progr Org Coatings 21:53
41. Ma Z, Chen M, Glass JE (1996) Colloids Surfaces A 112:163
42. Ma Z, Kaczmarski JP, Glass JE (1992) Polym Mater Sci Eng 66:23
43. Kaczmarski JP, Glass JE, Buchacek RJ (1993) Polym Mater Sci Eng 69:199

Progr Colloid Polym Sci (1998) 109:260–269
© Steinkopff Verlag 1998

J. Wieboldt
R. Zimehl
J. Ahrens
G. Lagaly

Some aspects of polymer colloids

III. Preparation and properties of different types of cationic latex particles

Received: 20 October 1997
Accepted: 27 October 1997

J. Wieboldt
PolymerLatex GmbH & Co. KG
D-45764 Marl
Germany

R. Zimehl (✉) · G. Lagaly
Institut für Anorganische Chemie
Universität Kiel
D-24098 Kiel
Germany

J. Ahrens
Hagenuk Telecom GmbH
Abteilung DEF
Westring 431
D-24118 Kiel
Germany

Abstract Different polymerization pathways concerning nucleation and growth mechanisms of anionic polymer lattices have been reported. However, limited information is available about cationic latices. We described preparation and properties of different types of anionic latex particles in a previous paper [Zimehl et al. (1990) Colloid Polym Sci 268: 924]. This part reports some possibilities to produce cationic latices by emulsion polymerization of styrene. Three types of polystyrene and polystyrene/comonomer lattices are distinguished by the structure of the boundary region which separates the latex particle from the surrounding medium. The mean particle size and the polydispersity index are obtained from quasi-elastic light scattering and transmission electron micrographs. Electron microscopy and the behavior of particles in a centrifugal force field are used as simple tools to gain an insight into particle morphology. The results were compared to the behavior of freeze-dried latex samples during nitrogen gas adsorption at 77 K. In the case of particles with extended boundary regions (core–shell particles or particles with hairy envelopes), film formation can reduce the specific surface area. Removal of soluble oligomers and polymers from the boundary region of these particles often increases the surface area.

Key words Emulsion polymerization – latices – light scattering – particle size – surface area

Introduction

The boundary region between hydrophobic polymer particles and the dispersion medium plays an important role during emulsion polymerization and influences the properties of the dispersion [1–4]. Monomer, initiator, and emulsifier not only determine the kinetics of particle formation but also the character of the particle interface and the properties of the latex particles [1, 5, 6]. Depending on the polymerization pathway different types of latex particles are obtained (Table 1, Fig. 1) [7, 8].

The "classical" latices are smooth, spherical polymer particles (B in Fig. 1). The stability of dispersions of this kind of particles (we call them type-1 latices) is mainly caused by fragments of initiator molecules accumulated at the particle surface and is in almost all cases of electrostatic nature.

Emulsion polymerization of a hydrophobic monomer, for instance styrene, in the presence of an emulsifier produces a more or less bristly particle surface with enhanced stability. The properties of this kind of dispersions (type-2 latices, Fig. 1C) are governed by a subtile balance of (i) interactions between the emulsifier ions adsorbed on the

Table 1 Different types of polystyrene latices

	Initiator	Emulsifier	Comonomer
Type-1 latices	ADMBA · 2 HCl[1)]	—	—
Type-2 latices	—	DPC	—
	—	OPC	—
	—	DTAB	—
	—	TBAC	—
	—	HY	—
	ADMBA · 2 HCl	DPC	—
	ADMBA · 2 HCl	OPC	—
	ADMBA · 2 HCl	DTAB	—
	ADMBA · 2 HCl	TBAC	—
	ADMBA · 2 HCl	BTAC	—
	ADMBA · 2 HCl	HY	—
Type-3 latices	ADMBA · 2 HCl	—	VTPC
	ADMBA · 2 HCl	—	VTAC

[1)] ADMBA · 2 HCl = azo bis-*N*,*N'*-dimethylene isobutyramidine dihydrochloride (see below); DPC = dodecyl pyridinium chloride; DTAB = dodecyl trimethylammonium bromide; HY = hyamin 1622; OPC = octadecyl pyridinium chloride; TBAC = tetradecyl dimethylbenzylammonium chloride; VTAC = vinylbenzyl trimethylammonium chloride; VTPC = vinylbenzyl triphenylphosphonium chloride.

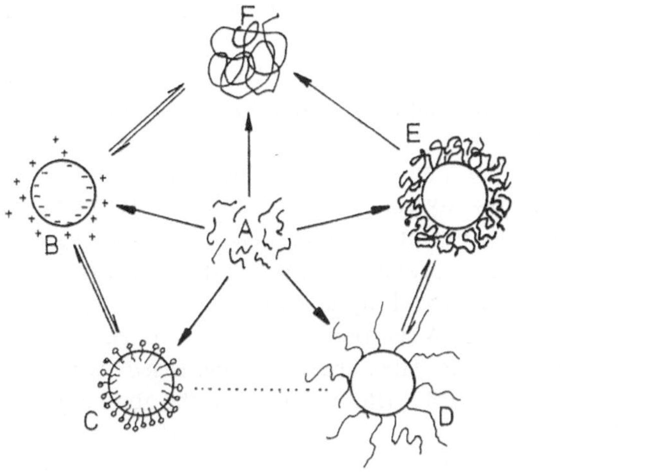

Fig. 1 Types of latex particles. **A**: monomer, **B**: particles with smooth surface (type-1 latices), **C**: particles with the surface bristly by adsorbed emulsifiers (type-2 latices), **D, E**: type-3 latices with hairy surface (D) or core-shell particles (E) and **F**: macromolecular coils

particle surface and (ii) interactions of the emulsifier with the dispersion medium.

When soluble or partially soluble unsaturated emulsifiers (comonomers, macromers) are added during polymerization, different types of boundary layers are created, and various particle structures and dispersion properties are obtained. In addition, the particles can be chemically modified, for instance by incorporating glycidyl methacrylate [9] or vinylbenzyl chloride [10]. Depending on the comonomer/monomer ratio, three types of particles are distinguished: (i) core-shell particles with a hydrophobic core and an hydrophilic envelope (Fig. 1E), (ii) particles surrounded by polyelectrolytes which protrude into the dispersion medium (Fig. 1D), and (iii) hydrophilic, highly swollen gel particles (microgels) which, in extreme case, resemble to macromolecules or polyelectrolytes in solution (Fig. 1F). Intermediate between particles E and F are multidomain structures (Fig. 4, Fig. 5G). All these types of copolymer particles are termed type-3 latices.

Examples of the different types of polystyrene latices with positive surface charges were prepared in laboratory scale (Table 1).

Material and methods

Preparation of latices

Type 1: The particles were prepared by emulsifier-free emulsion polymerization of styrene in the presence of azo bis-N,N'-dimethylene isobutyramidine dihydrochloride (ADMBA · 2 HCl):

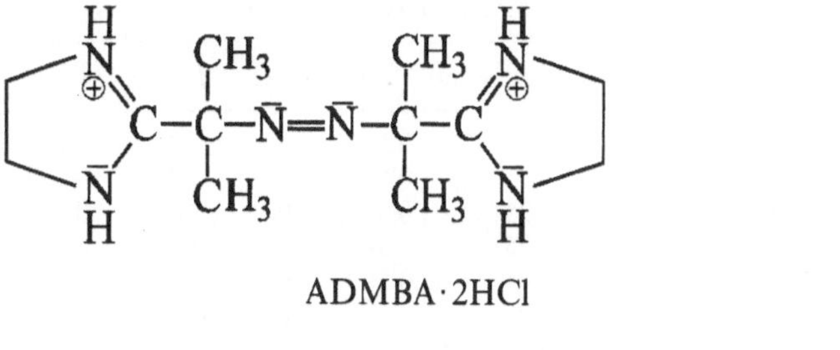

ADMBA · 2HCl

The desired amount of styrene (Table 2a) was dispersed in 220 ml of twice distilled water. The mixture was heated

Table 2a Preparation of type-1 polystyrene latices at 70 °C and 90 °C (without emulsifier)

Initiator/latex	M[1] (ml)	I (g)	T (°C)	t (h)	ϕ
ADMBA 4	2.5	0.24	70	23	0.0008
ADMBA 3	2.5	0.6	70	23	0.0008
ADMBA 7	25	0.6	70	23	0.016
ADMBA 2	1	0.24	90	23	0.002
ADMBA 8	1	0.6	90	23	0.004
ADMBA 1	2.5	0.24	90	23	0.003
ADMBA 5	2.5	0.6	90	23	0.008
ADMBA 100	25	1	90	7.5	0.021

[1] M: styrene; I: ADMBA · 2 HCl; total volume of emulsion: 250 ml except ADMBA 100 (500 ml); T: reaction temperature; t: polymerization time; ϕ: volume fraction of solid polymer.

Table 2d Preparation type-3 polystyrene latices. Initiator: ADMBA · 2 HCl, comonomer: VTPC

Comonomer/latex	M[1] (ml)	I (g)	C (g)	T (°C)	ϕ
VTPC 1	2.5	0.5	0.5	70	0.011
VTPC 5	2.5	0.5	0.5	100	0.004
VTPC 2	25	0.5	1.0	70	0.075
VTPC 8	25	0.5	0.5	90	0.051
VTPC 3	25	0.5	1.0	90	0.080
VTPC 4	25	0.5	2.0	90	0.097
VTPC 6	25	0.5	3.0	90	0.096
VTPC 7	25	0.5	5.0	90	0.104
VTPC 9	25	—	2.2	100	0.008

[1] M: styrene; I: ADMBA · 2 HCl; C: VTPC; total volume of emulsion: 250 ml; ϕ: volume fraction of solids.

Table 2b Preparation of type-2 polystyrene latices. Initiator: ADMBA · 2 HCl

Emulsifier/latex	I[1] (g)	E (g)	T (°C)	t (h)	ϕ
DPC 1	0.5	0.25	80	24	0.008
DPC 2	0.5	0.25	90	21	0.029
DPC 3	0.5	0.25	100	23	0.056
DCP 8	0.5	0.50	100	21	0.073
DCP 9	0.5	1.00	100	23	0.070
DCP 4	0.5	2.00	100	27	0.090
DCP 7	0.1	1.00	100	19	0.054
DCP 5	—	1.00	100	23	0.012
DCP 6	—	2.00	100	23	0.019
OPC 1	0.5	1.32	100	6	0.075
OPC 2	—	1.32	100	23	0.054
DTAB 1	0.5	1.11	100	4	0.081
DTAB 2	—	1.11	100	23	0.014
TBAC 1	0.5	0.75	100	5	0.075
TBAC 2	—	1.32	100	23	0.054
BTAC 1	0.5	1.32	100	24	0.048
HY 1	0.5	1.67	100	6	0.084
HY 2	—	1.67	100	23	0.052

[1] I: ADMBA · 2 HCl; E: emulsifier; T: polymerization temperature; t: polymerization time; ϕ: volume fraction of solids; styrene: 25 ml; total volume of emulsion: 250 ml.

Table 2c Preparation of type-3 polystyrene latices. Initiator: ADMBA · 2 HCl, comonomer: VTAC

Comonomer/latex	M[1] (ml)	I (g)	C (g)	T (°C)	ϕ
VTAC 1	2.5	0.5	0.50	70	0.011
VTAC 2	25	0.5	0.57	90	0.037
VTAC 3	25	0.5	1.00	90	0.089
VTAC 4	25	0.5	2.32	90	0.099
VTAC 5	25	0.5	4.06	90	0.113
VTAC 6	25	—	2.00	100	0.012

[1] M: styrene; I: ADMBA · 2 HCl; C: VTAC; total volume of emulsion: 250 ml; T: reaction temperature; ϕ: volume fraction of solids.

to 70 °C, and nitrogen was bubbled through the emulsion. ADMBA · 2 HCl dissolved in a small amount of water was added after 5 min. The total volume of the dispersion was 250 or 500 ml. The dispersion was held at 70 °C for 23 h or at 90 °C for 7.5 h, cooled, filtered through glass wool and outgased under reduced pressure. The volume fraction of solid polymer, ϕ, increased with the amount of initiator added to the emulsion (Table 2a).

Type 2: To prepare polystyrene latices of type 2, emulsions of styrene were stabilized by different surfactants in the presence or absence of the initiator ADMBA · 2 HCl (Table 2b). Most latices were obtained at 100 °C from 25 ml styrene and 0.25–2.0 g emulsifier in 225 ml water. When the initiator was added, the volume fraction of latex particles increased with the amount of DPC; the yield was lower in the absence of the initiator.

Type 3: Type-3 latices were obtained by copolymerization of (i) styrene and vinyl trimethylammonium chloride (VTAC) and (ii) styrene and vinyl triphenylphosphonium chloride (VTPC) (Table 2c, d). Emulsions of 2.5 ml styrene in 200 ml water or of 25 ml styrene in 180 ml water were heated to 70–100 °C before the desired amounts of VTAC, dissolved in 20 ml water, and ADMBA, dissolved in 25 ml water, were added. VTPC which is not sufficiently soluble in water was dispersed in the aqueous solution of the initiator. The total volume of emulsion was 250 ml. Generally, a blue opalescence was observed after 1–2 min. The period of polymerization was 4 h (24 h for runs VTPC 9 and VTAC 4). In most cases, polymerization yielded a milky, white dispersion. VTPC 1, 5, 9 and VTAC 1 were nearly translucent. The volume fraction of solid polymeric material, ϕ, increased with the comonomer content. The yield was modest in the absence of ADMBA · 2 HCl.

Quasi-elastic light scattering

Photon correlation spectroscopy (quasi-elastic light scattering, QELS [8, 11, 12]) measures the diffusion constant of freely moving particles in diluted dispersions. The hydrodynamic particle radius, R_{QUELS}, is obtained from

$$R_{\text{QUELS}} = q^2 kT \cdot t_c / 3\pi\eta \quad (1)$$

with $q = 4\pi n/\lambda \sin \Theta/2$; n and η are the refractory index and viscosity of the dispersion medium, respectively; Θ is the scattering angle, t_c a correlation time (details see [8]). To avoid artefacts caused by multiple light scattering, measurements were performed with a double beam optic at two different wavelengths [11, 12]. The polydispersity index σ given by

$$\sigma = \frac{\langle R^2 \rangle}{\langle R \rangle^2} - 1 \quad (2)$$

provides information on the width of the particle size distribution. Sample preparation for the QELS measurements was reported earlier [1].

Sedimentation experiments

Samples of 1 ml of the original latex dispersion were transfered to Eppendorf tubes (total volume of 1.5 ml) and centrifuged at 11 000 rpm (=11 634 g, Biofuge B, HERAEUS, Germany) for 24 h at about 40 °C. Immediately after centrifugation, the sediment volume V_{Sed} was determined with a cathetometer. The supernatant liquid was carefully withdrawn without disturbing the sediment. The volume fraction of the latex in the sediment, ϕ_{Sed}, was determined gravimetrically (drying at 70 °C for 24 h):

$$\phi_{\text{Sed}} = \frac{(m_{\text{Sed}} - m_{\text{H}_2\text{O}})/\rho}{V_{\text{Sed}}}, \quad (3)$$

where m_{Sed} is the mass of the sediment, $m_{\text{H}_2\text{O}}$ the mass of water in the sediment, ρ the density of the particles, $\rho = 1.06\ \text{g cm}^{-3}$, and V_{sed} the total volume of the sediment.

Electron microscopy

Electron micrographs of dried diluted dispersions were obtained with the scanning electron microscope PHILIPS 200. The particle size distribution (number and volume distribution) was determined by counting (TGZ-3, Karl Zeiss, Jena, Germany).

Surface charge titration

The surface charge of the particles was determined by the streaming current technique (PCD 02, Mütek, Germany) in combination with an automatic titration unit (DL 21, Mettler, Switzerland). The diluted latex dispersion (volume fraction about 2×10^{-3}) and its serum were titrated with 0.001 N solution of poly(sodium ethylene sulfonate). The specific charge of the latex particles was obtained as difference between the titration data of the latex and its serum.

Gas adsorption

The nitrogen gas adsorption isotherms of several latex samples were measured in an automatic gas adsorption apparatus [13, 14]. For sample preparation and analysis of the isotherms see [1]. On the basis of nitrogen adsorption studies on many latex samples a theoretical isotherm with a C value of 57 [25] was chosen as the standard isotherm for calculating t-plots. The C-value derived from the experimental adsorption isotherms ranges between 45 and 70. Due to the presence of micropores S_{BET} does not represent the real specific area but an apparent value. The similarity with S_{QELS} indicates that the difference between the true and the apparent specific surface area is modest.

Results and discussion

Size analysis by light scattering

The mean particle size of type-1 latices was virtually independent on the initiator concentration (Table 3a). ADMBA · 2 HCl was the only electrolyte added to the emulsion. The change of ionic strength by the amounts of initiator added was relatively small and had no significant influence on the particle size [15].

Increasing emulsifier concentration (at constant initiator level) decreased the mean particle radius of type-2

Table 3a Particle size of type-1 polystyrene latices

Initiator/latex	$10^3\ (I/M)$[1)]	R_{QUELS} (nm)	σ
ADMBA 7	8.5	79.5	0.35
ADMBA 100	15.2	65.2	0.22
ADMBA 1	33.9	63.9	0.32
ADMBA 4	33.9	73.5	0.14
ADMBA 2	84.3	61.8	0.16
ADMBA 5	84.8	60.2	0.14
ADMBA 3	84.8	87.5	0.24
ADMBA 8	210.9	72.3	0.20

[1)] I/M: molar ratio ADMBA/styrene; R_{QUELS}: mean hydrodynamic particle radius; σ: polydispersity index.

Table 3b Particle size of type-2 latices

Emulsifier/Latex	10^3 $(I/M)^{1)}$	10^3 (E/M)	T (°C)	R_{QUELS} (nm)	σ
DCP 5	—	18.7	100	198	0.47
DCP 6	—	37.5	100	143	0.51
DCP 3	7.9	4.7	100	115	0.01
DCP 8	7.9	9.4	100	188	0.38
DCP 9	7.9	18.7	100	32.9	0.11
DCP 4	7.9	37.5	100	20.8	0.24
DCP 5	—	18.7	100	198	0.47
DCP 7	1.6	18.7	100	67.8	0.24
DCP 9	7.9	18.7	100	32.8	0.11
DCP 1	7.9	4.7	80	86	0.47
DCP 2	7.9	4.7	90	39	0.41
DCP 3	7.9	4.7	100	115	0.01

1) I/M: molar ratio ADMBA/styrene; E/M: molar ratio DPC/styrene; R_{QUELS}: mean hydrodynamic particle radius; σ: polydispersity index.

Table 3c Influence of emulsifier type on the size of type-2 latices

Emulsifier/latex	R_{QUELS} (nm)	σ
Initiator monomer ratio: 10^3 $(I/M) = 7.9$		
Emulsifier monomer ratio: 10^3 $(E/M) = 1.87$		
DPC 9	32.9	0.11
OPC 1	21.8	0.39
DTAC 1	32.5	0.18
TBAC 1	22.5	0.35
BTAC 1	146	0.14
Without initiator		
Emulsifier monomer ratio: 10^3 $(E/M) = 1.87$		
DPC 5	198.0	0.47
OPC 2	56.0	0.24
DTAC 2	339.0	0.52
TBAC 2	75.4	0.26

latices (Table 3b). In the absence of initiator, however, high emulsifier concentrations produced large particles. At constant amounts of emulsifier, the mean particle size decreased with increasing initiator concentration. The structure of emulsifier governed particle size and particle size distribution (Table 3c).

Particle size determination of the copolymer dispersions (type-3 latices) was much more troublesome than of homopolymer type-1 or type-2 latices. The way the samples were prepared for the light scattering experiments played a decisive role in obtaining reliable results. In many cases, the particles became smaller when the dispersions were extensively purified by centrifugation/redispersion. For instance, the hydrodynamic particle radius of VTPC 3 latices decreased from 73.0 to 46.8 nm by centrifugation and redispersion, the polydispersity index remained constant ($\sigma = 0.15$). The radius of VTPC 4 latices changed from 46.8 to 31.8 nm, the index decreased from 0.63 to 0.21. For this reason, all type-3 dispersions were centrifuged and redispersed by a standardized procedure. The main cause of the changes is the strong aggregation tendency of type-3 copolymer latices at higher comonomer/monomer ratios (even at low particle volume fractions). Disintegration of the aggregates (secondary particles) during centrifugation produces smaller particles. After redispersion, the size of the primary particles is measured. Even certain core-shell particles can loose a part of their envelope and become smaller.

Electron microscopic studies

Scanning electron micrographs of cationic type-1 and type-2 latices showed dense, spherical particles with sharp boundaries. The mean diameter (number average) evaluated from scanning electron micrographs was similar to that obtained by light scattering. For type-3 latices agreement is certainly not expected. The surface layers consisting of polyelectrolytes or even of emulsifier ions can often be disintegrated by freeze-drying or degradated by the electron beam. Typical examples are latices HY 1 and VTPC 1. The type-2 latex particles HY 1 and type-3 particles VTPC 1 were comparable in size (radius $r = 19$ nm) as determined by light scattering. The electron micrographs showed particles HY 1 ($r = 12.5$ nm) being smaller than VTPC 1 ($r = 17$ nm). Similar results were obtained with anionic latices. It was shown that increasing amounts of anionic monomers produced increasingly disintegrated particles [1]. In the extreme case, only a small number of discrete particles of irregular shape were observed apart from amoeba-like forms with more or less extended tentacles.

Fig. 2 Specific surface charge of cationic latex particles as a function of comonomer added (cf. Table 2 c, d). (a) VTAC latex and (b) VTPC latex. **A**: theoretical charge, **B**: latex as received after preparation, **C**: latex after serum replacement, **D**: latex after centrifugation/redispersion

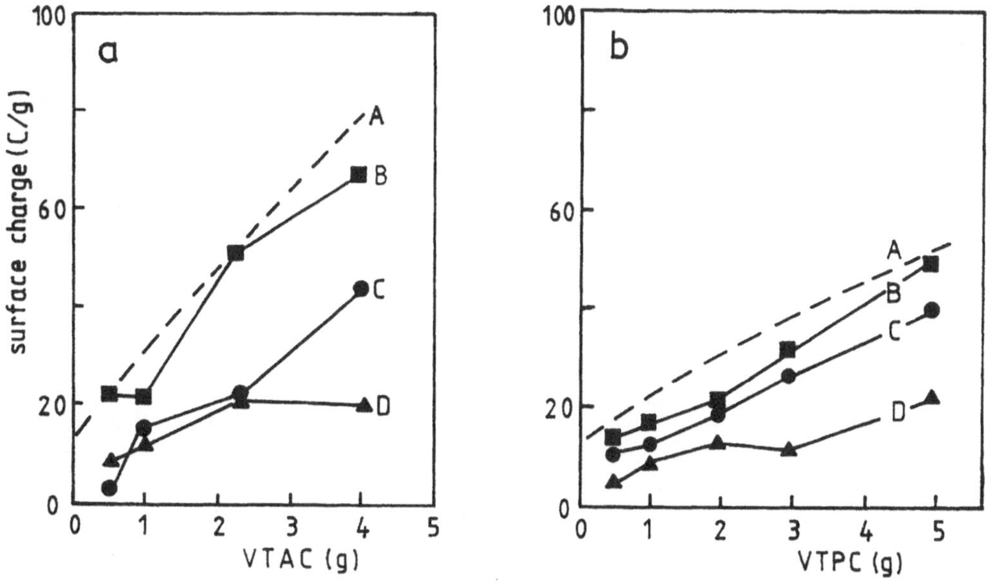

Titration

Polyelectrolyte titration of type-3 particles revealed a high specific surface charge in the range of 10–60 Cg^{-1}. The difference between the experimental and the theoretical charge was not too large (Fig. 2). The theoretical specific charge was calculated from the amount of initiator and styrene added, assuming that all positive charges are produced by decomposition of the azo initiator.

The charge density increased with increasing comonomer content. It was strongly reduced by centrifugation, and highly charged macromolecules with charges between 40 and 140 Cg^{-1} were found in the supernatant. This indicates that an appreciable amount of ionic comonomer was only physically adsorbed at the particle-solution interface. As shown in Fig. 2, the centrifugation/redispersion cleaning procedure was more effective than serum replacement.

The particles with highest comonomer content were multidomain structures and were disintegrated during centrifugation into hydrophobic, more lowly charged particles forming the sediment and hydrophilic, highly charged macroions remaining in solution [18].

Sedimentation

Centrifugation/redispersion cycles were very effective in cleaning latex dispersions. The procedure reduced not only the mean particle size of type-3 dispersions but also the titrable surface charge [19]. Gas adsorption was also largely influenced (see next section). For this reason, the results of centrifugation experiments are discussed in more detail.

Different types of sediments were formed by centrifugation at about 10000 g (Fig. 3). Type-1 and type-2 latices formed dense and compact white sediments (Fig. 3A) which could not be redispersed. The copolymer type-3 latices settled in different ways during centrifugation (Fig. 3B, C, Table 4). Often, translucent gel-like sediments developed with a sharp boundary to the clear liquid (Fig. 3C). Several of these gels showed iridescence and could be instantaneously redispersed in water. Turbid, gel-like sediments with formation of Liesegang rings were also observed. In other cases, a white opaque sediment at the bottom of the test tube was separated from the liquid phase by a translucent gel layer (Fig. 3B).

The transparency of the type-C sediments and the gel layers of B sediments is indicative of lattice-like structures. If particles are arranged in a regular way and in appropriate distances, destructive interferences occur which make the sediment transparent [20]. Many of the type-3 dispersions fit these requirements. Latices with higher comonomer content often form type-B sediments. Larger particles settle to a white opaque sediment; the smaller sized particles then aggregate forming a gel-like network.

Assuming that the particles in the sediment are closest packed (ccp) and that the mean particle size R_{QELS} is that of the particle core, the thickness of the interfacial layer surrounding the particles is estimated from the sediment volume. The effective radius of the particles in the sediment is related to the volume fraction for closest

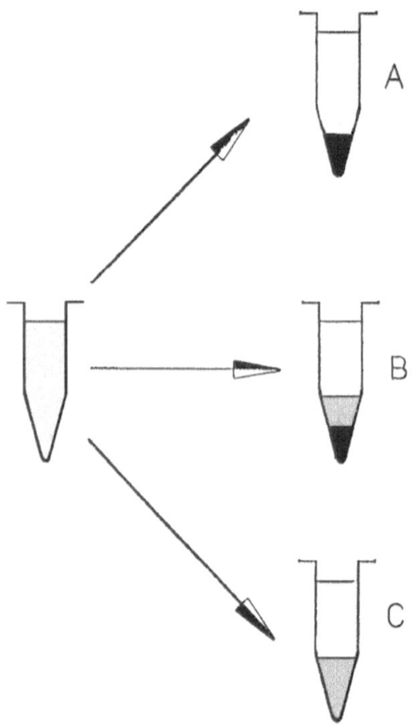

Fig. 3 Different types of sediments of latex particles. **A**: white, compact sediment, **B**: gel-like sediment with white, compact sublayer and **C**: voluminous, gel-like sediment

Table 4 Sedimentation of type-3 latex dispersions

Comonomer/latex	$C^{1)}$ (g)	R_{QUELS} (nm)	Sediment[2)]
VTAC 3	0.57	63	C
VTAC 2	1.00	39	C
VTAC 5	2.32	86	C
VTAC 6	4.06	35	B
VTPC 8	0.5	59	C
VTPC 3	1.0	34	C
VTPC 4	2.0	32	B
VTPC 6	3.0	137	B
VTPC 7	5.0	96	B

[1)] *C*: comonomer/25 ml styrene.
[2)] See Fig. 3.

packing ($\phi_{CP} = 0.7404$) and the measured volume fraction ϕ_{Sed}:

$$R_{eff} = R_{QELS}(\phi_{CP}/\phi_{Sed})^{1/3}. \quad (4)$$

The thickness of the interfacial layer is $\delta = R_{eff} - R_{QELS}$.

Particles forming C sediments carried envelopes of 1–13 nm thickness (Table 5). Similar values were obtained by microscopic investigation of sterically stabilized latices [16]. The values calculated in this way are only a crude approximation because the experimental hydrodynamic particle diameter may comprise a part of the interfacial region. VTAC particles were surrounded by thicker layers than VTPC latices: Incorporation of vinyl trimethylammonium chloride into the surface created thicker layers than copolymerization with the more hydrophobic vinyl triphenylphosphonium chloride.

Table 5 Layer thickness of type-3 particles forming gel-like sediments (Fig. 3C)

Comonomer/latex	$C^{1)}$ (g)	R_{QUELS} (nm)	R_{eff} (nm)	δ
VTAC 3	0.57	63	73	10
VTAC 2	1.00	39	44	5
VTAC 5	2.32	86	99	13
VTPC 8	0.5	59	61	2
VTPC 3	1.0	34	35	1

[1)] *C*: g comonomer/25 ml styrene.

Gas adsorption

The specific BET surface area of spherical type-1 and type-2 latex particles (anionic and cationic) is related to the mean particle diameter; *t*-plots did not reveal any microporosity [1]. The specific BET surface area of anionic type-3 latices was often smaller than that calculated from the average particle radius. One reason was a certain degree of film forming when the dispersion medium was removed by freeze-drying [1]. Soluble polymeric material or polyelectrolytes conglutinate the particles. Therefore, the specific surface area of many samples increased when the freeze-dried latices were dispersed in water, centrifuged and again freeze-dried. Similar observations were made with cationic type-3 latices, for instance VTAC 2 (Table 6). The specific BET surface area of this sample was 58 m^2/g. Centrifugation and redispersion increased the surface area to 76 m^2/g, a value identical with the area S_{QELS} calculated from the hydrodynamic radius. In contrast, redispersed particles VTPC 3 again glued together during centrifugation and freeze-drying so that redispersion did not increase the specific surface area, which was smaller than S_{QELS}.

The nitrogen adsorption isotherms were of type 4 (BDDT classification [21]). The *t*-plots [21] often indicated microporosity with micropore volumes of 3–5 ml/g. The nitrogen adsorption isotherms of several samples showed an unusual hysteresis loop extending to very low relative pressures. Similar observations were made with certain types of crystalline silicic acids [14]. The reason may be similar: at high relative pressure the nitrogen molecules are pressed into micropores between the latex particles and, because of some deformation of the particles, can not desorb when pressure is reduced.

Table 6 Specific area (m^2/g) of latex VTAC 2 and VTPC 3 after subsequent treatments

Comonomer/latex		S_{BET} [1)]	S_{QELS}	S_{eff}	C-BET
VTAC 2	freeze-dried sample	58			71
	Redispersion, centrifugation, freeze-drying	76	72	60	47
VTPC 3	freeze-dried sample	72			60
	Redispersion, centrifugation, freeze-drying	74	83	82	69

[1)] S_{BET}: apparent specific BET surface area; S_{QELS}, S_{eff}: specific surface area calculated from the hydrodynamic radius and the effective radius, resp.

Discussion

Preparation of type-1 and type-2 polystyrene latices represents the classical method of latex formation. In the absence of surface active agents the surface properties of type-1 latices are determined by the functional groups of the initiator. Free radicals formed by decomposition of the initiator react with the monomers forming oligomeric radicals. The reaction may be considered as a self-stabilizing, positive feed-back polymerization. In the beginning, micellar nucleation is the rate determining step [22]. According to calculations of Vanderhoff [23] for the system styrene/potassium peroxodisulfate small surface active oligomers are formed. They aggregate to spherical micelles with diameters and surface charge densities similar to sodium lauryl sulfate micelles. These micelles incorporate styrene molecules and swell considerably. During further polymerization these swollen micelles loose their individuality and form particles. The particle formation is then best fitted by a two step mechanism [24]. In the second step rather large particles with intermediate polydispersity form.

The role of emulsifiers in emulsion polymerization is manifold. The fundamental property of emulsifier molecules is their affinity to the interface of the growing latex particles. They are attached by van der Waals forces and can easily be desorbed. The interaction between the growing particle surface and the emulsifier competes with the interaction between adsorbed ions and molecules in the surface layer and the influence of the ions in the bulk phase on the adsorbed layer.

A coagulative mechanism is proposed for type-2 particles [26]. Surface activity and concentration of emulsifier are rate determining factors. At very low emulsifier concentration the particle surface charge mainly arises from initiator fragments and is independent on the emulsifier concentration. The coagulation rate is constant and also independent on the emulsifier concentration, so that a relatively small number of particles is formed. When the emulsifier concentration is increased, the surface charge density rises rapidly, the rate of coagulation decreases, and most of the small particles grow to larger ones. At still higher surfactant concentration, the particle surface becomes saturated and the surface charge density is again independent on the emulsifier concentration. Consequently, the rate of coagulation and the number of particles are independent of the emulsifier concentration. However, heterocoagulation (i.e. coagulation between large and small particles) can occur during the growing period. Small particles are captured by larger ones, and dispersions of particles with large polydispersity will be formed. At lower polymerization temperature the adsorption of emulsifier is enhanced, the period of seeding increases, a smaller number of primary seed particles coagulate and the size distribution becomes broader (Table 3b).

In the absence of initiator the seed periods are prolonged but the applied high emulsifier concentration stabilizes the seed particles. Low volume fractions of polydisperse particles are obtained.

Many methods [8, 27–30] are reported to incorporate functional monomers in latex particles by emulsion copolymerization. In most cases the comonomer is only used as a substitute of emulsifier and is added in small amounts. VTPC and VTAC monomers themselves are not surface active. During the initial steps of emulsion polymerization they copolymerize with dissolved styrene and form surface active oligomers ("polyelectrolyte snakes", Fig. 5B) which act as stabilizing agents for the growing latex particles. At low comonomer concentration (<0.01 mole fraction of the monomer), the particle size decreases linearly with increasing comonomer concentration. A narrow particle size distribution is obtained because a large number of highly charged, stable oligomeric radicals form in the aqueous phase. Even in this case, only a small proportion of these oligomers are incorporated into the particles. At higher comonomer concentrations the particle size does not necessarily decrease with the comonomer content [28, 29]. Production of large amounts

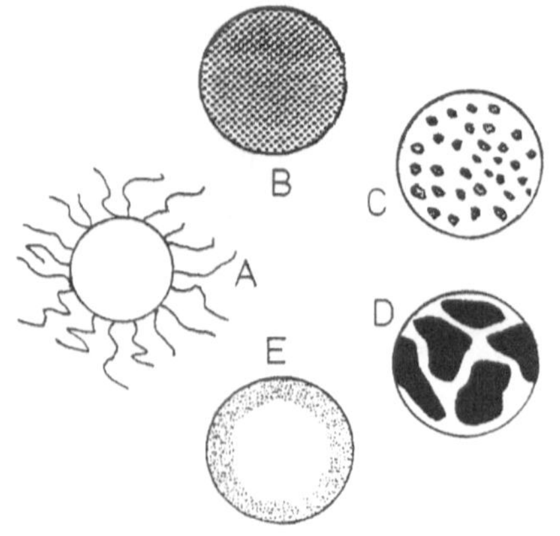

Fig. 4 Structure of copolymer latex particles. **A**: hairy particles: macromolecules form hairs at the surface of the particles, **B**: random distribution of polymer and copolymer within particles, **C, D**: multidomain particles and **E**: core–shell particles

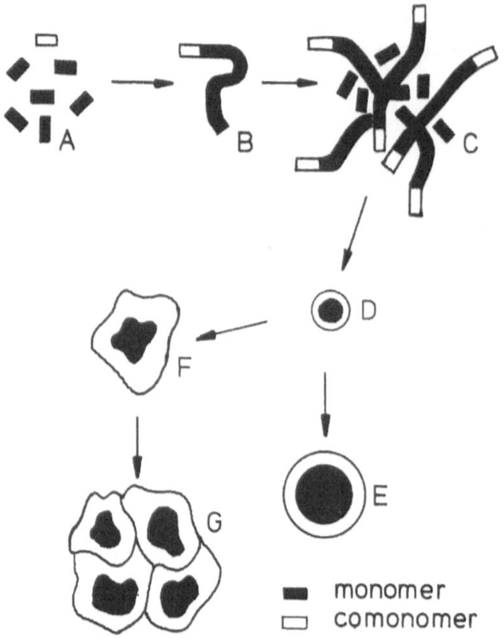

Fig. 5 Formation of cationic polystyrene particles. **A**: hydrophobic and hydrophilic monomers in solution, **B**: water soluble polyelectrolytes, **C**: polyelectrolytes aggregated and swollen by hydrophobic monomers, **D**: precursor particles, **E**: core–shell particles, **F**: copolymer particles of irregular shape and **G**: multidomain particles

of oligomers strongly affects nucleation and stabilization of the particles. In many cases, the oligomers grow to water soluble polymers. Low molecular weight polyelectrolytes were identified in the serum separated from the particles by ultracentrifugation [18, 19]. When the oligomers contain too high proportions of VTPC or VTAC, the charge increases, which reduces adsorption or makes the oligomers too hydrophilic to be surface active. In both cases the stabilizing effect on the particles is reduced and a broader particle size distribution is observed. Emulsion copolymerization with water soluble nonionic comonomers, mostly acrylic and methacrylic derivatives, shows similar trends. Kawaguchi et al. [31] observed a rapid consumption of the soluble comonomer at low degrees of conversion during emulsifier free emulsion copolymerization of styrene with nonionic acrylamide derivates with large N-substituents. He proposed a three step polymerization mechanism.

Typical are hairy particles or core-shell particles (Fig. 1D, E; Fig. 4A, E) because the charged segments are displaced from the hydrophobic core and enriched on the particle surface. Multidomain particles (Fig. 4) can also be formed. Low comonomer concentration leads to hydrophobic particles surrounded by more or less extended hydrophilic envelopes. At high dosages of hydrophilic comonomers, polyelectrolytes still free in the solution reduce the stability of the particles by bridging or depletion flocculation. The polymeric moieties tend to separate into two phases, hydrophobic/hydrophilic domains are formed, and eventually the particles assume a multidomain structure (Fig. 4C, D).

The micellar growing mechanism is schematically shown in Fig. 5A–E. The reaction starts in the aqueous phase by formation of comonomer rich "polyelectrolyte snakes" (Fig. 5B). The growing snakes become surface active, aggregate and solublize styrene (Fig. 5C). Polymerization of these styrene monomers yields precursor particles. The polyelectrolyte snakes adopt the role of emulsifier and stabilize the precursor particles which can grow by two mechanisms. When the monomers penetrate into the interior of the particles, the growing core remains enclosed in the hydrophilic envelope, and almost spherical particles form. This mechanism competes with template polymerization[1)] starting from the boundary region. If the concentration of comonomers and oligomers is still high, these molecules are enriched in the envelope of the precursor particles and react with monomers. The monomers are captured by the growing oligomers and, therefore, do not diffuse to the core. Particles of irregular shape form which

[1)] In a recently published paper [32], Miller et al. demonstrated that the number of polymer particles produced increased substantially when a small amount (about 0.05 wt%) of polystyrene ($M_n \approx 24\,400$ or $52\,600$, by GPC) was added.

easily aggregate to multidomain particles. As discussed above these particles can often be disintegrated by centrifugation.

It is evident from these studies that the type and size of oligomers play an important role. As stressed by El-Asser and Sudol [33] there is a need to separate and characterize the oligomers at various stages of the polymerization, particularly during the earliest stages.

Acknowledgements We are grateful to the Deutsche Forschungsgemeinschaft for financial support and thank Dr. Robbers (Hüls AG) for electron microscopic investigations.

References

1. Zimehl R, Lagaly G, Ahrens J (1990) Colloid Polym Sci 268:924
2. Zimehl R, Wieboldt J (1989) TIZ International Powder Magazine 113:629
3. Kluß BU, Zimehl R, Lagaly G, Hahn K (1993) Kurzfassungen 36. Hauptversammlung Kolloidgesellschaft, Jülich, p 75
4. Zimehl R, Priewe J (1996) Progr Colloid Polym Sci 101:116
5. Zimehl R, Lagaly G (1986) Progr Colloid Polym Sci 72:28
6. Maack A, Priewe J, Zimehl R (1995) Kurzfassungen, 11. Ulmer Kalorimetrietage, Freiberg/Sa., p 93
7. Arshady R (1992) Colloid Polym Sci 270:717
8. Lagaly G, Schulz O, Zimehl R (1997) Dispersionen und Emulsionen – Einführung in die Kolloidik feinerverteilter Stoffe einschließlich der Tonminerale, Steinkopff, Darmstadt
9. Schlund B, Pith T, Lambla M (1985) In: Guillot J, Pichit C (eds) Emulsion copolymerization. Hüthig Wepf, Basel, p 419
10. Loyd WG, Durocher TE (1963) J Appl Polym Sci 7:2025
11. Drewel M, Ahrens J, Podschus U (1990) J Opt Soc Am A 7:206
12. Ahrens J (1989) Thesis, University of Kiel
13. Kruse HH, Lagaly G (1988) GIT Fachz Lab 32:1096
14. Kruse HH, Beneke K, Lagaly G (1989) Colloid Polym Sci 267:844
15. Goodwin JW, Ottewill RH, Pelton R (1979) J Colloid Polym Sci 257:61
16. Paine AJ, Deslandes Y, Gerrior P, Henrissat B (1990) J Colloid Interface Sci 138:170
17. Zimehl R (1986) Thesis, University of Kiel
18. Zimehl R (1996) Kurzfassungen, DECHEMA-Jahrestagungen '96, Wiesbaden, p 291
19. Wieboldt J (1991) Thesis, University of Kiel
20. Meyer WH (1987) Chiu Z 21:59
21. Gregg SJ, Sing KSW (1982) Adsorption, surface area and porosity. Acad Press, London
22. Goodall AR, Wilkinson MC (1980) In: Fitch RM (ed) Polymer Colloids II. Plenum Press, New York
23. Vanderhoff JW (1985) J Polym Sci Polym Sym 72:161
24. Zhiqiang Song, Poehlein GW (1989) J Colloid Interface Sci 128:486; 501
25. Lecloux A, Pirard JP (1979) Colloid Interface Sci 70:265
26. Feeny PJ, Napper DH, Gilbert RG (1987) J Colloid Interface Sci 118:493
27. Guillot J, Pichit C (eds) (1985) Emulsion copolymerization, Hüthig and Wepf, Basel
28. Kim JH, Chainey M, El-Aasser MS, Vanderhoff JW (1989) J Polym Sci Part A: Polym Chem 27:3187
29. Guillot J, Guyot A, Pichot C (1990) In: Candau F and Ottewill RH (eds) Scientific Methods, Kluwer, Dordrecht, p 97
30. Daniels ES, Sudol ED, El-Aasser MS (eds) (1992) Polymer Latexes, ACS Symposium Series 492, Washington
31. Kawaguchi H, Sugi Y, Ohtsaka Y (1981) ACS Symp Series 165:145
32. Miller CM, Sudol ED, Silebi CA, El-Aasser MS (1995) Macromolecules 28:2754, 2765, 2772
33. El-Aasser MS, Sudol ED (1997) In: Lovell PA, El-Aasser MS (eds) Emulsion polymerization and emulsion polymers. Ch 2. Wiley, New York, p 38

Progr Colloid Polym Sci (1998) 109:270–277

C. Marburger
J. Kreuter

Production of nanoparticles: polymerization termination and molecular weights during radical-induced emulsifier-free emulsion polymerization (heterogeneous polymerization)

Received: 1 November 1996
Accepted: 14 April 1997

C. Marburger · J. Kreuter (✉)
Institut für Pharmazeutische Technologie
Biozentrum Niederursel
Johann Wolfgang Goethe-Universität
Marie-Curie-Straße 9
D-60439 Frankfurt am Main
Germany

Abstract Polymethylmethacrylate (PMMA) nanoparticles were produced by radical-induced emulsifier-free emulsion polymerization (heterogeneous polymerization) in aqueous solution using γ-rays, ammonium persulphate (APS) and potassium persulphate as initiators. The conditions of polymerization were varied with regard to the concentration of monomer and initiator, the temperature of reaction, and the properties of the reaction mixture. The molecular weights of the polymers were determined by gel permeation chromatography (GPC) and by viscosimetry. Increasing amounts of monomer increased the molecular weight up to a concentration of 3% MMA. Monomer concentrations higher than 3% led to a decrease in the molecular weight most likely due to the high viscosity and diffusion distances in the resulting polymer particles. The concentration of initiator (APS) up to a concentration of 0.3% decreased the molecular weight without significant change in the molecular weight distribution of the resulting polymers. Magnesium sulphate, potassium chloride as well as higher viscosities of the reaction media led to an increase in the molecular weight. In contrast, an increase in reaction temperature and addition of potassium nitrate decreased the molecular weights. The molecular weights determined by GPC were confirmed by a comparison with molecular weights resulting from viscosimetry.

Key words Heterogeneous polymerization – polymethylmethacrylate (PMMA) – molecular weights – gel permeation chromatography (GPC) – viscosimetry

Introduction

Emulsifier-free emulsion polymerization or heterogeneous polymerization represents a special case of emulsion polymerization [1–3]. This type of polymerization as well as emulsion polymerization with low amounts of emulsifiers were the most important methods so far for the production of nanoparticles [4]. Nanoparticles are defined as solid colloidal particles ranging in sizes from 10 to 1000 nm (1 μm) in which the drugs or other biologically active materials are dissolved, entrapped or encapsulated and/or to which the active principle is adsorbed or attached [4, 5]. Due to their small size, nanoparticles not only gain increased importance as drug delivery systems for intravenous, subcutaneous, and intramuscular injection and for targeting drugs to specific organs (drug targeting), but may also significantly improve drug bioavailability after peroral and ophthalmic administration [4].

Since polymer lattices with bound drugs produced by heterogeneous or emulsion polymerization represent the most widely used type of nanoparticles and since the molecular weight of these particles will have a significant influence on their biodegradation rate, it is very important to study the polymerization termination mechanism and the molecular weight resulting after polymerization.

A number of years ago it was proposed [6] that the termination of the polymer chain growth during γ-irradiation induced heterogeneous polymerization can occur by reaction of two radicals in both of the phases, the aqueous phase as well as in the polymer latex particles phase. It was further postulated that both of these processes take place simultaneously in a given polymerization mixture and that the ratio of the frequencies at which these processes occur is to a great extent governed by the aqueous phase solubility of the molecular species involved in the particle formation. These species include monomers, microradicals, soluble oligomer radicals, macroradicals, and polymers before their phase separation and/or their incorporation into polymer particles. The mentioned study [6] was based on the observation that after γ-ray induced polymerization the molecular weight was higher after addition of potassium chloride to the polymerization mixture than in an aqueous solution without electrolytes. This finding was explained as follows: in the electrolyte system, macroradicals are precipitated earlier than in distilled water. As a result, after precipitation, termination will take place mainly, if not exclusively, between macroradicals. In contrast, in distilled water the macroradicals remain for a longer time period than in the potassium chloride solution before separation by coagulation occurs. Hence, in distilled water the possibility that termination takes place by capture of a small radical is highly increased. Consequently, the average molecular weight will be lower after polymerization in water than after polymerization in an electrolyte solution [6].

The exclusive existence of only one of the two polymer growth termination mechanisms as earlier proposed by Baxendale, Evans, and Kilham [7–9] or by Fitch [1–3] was ruled out for the following reasons: Baxendale, Evans, and Kilham postulated that colloidal macroradicals coagulate prior to termination. However, due to the lower zeta potential in electrolyte solutions as well as to the salting-out effect, coagulation of the macroradicals would have to occur sooner in an electrolyte solution than in distilled water [10–12]. Since in this situation termination could take place only between two macroradicals in the coagulated phase, a larger molecular weight would have to be expected after polymerization in distilled water, the opposite to what was really observed. Fitch [3], on the other hand, postulated that chain radicals grow to macromolecular dimensions and undergo mutual termination while still in free solution. Particle formation and growth would occur subsequently by phase separation and coagulation of "dead" species. This mechanism, however, could not explain the observed difference in molecular weight because the possibility that the growing macroradical is hit by a radical is equal in both systems. The objective of the present study is to provide more insight into the mechanisms underlying the termination of polymerization in electrolyte-containing and viscous systems and to present different methods for monitoring the molecular weights of the resulting polymer lattices and nanoparticles.

Materials and methods

Reagents and chemicals

The following chemicals and reagents were used as obtained unless specified otherwise below: Methylmethacrylate (MMA) (Merck-Schuchardt, Hohenbrunn, Germany), ammonium persulphate (APS) (Hüls, Marl, Germany), potassium persulphate (KPS) (E. Merck, Darmstadt, Germany), tetrahydrofuran p.a. (E. Meck, Darmstadt, Germany), toluene p.a. (Fluka Chemie AG, Buchs, Switzerland), potassium chloride p.a. (E. Meck, Darmstadt, Germany), magnesium sulphate · 7 H_2O p.a. (Fluka Chemie AG, Buchs, Switzerland), potassium nitrate p.a. (E. Merck, Darmstadt, Germany), glycerol 85% (E. Merck, Darmstadt, Germany).

The water used for all steps of preparation was purified by ion exchange and boiled under a nitrogen stream to completely remove oxygen.

Polymerization procedure

Methylmethacrylate was purified from the polymerization inhibitor hydrochinone as described by Riddle [13] and Tessmar [14]. The purified monomer was then dissolved or, if its solubility was exceeded (above 1.5% MMA), dispersed in 50 ml distilled water, in an electrolyte solution or in diluted glycerol and then polymerized with 500 krad in a ^{60}Co source at a rate of 1.8–2.3 krad min^{-1}. As reaction vessels tightly closed 100 ml screw lid glasses were used (Schott, Mainz, Germany) (preparation nos. 4–7).

Different preparations were manufactured by altering the following variables:

1. The concentration of MMA was varied over a range of 0.5–5.0% (w/w) at an initiator concentration (APS) of 0.03% and a reaction temperature of 78 °C (Fig. 1).
2. The concentration of the initiator APS was varied between 0.01 and 1% (w/v) at a monomer concentration of 2%MMA and a reaction temperature of 78 °C (Fig. 2).
3. The reaction temperature was varied between 65 °C and 80 °C at monomer concentrations of 0.5, 0.75, 1.0, and

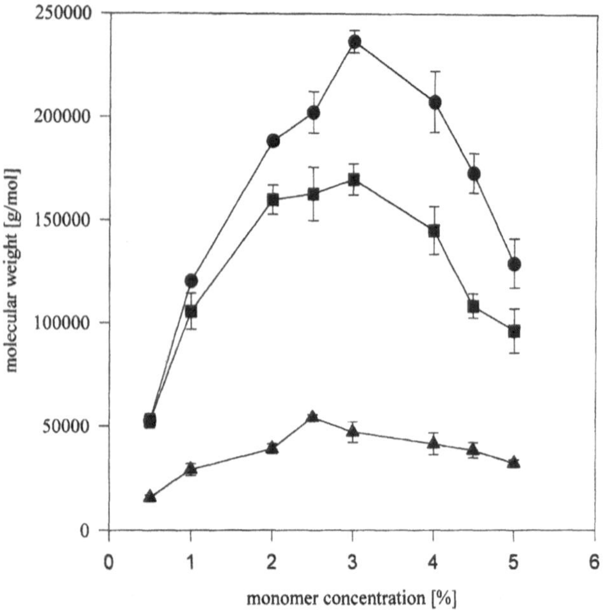

Fig. 1 Effect of the monomer concentration on the molecular weight of PMMA, shown as weight-average molecular weight (-●- M_w), molecular weight at the peak maximum (-■- M_p), and number-average molecular weight (-▲- M_n); Initiator concentration (APS): 0.03%. Standard deviations are indicated by vertical bars ($n = 3$)

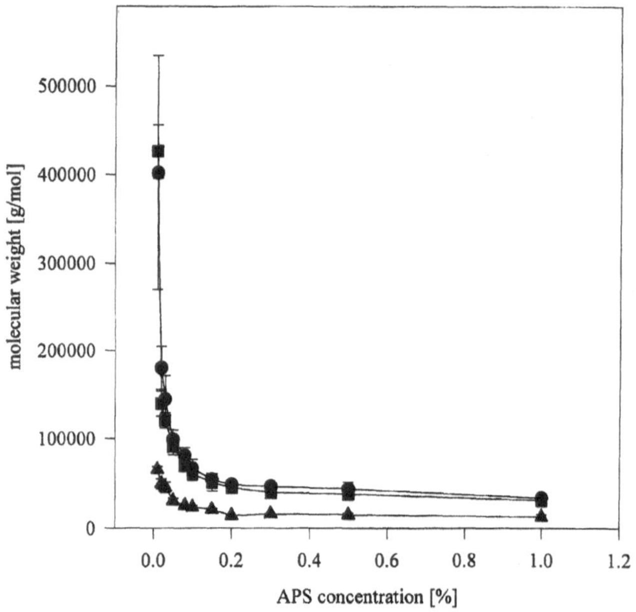

Fig. 2 Effect of initiator concentration (APS) on the molecular weight of PMMA, shown as weight-average molecular weight (-●- M_w), molecular weight at the peak maximum (-■- M_p), and number-average molecular weight (-▲- M_n); Monomer concentration (MMA): 2.0%. Standard deviations are indicated by vertical bars ($n = 3$)

1.5% MMA and an initiator concentration (KPS) of 0.1%, (Fig. 3).

4. The concentration of potassium chloride was varied between 0.25 and 3 mol/l at monomer concentration of 0.25 and 0.5% MA (Fig. 4).

5. The concentration of magnesium sulphate was varied between 0.25 and 2 mol/l at monomer concentrations of 0.25 and 0.5% MMA (Fig. 5).

6. The concentration of potassium nitrate was varied between 0.25 and 3 mol/l at monomer concentrations of 0.25 and 0.5% MMA (Fig. 6).

7. The content of glycerol was varied over a range of 5–42% (w/v) at monomer concentrations of 0.5 and 1.0% MMA (Fig. 7).

In the experiments with APS (preparation nos. 1 and 2) or KPS (preparations no. 3) as the reaction initiator the polymerization mixture were stired at 350 rpm (magnetic stirrer) for 24 h. The same reaction vessels as mentioned above were used. Addition of the initiator was performed when the reaction temperature was reached.

In all samples with electrolytes (preparation nos. 4–7) and with glycerol (preparation no. 7) polymerization was initiated by γ-irradiation at room temperature. No stirring was required in these experiments because the monomer solubility was not exceeded in any case.

The resulting particles were purified from the electrolytes by dialysis (preparations nos. 4–6). For this procedure

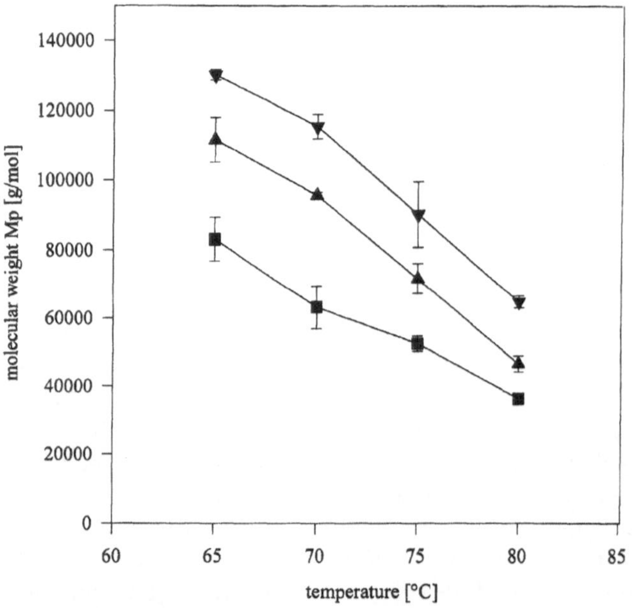

Fig. 3 Effect of reaction temperature on the molecular weight of PMMA for different monomer concentrations (-■- 0.75% MMA, -▲- 1.0% MMA, -▼- 1.5% MMA). Standard deviations are indicated by vertical bars ($n = 3$)

the particle suspensions were filled into a dialysis bag (exclusion limit 12 000–14 000, Medicell International LTD, London, England) and dialysed against tap water and after that against purified water both times for 48 h. In

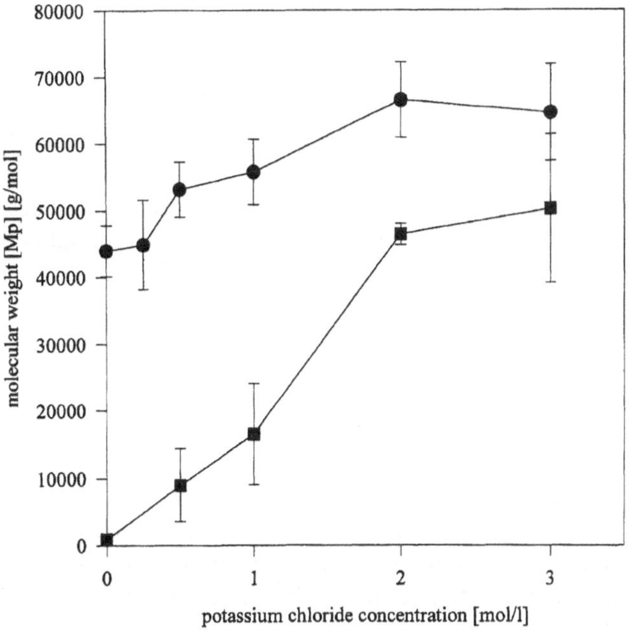

Fig. 4 Influence of potassium chloride on the molecular weight of PMMA, shown as the molecular weight at the peak maximum at monomer concentrations of 0.25% (-■-) and 0.5% (-●-). Standard deviations are indicated by vertical bars ($n = 6$)

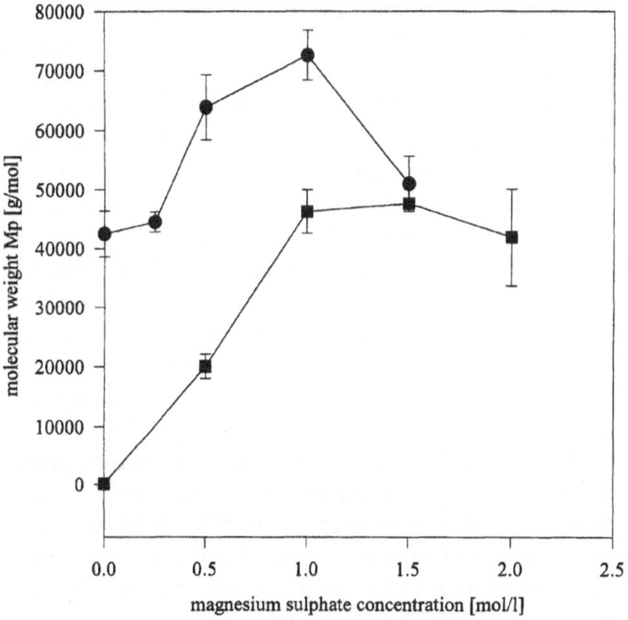

Fig. 5 Influence of magnesium sulphate on the molecular weight of PMMA, shown as the molecular weight at the peak maximum at monomer concentrations of 0.25% (-■-) and 0.5% (-●-). Standard deviations are indicated by vertical bars (0.25% MMA: $n = 4$; 0.5% MMA: $n = 6$)

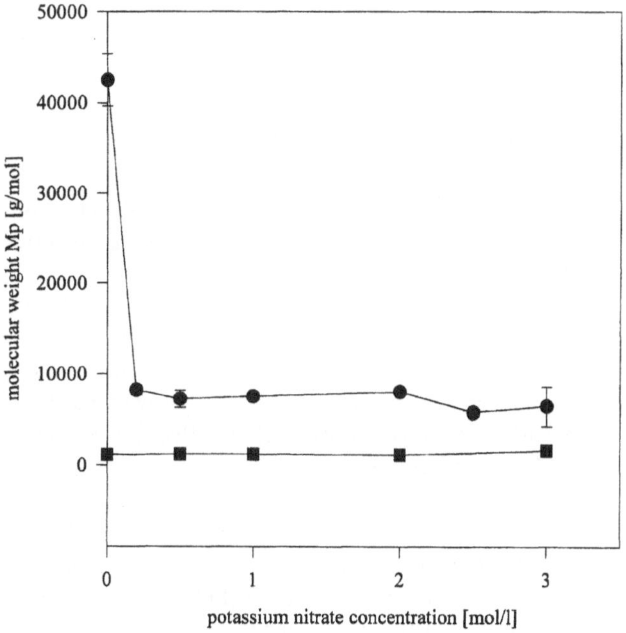

Fig. 6 Influence of potassium nitrate on the molecular weight of PMMA, shown as the molecular weight at the peak maximum at monomer concentrations of 0.25% (-■-) and 0.5% (-●-). Standard deviations are indicated by vertical bars (0.25% MMA: $n = 4$; 0.5% MMA: $n = 6$)

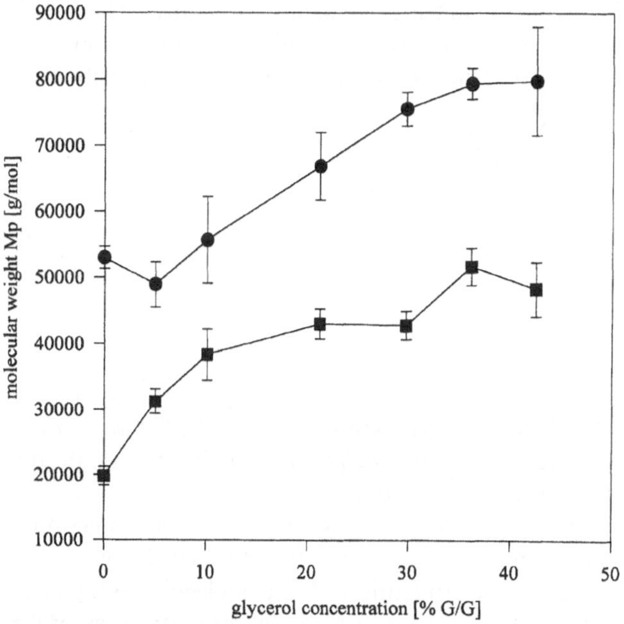

Fig. 7 Influence of glycerol on the molecular weight of PMMA, shown as the molecular weight at the peak maximum at monomer concentrations of 0.5% (-■-) and 1.0% (-●-). Standard deviations are indicated by vertical bars ($n = 4$)

order to remove the glycerol from the nanoparticles in preparation no. 7, the suspensions were diluted two-fold with distilled water. Then separation of the poly(methylmethacrylate) (PPMA) particles was performed by ultracentrifugation (30 min at 35 000 rpm, Beckmann Ultracentrifuge Optima L-80, Munich, Germany). This procedure was repeated three times. The purified particle suspensions were then lyophilizied in a Lyovac GT2 freeze-dryer

(Leybold-Heraeus, Hürth, Germany) for 24 h under vacuum (2×10^{-3} bar).

Determination of molecular weights

The molecular weights of the PMMA nanoparticles were determined using the methods of gel permeation chromatography (GPC) and viscosimetry. A GPC chromatograph equipped with a HPLC-chromatography pump (HPLC pump 64, Knae ur, Berlin, Germany) and a refraction index detector (Water 410 differential refractometer, Millipore, Eschborn, Germany) was used. 5μ Styragel 10^3 as well as 5 μ Styragel 10^5 (Polymer Standard Service, Mainz, Germany) served as columns. Tetrahydrofuran at a flow rate of 1 ml/min was used as eluent. Ten mg of the freeze dried PMMA nanoparticles were dissolved in 5 ml of tetrahydrofuran followed by filtration through a 0.45 μm filter of regenerated cellulose (Schleicher & Schüll, Dassel, Germany). Toluene at a concentration of 0.05% v/v was used as an internal standard. The injection volume was 100 μl. PMMA standards with molecular weights between 2030 and 10^6 (Polymer Standard Service, Mainz, Germany) were used for column calibration. The chromatograms were registered on a PC integration pack (Kontron Instruments, Neufahrn, Germany). The following equations [15] were used for the determination of the molecular weights and the molecular weight distributions:

$$M_n = \frac{\sum_{i=1}^{i=\infty} w_i}{\sum_{i=1}^{i=\infty} w_i/M_i}, \quad (1)$$

$$M_w = \frac{\sum_{i=1}^{i=\infty} w_i \cdot M_i}{\sum_{i=1}^{i=\infty} w_i}, \quad (2)$$

$$D = \frac{M_w}{M_n}, \quad (3)$$

M_w = weight-average molecular weight; w_i = amount of molecules with a certain molecular weight M_i; M_n = number-average molecular weight; D = molecular weight distribution (polydispersity index).

The molecular weight in the peak maximum ($M_{GPC} = M_p$) was taken from the chromatogram.

For the determination of the molecular weights by viscosimetry PMMA standards in concentrations of 0.5, 1.0, 2.0, 3.0% were dissolved in tetrahydrofuran for calibration. The density of these solutions was measured in a DMA 48 density meter (AP PAAR, Graz, Austria). The flow time of each solution was determined in a Cannon-Fenske viscometer ($k = 0.002897$, Schott, Hofheim, Germany). The reduced specific viscosity ($\eta_{sp} = \eta_{rel} - 1$; $\eta_{red} = \eta_{sp} c^{-1}$) was calculated [16] and plotted against the concentration of the solution followed by extrapolation to zero concentration. The resulting intercept represents the intrinsic viscosity. In order to establish the calibration curve, the logarithm of the intrinsic viscosity was plotted against the logarithm of the molecular weight of the PMMA standards.

The freeze dried nanoparticles were treated in the same way as the PMMA standards. Concentrations between 0.25 and 2.5% were employed. The molecular weights of the polymers (PMMA) were calculated by comparison with the calibration curve. All measurements were performed at a temperature of 20 °C.

Results and discussion

For the preparation of the poly(methylmethacrylate) (PMMA) nanoparticles methylmethacrylate (MMA) was polymerized in different reaction media by the method of heterogeneous free-radical polymerization either with γ-rays, or with ammonium persulphate or potassium persulphate as initiators [17, 18]. The temperature of reaction, the monomer concentration, and the initiator concentration, were also varied over a wide range to investigate their influence on the polymerization process.

The monomer concentration was varied over a range of 0.5–5.0% MMA (Fig. 1). For the first set of studies a fixed initiator concentration of 0.03% APS and a reaction temperature of 78 °C were chosen. The molecular weights were determined by using the method of gel permeation chromatography. The weight-average molecular weight, the number-average molecular weight, and the molecular weight distribution were calculated according to Eqs. (1)–(3). The molecular weight of PMMA increased with an increasing monomer concentration between 0.5 and 3% MMA (Fig. 1), For concentrations between 2.0 and 3% the differences were not significant (M_p). Monomer contents of more than 3% MMA led to a decrease in the molecular weight average [19]. In contrast to the molecular weight data, the polydispersity indexes showed no dependence on the monomer concentration over a range between 1 and 5% MMA (Table 1). Only a monomer concentration of 0.5% led to a more narrow molecular weight distribution. The results with concentrations below 2–3% initial monomer are in agreement with the scenario outlined in the introduction. Accordingly, chain growth and termination can occur in both phases, the polymer and the surrounding aqueous phase, and all types of species of radicals, i.e. micro-, oligo-, and macroradicals can react with each other. Due to the good solubility of the monomer in the polymer the dominant location of the polymerization and the termination will be the polymer phase. In this situation an increase in monomer leads to an increase in the molecular weight, since the number of radicals that can be generated by the decay of the initiator or by gamma-irridiation, is limited. At higher concentrations above 2%, and more pronounced above 3.0%, diffusion of radicals becomes limited by the viscosity of the growing polymer particles. The resulting reduction in the diffusion velocity, of course, is much lower with microradicals than with the macroradicals. Consequently, termination by two macroradicals becomes less frequent leading to a decrease in molecular weights. This scenario is similar to the Norrish-Trommsdorf-effect observed during bulk polymerization [20, 21].

The concentration of the initiator APS in the reaction mixture was varied from 0.01 to 1.0% at a fixed monomer concentration of 2% and a reaction temperature of 78 °C. An increasing amount of initiator led to a decrease in the molecular weight up to a concentration of 0.3% (Fig. 2). The molecular weight distribution was constant over a range of 0.02–1.0% APS (Table 1). The results of polymerization processes with higher initiator concentrations (≥0.0%) showed no further significant change in the average molecular weight. Low initiator concentrations led to a small number of radicals, resulting in a particulate system comprised of polymers of a high molecular weight. Higher APS concentrations (0.02–0.3%) increased the number of radicals resulting in a reduction of the molecular weight. For an initiator concentration above 0.3% the number of initiator radicals that were employed for polymerization reaction obviously remained largely constant.

Table 1 Molecular weight distributions (polymerization index: *D*) of PMMA by variation of monomer concentration (MMA) and initiator concentration (APS). SD = standard deviation

MMA content [%]	*D*	SD	APS content [%]	*D*	SD
0.5	3.38	0.06	0.01	6.17	1.06
1	4.18	0.41	0.02	3.73	0.68
2	4.81	0.31	0.03	3.23	0.24
2.5	3.92	0.75	0.05	3.31	0.13
3	5.05	0.54	0.08	3.22	0.06
4	5.07	0.93	0.1	2.89	0.06
4.5	4.54	0.47	0.15	2.62	0.42
5	4.02	0.22	0.2	3.50	0.03
			0.3	2.96	0.08
			0.5	2.93	0.11
			1	2.99	0.41

It has to be assumed that the excess starter molecules decayed without involvement in the polymerization process.

The variation of the reaction temperature over a range of 65 to 80 °C for monomer concentrations between 0.75 and 1.5% at a fixed initiator concentration of 0.1% KPS led to a linear decrease in the resulting molecular weights (Fig. 3). Under these reaction conditions, the number of generated polymerization initiating radicals is proportional to the reaction temperature, and consequently, the molecular weight at a given monomer concentration inversely related to these parameters, temperature, and number of active initiator molecules [1, 2, 22]. In addition, with an increase in the reaction temperature the contribution of disproportionation reactions to the overall termination as well as the extent and the rate of depolymerization increases [23]. Both effects augment the decrease in molecular weights. The molecular weight distribution (Table 2), on the other hand, was not significantly influenced by the reaction temperature.

The molecular weights of the PMMA nanoparticles polymerized using initiation with γ-rays in electrolyte solution were higher for magnesium sulphate and potassium choride and lower for potassium nitrate (for monomer concentrations ≥ 0.5% MMA) than after a corresponding polymerization in distilled water (Figs. 4–6). Magnesium sulphate and potassium chloride solution led to a poor solubility of the macroradicals in the aqueous phase and to a shift of the polymerization and termination reactions to the polymer phase. The increase of the molecular weight observed in these studies may be explained by the Norrish–Trommsdorff-effect [20]. In this situation termination to a great degree takes place between macroradicals as described above since the particle masses are still relatively low. Although with potassium nitrate similar results were expected, the opposite was observed. Potassium nitrate obviously strongly interferes with the polymerization: due to this interference not only the molecular weights were significantly reduced (0.5% MMA) or did not increase in comparison to the aqueous electrolyte free medium (0.25% MMA, Fig. 6) but also the polymer yield was very low (20–36%, depending on the electrolyte concentration). The molecular weight distributions (Tables 3 and 4) increased with increasing amounts of added magnesium sulphate

Table 2 Molecular weight distributions (polymerization index: *D*) of PMMA polymerized at different reaction temperatures

Reaction temperature [°C]	*D* 0.75% MMA	SD	*D* 1% MMA	SD	*D* 1.5% MMA	SD
65	4.53	0.11	4.19	0.65	5.49	0.04
70	3.05	0.61	4.58	0.07	5.00	0.36
75	5.26	1.37	3.37	0.16	5.14	0.19
80	2.88	0.40	3.89	0.34	4.68	0.07

Table 3 Molecular weight distributions (polymerization index: *D*) of PMMA polymerized in electrolyte solution at a monomer concentration of 0.25% MMA

Electrolyte content [mol/l]	*D* KCl	SD	*D* $MgSO_4$	SD	*D* KNO_3	SD
0	3.49	0.28	3.51	0.47	3.474	0.36
0.5	6.28	0.87	7.89	2.25	2.541	0.17
1	10.62	2.75	9.34	1.13	2.011	0.19
1.5	—	—	12.52	0.50	—	—
2	10.60	0.94	13.36	2.76	2.086	0.17
3	9.41	1.73	—	—	3.959	0.25

Table 4 Molecular weight distributions (polymerization index: *D*) of PMMA polymerized in electrolyte solution at a monomer concentration of 0.5% MMA

Electrolyte content [mol/l]	*D* KCl	SD	*D* $MgSO_4$	SD	*D* KNO_3	SD
0	10.52	1.19	9.90	2.68	9.87	1.23
0.25	10.59	2.57	8.01	1.29	4.77	0.12
0.5	10.56	2.11	11.70	2.80	3.88	0.67
1	11.28	3.17	14.82	5.34	4.61	1.48
1.5	—	—	10.65	4.41	—	—
2	11.02	2.06	—	—	5.03	1.39
2.5	—	—	—	—	3.55	0.24
3	11.84	2.02	—	—	5.31	1.20

and potassium chloride. In contrast, a general decrease of the molecular weight distribution was observed with potassium nitrate due to the reasons discussed above. Probably similar reasons that led to a decrease in molecular weight also decreased the molecular weight distribution.

Increasing amounts of glycerol over a range of 5–42% w/v led to higher molecular weights of the PMMA nanoparticles, produced by initiation with γ-rays at room temperature (Fig. 7). The polydispersity indexes, on the other hand, showed a tendency for lower values (Table 5). The increased viscosity of the reaction mixture definitely stabilized the macroradicals. The viscosity, in this preparation of course, still was by orders of magnitude lower than that in the bulk polymer phase within the polymer particles produced with high amounts of monomer (i.e. <2–3% monomer, preparation no. 1). In the polymer phase within the particles at later stages of polymerization macromolecule diffusion and, hence, termination by two macroradicals was impeded. In the viscous glycerol solution the viscosity by far was not high enough to impede this termination reaction. Therefore, in this situation viscosity rather favors termination by two macroradicals due its stabilizing effects. In addition to the increased viscosity, glycerol will alter the dielectric constant of the reaction

Table 5 Molecular weight distributions (polymerization index: *D*) of PMMA polymerized in glycerol solution at monomer concentrations of 0.5 and 1.0% MMA

Glycerol content [%]	*D* 0.5% MA	SD	*D* 1.0% MMA	SD
0	16.86	1.42	9.60	0.83
5	16.88	2.17	9.13	1.06
10	10.57	2.82	8.89	1.64
21	16.84	3.86	6.33	0.57
30	7.79	0.75	7.73	0.39
36	7.64	1.46	8.96	0.30
42	8.02	1.47	7.65	0.66

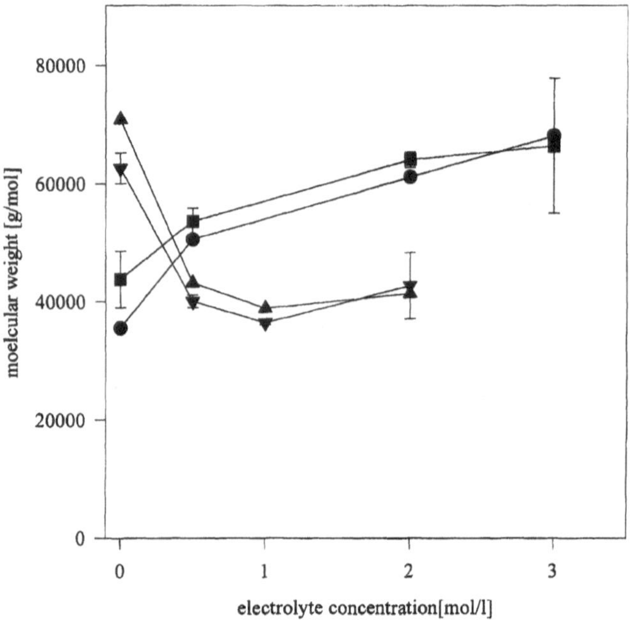

Fig. 8 Comparison of the molecular weights of PMMA determined by GPC (-■-) polymerization in potassium chloride solution, -▼- polymerization in potassium nitrate solution) and viscosimetry (-●- polymerization in potassium chloride solution, -▲- polymerization in potassium nitrate solution). Standard deviations are indicated by vertical bars ($n = 3$))

solution as well as the solubilities of the different reaction partners including the polymer. All these factors will further influence the resulting molecular weight. The increase in molecular weight due to the addition of glycerol was considerable and amounted to over 100% in the case of 0.5% MMA and to well over 50% in the case of a monomer concentration of 1% MMA.

In Fig. 8 GPC and viscosimetry as methods for the determination of the molecular weights were compared. The preparation of the polymers was performed by initiation with KPS at monomer concentrations of 0.5% MMA (potassium chloride) and 1.0% MMA (potassium nitrate) at a reaction temperature of 78 °C. As shown in Fig. 8 both methods yielded similar results.

This study largely confirms the assumptions made in an earlier study [6] that were outlined in the Introduction. It shows that the molecular weight of poly(methylmethacrylate) nanoparticles and probably other acrylic nanoparticles can be altered and, within certain limits, monitored by the addition of electrolytes, viscosity enhancers, and by the choice of polymerization initiation. As mentioned above the molecular weight of nanoparticles is very important because it governs their biodegradation rate. Poly(methylmethacrylate) nanoparticles in particular, due to their hydrophobic character [24, 25], represent promising adjuvants for vaccines. These particles did not only improve considerably the performance of vaccines [26–28] but they may also improve their stability [29].

References

1. Fitch RM (1973) Br Polym J 5:467–483
2. Fitch RM, Prenosil MB, Sprick KJ (1969) J Polym Sci C 27:95–118
3. Fitch RM, Tsai C (1970) Polym Lett 8:703–710
4. Kreuter J (1994) In: Kreuter J (ed) Colloidal Drug Delivery Systems. Marcel Dekker, New York, pp 219–342
5. Kreuter J (1994) Eur J Drug Metabol Pharmakokin 3:253–256
6. Kreuter J (1982) J Polym Sci, Polym Lett Ed 20:243–245
7. Baxendale JH, Evans MG, Kilham JK (1946) J Polym Sci 1:466–474
8. Baxendale JH, Evans MG, Kilham JK (1946) Trans Faraday Soc 42:668–675
9. Baxendale JH, Bywater S, Evans MG, (1946) Trans Faraday Soc 42:675–684
10. Riddich TM (1968) In: Control of Colloid Stability through Zeta Potential. Livingston, Wynneford, PA, pp 208–239
11. Verwey EJW (1940) Trans Farad Soc 36:192–203
12. Powis F (1941) Z Phys Chem 89: 186–212
13. Riddle EH (1954) In: Monomeric Acrylic Esters. Reinhold, New York, p 15
14. Tessmar K (1961) In: Müller E (ed) Methoden der organischen Chemie (Houben-Weyl) XIV/1. George Thieme, Stuttgart, p 1037
15. Waters Associates Inc (1974) DS 047, Milford, MA
16. Martin AN, Swarbrick JM, Cammarata A (1987) In: Physikalische Pharmazie: Wissenschaftliche Verlagsgesellschaft, Stuttgart, pp 455–456
17. Bentele V, Berg UE, Kreuter J (1982) Int J Pharm 13:109–113
18. Kreuter J (1983) Pharm Act Helv 56:196–209
19. Berg UE, Kreuter J, Speiser PP, Solvia M (1986) Pharm Ind 48:75–79
20. Wunderlich W, Strickler M (1985) Polym Sci Technol 31:505–516
21. Maschio G, Moutier C (1989) J Appl Polym Sci 37:825–840
22. Hamielec (1986) In: Notes on Polymerization Reaction Engineering. McMaster Univ
23. Ponnuswamy SR, Penilidis A, Kiparissides C (1988) Chem Eng J 39: 175–183
24. Kreuter J, Liehl E, Berg UE, Solvia M, Speiser PP (1988) Vaccine 6:253–256
25. Stieneker F, Kersten G, van Bloois L, Crommelin DJA, Hem SL, Löwer J, Kreuter J (1995) Vaccine 13:45–53
26. Kreuter J, Speiser P (1976) Infect Immunol 13:204–210
27. Kreuter, Mauler R, Gruschkau H, Speiser PP (1976) Exp Cell Biol 44: 12–19
28. Kreuter J, Liehl E (1978) Med Microbiol Immunol 165:111–117
29. Kreuter J, Liehl E (1981) J Pharm Sci 70:367–371

Progr Colloid Polym Sci (1998) 109:278–282

N.J. Marston
B. Vincent
N.G. Wright

The synthesis of spherical rutile titanium dioxide particles and their interaction with polystyrene latex particles of opposite charge

Received: 10 October 1997
Accepted: 13 October 1997

N.J. Marston · Prof. Dr. B. Vincent (✉)
N.G. Wright
University of Bristol
School of Chemistry
Cantock's Close
Bristol BS8 1TS
United Kingdom
E-mail: Brian.Vincent@bristol.ac.uk

Abstract This paper describes the preparation and characterization of monodisperse, cationic polystyrene latices and the preparation of monodisperse rutile titanium dioxide. Adsorption isotherms for the cationic latices onto the TiO_2 particles, as a function of NaCl concentration, are also be presented and discussed.

Key words Monodisperse rutile titanium dioxide particles – heteroaggregation – physical properties – novel apparatus – particle adsorption

Introduction

Several workers have studied the heteroaggregation between polymer latex particles of different sizes. For example, Cheung [1] studied the adsorption of 0.21 μm cationic latex particles onto 1.07 μm anionic polystyrene particles. High affinity adsorption isotherms were found. The number of small particles adsorbed by each large particle was found to increase with increasing electrolyte concentration. Vincent and Young [2, 3] reported results for similar systems, but with both sets of particles carrying physisorbed poly(vinylalcohol) (PVA) layers. Vincent and Harley [4] investigated heteroaggregate morphology as a function of relative particle sizes and electrolyte concentration, with and without adsorbed PVA, by the use of cryo-SEM.

In this paper the preparation and characterization of monodisperse, cationic polystyrene latices and the preparation of monodisperse rutile titanium dioxide will be described. The adsorption isotherms for the cationic latex particles onto the TiO_2 particles, as a function of NaCl concentration, will also be presented and discussed.

Experimental

Titanium dioxide dispersion preparations

The synthetic route used for the titanium dioxide dispersions is a modified version of that described by Barringer et al. [5–7]. In their method two separate ethanolic solutions are first prepared in a dry atmosphere: the first contains titanium (IV) ethoxide, $Ti(OC_2H_5)_4$, and the second a low concentration of water. On mixing, rapid hydrolysis of the $Ti(OC_2H_5)_4$ results, forming a dispersion containing monosized titanium dioxide (TiO_2) particles. A glove box was used for the entire preparation to prevent hydrolysis of the $Ti(OC_2H_5)_4$ prior to mixing the two solutions. In this work, however, an airtight apparatus was designed and constructed from glass (Fig. 1) that eliminated the need for a glove box.

The apparatus consisted of a round-bottomed reaction chamber and an upper dropping funnel. These two section were connected through a "Quickfit" joint. On separation both could be sealed by a valve (Valves 1 and 2). The glass taps (Taps 1–3) were to allow filling, and for evacuation of sections of the apparatus during the preparation. In addition, the reaction chamber had a "Quickfit" ground glass neck into which a "Subaseal" septum cap was inserted. Stirring was achieved using an Aldrich "High Shear" magnetic stirrer follower.

Prior to use the apparatus was thoroughly cleaned with concentrated sodium hydroxide followed by hydrochloric acid. The chambers were then rinsed with pure water before oven drying.

Fig. 1 Simplified representation of the reaction vessel used for TiO_2 synthesis

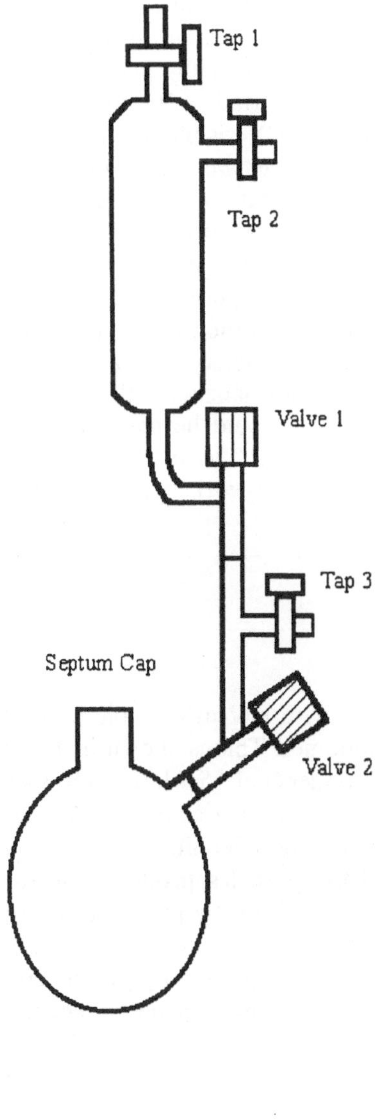

The ethanol used for preparing the initial $Ti(OC_2H_5)_4$ solutions was subjected to an extensive drying process. Approximately, 1 l of ethanol was placed in a flask over thoroughly dried alumina molecular sieves. These sieves had an average pore size of between 0.3 and 0.5 nm, and had been baked under a nitrogen flow for 12 h. This flask was then sealed with a "Subaseal", and purged with dry nitrogen. The ethanol was left in contact with the sieves for at least 1 week. At the end of this time approximately 100 cm^3 of the ethanol was transferred onto a sealed column of dry molecular sieves by the use of a double-ended needle. The ethanol was then forced through the column by an overpressure of dry nitrogen and passed through a 0.1 μm filter (to remove any fine particles washed from the sieves) and passed into the reaction vessel via a needle through the "Subaseal". Air was excluded at all times by use of an overpressure of dry nitrogen. The water concentration in the ethanol was determined at each stage of the drying process using a Karl Fischer 652KF Coulometer and was found to be reduced to <65 ppm.

The $Ti(OC_2H_5)_4$ was then drawn through a wide bore needle into the reaction flask by a vacuum line connected to Tap 3 (Fig. 2). The dropping funnel was then filled with a solution containing water, HCl and ethanol.

The two halves of the apparatus were connected and clamped in a retort stand. A vacuum line was then connected to Tap 3, and by opening Valve 2, the lower half of the apparatus was evacuated. The time that the vacuum was connected was kept to a minimum to prevent evaporation. Following closure of Tap 3, Taps 1 and 2 were opened prior to the withdrawal of Valve 1. This resulted in very rapid mixing of the two reactant solutions.

Details of the various titanium dioxide preparations are summarized in Table 1. Figure 2a and b show transmission electron micrographs of particles from a typical dispersion.

It was found that somewhat smaller particles could be produced if the reaction mixtures were precooled in ice water baths prior to mixing. In addition, the reaction vessel itself was similarly cooled for the duration of the preparation. This finding is presumably a result of the decreased rate of reaction.

Cationic polystyrene latex preparations

The basic emulsion polymerization method used is that described by Pelton [8]. All glassware used in the handling of positively charged latices was hydrophobed.

A four-necked, round-bottomed flask fitted with a water-cooled condenser was charged with electrolyte

a)

b)

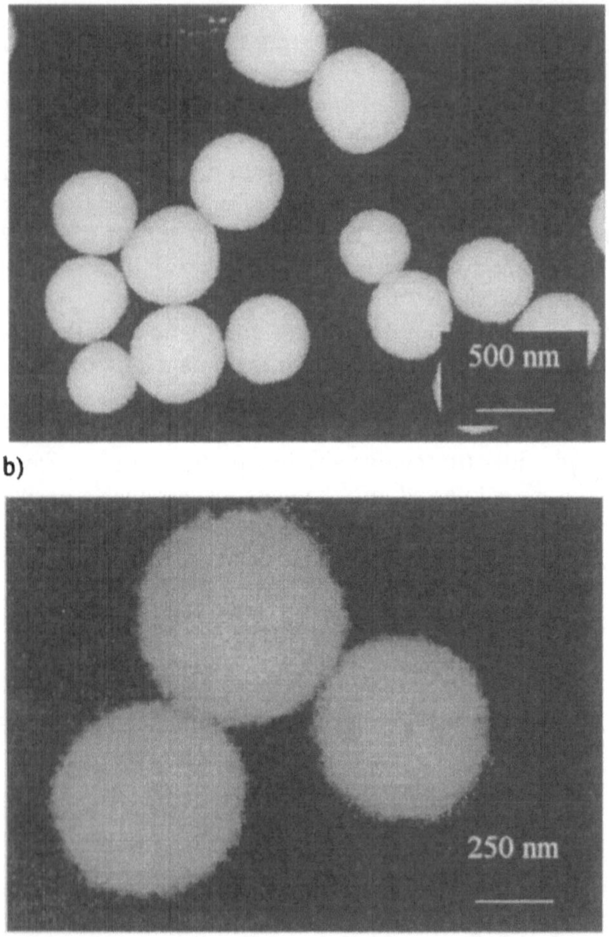

Fig. 2 Transmission electron micrographs of Dispersion TiO_2 #6

solution and heated to 70 °C in a thermostatted oil bath. After maintaining the mixture under a nitrogen atmosphere for 30 min, vacuum-distilled styrene was rapidly added via the fourth neck. After a further 30 min the initiator, azobis-(isobutylamidine)dihydrochloride (ABA·2HCl), dissolved in 10 cm^3 of pure water, was added. The onset of turbidity was typically observed after a further 45 min. The reaction was then allowed to proceed for at least a further 20 h.

Following removal from the reaction vessel the resulting latex was filtered through glass wool to remove coagulum. It was then exhaustively dialysed against distilled water.

Details of the latices prepared in this work are shown in Table 2. After purification the latex dispersions were stored in polyethene bottles, in a refrigerator, in order to minimize the hydrolysis of the amidine surface groups; this (acid- or base-catalyzed) reaction leads to the formation of carboxylic acid groups at the particle surface, which reduces their net positive surface charge density [9].

Electrophoresis measurements

A "Penkem System 3000" apparatus was used to study both the latex and TiO_2 particles. The short measurement times required is an advantage, as this minimizes any contamination of the cationic latex samples by polysilicates leached from the cell walls of the apparatus. Unfortunately, it was not possible to hydrophobe the walls of the cell, as this is an integral part of the equipment.

Adsorption of latex particles onto TiO_2 particles

The equilibrium extent of adsorption of PSL #2 particles onto rutile TiO_2 #6 particles was determined, for a range of NaCl concentrations. A 15 cm^3 aliquot of the latex dispersion, at the required electrolyte concentration, was placed in a "Sterilin" centrifuge tube, 0.25 cm^3 of stock TiO_2 #6 dispersion was then added. The final volume fraction of TiO_2 #6 in all cases was 4.7×10^{-5} and the pH of the dispersions was 6.2 ± 0.2, above the isoelectric point of the TiO_2 particles (pH ~ 5.6). The tube was then inverted to ensure homogeneous mixing, and placed on an end-over-end tumbler for 24 h. The TiO_2 particles were then allowed to sediment for a period of 10 days.

A Kontron "Uvikon 940" UV/visible spectrophotometer set at a wavelength of 500 nm was used to determine the concentration of the supernatant latex dispersion. A linear calibration of absorbance against particle volume fraction was established. The concentration of *unadsorbed* latex particles remaining in the supernatant was determined. This allowed calculation of the number adsorbed per TiO_2 particle (Γ). The fractional coverage, θ ($= \Gamma/\Gamma_{hcp}$), was then derived by normalizing with the maximum number of particles that could hexagonally close packed on each TiO_2 particle (Γ_{hcp}). This method has been used previously by Vincent and co-workers [2–4, 10–12].

Results and discussion

TiO_2 particles

The particle size decreased with increasing mole ratio of water to $Ti(OC_2H_5)_4$ monomer. This is shown in Fig. 3 along with the data of Barringer et al. [5, 6]. It can be seen that the two sets show similar decreases in particle size, although the absolute values are different.

The density of the TiO_2 particles produced by the ethoxide route was determined both by pycnometry and from sedimentation velocity measurements. The density of the particles was found to be 1.29 $g\,cm^{-3}$ by pycnometry and 1.35 $g\,cm^{-3}$ from the sedimentation measurements.

Table 1 Titanium dioxide dispersions prepared at room temperature

Reference number	Concentration of water [mol dm^{-3}]	Concentration of $Ti(OC_2H_5)_4$ [mol dm^{-3}]	Concentration of HCl [mol dm^{-3}]	Number average mean particle diameter* [nm]
TiO_2#1	0.29	8.2×10^{-2}	2.0×10^{-5}	616 ± 43
TiO_2#2	0.57	6.5×10^{-2}	4.0×10^{-5}	330 ± 35
TiO_2#3	0.37	2.4×10^{-2}	8.2×10^{-5}	200 ± 23
TiO_2#4	0.37	4.8×10^{-2}	8.2×10^{-5}	420 ± 31
TiO_2#5	0.33	4.3×10^{-2}	7.2×10^{-5}	450 ± 34
TiO_2#6	0.32	7.0×10^{-2}	7.3×10^{-5}	582 ± 45
TiO_2#7	0.37	8.6×10^{-2}	8.0×10^{-5}	479 ± 30

* From transmission electron microscopy.

Table 2 Cationic polystyrene latices

Latex reference number	Volume styrene [cm^3]	Volume water [cm^3]	Weight NaCl [g]	Weight of ABA·2HCl [g]	Number average mean diameter* [nm]
PSL#1	4.50	350	0.20	0.05	272 ± 12
PSL#2	12.75	750	0.50	0.57	276 ± 9

* From transmission electron microscopy.

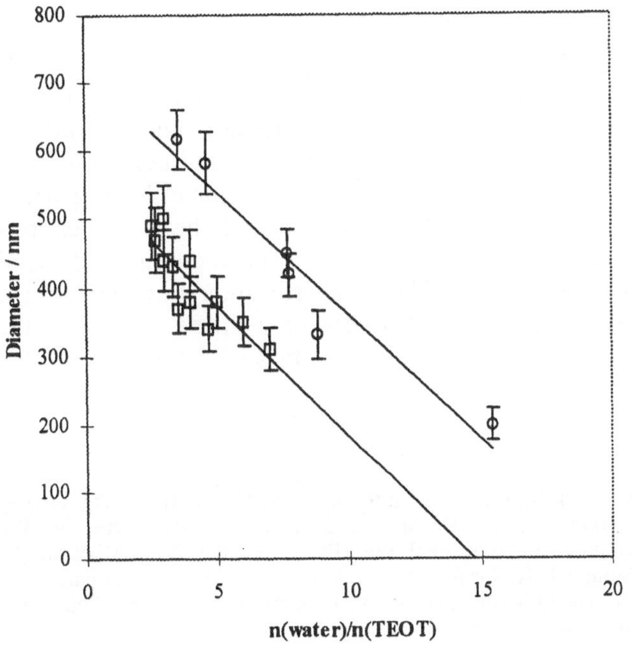

Fig. 3 Dependence of TiO_2 particle diameter on molar ratio of water to TEOT [$Ti(OC_2H_5)_4$] found in this work (○) compared to the findings of Barringer et al. [5, 6] (□)

These values are significantly lower than 3.35 g cm^{-3} determined by Barringer for particles produced by this route, which is in turn lower than the value for rutile TiO_2 of 4.25 g cm^{-3}.

It is clear from these data that the particles are likely to be highly porous, with the majority of their volume comprising of solvent. Electromicrographs of the particles seem to support this conclusion (see Fig. 2b). This observation has been made by previous authors [5–7].

Electrophoretic mobility of cationic latices and titanium dioxide

The electrophoretic mobilities of latices, PSL#2, and titanium dioxide dispersion, TiO_2#6 were determined as a function of pH, each at two NaCl concentrations. The results are shown in Fig. 4. The iso-electric point, IEP, of the latex dispersions (Fig. 4a) was pH 7.5 ± 0.5. The iso-electric points of TiO_2#6 was pH 5.6 ± 0.1 (Fig. 4b). This value is considerably higher than that for rutile TiO_2 pigment particles which exhibit an IEP in the region of pH 4. It has been proposed that this may be due to residual, unreacted ethoxide groups at the particle surface [7].

Adsorption of cationic latex on TiO_2

The adsorption isotherms, at pH 6.2 ± 0.2, are shown in Fig. 5.

All isotherms are of the "high-affinity" type, and display a plateau coverage, θ_{pl}. This type of isotherm has been observed for mixtures of oppositely charged polystyrene latices by Vincent and co-workers [1–4, 10–12].

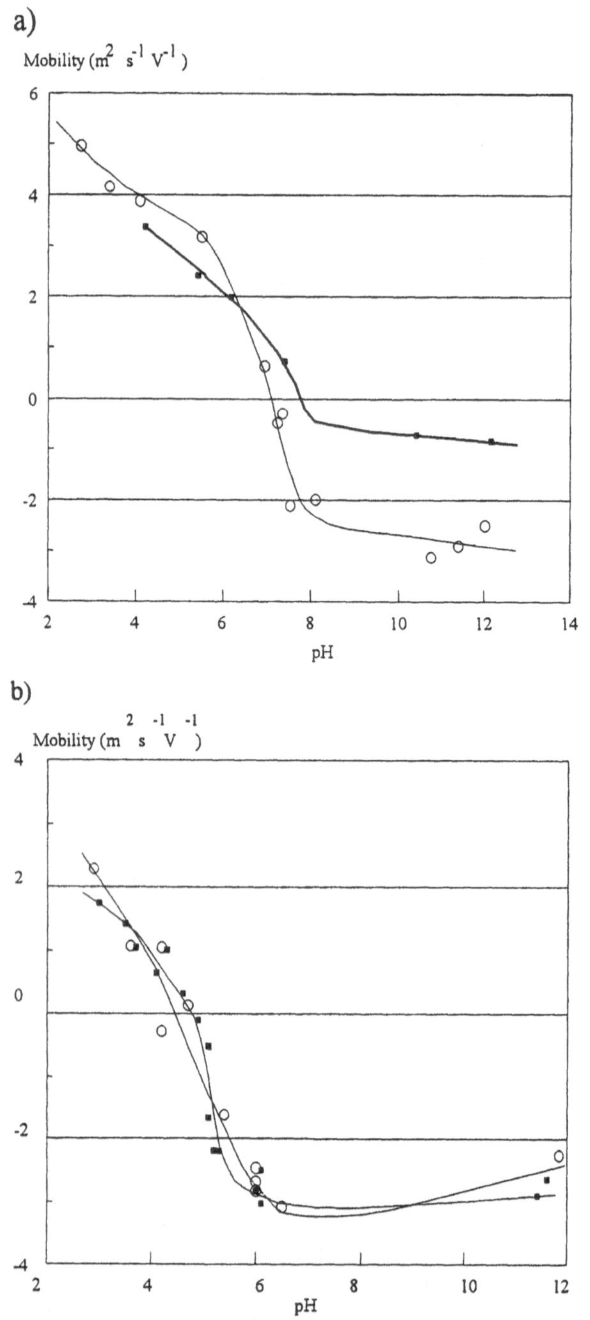

Fig. 4 A typical mobility vs. pH curve for (a) cationic polystyrene latex (PSL#2) and (b) titanium dioxide (TiO_2#6) in the presence of 10^{-2} mol dm^{-3} (■) and 10^{-4} mol dm^{-3} NaCl (○)

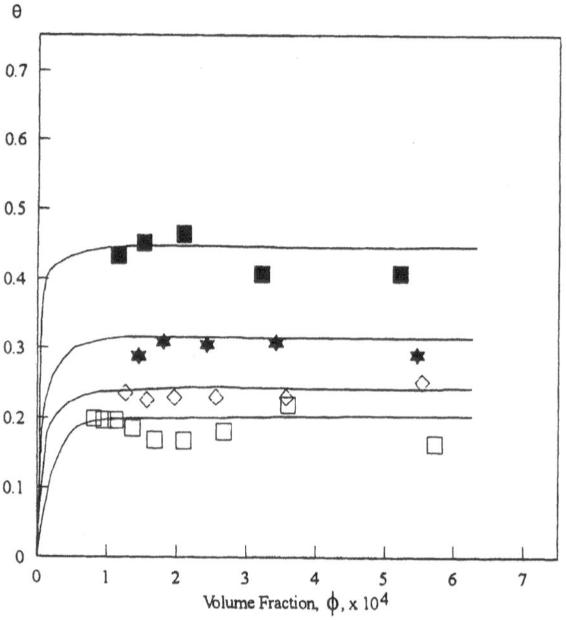

Fig. 5 Adsorbed amount (θ) versus Volume fraction (ϕ), for adsorption of PSL#2 onto rutile TiO_2 (TiO_2#6), in the presence of 10^{-5} mol dm^{-3} (□), 10^{-4} mol dm^{-3} (◇), 10^{-3} mol dm^{-3} (★) and 10^{-2} mol dm^{-3} (■) NaCl

The increase in plateau adsorbed amount, θ_{pl}, as the electrolyte concentration increases may be rationalised in terms of a balance between the normal electrostatic attraction, between the cationic latex particle and the negative TiO_2 particle, and the lateral electrostatic repulsion, between neighbouring adsorbed particles. As the NaCl concentration increases so the extent of both interactions is reduced. However, the reduction in the lateral repulsion allows the latex particles to pack more densely on the surface of the TiO_2 particles.

Acknowledgements The authors would like to thank Dr. Dudley Thompson of this Department for many valuable discussions. We would also like to thank the EPSRC and ICI Paints Division plc. (Slough, U.K.) for financial support of this work, and, in particular, Drs David Taylor and Simon Emmett from ICI for valuable discussions.

References

1. Cheung WK, PhD Thesis, Bristol, 1979
2. Vincent B, Young CA (1978) Faraday Disc Chem Soc 65:296
3. Vincent B, Young CA (1980) J Chem Soc Faraday Trans 1 76:665
4. Harley S, Vincent B (1992) Colloids Surf 62:163
5. Barringer EA, Bowen HK, Fegley B (1984) Commun Ceramics Soc C113
6. Barringer EA, Bowen HK (1985) Langmuir 1:414
7. Barringer EA, Bowen HK (1985) Langmuir 1:420
8. Pelton R (1976) PhD Thesis, Bristol
9. Obey T (1987) PhD Thesis, Bristol
10. Jafelicci M, Luckham PF, Tadros ThF, Vincent B (1980) J Chem Soc Faraday Trans 1 76:674
11. Luckham PF, Vincent B (1980) Colloids Surf 1:281
12. Luckman PF, Vincent B (1980) Colloids Surf 6:83

Progr Colloid Polym Sci (1998) 109:283

AUTHOR INDEX

SUBJECT INDEX